Biofuels

From Microbes to Molecules

Edited by

Xuefeng Lu

Key Laboratory of Biofuels
Shandong Provincial Key Laboratory of Energy Genetics
Qingdao Institute of Bioenergy and Bioprocess Technology
Chinese Academy of Sciences
Qingdao
China

Caister Academic Press

Caister Academic Press
Norfolk, UK

www.caister.com

British Library Cataloguing-in-Publication Data
A catalogue record for this book is available from the British Library

ISBN: 978-1-908230-45-4 (hardback)
ISBN: 978-1-908230-63-8 (ebook)

Cover design adapted from image courtesy of Xuefeng Lu

Contents

Current Books of Interest

Antibiotics: Current Innovations and Future Trends	2015
Leishmania: Current Biology and Control	2015
Acanthamoeba: Biology and Pathogenesis (2nd edition)	2015
Microarrays: Current Technology, Innovations and Applications	2014
Metagenomics of the Microbial Nitrogen Cycle: Theory, Methods and Applications	2014
Pathogenic *Neisseria*: Genomics, Molecular Biology and Disease Intervention	2014
Proteomics: Targeted Technology, Innovations and Applications	2014
Human Pathogenic Fungi: Molecular Biology and Pathogenic Mechanisms	2014
Applied RNAi: From Fundamental Research to Therapeutic Applications	2014
Halophiles: Genetics and Genomes	2014
Molecular Diagnostics: Current Research and Applications	2014
Phage Therapy: Current Research and Applications	2014
Bioinformatics and Data Analysis in Microbiology	2014
The Cell Biology of Cyanobacteria	2014
Pathogenic *Escherichia coli*: Molecular and Cellular Microbiology	2014
Campylobacter Ecology and Evolution	2014
Burkholderia: From Genomes to Function	2014
Myxobacteria: Genomics, Cellular and Molecular Biology	2014
Next-generation Sequencing: Current Technologies and Applications	2014
Omics in Soil Science	2014
Applications of Molecular Microbiological Methods	2014
Mollicutes: Molecular Biology and Pathogenesis	2014
Genome Analysis: Current Procedures and Applications	2014
Bacterial Toxins: Genetics, Cellular Biology and Practical Applications	2013
Bacterial Membranes: Structural and Molecular Biology	2014
Cold-Adapted Microorganisms	2013
Fusarium: Genomics, Molecular and Cellular Biology	2013
Prions: Current Progress in Advanced Research	2013
RNA Editing: Current Research and Future Trends	2013
Real-Time PCR: Advanced Technologies and Applications	2013
Microbial Efflux Pumps: Current Research	2013
Cytomegaloviruses: From Molecular Pathogenesis to Intervention	2013
Oral Microbial Ecology: Current Research and New Perspectives	2013
Bionanotechnology: Biological Self-assembly and its Applications	2013

Contributors

Norbert Ács
Department of Biotechnology
University of Szeged
Szeged
Hungary

acsn@brc.hu

Shota Atsumi
Department of Chemistry
University of California
Davis, CA
USA

satsumi@ucdavis.edu

Zoltán Bagi
Department of Biotechnology
University of Szeged
Szeged
Hungary

bagiz@brc.hu

Tamás Böjti
Department of Biotechnology
University of Szeged
Szeged
Hungary

tamas.bojti@gmail.com

Peter Dürre
Institute of Microbiology and Biotechnology
University of Ulm
Ulm
Germany

peter.duerre@uni-ulm.de

Dipankar Ghosh
Départment de Microbiologie et Immunologie
Université de Montréal
Montréal, QC
Canada

dipankarghosh.umontreal@gmail.com

Johann F. Görgens
Department of Process Engineering
Stellenbosch University
Stellenbosch
South Africa

jgorgens@sun.ac.za

Daniël C. la Grange
Department of Biochemistry, Microbiology and Biotechnology
University of Limpopo
Sovenga
South Africa

danie.lagrange@ul.ac.za

Riaan den Haan
Department of Microbiology
Stellenbosch University
Stellenbosch
South Africa

rdh@sun.ac.za

Patrick C. Hallenbeck
Département de Microbiologie et Immunologie
Université de Montréal
Montréal, QC
Canada; and
Life Sciences Research Center
Department of Biology
United States Air Force Academy
Colorado Springs, CO
USA

patrick.hallenbeck@umontreal.ca

Ryan E. Hill
LanzaTech NZ Ltd
Parnell
Auckland
New Zealand

ryan.hill@lanzatech.com

Rasmus Ø. Jensen
LanzaTech NZ Ltd
Parnell
Auckland
New Zealand

rasmus.jensen@lanzatech.com

Michael Köpke
LanzaTech NZ Ltd
Parnell
Auckland
New Zealand

michael.koepke@lanzatech.com

Etelka Kovács
Department of Biotechnology
University of Szeged
Szeged
Hungary

kovacse@brc.hu

Kornél L. Kovács
Departments of Biotechnology and Oral Biology and Experimental Dental Research
University of Szeged
Szeged
Hungary; and
Biological Research Center
Hungarian Academy of Sciences
Szeged
Hungary

kovacs.kornel@brc.mta.hu

Taek Soon Lee
Joint BioEnergy Institute
Emeryville, CA
USA; and
Physical Biosciences Division
Lawrence Berkeley National Laboratory
Berkeley, CA
USA

tslee@lbl.gov

Zhimin Li
Key Laboratory of Biofuels
Shandong Provincial Key Laboratory of Energy Genetics
Qingdao Institute of Bioenergy and Bioprocess Technology
Chinese Academy of Sciences
Qingdao
China

lizm@qibebt.ac.cn

Tiangang Liu
Key laboratory of Combinatorial Biosynthesis and Drug Discovery (Ministry of Education)
School of Pharmaceutical Science
Wuhan University
Wuhan
People's Republic of China; and
Hubei Engineering Laboratory for Synthetic Microbiology
Wuhan Institute of Biotechnology
Wuhan
People's Republic of China

liutg@whu.edu.cn

Xuefeng Lu
Key Laboratory of Biofuels
Shandong Provincial Key Laboratory of Energy Genetics
Qingdao Institute of Bioenergy and Bioprocess Technology
Chinese Academy of Sciences
Qingdao
China

lvxf@qibebt.ac.cn

Jordan T. McEwen
Department of Chemistry
University of California
Davis, CA
USA

jtmcewen@ucdavis.edu

Tingting Ning
Key laboratory of Combinatorial Biosynthesis and Drug Discovery (Ministry of Education)
School of Pharmaceutical Science
Wuhan University
Wuhan
People's Republic of China; and
Hubei Engineering Laboratory for Synthetic Microbiology
Wuhan Institute of Biotechnology
Wuhan
People's Republic of China

ttning@whu.edu.cn

Eugéne van Rensburg
Department of Process Engineering
Stellenbosch University
Stellenbosch
South Africa

eugenevrb@sun.ac.za

Orsolya Strang
Department of Biotechnology
University of Szeged
Szeged
Hungary

strangorsi@citromail.hu

Xiaoming Tan
Key Laboratory of Biofuels
Shandong Provincial Key Laboratory of Energy Genetics
Qingdao Institute of Bioenergy and Bioprocess Technology
Chinese Academy of Sciences
Qingdao
China

tanxm@qibebt.ac.cn

Yohei Tashiro
Department of Chemistry
University of California
Davis, CA
USA

ybtashiro@ucdavis.edu

Heinrich Volschenk
Department of Microbiology
Stellenbosch University
Stellenbosch
South Africa

volschenkh@sun.ac.za

Weihua Wang
Key Laboratory of Biofuels
Shandong Provincial Key Laboratory of Energy Genetics
Qingdao Institute of Bioenergy and Bioprocess Technology
Chinese Academy of Sciences
Qingdao
China

wangwh@qibebt.ac.cn

Roland Wirth
Department of Biotechnology
University of Szeged
Szeged
Hungary

roland.w@freemail.hu

Han Min Woo
Clean Energy Research Center
Korea Institute of Science and Technology
Seoul
Republic of Korea; and
Green School
Korea University
Seoul
Republic of Korea

hmwoo@kist.re.kr

Tao Zhu
Key Laboratory of Biofuels
Shandong Provincial Key Laboratory of Energy Genetics
Qingdao Institute of Bioenergy and Bioprocess Technology
Chinese Academy of Sciences
Qingdao
China

zhutao@qibebt.ac.cn

Willem H. van Zyl
Department of Microbiology
Stellenbosch University
Stellenbosch
South Africa

whvz@sun.ac.za

Preface

Energy demand, resource utilization and environmental protection have stimulated the research and development of renewable and sustainable biofuels. The diversified metabolic pathways related to biofuel molecules in nature make microbes offer great potential for applications in biofuel production. The fast developments of modern biotechnologies such as metabolic engineering, '-omics' techniques, and synthetic biology, have shown great promise for constructing highly efficient microbial cell factories with the optimized titre, yield and productivity of biofuels, to make microbial biofuels commercially viable.

This book provides general background, recent advances, technical challenges and future prospect for the production of different biofuel molecules including hydrogen, methane, ethanol, butanol, higher chain alcohols, isoprenoids and fatty acid derivatives by metabolically engineered microbes including bacteria, yeast and microalgae. Some newly emerging approaches for engineering microbes to enhance biofuel production with design and construction of biosynthetic pathways, optimization of genetic and regulatory processes, engineering of enzymes for modified substrate specificity and improved activity, and synthesis of protein scaffolds, are also discussed.

A panel of international experts in this area from Canada, China, Germany, Hungary, Korea, New Zealand, South Africa and the USA worked together on this book. Thanks to all of authors for their precious contributions and thanks to Caister Academic Press for providing me the opportunity to edit the book. I hope and I do believe that the book can be useful to students, scientists and engineers who are interested in microbial biofuel production.

Dr Xuefeng Lu

Metabolic Engineering: Key for Improving Biological Hydrogen Production

1

Dipankar Ghosh and Patrick C. Hallenbeck

Abstract

Hydrogen is a renewable, efficient and clean fuel. Photobiological and fermentative microorganisms hold great promise as a means for biological hydrogen production. Different metabolic pathways come into play for hydrogen evolution in various microbes depending upon the available substrate, enzyme activity, environmental conditions and growth kinetics of the microbial system. Known microbial systems are unable to achieve the theoretical maximal for either molar hydrogen yield or cumulative or specific hydrogen productivity. The major reasons for this are thermodynamic and physiological barriers related to the metabolisms of the different microorganisms. Here we review the development of metabolically engineered microorganisms for improved biohydrogen production.

1.1 Introduction

Biohydrogen is an attractive biofuel replacement for non-renewable fossil fuels. The joint challenges of climate change and diminishing fossil fuel reserves are driving intense research into alternative energy sources. Among the various biofuels options, biohydrogen is a promising future energy carrier due to its potentially higher efficiency of conversion to usable power, low to non-existent generation of pollutants and highest energy content per unit mass of any known fuel (143 GJ/t). Hydrogen can efficiently be converted to electricity by fuel cells; its biological production can utilize renewable energy resources, and finally this process usually operates at ambient temperature and atmospheric pressure (Hallenbeck *et al.*, 2009). Hydrogen can be produced by biological routes (Fig. 1.1); biophotolysis of water using algae and cyanobacteria; photofermentation of organic compounds by phototrophic bacteria; and dark fermentative hydrogen generation from organic substances (Hallenbeck and Ghosh, 2009). However, low hydrogen yields and productivity have been the key bottlenecks limiting the feasibility of biohydrogen production technology. In essence, the major question is 'Does a biological route for hydrogen production exists which can achieve complete substrate conversion (12 mol $H_2 \cdot mol^{-1}$ glucose equivalent)?'. Thermodynamic and metabolic constraints suggest that it would be impossible to find a native microorganism would be able to completely convert pentose- or hexose-based feed stocks to biohydrogen (Thauer *et al.*, 1977). Thus, a goal of microbial hydrogen production research has been to maximize reductant flow in native metabolic pathways to the hydrogen producing enzyme systems. To achieve this goal, metabolic engineering might be the best avenue to achieve or surpass stoichiometric barriers. Thus, the aim would be to use metabolic engineering for the construction, redirection and manipulation of cellular metabolism through the alteration of endogenous and/or incorporation of non-native metabolic pathways to achieve the desired hydrogen yields and productivities using non-food crops or industrial wastes. Extension of the range of substrate utilization is one pivotal issue in metabolic engineering of hydrogen producing microbes. However, except for glucose, the use of other biomass-derived sugars, in particular lignocellulosic biomass, has not been extensively studied so far. Cellulose and hemicellulose need

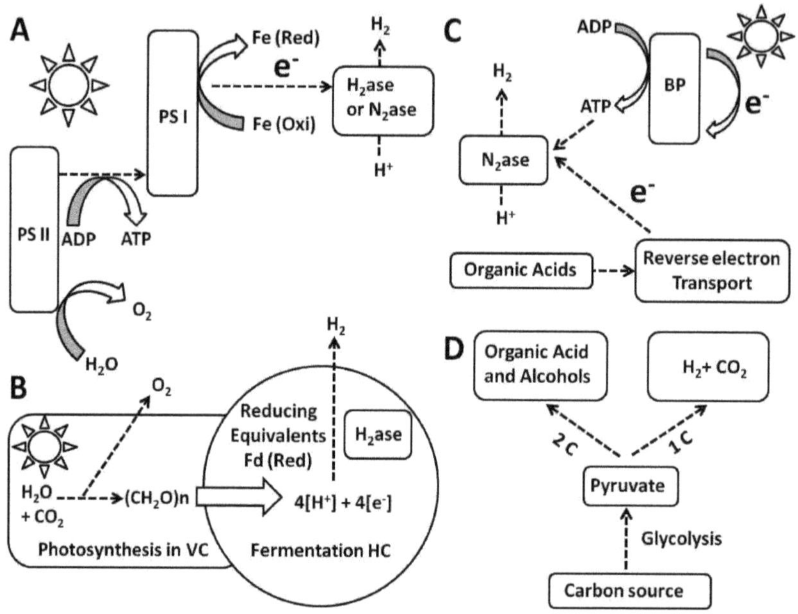

Figure 1.1 Basic routes of biological hydrogen production. (A) Direct biophotolysis. (B) Indirect biophotolysis (VC is vegetative cell while HC heterocyst cell). (C) Photofermentation from organic feedstock through reverse electron flow. (D) Dark fermentation by facultative or strict anaerobic hydrogen producers.

to be completely hydrolysed to fermentable sugars before they can be utilized by most hydrogen producers. The additional hydrolysis required may not be cost-effective and often results in the generation of inhibitory compounds such as furfurals or hydroxyl methyl furfurals (Abril and Abril, 2009). Thus, it would be beneficial if the hydrogen producing microbes can directly utilize cellulosic biomass without hydrolytic pretreatment. In the case of starch, which is often found in agricultural products and wastes, hydrolysis is relatively simple compared with lignocellulosic biomass. Therefore, the incorporation of amylase activity into hydrogen producing microorganisms can improve the economic feasibility of biological hydrogen production processes. This is just one example of improvements that can be made through metabolic engineering.

The major aim of this review is to deeply focus on metabolic pathways in different hydrogen producers. Finally, special emphasis will be focused on metabolic pathway alteration and/or incorporation of novel pathways towards complete conversion of reductant flow of substrate for achieving or surpassing stochiometric biohydrogen yields and current productivities (Fig. 1.2).

1.2 Metabolic engineering of bacterial systems for hydrogen production by dark fermentation

1.2.1 Dark fermentative microbial platforms

Biological hydrogen production by the dark fermentative mode has been studied over the past several decades. Different methodologies for improvement, mostly bioprocess based, have been examined including: reactor configuration, reverse micelle, immobilization, pretreatment of feedstock or industrial wastes, mathematical modelling with computer simulations, co-culture or two-stage technology, switching from batch to continuous mode operations (Hallenbeck and Ghosh, 2009). The main idea behind all of these approaches was to provide utilization of simple substrates (pentose and hexose sugars) by using existing or newly isolated microbial hydrogen producers; secondly, to mimic the natural adaptation while cultivating under laboratory conditions to achieve consistent proven operational modes; thirdly, to improve model systems of different hydrogen producers for scale

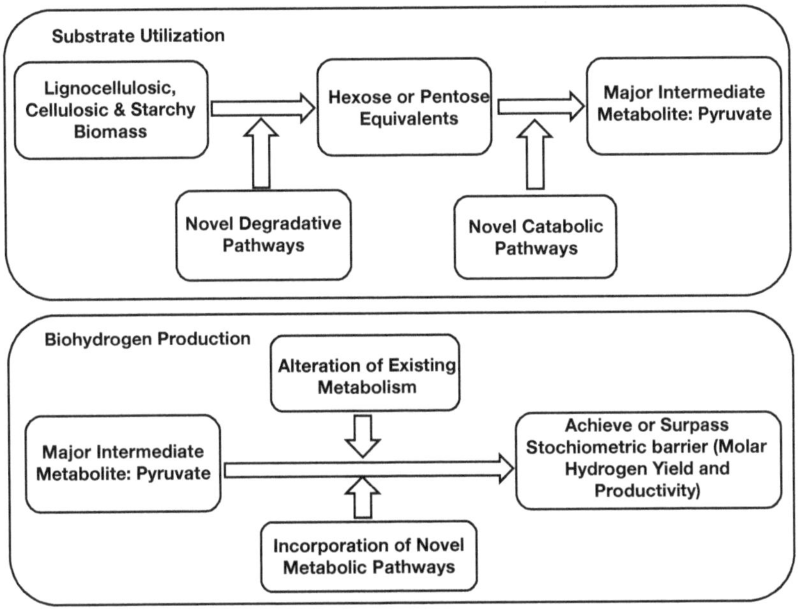

Figure 1.2 Central role of metabolic engineering for improving hydrogen production technology.

up towards commercialization. However, to date no single feasible and consistent technology for biohydrogen production has emerged and conversion efficiencies achieved so far by using these approaches are still suboptimal. In fact, no single native microbial platform has been found that can generate hydrogen consistently at a high enough yield or productivity for commercialization. The key reasons for this are mainly thermodynamic constraints and lack of sufficient knowledge of the different complex metabolic networks, enzyme systems, co-factors, reductant flow (ferredoxin, flavodoxin, NADH, NADPH, $FADH_2$), involved that couple the degradation of suitable substrates in different hydrogen producers. In fact, several different microbial platforms are available in which it is necessary to understand the complex metabolic pathways in order to carry out redirection of reductant flow to hydrogen. The minimum criteria in this regard for a potential microbial hydrogen producer platform would be availability of the complete genome sequence, capacity for broad substrate utilization, fast growth kinetics, and the ability to grow at ambient temperature and pressure.

So far *Escherichia coli* (a facultative anaerobic hydrogen producer) would appear to be the best microbial platform owing to its well-characterized genome, well-known metabolic pathways (Fig. 1.3) and its ability to utilize a wide range of carbon sources (Ghosh and Hallenbeck, 2009). In addition, *E. coli* has potential benefits over some other microbes since it exhibits rapid growth and has simple nutritional requirements as well as being tolerant to oxygen. A metabolic pathway database, *EcoCyc*, is available, allowing for metabolic network prediction (Keseler *et al.*, 2005). In *E. coli* a theoretical maximum hydrogen yield of 2 mol H_2/mol glucose can be achieved (Hallenbeck and Ghosh, 2012). *Enterobacter aerogenes* is another potential hydrogen producing microbial platform. *E. aerogenes* is a facultative anaerobe and can grow in the presence of oxygen, making it much easier to manipulate in bioprocesses than strict anaerobes. As well, it shows uninhibited growth properties in an atmosphere of 100% H_2 (Nakashimada *et al.*, 2002) and its high growth rate and high hydrogen production rate suggest that it has good potential for large-scale hydrogen production (Rachman *et al.*, 1998) (Fig. 1.4). Another potential candidate for hydrogen production are the *Clostridia*, strict anaerobic bacteria. Hydrogen is normally being generated through the action of PFOR and NFOR pathways. The maximum

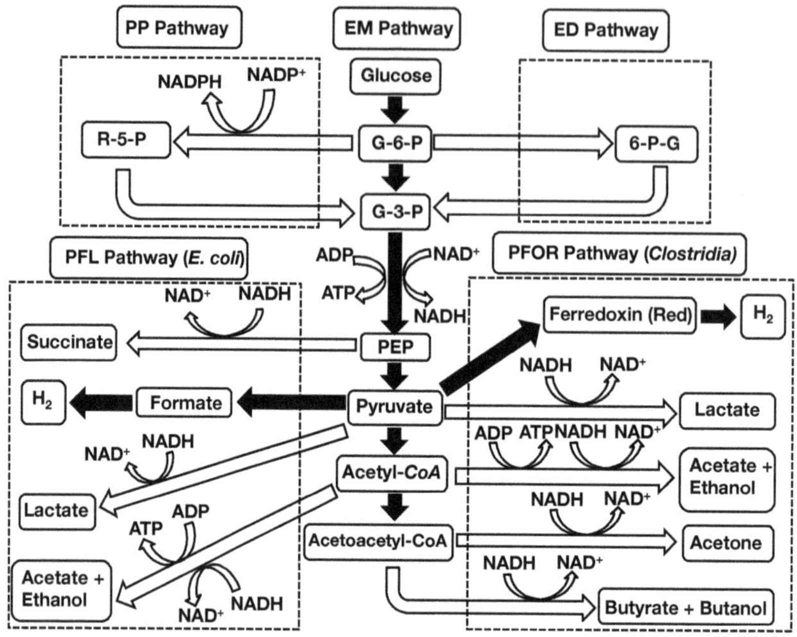

Figure 1.3 Metabolic pathways for hydrogen production in facultative (*E. coli*) and strict anaerobe (*Clostridia*) fermentative microbial systems.

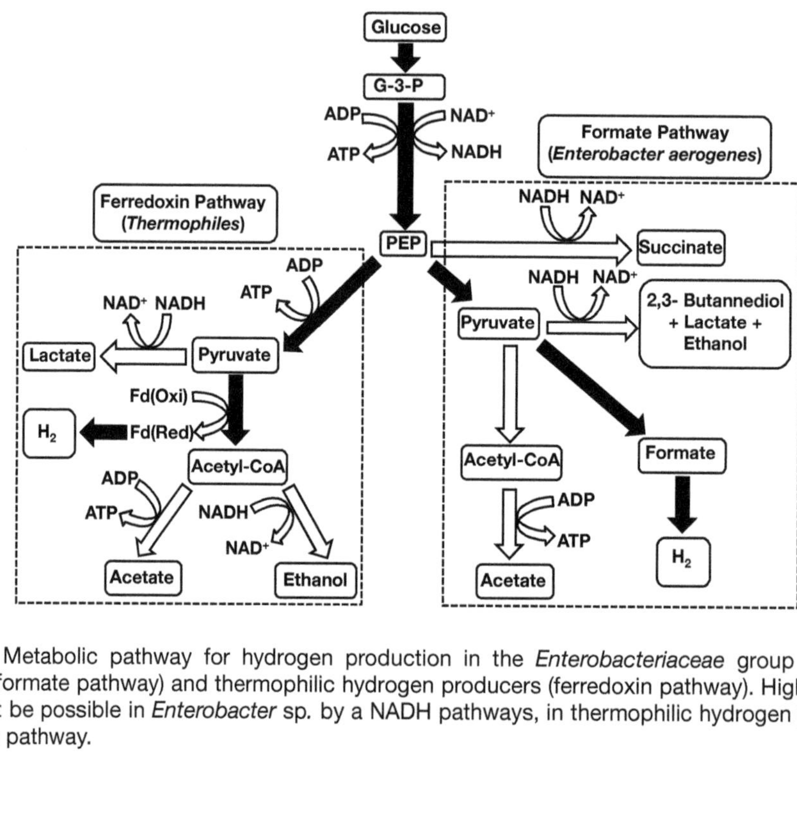

Figure 1.4 Metabolic pathway for hydrogen production in the *Enterobacteriaceae* group of hydrogen producers (formate pathway) and thermophilic hydrogen producers (ferredoxin pathway). Higher maximum yields might be possible in *Enterobacter* sp. by a NADH pathways, in thermophilic hydrogen producers by a ferredoxin pathway.

theoretical hydrogen yield that could be achieved is 4 mol H_2/mole glucose (Hallenbeck *et al.*, 2012). Degradation of glucose by central metabolism results in lactate, acetate, ethanol, butyrate, acetone, butanol, and butyrate as fermentation end metabolites (Fig. 1.3). The major advantage to using *Clostridia* is its capacity to produce hydrogen by removing excess reductant during growth on biomass materials (Kaji *et al.*, 1999; Morimoto *et al.*, 2005). Thermophilic bacteria could be interesting fermentative hydrogen producing platform in terms of hydrogen yield. Their higher temperature optimum also provides additional advantages relative to mesophilic hydrogen producers, including the ability to survive and grow under environmental conditions at which most contaminating hydrogen consuming microbial communities cannot. Elevated temperatures are also more favourable for biomass based feedstock pre-treatment, maximization of enzymatic reaction rates, and favour the equilibrium point of hydrogen generation pathway in the direction of hydrogen evolution (Veit *et al.*, 2008). Moreover, several species in this group, *Thermotoga* sp., *Thermoanaerobacter* sp. and *Caldicellulosiruptor* sp., are able to degrade complex polysaccharides and to assimilate pentose sugars (Van Niel *et al.*, 2002). Thermophilic glucose metabolism produces acetate as major metabolic end product and large amounts of NADPH (Fig. 1.4). The maximum yields of hydrogen that can be achieved through fermentation of glucose by thermophilic hydrogen producers are > 3 mol H_2/mol glucose (Jones, 2008).

1.2.2 Hydrogenases of dark fermentative hydrogen producers: potential targets for metabolic engineering

Hydrogenases are the key enzymes in microbial hydrogen producers responsible for the reduction of protons to hydrogen (Table 1.1). Crystal structures of several hydrogenases are available which unveil the molecular geometry and mode of action of their active centres (Fontecilla-Camps *et al.*, 2007). Extensive molecular phylogenetic investigations have revealed that microbial hydrogen producers harbouring genes encoding hydrogenases are ubiquitous in nature and are taxonomically diverse (Vignais and Billoud, 2007). These enzymes are utilized to produce energy, disposing of reducing equivalents produced by fermentative pathways, or, when carrying out hydrogen oxidation, generating reduced co-factors participating in many biochemical reactions central to cellular metabolism. Hydrogenases are classified into three distinct categories based on their active sites: Ni–Fe hydrogenases, Fe–Fe hydrogenases, and Fe-hydrogenases. Ni–Fe hydrogenases are only found in archaea and bacteria, promoting either hydrogen oxidation or generation *in vivo*, while Fe–Fe hydrogenases are distributed in bacteria and eukaryotes where they almost exclusively catalyse hydrogen production. In contrast, Fe-hydrogenases are only found in hydrogenotrophic methanogenic archaea where they use hydrogen to provide the reducing equivalents necessary for the sub-terminal steps of carbon dioxide reduction to methane production. All of these enzyme systems contain complex metallocentres that are very oxygen sensitive, with Fe–Fe hydrogenases being rapidly completely and irreversibly inactivated upon oxygen exposure. In contrast, Ni–Fe hydrogenases react with oxygen to produce intermediates that can be reactivated by reduction of the metal centres of the enzyme active site. Tolerance to oxygen can be a pivotal issue for the utilization of hydrogenases in some applications of metabolic engineering (Lyon *et al.*, 2004). One subclass of the Ni–Fe hydrogenases seen in nature are the [Ni–Fe–Se]-hydrogenases mostly found in sulfate-reducing bacteria. [Ni–Fe–Se]-hydrogenases are oligomeric, cytoplasmic or membrane-bound proteins generally involved in the oxidation of hydrogen (Garcin *et al.*, 1999).

Much effort is going into stable hydrogenase expression in heterologous microbial hosts in order to improve hydrogen yields and productivities. In addition, in many hydrogen producers hydrogen yields and productivities may be decreased due to the presence of uptake hydrogenases. Knocking out the genes encoding the uptake hydrogenases could be another possible route to increasing hydrogen production in metabolically engineered microorganisms.

Table 1.1 Hydrogenase enzyme systems as important segment on metabolic engineering of potential hydrogen producers

H_2-ase	Source	Microorganisms	Function	Localization
Ni–Fe H_2-ase (heterodimeric or multimeric)	Anaerobes, photosynthetic bacteria, cyanobacteria, sulfate-reducing bacteria, methanogens	**Group A** *Desulfovibrio gigas* *Desulfovibrio gigas* *Methanosarcina mazei* *Aquifex aeolicus* *Hydrogenobacter thermophilus* *Thiocapsa roseopercina* *Acidianus ambivalens* *Ralstonia eutropha* **Group B** *Methanothermobacter marburgensis* *Pyrococcus furiosis* *Ralstonia eutropha* *Thiocapsa roseopersicina* *Synechiocystis* sp. PCC6803 *Escherichia coli* **Group C** *Escherichia coli* *Rhodospirillum rubrum* *Rhodocyclus gelatinosus* *Carboxydothermus hydrogenoformans* *Methanosarcina barkeri* *Pyrococcus furiosus* **Group D** *Streptomyces* spp. **Group E*** *Anbaena* sp. PCC7120 *Bradyrhizobium japonicum* *Rhodobacter capsulatus* *Ralstonia eutropha* *Desulfovibrio fructosovorans*	Uptake of H_2	Membrane-bound, cytoplasmic, periplasmic
Fe–Fe H_2-ase (monomeric, dimeric, trimeric, and tetrameric)	Facultative anaerobes, strict anaerobes, thermophiles, Chlorophycean algae, sulfate-reducing bacteria, photosynthetic bacteria	*Clostridium pasteurianum* *Megasphaera elsdenii* *Desulfovibrio* spp. *Scenedesmus obliquus* *Shewanella oneidensis* *Clostridium pasteurianum* *Desulfovibrio desulfuricans* *Clostridium beijerinckii* *Thermotoga maritime* *Thermoanaerobacter tengcogensis* *Caldicellulosirupto saccharolyticum* *Eubacterium acidaminophilum* *Clostridium thermocellum* *Desulfovibrio fructosovorans* *Rhodobacter capsulatus*	Production of H_2 from reduced ferredoxin and formate. production of H_2 by Rnf complex (NADH, NADPH)	Cytoplasmic soluble, periplasmic, chloroplast, hydrogenosomes
Fe-H_2-ase (monomeric)	Hydrogenotrophic methanogenic Archaea	*Methanocaldococcus jannaschii* *Methanothemobacter marburgensis*	Production of H_2 by the dehydrogenation of methylene-tetrahydromethanopterin	Cytoplasmic
Ni–Fe–Se-H_2-ase (oligomeric)	Sulfate-reducing bacteria, methanogens	*Desulfomicrobium baculatum*	Oxidation of H_2	Membrane-bound, cytoplasmic

Ni–Fe hydrogenases

The Ni–Fe hydrogenases are heterodimeric proteins consisting of small (S) and large (L) subunits. The small subunit contains three iron–sulfur clusters while the large subunit contains the active site, a nickel–iron centre which is connected to the solvent by a molecular tunnel (Liebgott *et al.*, 2010). In some Ni–Fe hydrogenases, one of the Ni-bound cysteine residues is replaced by selenocysteine. On the basis of sequence similarity, however, the Ni–Fe and Ni–Fe–Se hydrogenases should be considered a single superfamily. To date, periplasmic, cytoplasmic, and membrane-bound hydrogenases have been found. The Ni–Fe hydrogenases, when isolated, are found to catalyse both H_2 evolution and uptake, with low-potential multi-haem cytochromes such as cytochrome c3 acting as either electron donors or acceptors, depending on their oxidation state. Generally speaking, however, Ni–Fe hydrogenases are more active in oxidizing H_2. Like Fe–Fe hydrogenases, Ni–Fe hydrogenases are known to be inactivated by molecular oxygen (O_2).

Recently, a novel hydrogenase participating in the hydrogen metabolism pathways of *Ralstonia eutropha* (Fig. 1.5) has been revealed to be oxygen tolerant (Burgdorf *et al.*, 2005). This finding has increased the hope that hydrogenases can one day be used in the photosynthetic production of molecular hydrogen via water splitting. The hydrogenases of *R. eutropha* are like other typical [Ni–Fe] hydrogenases in that they are made up of a large and small subunit. However, the hydrogenases of *R. eutropha* are different from typical [Ni–Fe] hydrogenases because they are tolerant to oxygen and are not inhibited by CO. One hydrogenase is a regulatory hydrogenase (RH) that signals to the cell that hydrogen is present. RH is a protein containing large and small [Ni–Fe] hydrogenase subunits attached to a histidine protein kinase subunit. Hydrogen gas is oxidized at the [Ni–Fe] centre in the large subunit and in turn reduces the [Fe–S] clusters in the small subunit. It is unknown whether the electrons are transferred from the [Fe–S] clusters to the protein kinase domain. The histidine protein kinase activates a response regulator. The response regulator is active in the dephosphorylated form. The dephosphorylated response regulator promotes the transcription of the membrane-bound hydrogenase and soluble

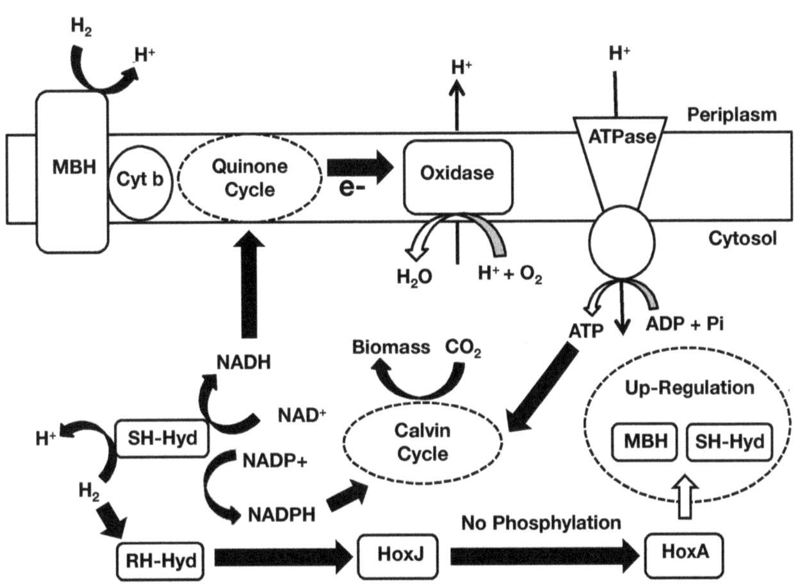

Figure 1.5 Hydrogen metabolism in *Ralstonia eutropha* involving MDH and SH hydrogenase (adopted and modified from Constant and Hallenbeck, 2013). The membrane-bound uptake hydrogenase (MBH) and soluble hydrogenase (SH) are under control of the H_2-sensing hydrogenase (RH). Cytb, cytochrome b; HoxJ, histidine protein kinase; HoxA, Ntrc-like response regulator.

hydrogenase. The membrane-bound hydrogenase (MBH) is linked to the respiratory chain through a specific cytochrome *b*-related protein in *R. eutropha.* Hydrogen gas is oxidized at the [Ni–Fe] active site of the large subunit and the electrons are shuttled through the [Fe–S] clusters in the small subunit to the cytochrome *b*-like protein. MBH is located on the outer cytoplasmic membrane. It produces energy for the cell by funnelling electrons into the respiratory chain and by increasing the proton gradient. *R. eutropha* MBH is not inhibited by CO and is tolerant to oxygen. The soluble hydrogenase (SH) creates NADH by oxidizing hydrogen gas. The SH is a heterodimeric protein with two subunits making up the large and small subunits of the [Ni–Fe] hydrogenase and the other two subunits comprising a protein similar to Complex I (as shown in the accompanying Fig. 1.5). The [Ni–Fe] active site oxidizes hydrogen gas, transferring electrons to a FMN cofactor, then to [Fe–S] clusters of the small hydrogenase subunit, then to another FMN cofactor and finally to NAD^+, thus providing reductant for carbon dioxide fixation when *R. eutropha* is growing autotrophically, or otherwise for biosynthetic reactions.

As previously mentioned, Ni–Fe hydrogenases are in general composed at a minimum of two structural subunits. The small β-subunit (~30 kDa) carries three Fe–S clusters which act to relay electrons between the active centre of the large subunit and the physiological electron mediator. The large subunit α (~60 kDa) has four highly conserved cysteine amino acid residues co-ordinating the Ni–Fe active site. The Ni–Fe active centre consists of two cyanide (CN^-) and one carbon monoxide (CO) ligands, important for hydrogen activation. Based on the consensus L1 and L2 amino acid signatures of the large α subunit, Ni–Fe hydrogenases are divided into nine groups (Constant and Hallenbeck, 2013).

Group A Ni–Fe hydrogenases allow microorganisms to use hydrogen as an energy source by coupling hydrogen oxidation to the reduction of carbon dioxide, sulfate, fumarate, nitrate, and iron in anoxygenic environments or under aerobic conditions. The enzyme couples directly with the electron transport chain, channelling electrons to the quinine pool through the membrane integral bis-Fe- cytochrome b.

Group B Ni–Fe hydrogenases are bi-directional heteromultimeric cytoplasmic associated proteins with modules capable interacting with cofactors such as NAD^+ or $NADP^+$ and F_{420}. Their activity is bidirectional *in vivo*. These enzymes participate in electron transport and ensure cellular redox balance.

Group C Ni–Fe hydrogenases are highly divergent from the other groups. These enzymes reduce protons to dispose of excess reducing equivalents produced during fermentation via the anaerobic oxidation of organic compounds. In general this group of enzymes is composed of four hydrophilic subunits and two integral membrane proteins which have a similar function to NADH::quinine oxidoreductase, suggesting that CO oxidation and hydrogen production reactions are coupled to the formation of an energy-conserving proton motive force across the cytoplasmic membrane.

Group D Ni–Fe hydrogenases are high-affinity Ni–Fe hydrogenases capable of oxidizing hydrogen gas at the minute concentrations at which it is found in the atmosphere. These high-affinity novel hydrogenases might in fact support specific ecosystems in nature, ensuring the stability of microbial communities and the ecological balance (Constant *et al.*, 2011). In addition, genes involved in maturation of Ni–Fe hydrogenases have been extensively studied in a variety of organisms; in *E. coli* (*hyp*-operon; *hyc*-operon; *hya*-operon) in *R. eutropha* (*hox*-operon) and in the methanogenic archaea (*frh*-operon; *vht*-operon *eha*-operon; *ehb*-operon) (Thauer *et al.*, 2010).

Fe–Fe hydrogenases

Fe–Fe hydrogenases are a distinct class of hydrogenase enzyme characterized by their catalytic H-cluster which contains iron as the only metal. The H-cluster consists of a [4Fe–4S] subcluster linked by a cysteine thiolate to a modified 2Fe subcluster with unique non-protein ligands, including CO and CN^-. The hydrogenase maturation enzymes HydF, HydE, HydG are together responsible for the synthesis of this 2Fe subcluster as well as the non-protein ligands. Fe–Fe hydrogenase requires at least one inorganic, toxic ligand, CO, as a necessary constituent of its active centre. As well, Fe–Fe hydrogenases also require CN^-. How these toxic ligands are produced by cellular metabolic pathway and integrated into the complex metal centres is of considerable interest. The CN^- of the active centre of Fe–Fe hydrogenases (H-cluster) is derived from the S-adenosylmethionine-dependent breakdown of tyrosine residue by HydG protein to produce *p*-cresol and cyanide (Kuchenreuther *et al.*, 2009). It is most likely that free CN^- is not liberated and that the CN-generated is bound immediately to HydF, which is thought to function as a scaffold for the assembly of at least part of the H-cluster (McGlynn *et al.*, 2008). Although Fe–Fe hydrogenases are of limited distribution, all of the enzymes in this category contain a basic domain capable of H-cluster binding. In addition, genomic sequence analysis has revealed a wide variety of different molecular structures (Meyer, 2007). Fe–Fe hydrogenases are found in four different forms; monomeric, dimeric, trimeric and tetrameric. Of course the main function of most of the members of this diverse protein family is probably hydrogen production through involvement with cellular fermentative metabolism. In general, this family of hydrogenases is thought to primarily use the pyruvate-ferredoxin oxidoreductase (PFOR) system to convert pyruvate to acetyl-CoA and carbon dioxide, producing reduced ferredoxin. Reduced ferredoxin donates electrons to Fe–Fe hydrogenase, enabling proton reduction to hydrogen (Fig. 1.6).

Fe hydrogenases

Fe-hydrogenases (Hmd) are homodimeric cytoplasmic enzymes with 38 kDa subunits. These enzymes utilize hydrogen and carbon dioxide to generate methane gas. This category of enzymes differs in several standpoints from the previously discussed Ni–Fe and Fe-hydrogenases. Fe-hydrogenases are devoid of Fe–S clusters and do not mediate the reduction of one electron acceptors (F_{420}, NAD^+, $NADP^+$ and methyl viologen). These enzymes do not even catalyse the exchange between hydrogen and protons of water in the absence of an electron mediator or the conversion of *para* and *ortho* hydrogen (Thauer *et al.*, 1996). The crystallographic analysis of the Fe-hydrogenase (Hmd) confirmed that an iron atom is co-ordinated by a reduced cysteine residue of the holoenzyme, two CO, and one guanylpyridol molecule, and a fifth ligand (Shima *et al.*, 2008). Two additional isozymes of Hmd exist, Hmd II, III. The relative amount of Hmd II transcript and protein was higher in cells grown under 5% hydrogen than for the 80% hydrogen grown cells, a level of hydrogen essential for Hmd III transcription (Afting *et al.*, 2000). The ecological benefit of Hmd may be related to a unique flexibility in hydrogen metabolism, providing the archaea with a high capacity system to exploit elevated hydrogen concentrations escaping from thermal vents or in close proximity to syntrophic hydrogen-producing bacteria. The maturation apparatus and corresponding mechanism of Hmd synthesis is not yet fully understood.

The distribution of Fe-hydrogenases is very restricted as they are found only in some hydrogenotrophic methanogenic archaea. They also feature a fundamentally different enzymatic mechanism (Fig. 1.7) in terms of redox partners and how electrons are delivered to the active site. In Ni–Fe and Fe–Fe hydrogenases, electrons travel over a long distance through a series of metallo-organic clusters and the active site structures remain unchanged during catalysis. In Fe-hydrogenases, however, electrons are directly delivered to the active site over a short distance. Methenyl-H_4MPT^+, a cofactor, directly accepts the hydride from H_2. Fe-hydrogenase is also known as H_2-forming methylenetetrahydromethanopterin (methylene-H_4MPT) dehydrogenase, because its function is the reversible reduction of methenyl-H_4MPT^+ to methylene-H_4MPT (Salomone-Stagnia *et al.*, 2010). The hydrogenation of a methenyl-H_4MPT^+ occurs instead of H_2 oxidation/production, which is the case for the other two types of hydrogenases. While the exact mechanism of catalysis is still under study, a recent finding suggests that molecular hydrogen is first heterolytically cleaved by Fe (II), followed by transfer of the hydride to the acceptor (Hiromoto *et al.*, 2009).

From a metabolic engineering standpoint these multifaceted hydrogenases are of fundamental

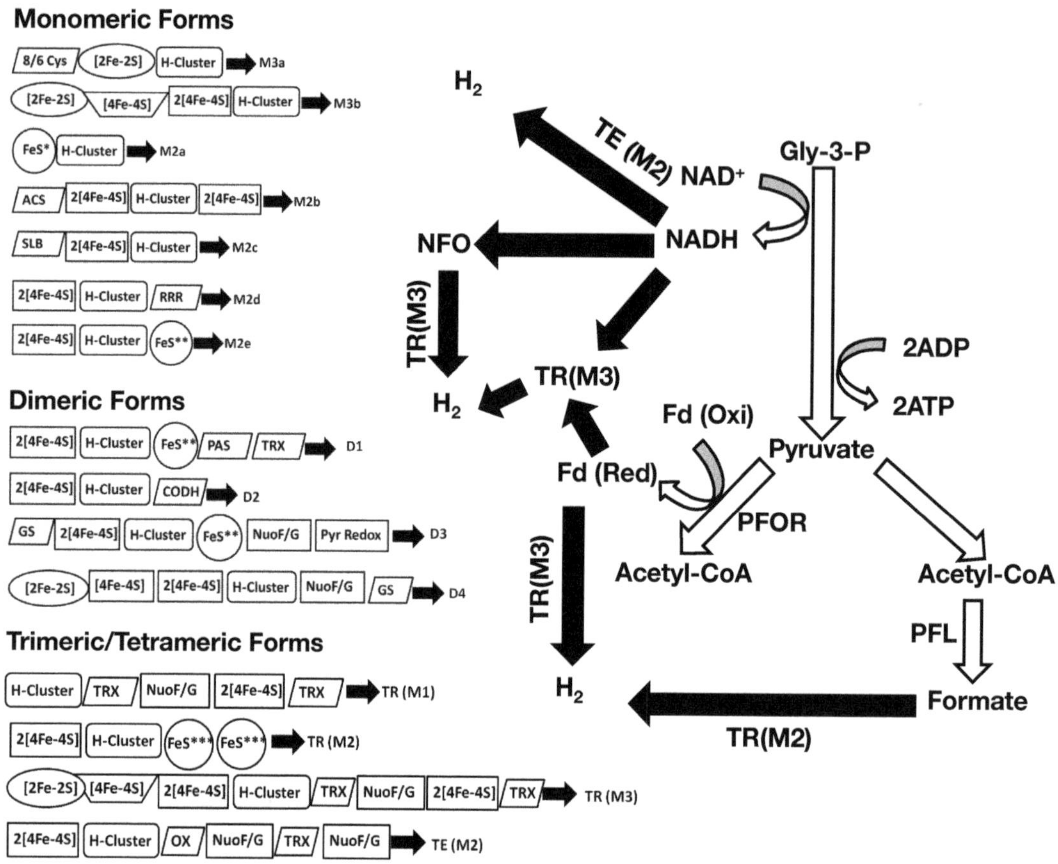

Figure 1.6 Modular domain structure of Fe–Fe hydrogenases with special emphasis on hydrogen metabolism mediated by Fe–Fe hydrogenase. Black arrows indicate the reductant flow for hydrogen production using different forms of Fe–Fe hydrogenases following the PFOR and PFL pathways. Many of the dimeric and trimeric forms contain Nuo subunits, suggesting a physiological role of NADPH in the function of these enzymes. Some clostridial species can contain up to five different hydrogenase forms. The well-known monomeric *C. pasteurianum*-type hydrogenase (type M3) catalyses H_2 production from reduced ferredoxin, either produced directly from pyruvate or produced by NADH oxidation by NADH: ferredoxin oxidoreductase (NFO). NFO could be the soluble activity described several decades ago or the more recently described membrane complex of the Rnf family. Some multimeric hydrogenases [TE (M2)], similar to Hnd from the sulfate reducers, could possibly evolve H_2 directly from NADPH at very low H_2 partial pressures. Finally, the newly described trimeric bifurcating hydrogenase [TR (M3)] could simultaneously use NADH and reduced ferredoxin to evolve H_2 with the excess energy available from ferredoxin oxidation driving the unfavourable oxidation of NADH.

interest in understanding how the basic process of hydrogen oxidation or proton reduction at the enzyme level is coupled to various types of microbial metabolic networks. Obviously, attempts to increase biological hydrogen yields and productivities will benefit from a greater understanding of the key enzymes involved. Fundamental knowledge about the reaction kinetics of metabolic pathways may be used to advantage to construct more efficient artificial catalytic systems or in developing hydrogenases active in hydrogen production in the presence of oxygen. Finally, metabolic engineering may increase hydrogen yields and productivity by creating engineered hydrogen producing strains containing hydrogenases linked to various native or synthetic cellular metabolisms.

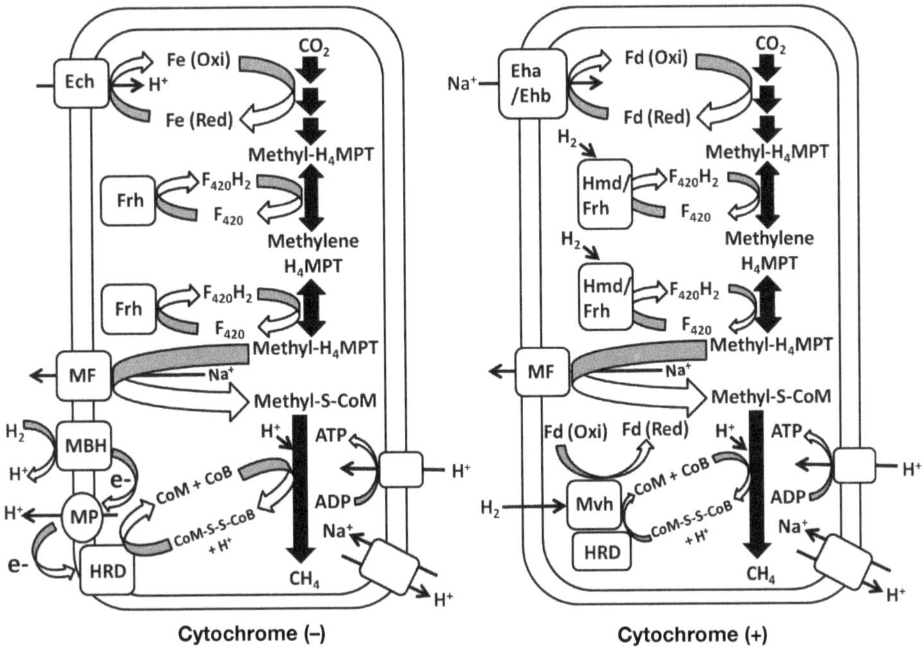

Figure 1.7 Hydrogen metabolism in methanogenic archaea (Adopted and modified from Thauer *et al.*, 2010). Nomenclature: (Group 4): energy-converting hydrogenase (Ech); (Group 3A): F_{420}-reducing hydrogenase (Frh); (Group 1): methyl transferase (MF); membrane-bound hydrogenase (MBH); (Group 4): heterodisulfide reductase (HRD); ferredoxin-reducing, energy-converting hydrogenases (Eha/Ehb); (group 3C): [Fe]-hydrogenase (Hmd); methylviologen reducing hydrogenase (Mvh).

1.2.3 Application of metabolic engineering to improving biological hydrogen production through dark fermentation

In general dark fermentative hydrogen production from pyruvate formed by central metabolism is via one of two basic pathways; the PFL (pyruvate–formate lyase) pathway (Fig. 1.8) or the PFOR (pyruvate–formate oxidoreductase) metabolic network (Fig. 1.9). However, these pathways do not permit the cell to surpass the stochiometric maximum of 2 or 4 mol H_2/mol glucose. The primary factors that restrict the amount of reductant that can be channelled into hydrogen in dark fermentative hydrogen producers are the incomplete oxidation of organic substrates, and the dissipation of reductant through alternative metabolic pathways that generate other by-products of fermentation. In addition to consuming electrons that might otherwise be used for hydrogen production, the production of these metabolic end products, acids and alcohols, also restrains growth (Kontur *et al.*, 2012). Thus, metabolic engineering offers a novel avenue to enhance hydrogen yields through the elimination of competing metabolic pathways or the introduction of novel metabolic pathways into potential hydrogen producers.

Escherichia coli

Metabolic engineering of native hydrogenases in *Escherichia coli*

E. coli is, in general, regarded as one of the best microbial platforms for metabolic engineering. *E. coli* carries four hydrogenase operons, *hya*, *hyb*, *hyc*, *hyf*, all encoding Ni–Fe hydrogenases. Functionally, *E. coli* hydrogenases can be divided into two classes, hydrogen oxidizing (H_2-ase 1 and H_2-ase 2) and energy-conserving, proton-reducing (H_2-ase 3 and H_2-ase 4) (Fig. 1.10). H_2-ase 1 and H_2-ase 2 are encoded by *hya*, *hyb* operons respectively, and function in the periplasmic uptake of hydrogen. H_2-ase 3, and potentially H_2-ase 4, is an important component of the cytoplasmically oriented formate hydrogen lyase complex. H_2-ase 3, encoded by the *hyc* operon, is a major part of

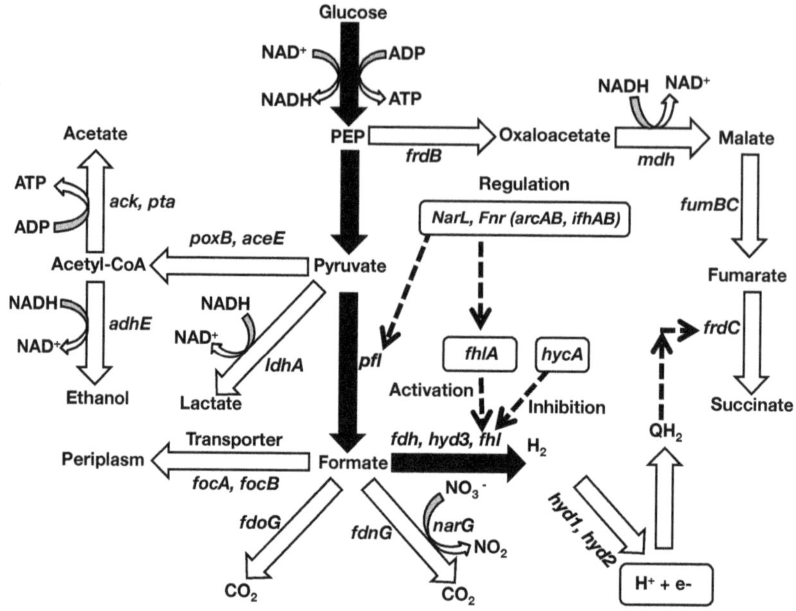

Figure 1.8 Metabolic engineering of the PFL pathway in facultative hydrogen producers (*Escherichia coli*). The maximum theoretical hydrogen yield that can be achieved is 2 moles of H_2 per mole of glucose if reductant that goes to several terminal metabolites can be redirected towards hydrogen evolution. Black arrows indicate the energy flow and dashed arrows indicate the global regulation of the PFL system. Here glucose is broken down to pyruvate through the glycolytic (Embden–Meyerhof–Parnas) pathway. The nomenclature is as follows *focA*, *focB*, Formate transporters; Phosphotransacetylase (*pta*); Acetate kinase (*ack*); pyruvate dehydrogenase (*aceE*); Pyruvate oxidase (poxB); NarL and Fnr, Global transcription regulators; lactate dehydrogenase (*ldhA*); fumarate reductase (*frdBC*); acetyl-CoA, Acetyl coenzyme A (*frdBC*); Phosphoenolpyruvate (PEP); Malate dehydrogenase (*mdh*); fumarate reductase (*frdC*); lactate dehydrogenase (*ldhA*); pyruvate dehydrogenase (*aceE*); pyruvate formate lyase (*pfl*); formate hydrogen lyase (*fhl*); Hydrogenase 3 (*hyd3*); formate dehydrogenase-H (*fdhH*); FHL activator (*fhlA*); FHL repressor HycA, Global regulators (Fnr, *arcAB*, *ihfAB*); Uptake hydrogenases; hydrogenase 1 (*hyd1*) and hydrogenase 2 (*hyd2*); Formate is metabolized by formate dehydrogenase-N (*fdnG*) which is linked with nitrate reductase A (*narG*) and formate dehydrogenase-O (*fdoG*); quinol (QH_2).

the formate hydrogen lyase complex and produces hydrogen from formate (FHL-1). H_2-ase 3 is active during mixed-acid fermentation at lower pHs. The *hyf* operon encoding H_2-ase 4 has been found to be cryptic and non-functional under normal conditions (Self *et al.*, 2004), therefore, all hydrogen production from formate is by the FHL complex formed by Hyd-3 and encoded by the *hyc* operon. Hydrogen produced from formate can be recycled by uptake hydrogenases Hyd-1 and Hyd-2. Thus, one possible approach to improving hydrogen production is to knock out the genes encoding uptake hydrogenases. Experiments have shown that a Hyd-2 hydrogenase mutation results in a high molar hydrogen yield while Hyd-1 knock out strains do not show an improved hydrogen yield. HycE (the large subunit of Hyd-3) has been subjected to the random protein engineering in an attempt to enhance the hydrogen yield (Maeda *et al.*, 2008a). HycE mutants having a 240 amino acid deletion at the carboxyl-terminus as well as strains carrying a 74 amino acid carboxy-terminal truncation show an improved hydrogen yield (30-fold). *fdnG* (α-formate dehydrogenase-N) and *fdoG* (α-formate dehydrogenase-O) encode formate dehydrogenases which consume formate and therefore might decrease hydrogen yields. Indeed, strains mutated in *fdoG* have been found to have improved hydrogen yield from formate as substrate (Maeda *et al.*, 2007). It is interesting to note that the various *E. coli* hydrogenases are differentially localized in the cell, permitting in

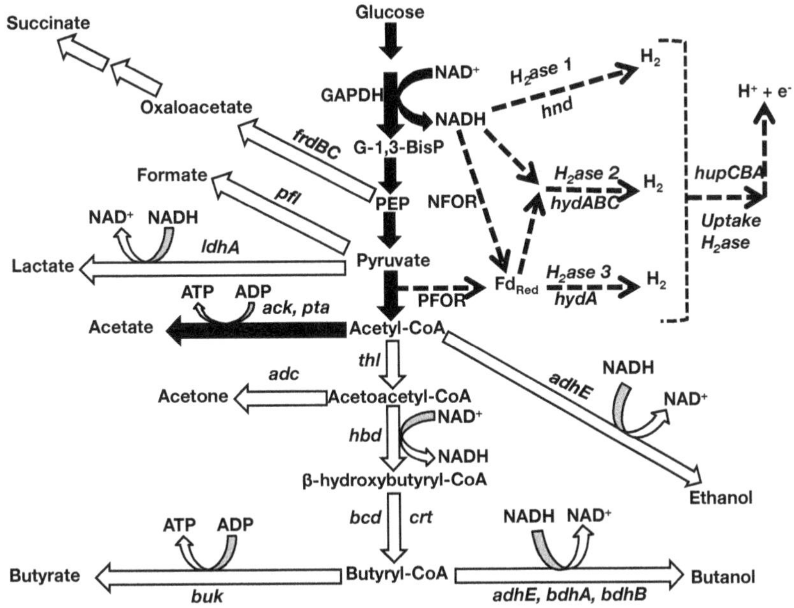

Figure 1.9 Metabolic engineering of strict anaerobic hydrogen producers (*Clostridia*). The PFOR, NFOR and PFL metabolic networks are involved in H_2 production and thus are targets for improving hydrogen yield. The maximum theoretical hydrogen yield that can be achieved is 4 moles of H_2 per mole of glucose without novel pathways (dashed lines) and if electron flow to several terminal metabolites can be redirected towards hydrogen evolution. Three different types of Fe–Fe hydrogenases have been described in *Clostridia*. HydA, Fd-dependent hydrogenase, Hnd, a NADPH-dependent hydrogenase, and HydABC, a novel bifurcating hydrogenase that reacts simultaneously with NADH and reduced ferredoxin; L-lactate dehydrogenase (*ldhA*); aldehyde-alcohol dehydrogenase (*adhE*); phosphotransacetylase (*pta*), thiolase (*thl*); acetoacetate decarboxylase (*adc*); b-hydroxybutyryl-CoA dehydrogenase (*hbd*); crotonase (*crt*); butyryl-CoA dehydrogenase (*bcd*); butanol dehydrogenase A and B (*bdhA* and *bdhB*); butyrate kinase (*buk*); uptake hydrogenase (*hupCBA*) fumarate reductase (*frdBC*); pyruvate formate lyase (*pfl*). Black arrows indicate the energy flow and dashed arrows indicate the novel pathways could be introduced to enhance reductant flow for hydrogen production.

theory hydrogen recycling which would develop a proton motive force across the cytoplasmic membrane. As mentioned before, *E. coli* Hyd-3, the hydrogenase responsible for hydrogen production is membrane bound and oriented towards the cytoplasm. Hyd-3 reduces protons to hydrogen with electrons derived from formate. The evolved hydrogen diffuses freely across the cytoplasmic membrane to the periplasmic space where it encounters Hyd-1. Hyd-1 then oxidizes the molecular hydrogen, producing protons in the periplasm and donating the extracted electrons to membrane-bound electron carriers which can reduce NAD^+ to NADH on the cytoplasmic side. The proton gradient thus formed can be used to drive ATP synthesis. The NADH is used to reduce acetyl-CoA to the more reduced metabolic end product ethanol to maintain redox balance (Hallenbeck *et al.*, 2011) (Fig. 1.10).

Metabolic engineering through heterologous hydrogenase overexpression in *Escherichia coli*

Another way to potentially increase hydrogen yields and productivities would be through the heterologous expression of foreign hydrogenases in *E. coli*. A stable Fe–Fe hydrogenase expression system has been created by co-expression of *hyd*FEG maturation system from the strict anaerobic bacterium *Clostridium acetobutylicum* (King *et al.*, 2006). Although *E. coli* is an efficient host expression system, it lacks NADH-dependent hydrogenase, creating hurdles for the engineering of hydrogen metabolism. There are different

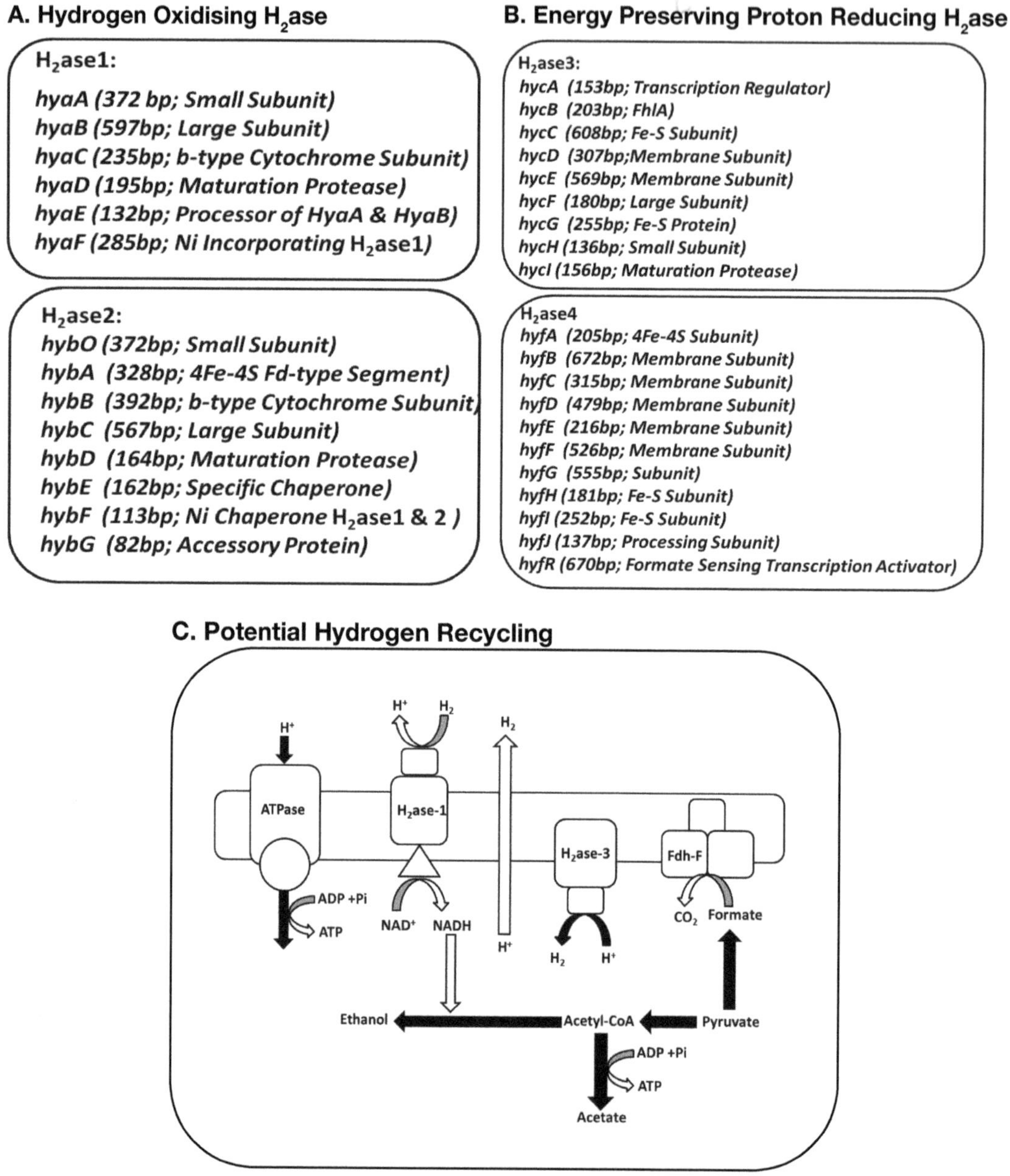

Figure 1.10 Genes of hydrogenase operon with potential hydrogen recycling in *E. coli* (C adopted and modified from Hallenbeck *et al.*, 2011). Detailed discussions are found in the text.

candidates for this kind of activity, for example, the NADP-dependent soluble Ni–Fe hydrogenase (SH) of *Ralstonia eutropha*, or Fe–Fe hydrogenases such as the Hnd (*hndA, hndB, hndC, hndD*) NADP-reducing hydrogenase from the sulfate-reducing bacterium *Desulfovibrio fructosovorans* (Malki *et al.*, 1995). A novel NADP-dependent hydrogenase from *Pyrococcus furiosus* has been overexpressed in *E. coli* and it was found that the native *E. coli* Ni–Fe maturation system could generate a functional hydrogenase when provided with only the genes encoding the hydrogenase subunits and a single protease from *P. furiosus*, the hydrogenase-specific protease necessary for

C-terminal cleavage during insertion of the Ni–Fe centre (Sun *et al.*, 2010). Ferredoxin-dependent [Fe–Fe] hydrogenase pathways have also been introduced along with the enzymes necessary to reduce ferredoxin with either NADH or NADPH. However, yields were disappointingly low: 0.025 (Y.M. Kim *et al.*, 2011), 0.04 (Cho *et al.*, 2011), 0.05 (Veit *et al.*, 2008) mol H_2/mol glucose. On the other hand, when a [Fe–Fe] hydrogenase was coupled with metabolism by the expression of a pyruvate: ferredoxin oxidoreductase yields as high as 1.46 (Akhtar and Jones, 2009) mol H_2/mol glucose was obtained. The heterologous expressions of [Ni–Fe] hydrogenases have been reported (J.Y. Kim *et al.*, 2011, 2012; Wells *et al.*, 2011; Weyman *et al.*, 2011; Zheng *et al.*, 2012). Recently, we introduced a [Ni–Fe] hydrogenase dependent pathway and showed that it is capable of higher yields than the previously characterized [Fe–Fe] hydrogenase dependent pathways. Heterologous expression of the soluble [Ni–Fe] hydrogenase from *Ralstonia eutropha* (soluble SH-*hyd*) demonstrated the successful introduction of a functional pathway capable of deriving hydrogen from NADH, giving an *in vivo* hydrogen yield of 2.11 mol H_2/mol glucose (Ghosh *et al.*, 2013). Of course, introduction of these types of hydrogenases would be even more useful if metabolic engineering is used to increase production of NADH or NADPH, through, for example, the pentose phosphate pathway (Y.M. Kim *et al.*, 2011).

Metabolic engineering involving *Escherichia coli* transporters

The hydrogen yield also depends on the functioning of transporter proteins. As discussed earlier, uptake hydrogenases can influence hydrogen yields, and their functional localization to the periplasm depends upon the twin-arginine translocation (Tat) system (Penfold *et al.*, 2006). TatA, TatB and TatC proteins form the large oligomeric Tat transporter (Berks *et al.*, 2003). Knockouts of *tatC* and *tatAE* (*ΔtatABCD*, *ΔtatE*) increase hydrogen production by almost 2-fold in both cases (Wexler *et al.*, 2000). The inactivity of respiratory dehydrogenases (FdHN, FdHO) and the of uptake hydrogenases (Hyd-1 and Hyd-2), which rely on Tat, leads to these increases in production rate. However, the growth rate also decreases, probably due to the disruption of the outer membrane in the engineered Tat knockout strains (Redwood *et al.*, 2008; Stanley *et al.*, 2001). The deletion of *tatC* gene does not have a significant effect on hydrogen production using sucrose as substrate as compared with the control strain, perhaps because if an excess of formate is present, formate production is no longer the rate-limiting step. This is corroborated by the fact that addition of plasmid pUR400, which encodes sucrose assimilation genes *ScrKYAB*, and a tetracycline resistance cassette did result in a significant increase in hydrogen gas evolution (Penfold *et al.*, 2004).

Metabolic engineering of alternative metabolite pathways in *Escherichia coli*

During the production of any product, side reactions can occur which decrease the yield of the desired product. Since *E. coli* carries out mixed acid fermentation, a variety of metabolic products are possible in addition to hydrogen. Thus, disruption of succinate- and lactate-producing pathways leads to an increased hydrogen yield, and disruption of *ldhA* and *frdBC*, which produce lactate and succinate respectively, gives an approximately 2-fold increase in hydrogen yield, up to 90% of the theoretical yield of hydrogen from glucose (Yoshida *et al.*, 2006a, 2007). An increase in the specific production rate as well as the volumetric production rate was also observed in the engineered strains (Yoshida *et al.*, 2006a). Another study suggested that knockouts of *dld*, *lldD*, *ldhA* (lactate dehydrogenases), two of which are membrane-bound flavoproteins, may help in increasing hydrogen production. Also, pyruvate is consumed by pyruvate dehydrogenase and pyruvate oxidase encoded by *aceE* and *poxB* respectively. Deletion of *aceE* resulted in an increased hydrogen production and yield (Maeda *et al.*, 2007). However, deletion of *poxB* was found to be inefficient in increasing hydrogen production, primarily because it is more active under aerobic conditions, and *E. coli* produces hydrogen only under anaerobic conditions (Abdel-Hamid *et al.*, 2001).

Table 1.2 Metabolic engineering on potential hydrogen producers for improving hydrogen production

Microorganism/strains	Knockout	Overexpression	H_2 yield (mol H_2/mol glucose)	Reference
Escherichia coli				
HD705	*ΔhycE*		ND	Redwood *et al.* (2008)
FTD22	*ΔhyaB*		0.800	
FTD67,	*ΔhybC*		1.024	
FTD89	*ΔhyaB, ΔhybC*		1.043	
HD701	*ΔhycA*		0.737	
B1LK0	*ΔtatC*		ND	Penfold *et al.* (2006)
DADE	*ΔtatAE*		ND	
FTD701	*ΔtatC+ΔhycA*		ND	
HD701	*ΔhycA*		ND	
SR15	*ΔldhA+ΔfrdBC*		1.82	Yoshida *et al.* (2006a)
SR13	*ΔhycA*	*fhlA*	0.95	
SR14	*ΔldhA+ΔfrdBC*	*fhlA*	1.87	
SR11	*ΔhycA*		ND	
SR12		*fhlA*	ND	
SR13	*ΔhycA*	*fhlA*	ND	
BL21 (DE3)pEMBTL-HJ2		*hupSL*	ND	Lee *et al.* (2010)
FTD701	*ΔtatC*		ND	Penfold and Macaskie (2004)
HD701/pUR400		*scrK+scrY+scrA+scrB*	ND	
FTD701/pUR400	*ΔtatC*	*scrK+scrY+scrA+scrB*	ND	
BL21 (DE3) pYdbK pAF	*MCS2*	*YdbK+CpFdx+hydA +hydF+hydG+hydE*	0.67	Akhtar and Jones (2009)
BL21 (DE3) ΔiscRpYdbK pAF	*ΔiscR+MCS2*	*YdbK+CpFdx+hydA +hydF+hydG+hydE*	1.46	
JWGH1 (JW135/pJWPH5)	*ΔadhE, ΔldhA, ΔarcA*	*Ni–Fe-hyd*	2.11	Ghosh *et al.* (2013)
ZF1	*ΔfocA*		0.54	Fan *et al.* (2009)
ZF2	*ΔhybC*		0.63	
ZF3	*ΔnarL*		0.7	
ZF4	*Δppc*		0.96	
ZF4	*ΔfocA+ΔhybC*		0.73	
DJT135	*Δhya-Km, Δhyb-Km, ΔldhA*	*fhlA-C*	2.5	Turcot *et al.* (2008)
Enterobacter aerogenes				
IAM1183 AO	*ΔhycA+ΔhybO*		1.36	Zhao *et al.* (2009)
TCC 13408/hydA		*hydA*	2.31	Zhao *et al.* (2010)
IAM1183 A (pMCL-fdhF)	*ΔhycA*	*fdhF*	1.19	Lu *et al.* (2009)
IAM1183 A (pMCL-fhlA)	*ΔhycA*	*fhlA*	1.15	
IAM1183 O (pMCL-fdhF)	*ΔhybO*	*fdhF*	1.13	
IAM1183 O (pMCL-fhlA)	*ΔhybO*	*fhlA*	1.09	
IAM1183 (pCOM 10-fdh1)	*ΔldhA*	*fdh1*	1.702	Lu *et al.* (2010)
pMCK		*ppk*	1.504	Lu *et al.* (2011)

Table 1.2 (Continued)

Microorganism/strains	Knockout	Overexpression	H_2 yield (mol H_2/mol glucose)	Reference
AY-2	alcohol dehydrogenase + Butadienol dehydrogenase		1.1	Rachman *et al.* (1998)
HGN-2	Acetate/butyrate pathway blocking		2.20	Jayasinghearachch *et al.* (2009)
HT34	acetate/butyrate Pathway blocking		2.23	
Enterobacter cloacae				
IIT-BT 08		*Hydrogen-promoting protein (hpp)*	2.55	Song *et al.* (2011)
***Clostridium* sp.**				
C. parapurtificum M-21 pJIR751		*hydA*	2.40	Morimoto *et al.* (2005)
C. acetobutylicum DSM 792 [pSOS]		*thl promoter*	1.77	Klein *et al.* (2010)
C. acetobutylicum DSM 792 [pSOS- hydACa]		*hydA*	1.81	
C. saccharoperbutylacetonicum	*ΔhupCBA operon*		ND	Nakayama *et al.* (2008)
C. tyrobutyricum PAK-Em	*Δack*		2.61	Liu *et al.* (2006)
C. tyrobutyricum ATCC 25755 PPTA-Em	*Δpta*		1.08	
Clostridium perfringens W11	*ΔldhA*		1.58 (Sucrose a carbon source)	R. Wang *et al.* (2011)
Clostridium perfringens W15	*ΔatoB*		ND	Yu *et al.* (2013)
Thermophiles				
Caldicellulosiruptor bescii DSM6725/pDCW121	*ΔldhA*		ND	Cha *et al.* (2013)
Caldicellulosiruptor bescii JWCB005	*ΔpyrF*		1.70	
Caldicellulosiruptor bescii JWCB017	*ΔldhA +ΔpyrF*		3.40	

ND, not determined.

Other enzymes and global transcription factors in *Escherichia coli*

A comprehensive study would provide insights into the effect of various knockouts and expressions, including expressions of global transcriptional regulators. Of course, various changes which increase formate hydrogen lyase activity will increase hydrogen production (Yoshida *et al.*, 2006b). Deletion of *focA* (formate transport gene) resulted in an accumulation of formate and an increase in yield of hydrogen gas (Fan *et al.*, 2009). Also, deletion of *ppc*, the gene encoding phosphoenolpyruvate carboxylase (PEPC), resulted in an increase in the molar yield. This is because the carboxylation of PEP is one of the most important steps in succinate production, and this step is abolished upon deletion of *ppc*. In addition, NarL (global transcription regulator) is a repressor of structural genes for FHL, PFL, and nickel transporters encoded by the *nik* operon. Inactivation of NarL gave a 2-fold increase in molar yield of hydrogen without compromising growth relative to the wild-type strain (Sawers, 2005). Interestingly, an increase of 63% was observed when the *selC* gene (for selenium metabolism) was overexpressed. Since FdhF is a selenocysteine-containing enzyme, this provides new insights into ways to improve cofactor production for the FHL enzyme system, which in turn improves hydrogen yield by increasing the use of formate. Overexpression of *fnr*, another global transcription regulator, resulted in a 5.5-fold increase in specific hydrogen yield. Moreover, this overexpression along with *focA* knockout resulted in very high specific hydrogen gas production in both M9 and TYP media (Sawers, 2005). Other global regulators, such as ArcAB and IhfAB, were found to have no significant impact on hydrogen yield (Sawers, 2005). Since the theoretical yield from FHL pathway is not currently economically viable, many authors have tried to enhance the molar yield by adopting a dual system approach (Redwood and Macaskie, 2006; Waks and Silver, 2009; Keasling *et al.*, 1998).

Metabolic engineering of *Escherichia coli* native metabolic pathways

Different native metabolic pathway alternations in *E. coli* have been attempted in order to increase hydrogen yields and productivities during dark fermentation over the past few decades. This is possible due to the establishment of chromosomal gene disruption technologies that allowed the creation of the 'Keio' and 'ASKA' collections. These valuable resources for metabolic engineering in *E. coli* have improved analysis of unknown gene function, gene regulatory networks, and genome-wide evaluation of mutational effects in a common background. Successive rounds of P1 viral transduction from the Keio *E. coli* K-12 metabolic and genomic library (Maeda *et al.*, 2007) has been performed to incorporate multiple, stable mutations into a single microbial platform (*E. coli*) in order to redirect metabolic flux and reductant flow towards hydrogen production. Some of these involved knockouts in the lactate and or succinate pathways to create *E. coli* SR15 (Δ*ldhA*, Δ*frdC*) which showed a considerably increased hydrogen yield (Yoshida *et al.*, 2006a). In addition, hydrogen production was increased by redirecting reductant flow or carbon flow towards increased formate production by knocking out the *hycA* gene (FHL repressor gene), allowing increased expression of the formate hydrogen lyase enzyme system. Similarly, *hycA* null *E. coli* strain HD701 was shown to have a 14-fold increase in hydrogen production (Penfold *et al.*, 2003). A stoichiometric hydrogen yield was obtained with *E. coli* DJT135 (Δ*hyaB*, Δ*hyb*, Δ*ldhA*, *fhlA**), a metabolically engineered strain which has a *ldhA* knockout and overexpression of FhlA for increased formate production (Bisaillon *et al.*, 2006). In another study, the regulation of FHL was altered by knocking out the repressor *hycA* and overexpression of *fhlA* activator, uptake hydrogenase activities were eliminated by deleting *hyaB* (hydrogenase 1) and *hybC* (hydrogenase 2), and hexose metabolism was redirected to formate using deletions of *fdnG*, *fdoG*, *narG*, *focA*, *focB*, *poxB* and *aceE*. Production of succinate and lactate were also eliminated in *E. coli* by knocking out *frdC*, *ldhA* for improving hydrogen production. The resulting metabolically engineered strain *E. coli* BW25113 (Δ*hyaB*, Δ*hybC*, Δ*hycA*, Δ*fdoG*, Δ*frdC*, Δ*ldhA*, Δ*aceE*) was shown to have a 4.6-fold increase in hydrogen production from glucose (Maeda *et al.*, 2008b).

Clostridium

Metabolic engineering of *Clostridium* with emphasis on gene knockout of existing metabolism and overexpression of foreign genes

Recently, metabolic engineering of strict anaerobic microbial systems, gene knockouts or insertions, and enzyme overexpression has been applied in attempts to increase hydrogen production. For example, *hydA* of *Clostridium perfringens* (encoding the Fe–Fe hydrogenase) has been extensively studied. It was found that hydrogen production was completely eliminated when the *hydA* gene is disrupted (Kaji *et al.*, 1999). Hydrogen production was increased 1.7-fold in *Clostridium parapurtificum* M-21 when *hydA* was overexpressed and, at the same time, there was a considerable reduction in lactate production. It was hypothesized that overexpression of *hydA* led to an over-oxidation of NADH to NAD^+ (Morimoto *et al.*, 2005). Similarly, *Clostridium butyricum* showed a higher specific hydrogenase activity with overexpression of *hydA*, but only a slight increase was seen when overexpression was made in *Clostridium acetobutylicum* (Klein *et al.*, 2010). From all these studies it can be concluded that hydrogen production is not limited by HydA activity, but that a more likely bottleneck is the rate at which electrons re-reduce oxidized ferredoxin. The overall balance in hydrogen production might in some cases be thermodynamically limited, a restriction that can at least be partially alleviated by constant removal of hydrogen gas, driving the forward chemical reaction. Hydrogen production was found to be increased 3.1-fold by down-regulation of the uptake hydrogenase (*hupCBA*) in *Clostridium saccharoperbutylacetonicum* N1–4 through anti-sense RNA technology (Nakayama *et al.*, 2007). In another study, *ack* (acetate kinase) gene deletion in *Clostridium tyrobutyricum* improved hydrogen yield up to 50% with an increase in hydrogenase and phosphotransacetylase activities (Liu *et al.*, 2006) while hydrogen production was found to be decreased slightly in *Clostridium tyrobutyricum* having a knock out in *pta* (encoding phosphotransacetylase). In *Clostridium tyrobutyricum* JM-1, homologous overexpression of Fe–Fe hydrogenase resulted in a 1.7-fold and a 1.5-fold increase in hydrogenase activity and hydrogen yield, respectively (Jo *et al.*, 2010) while lactate production was significantly decreased. Thus, it would appear that hydrogen production and lactate production in *Clostridia* are inversely related. These studies offer a number of insights into different gene disruptions that could affect hydrogen uptake as well as hydrogen productivity.

Recently, ClosTron, a group II intron homing-related mutagenesis system for *Clostridia*, was developed (Heap *et al.*, 2007). The bacterial group II introns are a relatively newly characterized type of mobile element that can be useful for the directed construction of stable mutations in the bacterial chromosome (Lambowitz and Zirnmerly, 2004). The mobile group II intron from the *ltrB* gene of *Lactococcus lactis* (*Ll.ltrB*) can insert into a specific target site via an RNA-mediated 'retro-homing' mechanism, achieved primarily by base-pairing between the excised intron lariat RNA and the target site. Therefore, introns can be reprogrammed to the appropriate target site by alteration of the DNA sequence of the appropriate part of the intron. Recently, several significant improvements have been made in the ClosTron system which allows the construction of strains with multiple mutations through marker recycling using FLP/FRT, or delivery of transgenes to the chromosome (Heap *et al.*, 2010). Automated ClosTron design bioinformatics and information about pMTL80000 modular plasmids are now available online (http://clostron.com). This group II intron delivery system has been successfully used in *C. difficile*, *C. acetobutylicum*, *C. sporogenes*, *C. botulinum*, *C. beijerinckii* and *C. perfringens*. In addition, a modular system of *Clostridium–E. coli* shuttle plasmids is now also available (Heap *et al.*, 2009). With different Gram-positive replicons, selectable markers, Gram negative replicons and specific application elements, a total of 400 different vectors can be constructed into which additional elements can be inserted using four unique rare restriction sites. This very flexible system allows researchers to identify combinations of components potentially suitable for particular clostridial hosts and applications. These newly developed and described tools open the door for the engineering of a variety of pathways to increase biofuels, including hydrogen, production. As yet, no studies to this effect have been reported.

Based on this novel approach, *Clostridium perfringens* W13 was metabolically engineered using the novel Targetron gene knockout system to remove lactate dehydrogenase (Δ*ldhA*) giving an increase in hydrogen yield and acetate and butyrate concentrations of 51%, 26%, and 57%, respectively (R. Wang *et al.*, 2011). Another approach used with *Clostridium perfringens* was to block the butyrate pathway. Acetyl-CoA acetyltransferase (*atoB*), the first enzyme participating in the butyrate pathway was knocked out, but this strain actually showed lower hydrogen production and greater ethanol accumulation. This study highlights the complexities in differential gene expression among the different proteins involved, including the electron transfer flavoprotein alpha subunit (*etfA*), electron transfer flavoprotein beta subunit (*etfB*), enoyl-CoA hydratase (*ert*), butyrate kinase (*buk*), and phosphate butyryl-transferase (*ptb*). Changes in the use of reducing equivalents (NADH) will have effects on hydrogen production (Yu *et al.*, 2013).

There are still several key pathways of central metabolic networks in *Clostridium* sp. that need to be understood before metabolic engineering can be fully employed. Most interestingly, the presence and completeness of tri-carboxylic acid (TCA) cycle in most anaerobic hydrogen producers remains to be elucidated (Lee *et al.*, 2008). For example, the whole genome annotation dataset of *Clostridium acetobutylicum* suggested the absence of a complete TCA cycle. However, a recent study has shown that this strictly anaerobic hydrogen producer has a complete, albeit bifurcated, TCA cycle (Amador-Noguez *et al.*, 2010). The major benefits of having bidirectional metabolic fluxes of carbon and electrons into succinate are unclear. One reason might be that a bifurcated TCA cycle ending with the succinate pathway plays a pivotal role in cellular redox balance; another could be that this phenomenon promotes the utilization of certain amino acids.

Metabolic engineering of *Enterobacter* through the alteration of existing metabolic networks and insertion of novel alternative pathways

Enterobacteriaceae are another group of potential hydrogen producers that can be subjected to metabolic engineering to improve molar hydrogen yields and productivities. The most two promising microorganisms in this group are *Enterobacter aerogenes* and *Enterobacter asburiae.* Hydrogenases enzymes belonging to these hydrogen producers have been studied (Ren *et al.*, 2005; Shin *et al.*, 2007) and a quantitative comparison of hydrogen evolution and hydrogen uptake activities has been made (Ren *et al.*, 2005), alleviating the necessity for development of novel approaches of controlling uptake activity to enhance hydrogen evolution capacity. *E. aerogenes* has been reported to have two distinct hydrogen-producing pathways: the pyruvate to formate pathway and the NADH oxidation pathway (Fig. 1.11). Formate generally comes from pyruvate by pyruvate formate lyase in the anaerobic respiration of glucose in facultative anaerobes such as *E. aerogenes*, and the formate thus produced can then be quickly converted to hydrogen by formate hydrogen lyase (FHL). However, the FHL system in *E. aerogenes* and its regulatory network are still not as clearly understood as that those of *E. coli*. However, the presence of FHL in *E. aerogenes* has been confirmed by isolating and cloning a 5431-bp DNA fragment, partially encoding the FHL gene cluster *hycABCDE* (Zhao *et al.*, 2009). Moreover the genes encoding the transcription factors of the FHL complex in *E. aerogenes* were cloned and characterized, including *fhlA* and *hycA*, which code for the FHL activator protein FhlA and the FHL repressor protein HycA, respectively (Lu *et al.*, 2009). There is some indication that hydrogen production though a NADH pathway might be possible in this organism (Zhao *et al.*, 2009; Lu *et al.*, 2010), but very little evidence is available on the enzymes and metabolic networks responsible for hydrogen generation in *Enterobacter aerogenes*. An important issue for the generation of hydrogen through a NADH pathway is the thermodynamic limit (E^O for H^+ to H_2 is −0.42 V; E^O for NAD^+ to NADH is −0.32 V; E^O for CO_2 to formate is −0.42 V). Therefore, hydrogen production through NADH is thermodynamically limited by the hydrogen partial pressure, which would be a key problem in the application NADH-dependent pathways in hydrogen production. At very low partial pressures (less than 42 Pa), the NADH and NAD^+ ratio is almost equal, and hydrogen production could be possible from NADH (Angenent *et al.*, 2004). Improvement of hydrogen production using *E. aerogenes* might be possible if the formate pathway, the genes related to the FHL complex and their regulatory proteins are manipulated. However, although deletion of *hycA* and *hybO*, encoding the FHL repressor and uptake hydrogenase, gave a 2-fold increase in hydrogen production, growth was drastically affected. This should be an area of focus for further scientific investigations (Zhao *et al.*, 2009). In another attempt to enhance the formate pathway in *E. aerogenes*, the formate dehydrogenase FDH-H gene (*fdhF*) and the FHL activator protein FhlA (*fhlA*) from *E. coli* were overexpressed in *E. aerogenes* at

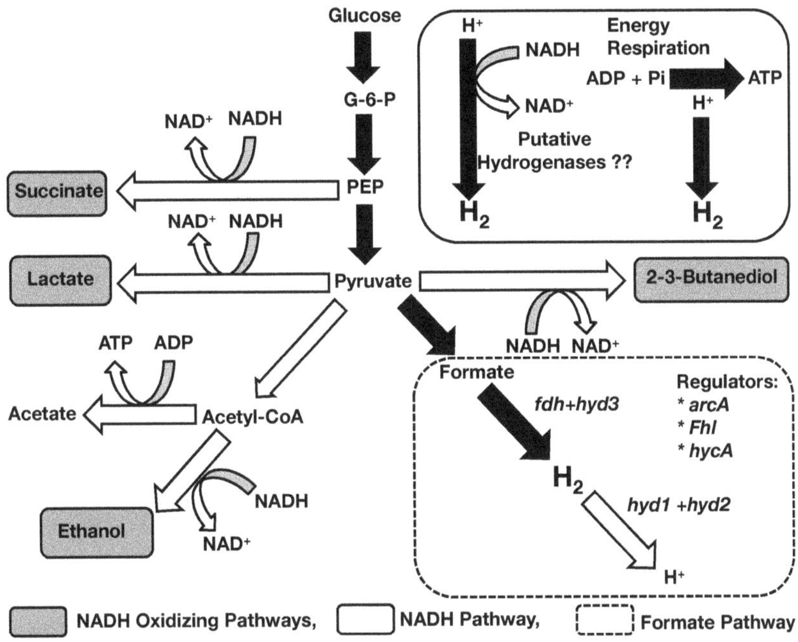

Figure 1.11 Metabolic engineering of the metabolic network of the *Enterobacteriaceae* group of hydrogen producers. Central metabolic (NADH and FHL) pathways in this group of organisms are presented as black arrows. Abbreviations: formate dehydrogenase (*fdh*); hydrogenase 3 (*hyd3*); Transcriptional regulators of the TCA cycle (*arcA*); formate hydrogen lyase (*Fhl*); Transcriptional regulator (*hycA*); uptake hydrogenase (*hyd1* and *hyd2*).

the same time that efforts were made to affect the intracellular NADH levels. Consuming pathways, such as lactate dehydrogenase, can reduce the NADH pool and thus hydrogen production if it is highly dependent on the intracellular NADH levels. Therefore, engineering such metabolic pathways would be expected to alter the metabolic flux and thus hydrogen evolution. *ldhA* disruption might be thought to be the best option to increase NADH pool, leading to an increased hydrogen yield. In practice, a deletion in *ldhA* results in an increase in the hydrogen yield but causes slight decrease in the hydrogen production rate (Lu *et al.*, 2010). Interestingly, introduction of an NADH regeneration system, namely the NAD^+-dependent formate dehydrogenase gene (*fdh1*) from *Candida boidini* which utilizes formate to generate NADH, resulted in an 86.8% increase in hydrogen yield per mole consumed glucose (Lu *et al.*, 2009).The polyphosphate kinase (*ppk*) of *E. aerogenes* was studied to better understand the global energy metabolism in order to improve hydrogen metabolic pathways. PPK is a key enzyme involved in energy storage and the release of energy in the form of polyphosphate. However, it has been found that polyphosphate kinase overexpression causes a readjustment of cellular oxidation which causes more electron flow to hydrogen production, improving the hydrogen yield per mole of glucose by 10.8% (Lu *et al.*, 2011). Even though there have been a lot of studies focusing on the improvement of hydrogen yield for *E. aerogenes*, the hydrogen yield is still far lower than the stochiometric maximum. To enhance hydrogen yields further, competing pathways for formic acid and NADH should be systematically knocked out (*ldhA, frd1, ppk*) in order to direct more electron flow towards hydrogen production. The maximum H_2 yield of 10 mol/mol glucose via NADH would be expected if NADH accumulated in aerobic conditions can be channelled to hydrogen production by the anaerobic metabolism. Thus one possibility is to screen for hydrogenases that are active during micro-aerobic cultivation of *E. aerogenes.* Secondly, reducing the partial pressure of hydrogen in the culture system would be another way of enhancing the hydrogen yield to overcome the thermodynamically limited NADH pathway.

Metabolic engineering of thermophilic hydrogen producers

Thermophilic hydrogen producing bacteria have received due to several advantages from a metabolic engineering standpoint. In these organisms, sugar (glucose) metabolism can proceed by a number of routes. The first pathway produces NADH during the conversion of glyceraldehyde-3-phosphate (Gly3P) to glycerate 1,3-bisphosphate (G1,3P) by an NAD^+ dependent glyceraldehyde-3-phosphate dehydrogenase (GAPDH) of the EMP pathway. Several different subsequent pathways might lead to H_2 production. The second route produces reduced ferredoxin (Fd_{red}) which can directly reduce hydrogenase to generate H_2 (Fig. 1.12). Experimental results have led to the proposal of an alternative route for *Thermotoga maritima,* in which a heterotrimeric [Fe–Fe] bifurcating hydrogenase simultaneously utilizes both Fd_{red} and NADH to produce hydrogen. It was proposed that as the reduction potential (E_0') of NADH is −320 mV, the trimeric [Fe–Fe] bifurcating hydrogenase utilizes the exergonic oxidation of ferredoxin (midpoint potential, E_m −453 mV) to drive the endergonic oxidation of NADH to produce hydrogen (E_0' −420 mV) (Schut and Adams, 2009). Metabolic data mining suggests that homologous enzymes may be fairly widespread. In *C. thermocellum* ATCC 27405, the existence of a ferredoxin-dependent energy conserving hydrogenase (Ech)-type Ni–Fe H_2-ase, two

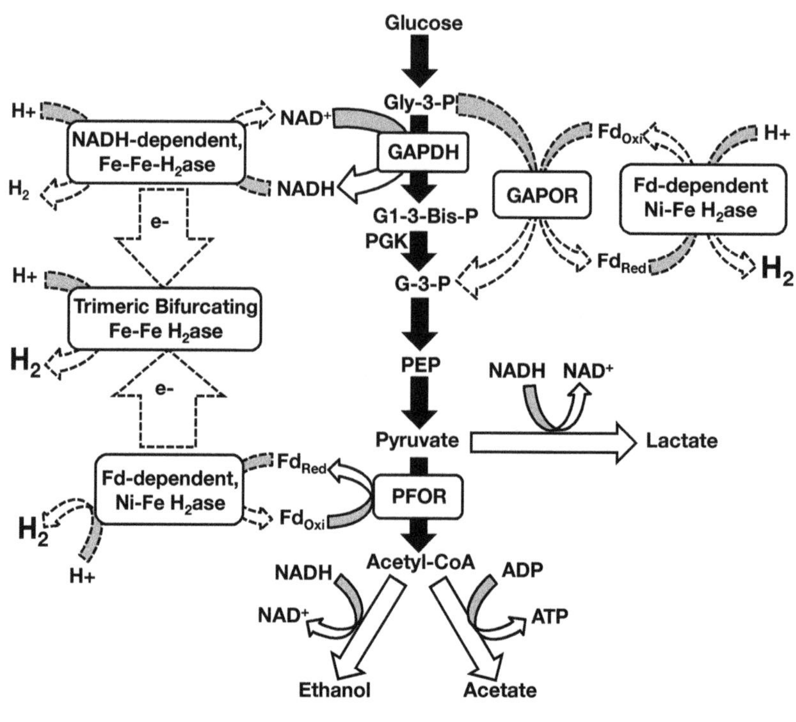

Figure 1.12 Metabolic engineering of the metabolic networks in thermophilic hydrogen producers. Thermophiles are able to produce a maximum of 3 to 4 mol hydrogen per mole glucose. Central metabolic hydrogen production pathways are represented as black arrows. Dashed arrows represent novel metabolic pathways that might be introduced with electron flow through reductant electron carriers such as; Ferredoxin (reduced), NADH. Finally, electrons will reduce protons to produce hydrogen via following novel enzymatic activities likely Trimeric Bifurcating Fe–Fe hydrogenase, Fe-dependent Ni–Fe hydrogenase, glyceraldehydes-3-phosphate dehydrogenase (GAPDH) and glyceraldehyde 3-phosphate oxidoreductase (GAPOR). Some abbreviation: pyruvate ferredoxin oxidoreductase (PFOR); phosphoglycerate kinase (PGK); phosphoenolpyruvate (PEP).

NADH-dependent, and one NADPH dependent Fe-only H_2-ases have been confirmed with transcriptomic data (Carere *et al.*, 2008). The recent transcriptomic analysis of *C. thermocellum* ATCC 27405 during cellulose fermentation suggested plausible roles in H_2 production and ATP synthesis for an NADH/ferredoxin oxidoreductase (NFOR) and four hydrogenases (H_2-ases) such as ferredoxin-dependent Ech-type Ni–Fe H_2-ase, two NADH-dependent Fe-only H_2-ases, and an NADPH-dependent Fe-only H_2-ase (Raman *et al.*, 2011). It was also proposed that the membrane-bound Fd-dependent Ni–Fe–H_2-ase, in association with NFOR, pumps H^+/Na^+ ions across the cell membrane and creates a proton gradient to synthesize ATP by ATP synthase and H^+/Na^+ transporting ATPase complexes (Raman *et al.*, 2011). Recent reports on *Clostridial* H_2-ases indicates that NADH-dependent Fe-only H_2-ases of *C. thermocellum, C. saccharolyticus* and *Anaerocellum thermophilum* are heterotrimeric [Fe–Fe] bifurcating hydrogenase (Calusinska *et al.*, 2010; Werken *et al.*, 2008; Willquist *et al.*, 2010). The recent hypothesis of electron bifurcation (Heap *et al.*, 2010) makes it quite reasonable to suggest that NADH dependent Fe-only H_2-ases of these strains might utilize both Fd_{red} and NADH synergistically to generate H_2 similar to that of the trimeric [Fe–Fe] bifurcating hydrogenase of *T. maritima*. Nevertheless, the NADH-dependent Fe-only H_2-ases of *C. thermocellum, C. saccharolyticus*, and *A. thermophilum* need to be characterized for their physiological substrates in order to understand clearly their function since the purified membrane-bound Fd-dependent Ech-type Ni–Fe H_2-ase, and an NADH dependent Fe-only H_2-ase of *Thermoanaerobacter tengcongensis* were shown to produce H_2 from Fd_{red} and NADH, respectively (Soboh *et al.*, 2004).

The most promising thermophilic hydrogen producers, *Thermotoga maritama* DSM 3109, *Thermotoga neapolitana* DSM 4359, *Caldicellulosiruptor owensensis* DSM 13100, *Thermotoga elfii* DSM 9442, *Caldicellulosiruptor saccharolyticus* DSM 8903, *Thermococcus kodakarensis* KOD1, *Clostridium thermocellum* ATCC 27405, *C. thermocellum* JN4, *Thermoanaerobacterium thermosaccharolyticum* GD17, *Caldicellulosiruptor bescii* sp. nov. DSM 6725 generate hydrogen utilizing different carbon sources: glucose-, xylose-, starch-, hemicellulose- and cellulose-based feed stocks, at temperatures ranging from 65°C to 80°C with the production of acetate, lactate and butyrate and ethanol as major metabolic end products. Although thermophilic organisms might pay a price energetically for maintaining enzyme functionality at high temperatures, they have a metabolic advantage in that the net hydrogen energy gain, 656.35 kJ, is higher than that of mesophilic hydrogen producers, 609.75 kJ (Mohan Raj *et al.*, 2012). Thus, elevated temperature offers the advantages of surpassing thermodynamic hurdles and improving metabolic reaction rates (Nguyen *et al.*, 2008). Metabolic engineering of cellulolytic pathways in *Clostridium thermocellum* has improved hydrogen production by promoting more rapid cellulose degradation (Levin *et al.*, 2004). A *Thermoanaerobacterium aotearoense* ldhA mutant was created which showed considerably higher hydrogen yields without any lactate production after 48 h of cultivation (S. Li *et al.*, 2010). Recently a novel gene disruption system has been developed for *Clostridium thermocellum* using a native dual selectable gene marker based on pyrF. The ΔpyrF strain, a uracil auxotroph, can be restored to prototrophy through the ectopic expression of pyrF from a plasmid, providing a positive genetic screening. A toxic uracil analogue, 5-fluoro-orotic acid can be used to screen against plasmid expressed pyrF, generating a selection pressure for plasmid curing. This approach was used with *C. thermocellum* to create a mutant having a deletion in the phosphotransacetylase (Δpta) enzyme system to eliminate acetate and direct metabolic flux towards hydrogen production (Tripathi *et al.*, 2010). Very recently, a metabolically engineered *Caldicellulosiruptor bescii* was generated carrying a lactate dehydrogenase gene knock out (ΔldhA) to improve hydrogen production from lignocellulosic biomass. This engineered strain grew to a higher maximum optical density than the wild type on maltose and cellobiose, consistent with the prediction that the mutant would gain additional ATP with increased acetate production via the Embden–Meyerhof glycolytic pathway (Cha *et al.*, 2013).

Potential candidates for metabolic engineering with emphasis on halophiles, amino acid degrading and nitrogen-fixing facultative bacteria

Most of the metabolic engineering research for improving hydrogen production has been done with organisms whose complete genome sequence is known, *E. coli*, *Clostridium* and *Enterobacter* sp. However, many other microorganisms show a high capacity for hydrogen production. Major candidates for expanding metabolic engineering for hydrogen would be halophilic microorganisms capable of surviving at high salt concentrations. *Halanaerobium saccharolyticum* subsp. *Saccharolyticum* and *H. saccharolyticum* subsp. *snegalesis* have been shown to produce hydrogen from glycerol using the 1,3-propane-diol pathway (Kivisto *et al.*, 2010). In addition, several other newly described microbes have shown the ability to produce hydrogen; *Megasphaera cerevisiae, Ethanoligenens harbinense, Anaerotruncus* sp., *Klebsiella* sp., *Bacillus* sp., *Bifidobacterium* sp., *Streptococcus* sp. etc. (Hung *et al.*, 2011a,b). It has also been shown that *Klebsiella pneumoniae*, a facultative anaerobic nitrogen-fixing bacterium, is able to produce hydrogen, with the major metabolic end product being acetate. A hydrogen yield of 2.83 mole H_2 per mole of glucose was obtained (Chen *et al.*, 2006). Here, hydrogen production from glucose is carried out by the nitrogenase enzyme system which requires ATP as energy source and reduces protons to hydrogen in the absence of molecular nitrogen (Fig. 1.13). It would appear that NADH is the most promising reductant for hydrogen production in this microorganism which could be subjected to metabolic engineering for improvement of hydrogen yield. Alternatively, various other pathways involved in acetate, ethanol, and CO_2 production could be a standpoint for metabolic engineering to ameliorate hydrogen production in *Klebsiella pneumoniae*. The glutamate-fermenting microorganism *Acidaminococcus fermentans* could in the future be modified to increase hydrogen production. This would involve using the key enzymes, Na^+-dependent NAD^+-dehydrogenase

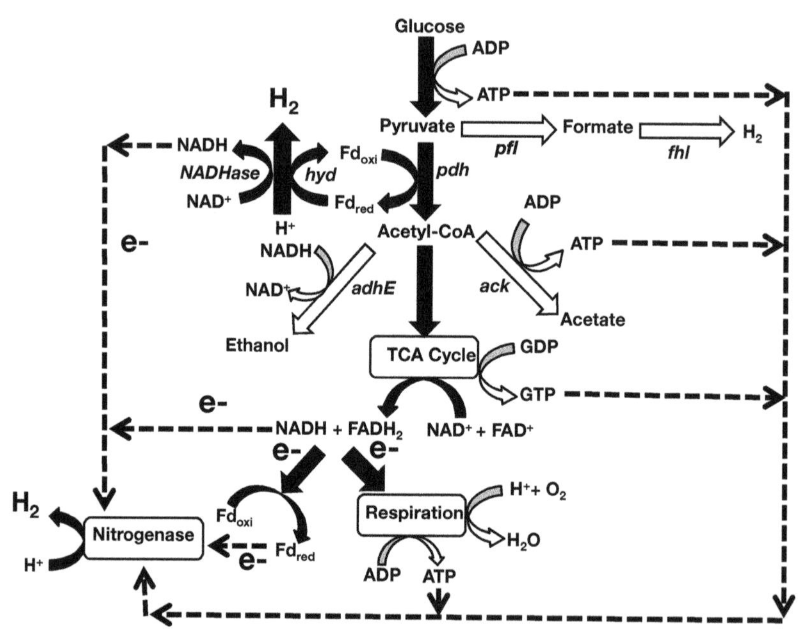

Figure 1.13 Metabolic pathways involved in hydrogen production in *Klebsiella pneumoniae. K. pneumoniae* produces hydrogen through three major pathways. Energy for hydrogen metabolism in the form of ATP, GTP is provided by respiration and the TCA cycle as well as glycolysis. Electron flow from NADH, $FADH_2$, and reduced ferredoxin is used to produce hydrogen. Abbreviations: hydrogenase (*hyd*); pyruvate dehydrogenase (*pdh*); pyruvate formate lyase (*pfl*); formate hydrogen lyase (*fhl*); acetate kinase (*ack*); alcohol dehydrogenase (*adhE*); NADH dependent oxidoreductase (*NADHase*).

and glutaconyl-CoA decarboxylase. When external Na^+ concentrations are lower than 10 mM, glutaconyl-CoA decarboxylase exchanges H^+ from the outside with Na^+ from the inside. The NADH that is formed is used exclusively to reduce crotonyl-CoA to butyryl-CoA, leading to an acetate/butyrate ratio of 2.0. The electrochemical Na^+ gradient is probably used for glutamate uptake. On the other hand, when the external Na^+ concentration is high (≥ 10 mM) H^+ uptake from the outside by glutaconyl-CoA decarboxylase is suppressed. Now H^+ comes from the inside, leading to an electrochemical H^+ gradient that drives hydrogen formation via an NADH dehydrogenase and a hydrogenase. Less butyrate is produced, which increases the acetate/butyrate ratio to 3.0. These observations may open a new avenue for metabolic engineering involving novel hydrogenases that are dependent on the Na^+ gradient across the cytoplasmic membrane (Härtel *et al.*, 1996) of the glutamate-fermenting microorganism *Acidaminococcus fermentans* (Fig. 1.14).

1.3 Metabolic engineering of green algae, cyanobacteria, and bacteria for improving hydrogen production

Hydrogen production through either a photoautotrophic or a photoheterotropic mode is of great interest due to the potential for a higher stochiometric hydrogen yield. Three major options are available for hydrogen production using light energy: direct biophotolysis, indirect biophotolysis and photofermentation. Biophotolysis is the use of water-splitting oxygenic photosynthesis for direct proton reduction to hydrogen production. In direct biophotolysis (Fig. 1.15), green algae or cyanobacteria absorb photons through their two photosystems, PSII (680 nm) and PSI (700 nm). The captured solar energy generates a strong oxidant that can oxidize water (producing oxygen) giving protons and electrons which can be used to reduce ferredoxin, which in turn will be used to form hydrogen using either a Fe–Fe hydrogenase and or nitrogenase. Direct biophotolysis phenomenon has been mostly studied in green algae and

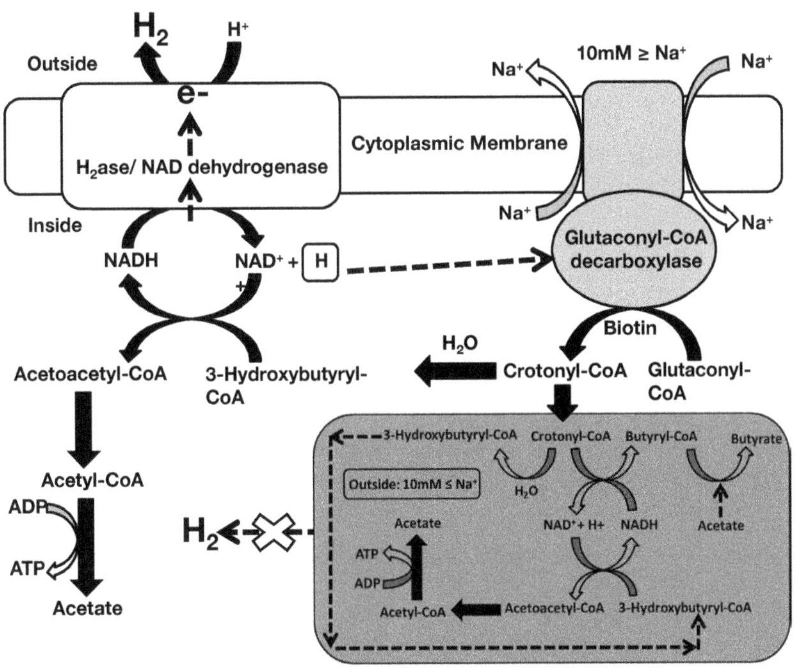

Figure 1.14 Na^+ ion-dependent hydrogen production in *Acidaminococcus fermentans*. Sodium ion concentrations greater than 10mM outside the cytosolic membrane are effective in driving hydrogen production in glutamate utilizing microorganism. Three major enzymes are responsible; H_2-ase, NADH-dependent dehydrogenase, and glutaconyl-CoA decarboxylase (in with biotin as major co-factor).

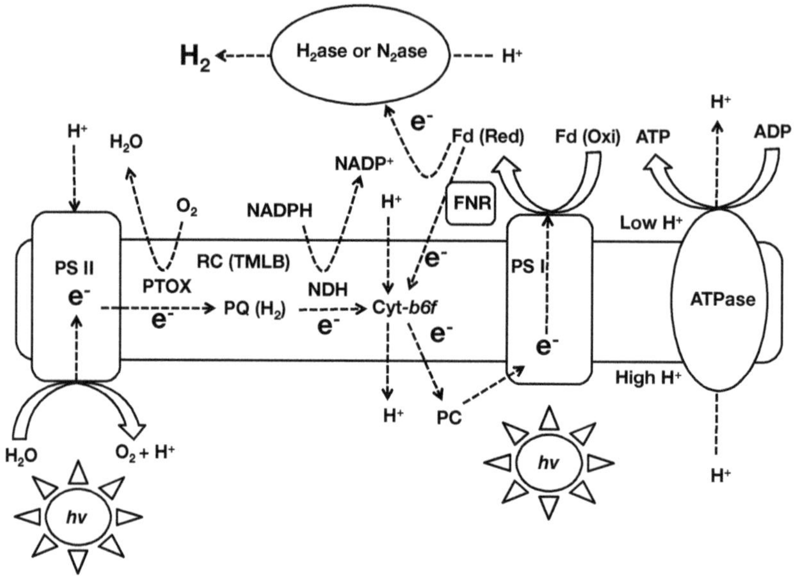

Figure 1.15 Molecular mechanism of direct biophotolysis using oxygenic photosynthesis for hydrogen production. The overall reaction occurs via membrane (lipid bilayer of the thylakoid membrane) bound enzyme systems (ATPase, ATP synthase; Cyt b6f, cytochrome b6f complex; Fd, ferredoxin; FDP, flavodiiron protein; FNR, ferredoxin NADP+ reductase; H_2-ase, hydrogenase; NDH, NAD (P)H dehydrogenase; PC, plastocyanin; PQ, plastoquinones; PS, photosystem; PTOX, plastid terminal oxidase.). Black dashed arrows indicate the multiple reactions.

cyanobacteria (Brentner *et al.*, 2010). In indirect biophotolysis (Fig. 1.16), the electrons or reducing equivalents from water are used for carbon dioxide fixation and the fixed carbon is stored as endogenous storage carbohydrates $(CH_2O)_n$ before being utilized to produce hydrogen. Finally this reserve carbohydrate is broken down and the electrons which are extracted are used to reduce nitrogenase or hydrogenase, thus producing hydrogen. Indirect biophotolysis has been mostly studied with blue green algae or cyanobacteria (Antal and Lindblad, 2005). Although these biophotolysis approaches are very attractive given the very abundant and inexpensive feedstock (water) and the capacity to capture solar energy and use this electron flux to reduce protons to hydrogen, they suffer from several serious bottlenecks. Generation of molecular oxygen is of course a major issue given the oxygen sensitivity of hydrogenases (green algae) (Brentner *et al.*, 2010). Furthermore, light conversion efficiencies are low, less than 2% (Lindblad *et al.*, 2002). Some are attempting to address the oxygen problem through engineering of algal hydrogenases, a very difficult proposition since algal hydrogenases have a complicated molecular structure, and genetic toolboxes for microalgae have not yet been well developed. Photofermentation is another avenue through which hydrogen can be produced using light energy. Photofermentation is a photoheterotrophic mode of hydrogen production carried out mostly by purple non-sulfur bacteria (Fig. 1.17). Under anaerobic conditions, the electrons scavenged from organic acids are transferred via reverse electron flow to oxidized ferredoxin through a series of membrane-bound, electron transport carrier molecules, the Rnf or FixABC systems (McKinlay *et al.*, 2010). Reduced ferredoxin in turn gives its electrons to nitrogenase which, when it is present, reduces molecular nitrogen to ammonia. Finally, in the absence of molecular nitrogen, nitrogenase catalyses the reduction of protons to hydrogen (Harwood, 2008). The electrochemical gradient generated by photosynthesis is utilized for ATP synthesis to supply the necessary energy to nitrogenase. This metabolic network, with its membrane-bound enzymes and electron carriers, has not yet been

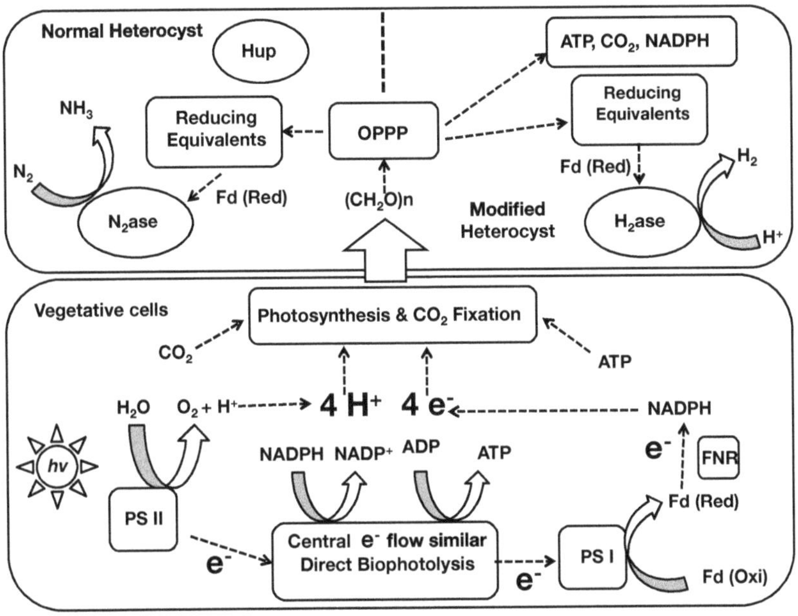

Figure 1.16 Molecular mechanism of indirect biophotolysis which couples oxygenic photosynthesis with hydrogen production. Black dashed arrows indicate the multiple reactions. Nomenclature: FNR, ferredoxin $NADP^+$ reductase; $(CH_2O)n$, starch or carbohydrates; hv, photon energy; OPPP, oxidative pentose phosphate pathways.

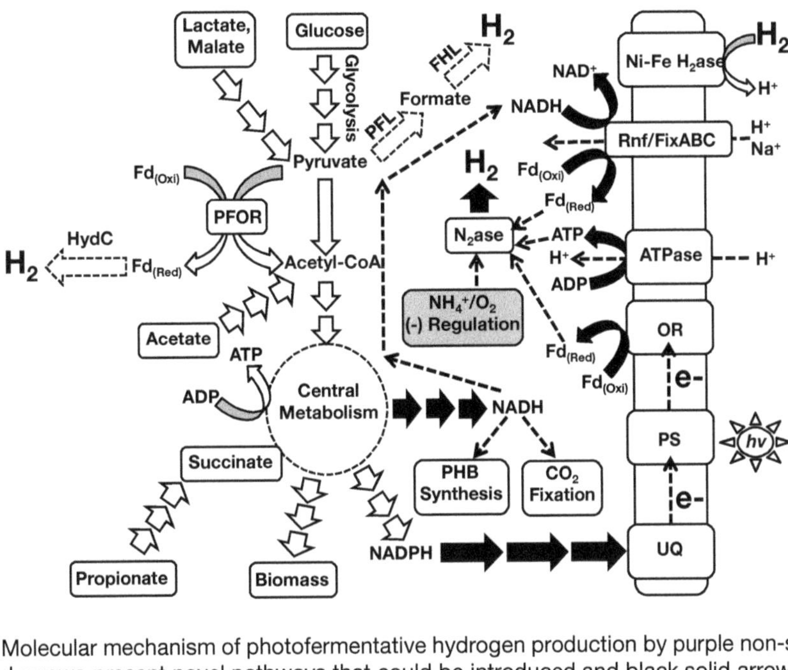

Figure 1.17 Molecular mechanism of photofermentative hydrogen production by purple non-sulfur bacteria upon. Dashed arrows present novel pathways that could be introduced and black solid arrows indicate the central electron or reductant flow towards hydrogen production by the N_2-ase enzyme system. Abbreviations: pyruvate:ferredoxin oxidoreductase (PFOR); ferredoxin-dependent hydrogenase (HydC); pyruvate formate lyase (PFL); formate hydrogen lyase (FHL); poly-beta-hydroxybutyrate (PHB); ubiquinone (UQ); photosystem (PS); oxidoreductase enzyme (OR); ferredoxin (Fd).

fully characterized. To understand and engineer the metabolic networks in photofermentative hydrogen production will first require detailed studies including 'omics' and detailed biochemical and genetic characterization. The photoheterotrophic hydrogen production approach has several advantages including (a) high theoretical conversion yields, (b) no oxygen evolution, (c) the ability to use a wide spectrum of light, and (d) the ability to use organic substrates, organic acids, glucose equivalents, glycerol, derived from wastes. However, this process also suffers from different drawbacks, including low light conversion efficiencies, low volumetric rates of hydrogen production, and having several competing electron consuming pathways (Hallenbeck, 2011).

1.3.1 Metabolic engineering for improving photosynthetic microbial platforms for hydrogen production

As discussed above, light-driven hydrogen production is dominated by three photosynthetic hydrogen producing microbial platforms. These photoautotrophs and photohetertrophs include microalgae, cyanobacteria, and anaerobic photosynthetic bacteria (mostly Gram negative non-sulfur purple bacteria) (Table 1.3). From a metabolic standpoint, the hydrogen production capacities of these groups differ greatly in several respects: (a) carbon source or source of reducing energy, (b) light-harvesting pigments and photosystems and (c) hydrogen-producing enzymes with variable sensitivities.

The model green alga for hydrogen production is *Chlamydomonas reinhardtii*. Anoxic conditions lead to the activation of several fermentative pathways which produce a number of end metabolites including formate, malate, and acetate along with ethanol, carbon dioxide and hydrogen (Fig. 1.18). Many of the efforts to improve yields have focused on *C. reinhardtii* due to its good basal levels of H_2 production (Timmins *et al.*, 2009; Melis *et al.*, 2000), the availability of completely sequenced and transformable mitochondrial, chloroplast and nuclear genomes, detailed information about its metabolic pathways, and an extensive library of expressed sequence tags (Merchant *et al.*, 2007). *Chlamydomonas* is known to ferment starch when deprived of sulfur (which creates the anoxic conditions necessary for hydrogen production), creating a range of products that accumulate in direct competition with H_2 production (Matthew *et al.*, 2009). Sulfur deprivation is one approach to overcoming the inhibition of hydrogenase by oxygen evolved during photosynthesis. In essence, sulfur depletion separates the growth and hydrogen generation phases. On a molecular level, the rate of oxygen evolution, catalysed by photosystem II, drops below respiration, driving sealed cultures into anaerobic conditions which allow sustained hydrogen production for long periods of time. Analysis of the *C. reinhardtii* genome sequence as well as inhibitor studies and determination of the products of fermentation suggests the presence of a three branched pathway (Fig. 1.18) (Hemschemeier *et al.*, 2008a; Mus *et al.*, 2007; Grossman *et al.*, 2007) in which pyruvate can be broken down to produce ethanol, formic acid or hydrogen along with acetic and malic acid further downstream (Gfeller *et al.*, 1984). The predominant fermentative pathway is thought to involve the conversion of pyruvate to acetyl-CoA and formic acid by the enzyme pyruvate formate lyase (PFL1) (Hemschemeier *et al.*, 2008b) which has subsequently been identified as a good candidate for a reverse genetic approach to improving hydrogen production. To date, problems transforming the *C. reinhardtii* nuclear genome by homologous recombination (Zorin *et al.*, 2008), and transcriptional silencing of conventional RNAi constructs, consisting of long dsRNAs (Rohr *et al.*, 2004), have hampered the ability to switch off or down regulate fermentative pathways. However, recent advances in artificial mi-RNA (a mi-RNA) technology have opened up the potential for stable metabolic engineering in *Chlamydomonas* (Molnar *et al.*, 2009; Tao *et al.*, 2009).

Synechocystis sp. and *Synechococcus* sp. are the best model cyanobacterial systems for metabolic engineering due to the availability of their genome sequences and their capacity for natural transformation (Kaneko *et al.*, 1996; Eaton-Rye, 2004). In general, the cyanobacteria are a physiologically and evolutionarily diverse group of prokaryotic oxygenic photoautotrophs which are tolerant to widely variable environmental conditions, including; pH, temperature, and salinity.

Table 1.3 General outline for photosynthetic microbial platforms for metabolic engineering towards improving biohydrogen generation

Groups	Hydrogen producer	PS (LHP)	Source of Reducing energy	Enzymes	Benefits	Drawbacks
Green algae (aerobic, oxygenic, eukaryotes)	*Chlamydomonas reinhardtii* *Chlorella sorokiniana* *Platymonas subcordiformis*	Chlorophyll a, b and carotenoids (PSI, PSII)	H_2O and/or organic acids (acetate), CO_2	H_2-ase	a. Energy production is carbon neutral b. Stochiometric energy efficiency is comparative higher for hydrogen production than do the biomass production	a. Oxygen sensitivity of H_2-ase
Blue-green algae or Cyanobacteria (aerobic, oxygenic, Gram+ve prokaryotes)	*Anabaena variabilis** *Anabaena cylindrica** *Aphanocapsa montana** *Gleobacter* sp.* *Aphanothece halophytico*** *Chroococcidiopsis thermalis*** *Spirulina platensis*** *Microcystis* sp.** *Oscillotoria* sp.** *Synechococcus* sp.** *Synechocystis* sp.**	Chlorophyll a, Carotenoids, and Phycobilisome (PSI, PSII)	H_2O and/or organic acids (acetate), CO_2 and air	Bi-directional H_2-ase and N_2-ase	a. Biohydrogen production is compartmentalized from oxygen evolution in heterocystous nitrogen-fixing cyanobacteria b. Non-heterocystous cyanobacteria are separated temporally (dark or light)	a. Biosynthesis and cellular maintenance of Heterocystous form b. N_2-ase required higher ATP for cellular energy metabolisms c. Re-oxidation of generated hydrogen by uptake H_2-ase
Photosynthetic bacteria (anaerobic, anoxygenic, Gram–ve prokaryotes)	*Thiocapsa roseopersicina*† *Chromatium* sp.† *Rhodobacter capsulatus*†† *Rhodobacter sphaeroides*†† *Rhodopseudomonas palustris*†† *Rhodospirillum rubnum*††	Bacterial chlorophyll a/b and carotenoids (single photosystem homologous to PSII)	†H_2S, S^O, $S_2O_3^{2-}$, thiosulfate, succinate ††Organic acids (malate)	†N_2-ase ††N_2-ase and/ or Bbidirectional H_2-ase		a. Required nitrogen limiting environment to produce hydrogen gas b. Industrial waste pretreatment requires as it may show toxic effects on cellular growth or reductant flow c. Re-oxidation of generated hydrogen by uptake H_2-ase

*Heterocystous.
**Non-heterocystous.
†Purple sulfur.
††Non-purple sulfur.
PS (LHP), photosystem (light-harvesting pigments); H_2-ase, hydrogenase enzyme; N_2-ase, nitrogenase enzyme.

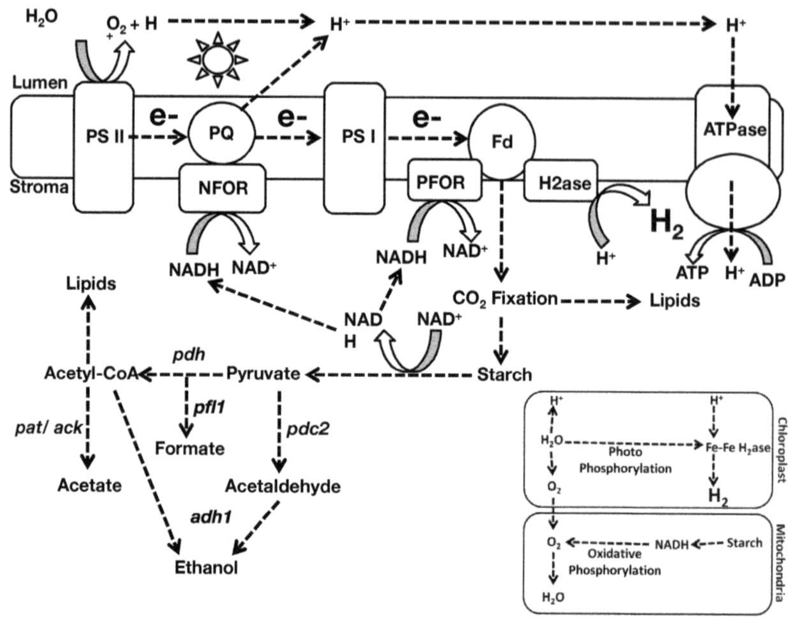

Figure 1.18 Metabolic pathways involved in the green alga *Chlamydomonas reinhardtii.* On the molecular level, ferredoxin (Fd) is reduced by interaction with electrons from either water oxidation with the help of photosystem II (PS II) or by non-photochemical reduction of plastoquinone (PQ). The reduced Fd subsequently reduces ferredoxin-dependent NADH or NADPH oxidoreductase (NFOR) to produce more reducing equivalents (NADH or NADPH); finally used in CBB pathway for carbon dioxide fixation leading to starch and lipid biosynthesis. Fd can reduce H_2-ase to produce hydrogen during anaerobiosis. The oxidation of pyruvate during glycolysis; catalysed by either the pyruvate dehydrogenase (*pdh*) complex under aerobic environments or pyruvate–ferredoxin oxidoreductase (PFOR) under anaerobic conditions, can be utilized to produce acetyl-CoA for lipid biosynthesis. Alternative pathways leading to acetyl-CoA from pyruvate compete with hydrogen metabolism for reducing equivalents (NADH); acetate pathways (*pat* and *ack*); ethanol pathways (*adh1* and *pdc2*) and formate pathway (*pfl1*). Reduced ferredoxin resulting from the activity of PFOR can be used to reduce H_2-ase. (Modified from Posewitz *et al.*, 2009.)

Many cyanobacteria are facultative anaerobes with the capability to carry out fermentation and produce hydrogen as a major by-product of the dark anaerobic catabolism of photosynthetically derived, energy storage substances (e.g. glycogen) (Stal and Moezelaar, 1997). Thus, there are multiple driving forces for focussing on these microorganisms as potential sources for the sustainable and renewable production of hydrogen. If hexoses or pentoses equivalents enter catabolic metabolic networks under anoxic conditions, the carbon flux will passes through several metabolic pathways downstream of pyruvate, including hydrogen production and uptake. Pyruvate and reductant formed by glycolysis and/or the pentose phosphate pathway in cyanobacteria can be transformed into lactate, ethanol, acetate, formate, carbon dioxide and molecular hydrogen gas (Stal and Moezelaar, 1997). Electron flow is through photosynthesis or the respiratory electron transport chain allowing for interconversion of $NADP^+$ to NADPH or reduction of hydrogenase (Fig. 1.19). Additionally, the TCA cycle can affect the electron balance. Although the TCA cycle in cyanobacteria was long thought to be incomplete, the recent demonstration of genes encoding novel 2-oxoglutarate decarboxylase and succinic semialdehyde dehydrogenase were identified in most cyanobacteria, demonstrating that they too have a complete TCA cycle (Zhang and Bryant, 2011). Although nitrogen assimilation is redox neutral in some cases, e.g. the production of alanine by the reductive amination of pyruvate, in most cases nitrogen assimilation will consume electrons that could in principle be used for hydrogen production (Kengen *et al.*, 1994;

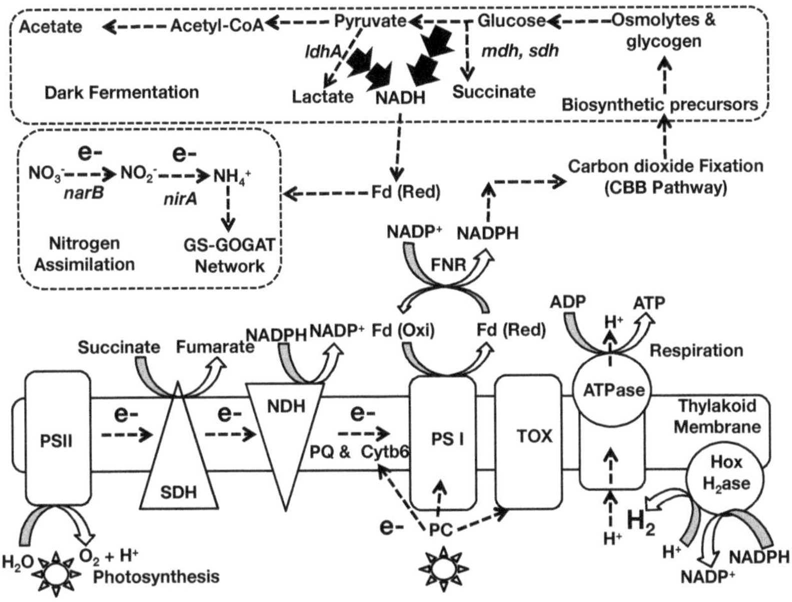

Figure 1.19 Metabolic pathways involved in hydrogen evolution by cyanobacteria. This model system is a combined form taking into account two systems of hydrogen production, *Synechocystis* sp. PCC 6803 and *Spirulina maxima.* Hydrogen metabolism and its relationships with photosynthesis, respiration, dark fermentation and nitrate assimilatory pathways (nitrate is being transported by Nrt, a nitrate transporter) are shown. Electrons derived from the hydrolysis of water in photosynthesis reduce Fd and subsequently NADPH. NADPH equivalents are oxidized to produce hydrogen or oxidized by competing metabolic networks; CBB pathways (carbon dioxide fixation to make glycerol-dehyde-3-phoshphate) and biosynthesis of precursors; respiration (oxidative phosphorylation), or by nitrate assimilation via ferredoxin for protein biosynthesis. In the absence of oxygen, stored energy compounds made by carbon fixation (such as osmolytes and glycogen) can be fermented to produce ATP for the cell and subsequently make NADH, which can then be reoxidized to NAD^+ through production of organic acids and alcohols, through nitrate assimilation via ferredoxin, or to produce hydrogen. Abbreviations: PSII (Photosystem II); SDH (succinate dehydrogenase); NDH (Type I NADPH-dehydrogenase); PQ (Plastiquinone pool); Cyt b6 (cytochrome b6); PC (plastocyanin); PSI (Photosystem I); TOX (terminal oxidase); ATPase (ATP synthase); Fd (ferredoxin); FNR (Ferredoxin NADPH Reductase); GS-GOGAT (glutamine synthetase-Glutamate synthase). Hox (hox-encoded bidirectional hydrogenase). (Modified from Baebpraserta *et. al.*, 2011; Carrieri *et al.*, 2011; Ananyev *et al.*, 2008.)

O¨rlygsson *et al.,* 1995). Thus, nitrate assimilation, which competes with hydrogen production for electrons, is a target for metabolic engineering and hydrogen production has already been shown to increase in *Synechocystis* sp. under nitrate limitation (Oxelfelt *et al.,* 1995; Axelsson and Lindblad, 2002; Antal and Lindblad, 2005; Antal *et al.,* 2006). Nitrate is transported into the cell by an active transport system and reduced to nitrite by nitrate reductase (encoded by *narB*) followed by reduction to ammonium by nitrite reductase (*nirA*) (Incharoensakdi, 2006). Nitrate reduction to ammonium ions via these two sequential steps requires two electrons and six electrons for a total of eight electrons (Flores *et al.,* 2005). Given the diversity of the environments in which various cyanobacteria live, it is perhaps unsurprising that several species have evolved different strategies for tolerating, or even thriving in, anaerobic or microaerobic environments. With this in mind, fermentative pathway map for products that result from NADPH regeneration after oxidation of carbon storage compounds (such as glycogen or osmolyte sugars) to pyruvate has been assembled in the form at the Joint Genome Institute's Integrated Microbial Genomes (IMG) database (www.img.jgi.doe.gov), where 58 cyanobacterial genome sequences are stored along with a notation of the presence or absence of fermentative enzymes.

Rhodobacter capsulatus, *Rhodobacter sphaeroides*, *Rhodopseudomonas palustris* and *Rhodospirillum rubrum* are the most widely studied photosynthetic bacteria as potential hydrogen producers. The photosynthetic bacteria consist of purple sulfur bacteria, obligate anaerobic autotrophs which utilize H_2, H_2S, and elemental sulfur, whereas the non-sulfur purple photosynthetic bacteria do not use sulfur and are capable of growing aerobically on organic substrates in absence of light. In these organisms, H_2 production begins after the exhaustion of ammonium, a production that is associated with the photofermentative metabolism of organic substances in the absence of significant growth but with a nearly stoichiometric release of H_2 (Ormerod and Gest, 1962) due to the presence of a highly active anaerobic citric acid cycle (Fig. 1.17). All photosynthetic bacteria use H_2 as a reductant for the fixation of CO_2 and are capable of fixing molecular nitrogen. It has been known for a long time that in these organisms nitrogenase has the dual activities of nitrogen reduction as well as ATP-dependent H_2 evolution (Bulen *et al.*, 1965; Nandi and Sengupta, 1998).

1.3.2 Hydrogenases and nitrogenase of phototrophs and the potential for metabolic engineering

The hydrogenase enzymes involved in photoautotrophic or photoheterotrophic mode have been classified in general as follows.

- **Group A:** This is characteristically the cytoplasmic membrane-bound Ni–Fe uptake hydrogenase located in cyanobacterial heterocysts, a differentiated cell type specialized for nitrogen fixation in which cell division and photosynthesis are arrested, creating the microaerobic conditions necessary for the functioning of the oxygen sensitive nitrogenase (Bothe *et al.*, 2010). In the vegetative cells of *Anabaena* sp. PCC 7120, *hupL*, coding for the large subunit of the Group A hydrogenase is interrupted by a 9.5-kb DNA fragment specifying for a recombinase enzyme encoded by *xisC* and other genes (Carrasco *et al.*, 2005). This group of hydrogenases has the dual functions of protecting nitrogenase against oxygen inactivation and recycling the hydrogen produced as an obligate by-product of nitrogen fixation (Peterson and Burris, 1978)
- **Group B:** This is mostly the bi-directional hydrogenase found in *Synechocystis* sp. PCC 6803 and the purple sulfur photosynthetic bacteria *Thiocapsa roseopersicina*. Group B hydrogenases are potentially active under anaerobic conditions (Ra′khely *et al.*, 2004; Cournac *et al.*, 2004). Recently bioengineering tools have been used to improve biohydrogen generation in cyanobacteria by directing electrons towards the bidirectional hydrogenase and away from the other electron-competing metabolic networks (Tamagnini *et al.*, 2007).
- **Group C:** this group of hydrogenases are distributed in diverse facultative chemolithoautotrophic bacteria which use CO as the sole energy and carbon source under anaerobic conditions. This metabolic flexibility has been revealed in a few purple non-sulfur bacteria, namely *Rhodospirillum rubrum* and *Rhodocyclus gelatinosus*. In these photosynthetic bacteria, the presence of CO induces expression of CO-hydrogenase and hydrogenase which together catalyse the evolution of molecular hydrogen from CO (Bonam *et al.*, 1989).

In the green algae and cyanobacteria, photosynthetic water splitting is functionally linked to H_2 production by the activity of either hydrogenase or nitrogenase enzymes. Interestingly, each of these organisms contains only one of two major types of hydrogenases, Fe–Fe or Ni–Fe enzymes, which are phylogenetically distinct but perform the same catalytic reaction, suggesting convergent evolution. Green algal hydrogenases belong to the class of Fe–Fe-hydrogenases, which are also found in strict anaerobes, fungi, and protists (Boichenko *et al.*, 2004). The sequence similarity between different Fe–Fe-hydrogenase proteins is very high, around 50% (Florin *et al.*, 2001). This is due to conservation of the residues involved in binding or in providing the appropriate structure and environment for the metallocatalytic cluster (based on L1, L2 and L3 motifs). Fe–Fe-hydrogenases contain only Fe and S in their catalytic site, and are typically involved in H_2 production rather than H_2 oxidation. The metallocatalytic cluster of Fe–Fe-hydrogenases, the H-cluster, is unique. It consists of a 4Fe–4S cubane linked through a protein cysteine residue to a 2Fe subcluster. The iron atoms of the 4Fe–4S centre are coordinated to the protein structure by three additional conserved cysteine residues, which are found in motifs L1–L3. Except for the bridging cysteine, the iron atoms of the 2Fe centre are coordinated to the non-protein carbon monoxide (CO) and cyanide (CN) ligands (Tye *et al.*, 2005). In green algal hydrogenases, electrons are delivered directly to the H-cluster by reduced ferredoxin. All other hydrogenases, however, contain electron relays composed of additional FeS centres either 2Fe–2S or 4Fe–4S, the F-clusters, that may be involved in electron transport from an external donor to the H-cluster (Cammack, 1999). At present, Fe–Fe-hydrogenase genes have been identified and sequenced from several green algal species namely *hydA1* and *hydA2* (*Chlamydomonas reinhardtii*), *hydA1* (*Scenedesmus obliquus*), *hydA* (*Chlorella fusca*). Moreover, the maturation proteins involved in Fe–Fe-hydrogenase assembly were initially discovered in *C. reinhardtii,* where two novel radical S-adenosylmethionine (Radical-SAM or AdoMet radical) proteins, HydEF and HydG, are required for Fe–Fe-hydrogenase activity (Posewitz *et al.*, 2004).

In cyanobacteria hydrogen production is mainly associated with nitrogenase. These organisms also contain an uptake hydrogenase, which catalyses the consumption of the hydrogen produced by nitrogenase, and a bidirectional hydrogenase, which has the ability to carry out either hydrogen production or uptake (Tamagnini *et al.*, 2002). The nitrogenase complex consists of two proteins: dinitrogenase (Mo–Fe protein or protein I) and dinitrogenase reductase (Fe protein or protein II). Dinitrogenase is a $\alpha_2\beta_2$ heterotetramer of about 220 to 240 kDa, and the α and β subunits are encoded by *nifD* and *nifK,* respectively. Dinitrogenase reductase, encoded by *nifH,* is a homodimer of about 60 to 70 kDa and specifically mediates the transfer of electrons from the external electron donor, a ferredoxin or a flavodoxin, to dinitrogenase (Masepohl *et al.*, 1997). The reduction of nitrogen to ammonia mediated by nitrogenase is a highly endergonic reaction requiring metabolic energy in the form ATP. Cyanobacterial nitrogenases, like all others, are highly oxygen sensitive. Therefore, all cyanobacterial nitrogenases require some mechanism of protection from the detrimental effects of molecular oxygen. There are several alternative nitrogenases present in the heterocystous cyanobacterium *A. variabilis* ATCC 29413; Mo-containing nitrogenases (encoded by *nif1* and *nif2*), a V-containing nitrogenase (encoded by *vnfDGK*) and a Fe-containing nitrogenases (encoded by *anf* gene) (Tamagnini *et al.*, 2002). These alternative nitrogenases differ to some degree from the conventional Mo enzyme physically, chemically and catalytically. Specifically, research has shown that these enzymes allocate a higher proportion of electrons to the reduction of protons to molecular hydrogen than do the Mo nitrogenases (Almon and Böger, 1988). However, this is not really a useful property for producing hydrogen as their overall catalytic activity is much lower.

As mentioned above, cyanobacteria carry two other hydrogenases, an uptake Ni–Fe hydrogenase, and a bidirectional hydrogenase enzyme. Uptake hydrogenases (*hupSL*) are mostly involved in hydrogen oxidation and found in all nitrogen-fixing bacteria (Hansel and Lindblad, 1998). On the other hand, the cyanobacterial bidirectional hydrogenases (encoded by *hoxEFUYHW* genes) are soluble or loosely membrane-bound Ni–Fe hydrogenases found in both nitrogen fixing and non-nitrogen fixing cyanobacteria. This novel bidirectional hydrogenase has been partially purified and characterized in *A. variabilis* ATCC 29413 and found to differ from uptake hydrogenases from both a physical and catalytic standpoint (Serebryakova *et al.*, 1996). Although hydrogenase genes have been identified and characterized in a number of different cyanobacteria, very little is known about their maturation systems, typically encoded by *hyp*, e.g. *hypABCDEF* in *Synechococcus* sp. PC6803, having several functions; Ni ion incorporation, chaperone proteins, CO/CN synthesis, etc. (Tamagnini *et al.*, 2002).

Photosynthetic bacteria possess a diverse and evolutionarily ancient metabolism, which is reflected to the different ways in which they can metabolize H_2. Three enzymes have been implicated in H_2 metabolism in these organisms: (a) nitrogenase, which catalyses a unidirectional, ATP-dependent H_2 evolution and can function either in the light, or in the dark under anaerobic or micro-aerobic conditions, (b) uptake hydrogenase (*hupSL*), which is membrane-bound and, functions physiologically in the direction of H_2 oxidation, and (c) a fermentative Ni–Fe hydrogenase, which may be either soluble or membrane bound and functions mainly during dark anaerobic fermentation (Vignais *et al.*, 1985). Most nitrogen-fixing bacteria encode only a molybdenum nitrogenase (Mo nitrogenase). However, some, including *Rhodobacter capsulatus* and *Rhodospirillum rubrum* encode in addition an iron nitrogenase (Fe nitrogenase) or a vanadium nitrogenase (V nitrogenase). Alternative nitrogenases have been proposed to serve as a route for nitrogen fixation in situations where molybdenum is limited in the environment (Richaud *et al.*, 1990; Oda *et al.*, 2005).

The *R. palustris* CGA009 genome is predicted to encode three different nitrogenases, each with a different transition metal-containing cofactor at its active site (Larimer *et al.*, 2004). Mo nitrogenases consist of two components, designated the dinitrogenase reductase (encoded by *nifH*) and the dinitrogenase or Mo–Fe protein (encoded by *nifDK*) (Fisher *et al.*, 2002). The Mo–Fe protein has an iron molybdenum cofactor at its active site. Reduced NifH serves as the electron donor to the Mo–Fe protein, which contains the site of nitrogen reduction. The electron transfer reaction from NifH to NifDK is accompanied by the hydrolysis of ATP. Nitrogen is progressively reduced at the active site to produce partially reduced intermediates until ammonium is finally formed and released along with hydrogen. Alternative V and Fe nitrogenases comprise homologous VnfHDK and AnfHDK subunits. These enzymes also include VnfG and AnfG subunits as additional structural components of the dinitrogenase (Masepohl *et al.*, 2002).

1.3.3 Application of metabolic engineering to existing microorganisms to improve light-dependent hydrogen production

Phototrophic organisms depend in general upon captured light energy for hydrogen production. Hence one way to enhance the photoproduction of hydrogen is to increase photosynthetic efficiencies. One approach to this is to modify the size of the light harvesting antenna complexes responsible for capturing solar energy. Large antenna complexes of cyanobacteria and green algae help them to grow under low light condition, but results in dissipation and loss of excess photons as fluorescence or heat under high light conditions. It is debatable whether the LHC II b (major peripheral antenna), or CP26 and CP29 (minor antenna) or another component is critical for non-photochemical quenching (NPQ) that protects the PSII in *Chlamydomonas reinhardtii* (Elrad *et al.*, 2002). Reduction in pigment content can lead to better penetration of light inside the reactor and reduce the wastage of light energy. It has been observed that hydrogen production is increased in a mutant algal strain having a reduced antenna (Polle *et al.*, 2003). It is also expected that cyanobacteria deficient in phycobilisomes, which implies an uncoupling of electron transport and preferential use of excitation by PSI, might show an enhanced rate of hydrogen production (Bernat *et al.*, 2009). Another promising technology, RNA interference (RNAi), has been used to down-regulate the entire family of light harvesting complexes (LHC) in *Chlamydomonas reinhardtii* (Mussgnug *et al.*, 2007). Using a *Rhodobacter sphaeroides* mutant MTP4 in a plate-type reactor, 50% more hydrogen was found to be produced than with its wild type counterpart *R. sphaeroides* RV (Kondo *et al.*, 2002). Another mutant (P3 mutant) with a 2.7-fold decrease in core antenna (LH1) content and a 1.6-fold increase in peripheral antennal (LH2) content showed accelerated H_2 production compared to the wild type (Vasilyeva *et al.*, 1999).

Another potentially major problem for producing hydrogen using biological systems is the extreme sensitivity of hydrogen-producing enzymes to oxygen. Cyanobacteria and green algae produce oxygen due to the water splitting by PSII, potentially inactivating any hydrogenase that is present. Under normal growth conditions, the rate of photosynthesis is 4- to 7-fold higher than the respiration rate. By using attenuated photosynthesis/respiration ratio (P/R ratio) mutants (*apr* mutants) of *Chlamydomonas reinhardtii*, the P/R ratio drops below one, thereby establishing anaerobic conditions which mimic the physiological status of sulfur-deprived cells but avoids the cell growth inhibition and death that can be brought about by sulfur deprivation (Melis, 2007; Ruhle *et al.*, 2008). It has also been reported that a sulfur-deprived *C. reinhardtii* D1 mutant that carries a double amino acid substitution is superior to the wild type for hydrogen production. The leucine residue L159 in the D1 protein was replaced by isoleucine, and the asparagine N230 was replaced by tyrosine (L159I-N230Y). This strain is very efficient in prolonged H_2 production and as well as having lower chlorophyll content and higher respiration rate, both of which contribute to a higher net yield (Torzillo *et al.*, 2009). The actual activity of the hydrogen evolving enzymes might be another important factor limiting the rate of photoproduction of hydrogen. Hydrogen production can be increased in photosynthetic bacteria by improving the efficiency of nitrogenase. The problem of repression of nitrogenase by ammonium ions has been overcome in *R. sphaeroides* by knocking out *glnB* and *glnK*, genes encoding P II-like proteins (Kim *et al.*, 2008). Another possibility in the same vein is the heterologous overexpression of a more efficient enzyme. Cyanobacteria have a bidirectional Ni–Fe-hydrogenase which may be inefficient for hydrogen production and hydrogen is mostly produced by nitrogenase. Interestingly a Fe–Fe-hydrogenase (*hydA*) from *Clostridium pasteurianum* has been introduced into the cyanobacterium *Synechococcus elongates*, expressed and found to be functional without the co-expression of maturation proteins (Miyake and Asada 1997; Asada *et al.*, 2000). Attempts have been made in *R. rubrum* to both overexpress a native hydrogenase and to carry out the heterologous expression of a foreign hydrogenase (Kim *et al.*, 2008).

There is an additional obstacle to achieving the theoretical maximum value for hydrogen production in cyanobacteria and photosynthetic bacteria; the produced hydrogen can be re-oxidized by an uptake hydrogenase (Ni–Fe-hydrogenase). To date significant research has been performed to inhibit the uptake hydrogenase activity using different approaches. Mutants deficient in uptake hydrogenase activity show a dramatically increased net production of hydrogen. Inactivation of all hydrogenases, including all uptake activities, in *Thiocapsa roseopersicina* (*hypF*-deficient mutant) caused a significant increase in the hydrogen evolution capacity under nitrogen-fixing conditions (Fodor *et al.*, 2001). It has also been reported that a *R. rubrum* mutant deleted for *hupL*, encoding the large subunit of the uptake hydrogenase, produced increased amounts of H_2 (Ruiyan *et al.*, 2006). In another study, a suicide vector was used for site directed mutagenesis of the *R. sphaeroides* uptake hydrogenase (*hupSL*), giving 20% greater H_2 production (Gokhan *et al.*, 2008). In addition to external electron donors like H_2O for cyanobacteria and green algae, and sulfur containing inorganic compounds or organic acids for photosynthetic bacteria, reducing equivalents can be derived from reserve carbon compounds. In cyanobacteria and green algae, this alternative metabolic process can produce hydrogen via fermentative reactions or can proceed by reducing plastoquinone (PQ) by Ndh to maintain the NAD^+/NADH balance and ATP supply (Gfeller and Gibbs, 1984; Hemschemeier and Happe, 2005). After random gene insertion in *C. reinhardtii*, a strain named Stm6 with modified respiratory metabolism was isolated. This strain is able to accumulate large amounts of starch, enabling it to produce 5–13 times more hydrogen than the wild type (Kruse *et al.*, 2005). Obviously, supplying a fixed carbon source like acetate or glucose permits greater hydrogen production than growth on inorganic medium alone. *Chlorella* has a hexose uptake protein that is involved in transporting external carbohydrate into the cell. Recently, the HUP1 (hexose uptake protein) from *Chlorella kessleri* was introduced into *C. reinhardtii* to increase the supply of external glucose into the cell. Hydrogen production capacity was increased about 150% by supplying 1 mM glucose to a derivative of Stm6 in which HUP1 had been inserted (Doebbe *et al.*, 2007).

In the green alga *Chlamydomonas reinhardtii* the main competitor for hydrogenase for reducing equivalents is the Calvin–Benson–Bassham (CBB) pathway in which RuBisCO plays a key role. RuBisCO (Ribulose-1,5-bisphosphate carboxylase oxygenase) is involved in the first major step of carbon fixation, a process by which atmospheric carbon dioxide is converted by green algae to energy-rich molecules such as glucose. In chemical terms, it catalyses the carboxylation of ribulose-1,5-bisphosphate. A RuBisCO mutant *C. reinhardtii*, Y67A, has been shown to have a 10- to 15-fold higher hydrogen production (Pinto *et al.*, 2013). This result clearly shows that when the CBB pathway is partially or completely knocked out, more reductant generated by the light driven reactions of photosynthesis may become available to the Fe–Fe hydrogenase present in *C. reinhardtii.*

The photosynthetic bacterium *R. capsulatus* uses the CBB pathway for intracellular redox homeostasis under photoheterotrophic growth conditions (Tichi and Tabita, 2001). Normally, the highly regulated CBB pathway is used to assimilate atmospheric CO_2, the sole source of cellular carbon, during photo and chemoautotrophic growth. However during photoheterotrophic growth, the CBB pathway plays a role in redox balance by allowing CO_2 to serve as an electron sink rather than serving as a major means for generating organic carbon. RuBisCO and phosphoribulokinase (PRK), key enzymes of the CBB cycle, are produced in much lower yet substantial levels for this purpose under photoheterotrophic growth conditions (Tichi and Tabita, 2001). In the absence of a functional CBB pathway (inactivation of RuBisCO or phosphoribulokinase), *R. capsulatus* strains fail to grow photoheterotrophically in the presence of ammonium unless an exogenous electron acceptor such as dimethylsulfoxide (DMSO) is provided (Tichi and Tabita, 2001; Hallenbeck *et al.*, 1990). In the presence of ammonium and the absence of external electron acceptors, RuBisCO or PRK deficient strains give photoheterotrophically competent pseudo-revertants by de-repressing the synthesis of nitrogenase. Nitrogenase permits growth through the dissipation of excess reducing equivalents via proton reduction to generate hydrogen (Öztürk *et al.*, 2012). Recently it was shown that the poor growth of RuBisCO (*ΔcbbM*) mutants of the related photosynthetic bacterium *Rhodospirillum rubrum* is due to the accumulation of ribulose-1,5-bisphosphate (RuBP) under photoheterotrophic growth conditions (D. Wang *et al.*, 2011). The mutational elimination of RuBisCO in a *cbbM* mutant would lead to the accumulation of RuBP, while the disruption of *cbbP* (phosphoribulokinase) or *cbbF* (fructose-1,6-bisphosphatase) should block this which would explain why the *cbbM, cbbP* and *cbbM,* cbbF double mutants grew well in nitrogen-limited malate glutamate medium. It was also shown that there was a redox effect on growth, but it was apparently an indirect effect on the accumulation of RuBP by the regulation of the activities of enzymes involved in RuBP regeneration (D. Wang *et al.*, 2011). Similarly, a recent study of an ammonium insensitive pseudo-revertant of a CBB deficient strain of *R. capsulatus* (YO7-R3) found a 56% higher production rate under nitrogen-limiting environment (Ozturk and Gokce, 2012).

Recently hydrogen production was increased around 140-fold (to 186 nmol hydrogen/mg chlorophyll a/hour) using the metabolically engineered cyanobacterium *Synechocystis* sp. PCC6803 where the electron supply was redirected by knocking out nitrate assimilation (Baebpraserta *et al.*, 2011) since nitrate assimilation functions as an electron sink (Antal *et al.*, 2006; Gutthann *et al.*, 2007; Baebprasert *et al.*, 2010; Carrieri *et al.*, 2011). Nitrate is transported into the cell by an active transport system and reduced to nitrite by nitrate reductase (encoded by *narB*) followed by the action of nitrite reductase (encoded by *nirA*) to reduce nitrite to ammonium (Incharoensakdi, 2006). Nitrate reduction to ammonium through the two sequential reactions of nitrate reductase and nitrite reductase requires 2 electrons and 6 electrons, respectively (Flores *et al.*, 2005).

Hydrogen production in PNSB is often limited by low nitrogenase activity (Hallenbeck and Benemann, 2002; Harwood, 2008). To improve cellular nitrogenase activity, this enzyme can be engineered to have a higher catalytic turnover number or attempts can be made to increase the cellular expression level of the enzyme. Most of the studies thus far have focused on improving the expression level of nitrogenase rather than engineering it. The expression of nitrogenase is strictly controlled at the transcriptional and post-transcriptional levels in response to the availability of fixed nitrogen, particularly NH_4^+. In the presence of NH_4^+, two PII-like proteins, GlnB and GlnK, activate DraT, which inhibits nitrogenase by ADP-ribosylation. The two PII-like proteins also control the activity of NifA, the transcriptional activator of the nitrogenase structural genes. Furthermore, under nitrogen-replete conditions, GlnB inhibits a sensor kinase, NtrB, which is involved in dephosphorylation of the transcriptional activator NtrC and subsequently inactivates the transcription of *nifA* (Kim *et al.*, 2008). At 20 mM NH_4^+ nitrogenase is fully repressed and H_2 production becomes negligible (Koku *et al.*, 2002). Many efforts have been directed toward preventing this repression. For example, a *glnB-glnK* double mutant of *R. capsulatus* has been reported where the synthesis of nitrogenase, and thus its enzymatic activity, is enhanced greatly during cultivation in the presence of 200 mM NH_4^+ (Drepper *et al.*, 2003). H_2 production of the mutant strain was 1.5-fold higher than that of the wild-type strain. In another PNSB, *R. sphaeroides* KCTC 12085, the deletion of both *glnB* and *glnK* genes alleviates the repression of nitrogenase genes by NH_4^+ and hence improves H_2 production (Kim *et al.*, 2008). The effect of Δ*glnA1*, a deletion of glutamine synthase, key enzyme in the assimilation pathways of NH_4^+, on nitrogenase activity and H_2 production was also studied in *R. sphaeroides* 6016 in the presence of glutamine and NH_4^+ as nitrogen sources (X. Li *et al.*, 2010). The mutant *R. sphaeroides* produced a substantial amount of H_2 at relatively high NH_4^+concentrations (15 to 40 mM), as the elimination of glutamine synthase prevented NH_4^+ assimilation. Hydrogen production in the PNSB can also be improved by deleting uptake hydrogenases (Hup). H_2 production in the *hup*-negative *Rhodopseudomonas palustris* was improved greatly under nitrogen-fixing conditions with malate as the carbon source (Rey *et al.*, 2006). Two mutants, Δ*hupV* and Δ*hupS*, produced 120 and 110 mmol H_2/mg protein, respectively, while the wild type produced less than 1 mmol H_2/mg protein. Other *Rhodobacter* sp., such as *R. capsulatus* MT1131 (Ozturk *et al.*, 2006) and *R. sphaeroides* RV (Franchi *et al.*, 2004), also showed improved rates of H_2 production, around 36%, when Hup was disrupted. Two mutant strains of *R. sphaeroides* 6016, one without uptake hydrogenase HupSL, and the other lacking its positive regulator protein HupR gave a high H_2 production yield (Liu *et al.*, 2010). A similar result has been reported for the Δ*hupS* mutant of *R. sphaeroides* O.U. 001 (Kars *et al.*, 2009). In another study, the double-mutant *R. sphaeroides* KD131, which lacks both uptake hydrogenase (Hup) and poly-β-hydroxybutyrate (PHB) synthase (Phb), was studied (Kim *et al.*, 2006). This mutant strain showed a much higher H_2 production (3.34 ml H_2/mg cell on a dry weight basis) than that of the wild-type parent strain (1.32 ml H_2/mg cell). Poly-β-hydroxybutyrate synthase, which is responsible for the accumulation of the carbon storage material PHB in many PNSBs, is a major competitor of nitrogenase for electrons (Harwood *et al.*, 2008; Koku *et al.*, 2002). It seems that reduction of PHB synthesis is an efficient method for increasing H_2 yield in PHB-synthesizing PNSB.

1.4 Future directions

Metabolic engineering approaches have shown enormous potential for understanding and manipulating biological hydrogen metabolism either directly by affecting hydrogen production pathways or indirectly by changing alternative competing pathways. Greater knowledge is required to understand how to fine tune the electron flow of reducing equivalents (NADH, NADPH or ferredoxin) towards hydrogen production. More advanced research is required to make full use of metabolic engineering in the development of existing or even novel microbial platforms for hydrogen production. This will be necessary since enhanced hydrogen yields and productivities are required to make biohydrogen technology feasible on a commercial industrial level. Metabolic engineering along with two stage or co-culture approaches and electrochemical fuel cells or improved bioreactor design may solve some of the bottlenecks and accelerate hydrogen production technology by 2040 (Hallenbeck *et al.*, 2009; Hallenbeck and Ghosh, 2009). Additionally, bioprocess development could

Table 1.4 Improving hydrogen productivity by metabolically engineered light dependent hydrogen producers

Microorganisms	Target	Effects	References
Chlamydomonas reinhardtii	Pigment Reduction	Greater conversion efficiencies, productivity under high light	Polle *et al.* (2003), Mussgnug *et al.* (2007)
Synechocystis sp. PCC 6803		Greatly increased linear photosynthetic electron transport	Bernat *et al.* (2009)
Rhodobacter sphaeroides (mutant MTP4 and P3		50% greater H_2 production rates and yields	Kondo *et al.* (2002), Vasilyeva *et al.* (1999)
C. reinhardtii (arp mutant)	Modification of PSII, decrease in P/R	2–3-fold higher H_2 production	Melis (2002), Ruhle *et al.* (2008)
C. reinhardtii (D1 protein mutant)		10-fold higher H_2 production	Torzillo *et al.* (2009)
Synechococcus elongates	Introduction of foreign hydrogenase	Small in vivo activity, uncertain maturation process (HydE, F and G absent)	Miyake and Asada (1997), Asada *et al.* (2000)
Rhodospirillum rubrum			Kim *et al.* (2008)
Thiocapsa roseopersicina (*hypF*-deficient mutant)	Elimination of uptake hydrogenase activity	100-fold higher H_2 production	Fodor *et al.* (2001)
Rhodospirillum rubrum (*hupL*-deficient mutant)		2-fold higher H_2 production	Ruiyan *et al.* (2006)
R. sphaeroides (*hupSL*-deficient mutant)		20% higher H_2 production	Gokhan *et al.* (2008)
Rhodobacter capsulatus JP91 (*hup–*)		5.5 mol H_2/mol glucose	Ghosh *et al.* (2012a)
C. reinhardtii (Stm6 strain)	Enhancing the capacity of deriving electron from carbohydrates	5-fold greater rates and efficiency of H_2 production, incorporation of hexose transporter increases and additional 59%	Kruse *et al.* (2005), Doebbe *et al.* (2007)
Synechocystis sp. strain PCC 6803	Knocking out nitrate assimilation pathways	50–150-fold higher H_2 production	Baebpraserta *et al.* (2011)
R. capsulatus	Inhibition of CBB Pathways (RuBisCo mutation)	Decrease in H_2 –N, greatly increased H_2 +N	Ozturk *et al.* (2012)

become an important field where multi-process parameter optimization is applied to engineered hydrogen producers for stable hydrogen productivity with high yields of hydrogen. Ideally, the hydrogen would be produced with biophotolysis and water as a feedstock or with dark or photofermentative hydrogen production using various feedstocks derived from industrial wastes which would not compete directly with food production (Ghosh and Hallenbeck, 2010; Ghosh *et al.*, 2012a,b). Thus, metabolic engineering and bioprocess engineering combined will be key for making hydrogen production technology feasible at the commercial and industrial level.

Acknowledgements

Hydrogen research in the laboratory of PCH is supported by NSERC (Natural Science and Engineering Research Council of Canada) and FQRNT (Fonds québécois de la recherche sur la nature et les technologies). DG was supported by a fellowship from FQRNT.. Part of this work was performed while PCH held a National Research Council Research Associateship Award at the Life Sciences Research Center, Department of Biology, United States Air Force Academy.

References

Abdel-Hamid, A.M., Attwood, M.M., and Guest, J.R. (2001). Pyruvate oxidase contributes to the aerobic growth efficiency of *Escherichia coli*. Microbiology *147*, 1483–1498.

Abril, D., and Abril, A. (2009). Ethanol from lignocellulosic biomass. Cien. Inv. Agr. *36*, 177–190.

Afting, C., Kremmer, E., Brucker, C., Hochheimer, A., and Thauer, R.K. (2000). Regulation of the synthesis of H2-forming methylenetetrahydromethanopterin dehydrogenase (Hmd) and of HmdII and HmdIII in *Methanothermobacter marburgensis*. Arch. Microbiol. *174*, 225–232.

Akhtar, M.K., and Jones, P.R. (2009). Construction of a synthetic YdbK-dependent pyruvate:H2 pathway in *Escherichia coli* BL21 (DE3). Metabol. Eng. *11*, 139–147.

Almon, H., and Böger, P. (1988). Nitrogen and hydrogen metabolism: induction and measurement. Methods Enzymol. *167*, 459–467.

Amador-Noguez, D., Feng, X.J., Fan, J., Roquet, N., Rabitz, H., and Rabinowitz, J.D. (2010). Systems-level metabolic flux profiling elucidates a complete, bifurcated tricarboxylic acid cycle in *Clostridium acetobutylicum*. J. Bacteriol. *192*, 4452–4461.

Ananyev, G., Carrieri, D., and Dismukes, G.C. (2008). Optimization of metabolic capacity and flux through environmental cues to maximize hydrogen production by the cyanobacterium *Arthrospira* (*Sprilulina*) *maxima*. Appl. Environ. Microbiol. *74*, 6102–6113.

Angenent, L.T., Karim, K., Al-Dahhan, M.H., and Domiguez-Espinosa, R. (2004). Production of bioenergy and biochemicals from industrial and agricultural wastewater. Trends Biotechnol. *22*, 477–485.

Antal, T.K., and Lindblad, P. (2005). Production of H_2 by sulfur-deprived cells of the unicellular cyanobacteria Gloeocapsaalpicola and *Synechocystis* sp. PCC 6803 during dark incubation with methane or at various extracellular pH. J. Appl. Microbiol. *98*, 114–120.

Antal, T.K., Oliveira, P., and Lindblad, P. (2006). The bidirectional hydrogenase in cyanobacterium *Synechocystis* sp. PCC 6803. Int. J. Hydrogen Energy *31*, 1439–1444.

Asada, Y., Koike, Y., Schnackenberg, J., Miyake, M., Uemura, I., and Miyake, J. (2000). Heterologous expression of clostridial hydrogenase in the cyanobacterium *Synechococcus* PCC7942. Biochim. Biophys. Acta *1490*, 269–278.

Axelsson, R., and Lindblad, P. (2002). Transcriptional regulation of Nostoc hydrogenases: effects of oxygen, hydrogen and nickel. Appl. Environ. Microbiol. *68*, 444–447.

Baebprasert, W., Lindblad, P., and Incharoensakdi, A. (2010). Response of H2 production and Hox-hydrogenase activity to external factors in the unicellular cyanobacterium *Synechocystis* sp. PCC 6803. Int. J. Hydrogen Energy *35*, 6611–6616.

Baebpraserta, W., Jantaro, S., Khetkorn, W., Lindblad, P., and Incharoensakdi, A. (2011). Increased H2 production in the cyanobacterium Synechocystis sp. Strain PCC 6803 by redirecting the electron supply via genetic engineering of the nitrate assimilation pathway. Metab. Eng. *13*, 610–616.

Beer, L.L., Boyd, E.S., Peters, J.W., and Posewitz, M.C. (2009). Engineering algae for biohydrogen and biofuels production. Curr. Opin. Biotechnol. *20*, 264–271.

Berks, B.C., Palmer, T., and Sargent, F. (2003). The Tat protein translocation pathway and its role in microbial physiology. In Advances in Microbial Physiology (Academic Press, Waltham, MA, USA), pp. 187–254.

Bernat, G., Waschewski, N., and Rogner, M. (2009). Towards efficient hydrogen production: the impact of antenna size and external factors on electron transport dynamics in *Synechocystis* PCC 6803. Photosynthetic Res. *99*, 205–216.

Bisaillon, A., Turcot, J., and Hallenbeck, P.C. (2006). The effect of nutrient limitation on hydrogen production by batch cultures of *Escherichia coli*. Int. J. Hydrogen Energy *31*, 1504–1508.

Boichenko, V.A., Greenbaum, E., and Seibert, M. (2004). Hydrogen production by photosynthetic microorganisms. In: MD Archer, J Barber (eds.). Photoconversion of Solar Energy: Molecular to Global Photosynthesis. London: Imperial College Press, pp. 397–452..

Bonam, D., Lehman, L., Roberts, G.P., and Ludden, P.W. (1989). Regulation of carbon monoxide dehydrogenase and hydrogenase in Rhodospirillum rubrum: effects of CO and oxygen on synthesis and activity. J. Bacteriol. *171*, 3102–3107.

Bothe, H., Schmitz, O., Yates, M.G., and Newton, W.E. (2010). Nitrogen fixation and hydrogen metabolism in cyanobacteria. Microbiol. Mol. Biol. Rev. *74*, 529–551.

Brentner, L.B., Jordan, P.A., and Zimmerman, J.B. (2010). Challenges in developing biohydrogen as a sustainable energy source: Implications for a research agenda. Environ. Sci. Technol. *44*, 2243–2254.

Bulen, W.A., Burns, R.C., and Le Combe, J.R. (1965). Nitrogen fixation: hydrosulfite as electron donor with cell free preparations of *Azotobacter vinelandii* and *Rhodospirillum rubrum*. Proc. Natl. Acad. Sci. U.S.A. *53*, 532–541.

Burgdorf, T., Buhrke, T., van der Linden, E., Jones, A., Albracht, S., and Friedrich, B. (2005). [NiFe]-Hydrogenases of *Ralstonia eutropha* H16: modular enzymes for oxygen-tolerant biological hydrogen oxidation. J. Mol. Microbiol. Biotechnol. *10*, 181–196.

Calusinska, M., Happe, T., Joris, B., and Wilmotte, A. (2010). The surprising diversity of clostridial hydrogenases: a comparative genomic perspective. Microbiology *156*, 1575–1588.

Cammack, R. (1999). Hydrogenase sophistication. Nature *397*, 214–215.

Carere, C.R., Kalia, V., Sparling, R., Cicek, N., and Levin, D.B. (2008). Pyruvate catabolism and hydrogen synthesis pathway genes of *Clostridium thermocellum* ATCC 27405. Ind. J Microbiol. *48*, 252–266.

Carrasco, C.D., Holliday, S.D., Hansel, A., Lindblad, P., and Golden, J.W. (2005). Heterocyst-specific excision of the *Anabaena* sp. strain PCC 7120 hupL element requires xisC. J. Bacteriol. *187*, 6031–6038.

Carrieri, D., Wawrousek, K., Eckert, C., Yu, J., and Maness, P.C. (2011). The role of the bidirectional hydrogenase in cyanobacteria. Bioresour. Technol. *102*, 8368–8377.

Cha, M., Chung, D., Elkins, J.G., Guss, A.M., and Westpheling, J. (2013). Metabolic engineering of *Caldicellulosiruptor bescii* yields increased hydrogen production from lignocellulosic biomass. Biotechnol. Biofuels *6*, 85.

Chen, X., Sun, Y., Xiu, X.-L., Li, X., and Zhang, D. (2006). Stochiometric analysis of biological hydrogen production by fermentative bacteria. Int. J. Hydrogen Energy *31*, 539–549.

Cho, H.S., Kim, Y.M., Min, B.E., Jung, G.Y., and Park, J.M. (2011). Improvement of hydrogen production yield by rebalancing NADPH/NADH ratio in a recombinant *Escherichia coil.* J Nanoelectronics and Optoelectronics *6*, 343–347.

Constant, P., and Hallenbeck, P.C. (2013). Hydrogenase. Biohydrogen. Edited by A. Pandey, J.S. Chang, P.C. Hallenbeck and C. Larroche. Elsevier Pvt. Ltd., pp. 75–102.

Constant, P., Chowdhury, S.P., Hesse, L., Pratscher, J., and Conrad, R. (2011). Genome data mining and soil survey for the novel group 5 [NiFe]-hydrogenase to explore the diversity and ecological importance of presumptive high affinity H_2-oxidizing bacteria. Appl. Environ. Microbiol. 77, 6027–6035.

Cournac, L., Guedeney, G., Peltier, G., and Vignais, P.M. (2004). Sustained photoevolution of molecular hydrogen in a mutant of *Synechocystis* sp. strain PCC 6803 deficient in the type I NADPH-dehydrogenase complex. J. Bacteriol. *186*, 1737–1746.

Doebbe, A., Rupprecht, J., Beckmann, J., Mussgnug, J.H., Hallmann, A., and Hankamer, B. (2007). Functional integration of the HUP1 hexose symporter gene into the genome of *C. reinhardtii* impacts on biological H_2 production. J Biotechnol. *131*, 27–33.

Drepper, T., Gross, S., Yakunin, A.F., Hallenbeck, P.C., Masepohl, B., and Klipp, W. (2003). Role of GlnB and GlnK in ammonium control of both nitrogenase systems in the phototrophic bacterium *Rhodobacter capsulatus*. Microbiology *149*, 2203–2212.

Eaton-Rye, J.J. (2004). The construction of gene knockouts in the cyanobacterium *Synechocystis* sp. PCC 6803. In: R. Carpentier (ed.), Photosynthesis Research Protocol. Humana Press Inc., New Jersey, pp. 309–324.

Elrad, D., Niyogib, K.K., and Grossmanc, A.R. (2002). A major light harvesting polypeptide of photosystem II functions in thermal dissipation. Plant Cell *14*, 1801–1816.

Fan, Z., Yuan, L., and Chatterjee, R. (2009). Increased hydrogen production by genetic engineering of *Escherichia coli*. PLoS ONE 4, e4432.

Fisher, K., and Newton, W.E. (2002). Nitrogen fixation – a general overview. In: G.J. Leigh (ed.), Nitrogen Fixation in the Millennium. Elsevier, Amsterdam, pp. 1–34.

Flores, E., Frı´as, J.E., Rubio, L.M., and Herrero, A. (2005). Photosynthetic nitrate assimilation in cyanobacteria. Photosynth. Res. *83*, 117–133.

Florin, L., Tsokoglou, A., and Happe, T. (2001). A novel type of [Fe]-hydrogenase in the green alga *Scenedesmus obliquus* is linked to the photosynthetic electron transport chain. J. Biol. Chem. *276*, 6125–6132.

Fodor, B., Rakhely, G., Kovacs, A.T., and Kovacs, K.L. (2001).Transposon mutagenesis in purple sulfur photosynthetic bacteria: identification of hypF, encoding a protein capable of processing [NiFe] hydrogenases in alpha, beta, and gamma subdivisions of the proteobacteria. Appl. Env. Microbiol. *67*, 2476–2483.

Fontecilla-Camps, J.C., Volbeda, A., Cavazza, C., and Nicolet, Y. (2007). Structure/function relationships of [NiFe]- and [FeFe]-hydrogenases. Chem. Rev. *107*, 4273–4303.

Franchi, E., Tosi, C., Scolla, G., Della Penna, G., Rodriguez, F., and Pedroni, P.M. (2004). Metabolically engineered *Rhodobacter sphaeroides* RV strains for improved biohydrogen photoproduction combined with disposal of food wastes. Marine Biotechnol. (NY) 6, pp. 552–565.

Garcin, E., Vernede, X., Hatchikian, E.C., Volbeda, A., Frey, M., and Fontecilla-Camps, J.C. (1999). The crystal structure of a reduced [Ni–Fe–Se] hydrogenase provides an image of the activated catalytic centre. Struct. Fold. Des. *9*, 557–566.

Gfeller, R.P., and Gibbs, M. (1984). Fermentative metabolism of *Chlamydomonas reinhardtii*. I. Analysis of fermentative products from starch in dark and light. Plant Physiol. *75*, 212–218.

Ghosh, D., and Hallenbeck, P.C. (2009). Fermentative hydrogen yields from different sugars by batch cultures of metabolically engineered *Escherichia coli* DJT135. Int. J. Hydrogen Energy *34*, 7979–7982.

Ghosh, D., and Hallenbeck, P.C. (2010). Response surface methodology for process parameter optimization of hydrogen yield by the metabolically engineered strain *Escherichia coli* DJT135. Bioresour. Technol. *101*, 1820–1825.

Ghosh, D., Sobro, I.F., and Hallenbeck, P.C. (2012a). Optimization of the hydrogen yield from single-stage photofermentation of glucose by *Rhodobacter capsulatus* JP91 using response surface methodology. Bioresour. Technol. *123*, 199–206.

Ghosh, D., Sobro, I.F., and Hallenbeck, P.C. (2012b). Stoichiometric conversion of biodiesel derived crude glycerol to hydrogen: response surface methodology study of the effects of light intensity and crude glycerol and glutamate concentration. Bioresour. Technol. *106*, 154–160.

Ghosh, D., Bisaillon, A., and Hallenbeck, P.C. (2013). Increasing the metabolic capacity of *Escherichia coli* for hydrogen production through heterologous expression of the *Ralstonia eutropha* SH operon. Biotechnol. Biofuels *6*, 122.

Grossman, A.R., Croft, M., Gladyshev, V.N., Merchant, S.S., Posewitz, M.C., Prochnik, S., and Spalding, M.H. (2007). Novel metabolism in *Chlamydomonas* through the lens of genomics. Curr. Opin. Plant Biol. *10*, 190–198.

Gutthann, F., Egert, M., Marques, A., and Appel, J. (2007). Inhibition of respiration and nitrate assimilation enhances photohydrogen evolution under low oxygen concentration in *Synechocystis* sp. PCC 6803. Biochim. Biophys. Acta. *1767*, 161–169.

Hallenbeck, P.C. (2011) Microbial paths to renewable hydrogen production. Biofuels 2, 285–302.

Hallenbeck, P.C., and Benemann, J.R. (2002). Biological hydrogen production; fundamentals and limiting processes. Int. J. Hydrogen Energy *27*, 1185–1193.

Hallenbeck, P.C., and Ghosh, D. (2009). Advances in fermentative biohydrogen production: the way forward? Trends Biotechnol. *27*, 287–297.

Hallenbeck, P.C., and Ghosh, D. (2012). Improvements in fermentative biological hydrogen production through metabolic engineering. J. Environ. Manag. *95*, S360–S364.

Hallenbeck, P.L., Lerchen, R., Hessler, P., and Kaplan, S. (1990). Phosphoribulokinase activity and regulation of CO_2 fixation critical for photosynthetic growth of *Rhodobacter sphaeroides*. J Bacteriol. *172*, 1749–1761.

Hallenbeck, P.C., Ghosh, D., Skonieczny, M.T., and Yargeau, V. (2009). Microbiological and engineering aspect of biohydrogen production. Ind. J. Microbiol. *49*, 48–59.

Hallenbeck, P.C., Ghosh, D., Abo-Hasesh, M., and Wang, R. (2011). Metabolic engineering for enhanced biofuels production with emphasis on the biological production of hydrogen. Adv. Chem. Res. 6, 125–154.

Hallenbeck, P.C., Abo-Hashesh, M., and Ghosh, D. (2012). Strategies for improving biological hydrogen production. Bioresour. Technol. *110*, 1–9.

Hansel, A., and Lindblad, P. (1998). Towards optimization of cyanobacteria as biotechnologically relevant producers of molecular hydrogen, a clean and renewable energy source. Appl. Microbiol. Biotechnol. *50*, 153–160.

Härtel, U., and Buckel, W. (1996). Sodium ion-dependent hydrogen production in *Acidaminococcus fermentans*. Arch. Microbiol, *166*, 350–356.

Harwood, C.S. (2008). Nitrogenase-catalyzed hydrogen production by purple non-sulfur photosynthetic bacteria. In: J.D. Wall, C.S. Harwood, and A. Demain (eds.), Bioenergy. ASM Press, Washington, DC, pp. 259–271.

Heap, J.T., Pennington, O.J., Cartman, S.T., Carter, G.P., and Minton, N.P. (2007). The ClosTron: A universal gene knock-out system for the genus *Clostridium*. J. Microbiol. Methods *70*, 452–464.

Heap, J.T., Pennington, O.J., Cartman, S.T., and Minton, N.P. (2009). A modular system for *Clostridium* shuttle plasmids. J. Microbiol. Methods *78*, 79–85.

Heap, J.T., Kuehne, S.A., Ehsaan, M., Cartman, S.T., Cooksley, C.M., Scott, J.C., and Minton, N.P. (2010). The ClosTron: mutagenesis in *Clostridium* refined and streamlined. J. Microbiol. Methods *80*, 49–55.

Hemschemeier, A., and Happe, T. (2005). The exceptional photofermentative hydrogen metabolism of the green alga *Chlamydomonas reinhardtii*. Biochem. Soc. Trans. *33*, 39–41.

Hemschemeier, A., Fouchard, S., Cournac, L., Peltier, G., and Happe, T. (2008a). Hydrogen production by *Chlamydomonas reinhardtii*: an elaborate interplay of electron sources and sinks. Planta *227*, 397–407.

Hemschemeier, A., Jacobs, J., and Happe, T. (2008b). Biochemical and physiological characterization of the pyruvate formate-lyase Pfl1 of *Chlamydomonas reinhardtii*, a typically bacterial enzyme in a eukaryotic alga. Eukaryot.Cell 7, 518–526.

Hiromoto, T., Warkentin, E., Moll, J., Ermler, U., and Shima, S. (2009). Iron-chromophore circular dichroism of [Fe]-hydrogenase: the conformational change required for H_2 activation. Angew. Chem. Int. 48: 6457–6460.

Hung, C.H., Chang, Y.T., and Chang, Y.J. (2011a). Roles of microorganisms other than *Clostridium* and *Enterobacter* in anaerobic fermentative biohydrogen production systems – a review. Bioresour. Technol. *102*, 8437–8444.

Hung, C.-H., Cheng, C.-H., Guan, D.-W., Wang, S.-T., Hsu, S.-C., Liang, C.-M., and Lin, C.-Y. (2011b). Interactions between *Clostridium* sp. and other facultative anaerobes in a self-formed granular sludge hydrogen-producing bioreactor. Int. J. Hydrogen Energy *36*, 8704–8711.

Incharoensakdi, A. (2006). Nitrogen metabolism in cyanobacteria under osmotic stress. In: A.K. Rai and T. Takabe (eds.), Abiotic Stress Tolerance in Plants. Springer, Dordrecht, pp. 195–212.

Jo, J.H., Jeon, C.O., Lee, S.Y., Lee, D.S., and Park, J.M. (2010). Molecular characterization and homologous overexpression of [FeFe]-hydrogenase in *Clostridium tyrobutyricum* JM1. Int. J. Hydrogen Energy *35*, 1065–1073.

Jones, P.R. (2008). Improving fermentative biomass-derived H2-production by engineering microbial metabolism. Int. J. Hydrogen Energy *33*, 5122–5130.

Kaji, M., Taniguchi, Y., Matsushita, O., Katayama, S., Miyata, S., Morita, S., and Okabe, A. (1999). The hydA gene encoding the H_2-evolving hydrogenase of *Clostridium perfringens*: molecular characterization and expression of the gene. FEMS Microbiol. Lett. *181*, 329–336.

Kaneko, T., Sato, S., Kotani, H., Tanaka, A., Asamizu, E., Nakamura, Y., Miyajima, N., Hirosawa, M., Sugiura, M., Sasamoto, S., *et al.* (1996). Sequence analysis ofthe genome of the unicellular cyanobacterium *Synechocystis* sp. strain PCC6803. II. Sequence determination of the entire genome and assignment of potential protein-coding regions. DNA Res. 3, 109–136.

Kars, G., Gündüz, U., Rákhely, G., Yücel, M., Eroglu, I., and Kovács, K.L. (2008). Improved hydrogen production by uptake hydrogenase deficient mutant strain of *Rhodobacter sphaeroides* O.U.001. Int. J. Hydrogen Energy 33, 3056–3060.

Kars, G., Gunduz, U., Yucel, M., Rakhely, G., Kovacs, K.L., and Eroglu, I. (2009). Evaluation of hydrogen production by *Rhodobacter sphaeroides* O.U.001 and its hupSL deficient mutant using acetate and malate as carbon sources. Int. J. Hydrogen Energy *34*, 2184–2190.

Keasling, J.D., Benemann, J., Pramanik, J., Carrier, T.A., Jones, K.L., and Van Dien, S.J. (1998). A toolkit for metabolic engineering of bacteria. In: O.R. Zaborski, J.R. Benemann, T. Matsunaga, J. Miyake, and A. San Pietro (eds.). Bio Hydrogen. Plenum Press, USA. pp. 87–97.

Kengen, S.W.M., and Stams, A.J.M. (1994). Formation of L-alanine as a reduced end product in carbohydrate fermentation by the hyperthermophilic archaeon *Pyrococcus furiosus*. Arch. Microbiol. *161*, 168–175.

Keseler, I.M., Collado-Vides, J., Gama-Castro, S., Ingraham, J., Paley, S., Paulsen, I.T., Peralta-Gil, M., and Karp, P.D. (2005). EcoCyc: a comprehensive database resource for *Escherichia coli*. Nucleic Acids Res. *33*, 334–337.

Kim, E.J., Lee, M.K., Kim, M.S., and Lee, J.K. (2008). Molecular hydrogen production by nitrogenase of *Rhodobacter sphaeroides* and by Fe-only hydrogenase of *Rhodospirillum rubrum*. Int. J. Hydrogen Energy 33, 1516–1521.

Kim, J.Y.H., Jo, B.H., and Cha, H.J. (2011). Production of biohydrogen by heterologous expression of oxygen-tolerant *Hydrogenovibrio marinus* NiFe -hydrogenase in *Escherichia coli*. J. Biotechnol. *155*, 312–319.

Kim, J.Y.H., Jo, B.H., Jo, Y., and Cha, H.J. (2012). Improved production of biohydrogen in light-powered *Escherichia coli* by co-expression of proteorhodopsin and heterologous hydrogenase. Microb. Cell Fact. *11*, 2.

Kim, M.S., Baek, J.S., and Lee, J.K. (2006). Comparison of H_2 accumulation by *Rhodobacter sphaeroides* KD131 and its uptake hydrogenase and PHB synthase deficient mutant. Int. J. Hydrogen Energy *31*, 121–127.

Kim, Y.M., Cho, H.-S., Jung, G.Y., and Park, J.M. (2011). Engineering the pentose phosphate pathway to improve hydrogen yield in recombinant *Escherichia coli*. Biotechnol. Bioeng. *108*, 2941–2946.

King, P.W., Posewitz, M.C., Ghirardi, M.L., and Seibert, M. (2006). Functional studies of [FeFe] hydrogenase maturation in an *Escherichia coli* biosynthetic system. J. Bacteriol. *188*, 2163–2172.

Kivisto, A., Santala, V., and Karp, M. (2010). Hydrogen production from glycerol using halophilic fermentative bacteria. Bioresour. Technol. *101*, 8671–8677.

Klein, M., Ansorge-Schumacher, M.B., Fritsch, M., and Hartmeier, W. (2010). Influence of hydrogenase overexpression on hydrogen production of *Clostridium acetobutylicum* DSM 792. Enzyme Microb. Technol. *46*, 384–390.

Koku, H., Eroglu, I., Gunduz, U., Yucel, M., and Turker, L. (2002). Aspects of the metabolism of hydrogen production by *Rhodobacter sphaeroides*. Int. J. Hydrogen Energy 27, 1315–1329.

Kondo, T., Arakawa, M., Hirai, T., Wakayama, T., Hara, M., and Miyake, J. (2002). Enhancement of hydrogen production by a photosynthetic bacterium mutant with reduced pigment. J. Biosci. Bioeng. *93*, 145–150.

Kontur, W.S., Noguera, D.R., and Donohue, T.J. (2012). Maximizing reductant flow into microbial H_2 production. Curr. Opin. Biotechnol. 23. 382–389.

Kruse, O., Rupprecht, J., Bader, K., Thomas-Hall, S., Schenk, P.M., Finazzi, G., *et al.* (2005). Improved photobiological H2 production in engineered green algal cells. J. Biol. Chem. *280*, 34170–34177.

Kuchenreuther, J.M., Stapleton, J.A., and Swartz, J.R. (2009). Tyrosine, cysteine, andS-adenosyl methionine stimulatein vitro [FeFe] hydrogenase activation. PLoS One 4(10), e7565.

Lambowitz, A.M., and Zirnmerly, S. (2004). Mobile group II introns. Annu. Rev. Genet. *38*, 1–35.

Larimer, F.W., Chain, P., Hauser, L., Lamerdin, J., Malfatti, S., Do, L., Land, M.L., Pelletier, D.A., Beatty, J.T., Lang, A.S., *et al.* (2004). Complete genome sequence of the metabolically versatile photosynthetic bacterium *Rhodopseudomonas palustris*. Nat. Biotechnol. *22*, 55–61.

Lee, J., Yun, H., Feist, A.M., Palsson, B.O., and Lee, S.Y. (2008). Genome-scale reconstruction and in silico analysis of the *Clostridium acetobutylicum* ATCC 824 metabolic network. Appl. Microbiol. Biotechnol. *80*, 849–862.

Levin, D.B., Pitt, L., and Love, M. (2004). Biohydrogen production: prospects and limitations to practical application. Int. J. Hydrogen Energy *29*, 173–185.

Li, S., Lai, C., Cai, Y., Yang, X., Yang, S., Zhu, M., Wang, J., and Wang, X.. (2010). High efficiency hydrogen production from glucose/xylose by the ldh-deleted *Thermoanaerobacterium* strain. Bioresour. Technol. *101*, 8718–8724.

Li, X., Liu, T., Wu, Y., Zhao, G., and Zhou, Z. (2010). Derepressive effect of NH4+on hydrogen production by deleting the glnA1 gene in *Rhodobacter sphaeroides*. Biotechnol. Bioeng. *106*, 564–572.

Liebgott, P.P., Leroux, F., Burlat, B., Dementin, S., Baffert, C., Lautier, T., Fourmond, V., Ceccaldi, P., Cavazza, C., Meynial-Salles, I., *et al.* (2010). Relating diffusion along the substrate tunnel and oxygen sensitivity in hydrogenase. Nat. Chem. Biol. *6*, 63–70.

Lindblad, P., Christensson, K., Lindberg, P., Pinto, F., and Tsygankov, A. (2002). Photoproduction of H2 by wild type *Anabaena* PCC 7120 and a hydrogen uptake deficient mutant: from laboratory experiments to outdoor culture. Int. J. Hydrogen Energy *27*, 1271–1281.

Liu, T., Li, X., and Zhou, Z. (2010). Improvement of hydrogen yield byhupRgene knock-out and nifA gene overexpression in *Rhodobacter sphaeroides* 6016. Int. J. Hydrogen Energy 35, 9603–9610.

Liu, X., Zhu, Y., and Yang, S.T. (2006). Construction and characterization of ack deleted mutant of *Clostridium tyrobutyricum* for enhanced butyric acid and hydrogen production. Biotechnol. Prog. 22, 1265–1275.

Lu, Y., Zhao, H.X., Zhang, C., Lai, Q.H., and Xing, X.H. (2009). Perturbation of formate pathway for hydrogen production by expressions of formate hydrogen lyase and its transcriptional activator in wild *Enterobacter aerogenes* and its mutants. Int. J. Hydrogen Energy *34*, 5072–5079.

Lu, Y., Zhao, H.X., Zhang, C., Lai, Q.H., Wu, X., and Xing, X.H. (2010). Alteration of hydrogen metabolism of ldh-deleted *Enterobacter aerogenes* by overexpression of NAD^+-dependent formate dehydrogenase. Appl. Microbiol. Biotechnol. *86*, 255–262.

Lu, Y., Zhang, C., Lai, Q., Zhao, H., and Xing, X.H. (2011). Improved hydrogen production under microaerophilic conditions by overexpression of polyphosphate kinase in *Enterobacter aerogenes*. Enzyme Microb. Technol. *48*, 187–192.

Lyon, E.J., Shima, S., Buurman, G., Chowdhuri, S., Batschauer, A., Steinbach, K., and Thauer, R.K. (2004). UV-A/blue-light inactivation of the 'metal-free' hydrogenase (Hmd) from methanogenic archaea. Eur. J. Biochem. *271*, 195–204.

McGlynn, S.E., Shepard, E.M., Winslow, M.A., Naumov, A.V., Duschene, K.S., Posewitz, M.C., Broderick, W.E.;

Broderick, J.B.; Peters, J.W.. (2008). HydF as a scaffold protein in [FeFe] hydrogenase H-cluster biosynthesis. FEBS Lett. *582*, 2183–2187.

McKinlay, J.B., and Harwood, C.S. (2010). Photobiological production of hydrogen gas as a biofuel. Curr. Opin. Biotechnol. *21*, 244–251.

Maeda, T., Sanchez-Torres, V., and Wood, T. (2007). Enhanced hydrogen production from glucose by metabolically engineered *Escherichia coli*. Appl. Microbiol. Biotechnol. *77*, 879–890.

Maeda, T., Sanchez-Torres, V., and Wood, T. (2008a). Protein engineering of hydrogenase 3 to enhance hydrogen production. Appl. Microbiol. Biotechnol. *79*, 77–86.

Maeda, T., Sanchez-Torres, V., and Wood, T.K. (2008b). Metabolic engineering to enhance bacterial hydrogen production. Microb.Biotechnol. 1, 30–39.

Malki, S., Saimmaime, I., De Luca, G., Rousset, M., Dermoun, Z., and Belaich, J.P. (1995). Characterization of an operon encoding an NADP-reducing hydrogenase in *Desulfovibrio fructosovorans*. J. Bacteriol. *177*, 2628–2636.

Masepohl, B., Schoelisch, K., Goerlitz, K., Kutzki, C., and Böhme, H. (1997). The heterocyst-specific fdxH gene product of the cyanobacterium *Anabaena* sp. PCC 7120 is important but not essential for nitrogen fixation. Mol. Gen. Genet. *253*, 770–776.

Masepohl, B., Schneider, K., Drepper, T., Müller, A., and Klipp. W. (2002). Alternative nitrogenases. In G.J. Leigh (ed.), Nitrogen fixation at the millennium. Elsevier, Amsterdam, pp. 191–222.

Matthew, T., Zhou, W., Rupprecht, J., Lim, L., Thomas-Hall, S.R., Doebbe, A., Kruse, O., Hankamer, B., Marx, U.C., Smith, S.M., *et al.* (2009). The metabolome of *Chlamydomonas reinhardtii* following induction of anaerobic H_2 production by sulfur depletion. J. Biol. Chem. *284*, 23415–23425.

Melis A. (2007). Photosynthetic H_2 metabolism in *Chlamydomonas reinhardtii* (unicellular green algae). Planta *226*, 1075–1086.

Melis, A., Zhang, L., Forestier, M., Ghirardi, M.L., and Seibert, M. (2000). Sustained photobiological hydrogen gas production upon reversible inactivation of oxygen evolution in the green alga *Chlamydomonas reinhardtii*. Plant Physiol. *122*, 127–136.

Merchant, S.S., Prochnik, S.E., Vallon, O., Harris, E.H., Karpowicz, S.J., Witman, G.B., Terry, A., Salamov, A., Fritz-Laylin, L.K., and Marechal-Drouard, L. (2007). The *Chlamydomonas* genome reveals the evolution of key animal and plant functions. Science *318*, 245–250.

Meyer, J. (2007). [FeFe] hydrogenases and their evolution: a genomic perspective. Cell. Mol. Life Sci. *64*, 1063–1084.

Miyake, M., and Asada, Y. (1997). Direct electroporation of clostridial hydrogenase into cyanobacterial cells. Biotechnol. Tech. *11*, 787–790.

Mohan Raj, S., Talluri, S., and Christopher, L.P. (2012). Thermophilic hydrogen production from renewable resources: current status and future perspectives. Bioenerg. Res. *5*, 515–531.

Molnar, A., Bassett, A., Thuenemann, E., Schwach, F., Karkare, S., Ossowski, S., Weigel, D., and Baulcombe, D. (2009). Highly specific gene silencing by artificial microRNAs in the unicellular alga *Chlamydomonas reinhardtii*. Plant J. *58*, 165–174.

Morimoto, K., Kimura, T., Sakka, K., and Ohmiya, K. (2005). Overexpression of a hydrogenase gene in *Clostridium paraputrificum* to enhance hydrogen gas production. FEMS Microbiol. Lett. *246*, 229–234.

Mus, F., Dubini, A., Seibert, M., Posewitz, M.C., and Grossman, A.R. (2007). Anaerobic acclimation in *Chlamydomonas reinhardtii*: Anoxic gene expression, hydrogenase induction and metabolic pathways. J Biol. Chem. 282, 25475–25486.

Mussgnug, J.H., Thomas-Hall, S., Rupprecht, J., Foo, A., and Klassen, V. (2007). Engineering photosynthetic light capture: impacts on improved solar energy to biomass conversion. Plant Biotechnol. J. 5, 802–814.

Nakashimada, Y., Rachman, M.A., Kakizono, T., and Nishio, N. (2002). Hydrogen production of *Enterobacter aerogenes* altered by extracellular and intracellular redox states. Int. J. Hydrogen Energy *27*, 1399–1405.

Nakayama, S.-I., Kosaka, T., Hirakawa, H., Matsuura, K., Yoshino, S., and Furukawa, K. (2008). Metabolic engineering for solvent productivity by down regulation of the hydrogenase gene cluster in *Clostridium saccharoperbutylacetonicum* strain N1–4. Appl. Microbiol. Biotechnol. *78*, 483–493.

Nandi, R., and Sengupta, S. (1998). Microbial production of hydrogen: an overview. Crit. Rev. Microbiol. *24*, 61–84.

Nguyen, T.A.D., Kim, J.P., Kim, M.S., Oh, Y.K., and Sim, S.J. (2008). Optimization of hydrogen production by hyperthermophilic eubacteria, *Thermotoga maritima* and *Thermotoga neapolitana* in batch fermentation. Int. J. Hydrogen Energy 33, 1483–1488.

Oda, Y., Samanta, S.K., Rey, F.E., Wu, L., Liu, X., Yan, T., Zhou, J., and Harwood, C.S. (2005). Functional genomic analysis of three nitrogenase isozymes in the photosynthetic bacterium *Rhodopseudomonas palustris*. J. Bacteriol. *187*, 7784–7794.

Örlygsson, J., Anderson, R., and Svensson, B.H. (1995). Alanine as an end product during fermentation of monosaccharides by *Clostridium* strain P2. Antonie Van Leeuwenhoek *68*, 273–280.

Ormerod, J.G., and Gest, H. (1962). Symposiumon metabolism of inorganic compounds. IV. Hydrogen photosynthesis and alternative metabolic pathways in photosynthetic bacteria. Bacteriol. Rev. *26*, 51.

Oxelfelt, f., Tamagnini, P., Salema, R., and Lindblad, P. (1995). Hydrogen uptake in Nostoc strain PCC 73102: effects of nickel, hydrogen, carbon and nitrogen. Plant Physiol. Biochem. *33*, 617–623.

Ozturk, Y., and Gokce, A. (2012). Characterization of hydrogen production by an ammonium insensitive pseudo-revertant of a CBB deficient *Rhodobacter capsulatus* strain. Int. J. Hydrogen Energy 37, 8811–8819.

Öztürk, Y., Yücel, M., Daldal, F., Mandacı, S., Gündüz, U., Türker, L., Eroğlu, İ. (2006). Hydrogen production by using *Rhodobacter capsulatus* mutants with genetically modified electron transfer chains. Int. J. Hydrogen Energy *31*, 1545–1552.

Öztürk, Y., Gökçe, A., Peksel, B., Gürgan, M., Özgür, E., Gündüz, U., Eroğlu, İ., Yücel, M.. (2012). Hydrogen production properties of *Rhodobacter capsulatus* with genetically modified redox balancing pathways. Int. J. Hydrogen Energy *37*, 2014–2020.

Penfold, D.W., and Macaskie, L.E. (2004). Production of H_2 *Escherichia coli* strains carrying the pUR400 plasmid, which encodes invertase activity. Biotechnol. Lett. *26*, 1879–1883.

Penfold, D.W., Forster, C.F., and Macaskie, L.E. (2003). Increased hydrogen production by *Escherichia coli* strain HD701 in comparison with the wild-type parent strain MC4100. Enzyme Microb. Technol. *33*, 185–189.

Penfold, D.W., Sargent, F., and Macaskie, L.E. (2006). Inactivation of the *Escherichia coli* K-12 twin-arginine translocation system promotes increased hydrogen production. FEMS Microbiol. Lett. *262*, 135–137.

Peterson, R.B., and Burris, R.H. (1978). Hydrogen metabolism in isolated heterocysts of Anabaena 7120. Arch. Microbiol. *116*, 125–132.

Pinto, T.S., Malcata, F.X., Arrabaca, J.D., Silva, J.M., Spreitzer, R.J., and Esquivel, M.G. (2013). Rubisco mutants of *Chlamydomonas reinhardtii* enhance photosynthetic hydrogen production. Appl. Microbiol. Biotechnol. 97, 5635–5643.

Polle, J.E.W., Kanakagiri, S., and Melis, A. (2003). tla1, a DNA insertional transformant of the green alga *Chlamydomonas reinhardtii* with a truncated light-harvesting chlorophyll antenna size. Planta *217*, 49–59.

Posewitz, M.C., King, P.W., Smolinski, S.L., Zhang, L., Seibert, M., and Ghirardi, M.L. (2004). Discovery of two novel radical S-adenosylmethionine proteins required for the assembly of an active [Fe] hydrogenase. J. Biol. Chem. *279*, 25711–25720.

Rachman, M.A., Nakashimada, Y., Kakizono, T., and Nishio, N. (1998). Hydrogen production with high yield and high evolution rate by self-flocculated cells of *Enterobacter aerogenes* in a packed-bed reactor. Appl. Microbiol. Biotechnol. *49*, 450–454.

Rákhely, G., Kovács, A.T., Maróti, G.,. Fodor, B.D., Csanádi, G., Latinovics, D., and Kovács, K.L. (2004). Cyanobacterial-type, heteropentameric, NAD+-reducing NiFe hydrogenase in the purple sulfur photosynthetic bacterium *Thiocapsa roseopersicina*. Appl. Environ. Microbiol. *70*, 722–728.

Raman, B., McKeown, C.K., Rodriguez, M. Jr., Brown, S.D., and Mielenz, J.R. (2011). Transcriptomic analysis of *Clostridium thermocellum* ATCC 27405 cellulose fermentation. BMC Microbiol. *11*, 1471–2180.

Redwood, M.D., and Macaskie, L.E. (2006). A two-stage, two-organism process for biohydrogen from glucose. Int. J. Hydrogen Energy *31*, 1514–1521.

Redwood, M.D., Mikheenko, I.P., Sargent, F., and Macaskie, L.E. (2008). Dissecting the roles of *Escherichia coli* hydrogenases in biohydrogen production. FEMS Microbiol. Lett. *278*, 48–55.

Ren, Y., Xing, X.H., Zhang, C., and Gou, Z. (2005). A simplified method for assay of hydrogenase activities of evolution and uptake in *Enterobacter aerogenes*. Biotechnol. Lett. *27*, 1029–1033.

Rey, F.E., Oda, Y., and Harwood, C.S. (2006). Regulation of uptake hydrogenase and effects of hydrogen utilization on gene expression in *Rhodopseudomonas palustris*. J. Bacteriol. *188*, 6143–6152.

Richaud, P., Vignais, P.M., Colbeau, A., Uffen, R.L., and Cauvin, B. (1990). Molecular biology studies of the uptake hydrogenase of *Rhodobacter capsulatus* and *Rhodocyclus gelatinosus*. FEMS Microbiol. Rev. *7*, 413–418.

Rohr, J., Sarkar, N., Balenger, S., Jeong, B.-R., and Cerutti, H. (2004). Tandem inverted repeat system for selection of effective transgenic RNAi strains in *Chlamydomonas*. Plant J. *40*, 611–621.

Ruhle, T., Hemschemeier, A., Melis, A., and Happe, T. (2008). A novel screening protocol for the isolation of hydrogen producing *Chlamydomonas reinhardtii* strains. BMC Plant Biol. *8*, 107.

Ruiyan, Z., Di, W., Yaoping, Z., Jilun, Li. (2006). Hydrogen production by draTGB hupL double mutant of *Rhodospirillum rubrum* under different light conditions. Chin. Sci. Bull. *51*, 2611–2618.

Salomone-Stagnia, M., Stellatob, F., Whaleyc, C.M., Vogtd, S., Moranteb, S., Shimad, S., Rauchfussc, T.B., and Meyer-Klaucke, W. (2010). The iron-site structure of [Fe]-hydrogenase and model systems: an X-ray absorption near edge spectroscopy study. Dalton Transactions *39*, 3057–3064.

Sawers, R.G. (2005). Formate and its role in hydrogen production in *Escherichia coli*. Biochem. Soc. Trans. *33*, 42–46.

Schut, G.J., and Adams, M.W.W. (2009). The iron-hydrogenase of *Thermotoga maritima* utilizes ferredoxin and NADH synergistically: a new perspective on anaerobic hydrogen production. J Bacteriol. *191*, 4451–4457.

Self, W.T., Hasona, A., and Shanmugam, K.T. (2004). Expression and regulation of a silent operon, hyf, coding for hydrogenase4 isoenzyme in *Escherichia coli*. J. Bacteriol. *186*, 580–587.

Serebryakova, L.T., Medina, M., Zorin, N.A., Gogotov, I.N., and Cammack, R. (1996). Reversible hydrogenase of *Anabaena variabilis* ATCC 29413: catalytic properties and characterization of redox centres. FEBS Lett. *383*, 79–82.

Shima, S., Pilak, O., Vogt, S., Schick, M., Stagni, M.S., Meyer-Klaucke, W., *et al.* (2008). The crystal structure of [Fe]-hydrogenase reveals the geometry of the active site. Science *321*, 572–575.

Shin, J.-H., Yoon, J.H., Ahn, E.K., Kim, M.-S., Sim, S.J., and Park, T.H. (2007). Fermentative hydrogen production by the newly isolated *Enterobacter asburiae* SNU-1. Int. J. Hydrogen Energy *32*, 192–199.

Soboh, B., Linder, D., and Hedderich, R. (2004). A multi subunit membrane-bound [NiFe] hydrogenase and an NADH-dependent Fe-only hydrogenase in the fermenting bacterium Thermoanaerobacter tengcongensis. Microbiology *150*, 2451–2463.

Stal, L.J., and Moezelaar, R. (1997). Fermentation in cyanobacteria. FEMS Microbiol. Rev. *21*, 179–211.

Stanley, N.R., Findlay, K., Berks, B.C., and Palmer, T. (2001). *Escherichia coli* strains blocked in

Tat-dependent protein export exhibit pleiotropic defects in the cell envelope. J. Bacteriol. *183*, 139–144.

Sun, J., Hopkins, R.C., Jenney Jr, F.E., McTernan, P.M., and Adams, M.W. (2010). Heterologous expression and maturation of an NADP-dependent [NiFe] hydrogenase: a key enzyme in biofuel production. PLOS One 5(5), e10526.

Tamagnini, P., Axelsson, R., Linberg, P., Oxelfelt, F., Wunchiers, R., and Lindblad, P. (2002). Hydrogenase and hydrogen metabolism of Cyanobacteria. Microbiol. Mol. Biol. Rev. *66*, 1–20.

Tamagnini, P., Leitão, E., Oliveira, P., Ferreira, D., Pinto, F., Harris, D.J., Heidorn, T., and Lindblad, P. (2007). Cyanobacterial hydrogenases: diversity, regulation and applications. FEMS Microbiol. Rev. *31*, 692–720.

Tao, Z., Wei, W., Xue, B., and Yijun, Q. (2009). Gene silencing by artificial microRNAs in *Chlamydomonas*. Plant J. *58*, 157–164.

Thauer, R.K., Jungermann, K., and Decker, K. (1977). Energy conservation in chemotrophic anaerobic bacteria. Bact. Rev. *41*, 100–180.

Thauer, R.K., Klein, A.R., and Hartmann, G.C. (1996). Reactions with molecular hydrogen in microorganisms: evidence for a purely organic hydrogenation catalyst. Chem. Rev. *96*, 3031–3042.

Thauer, R.K., Kaster, A.-K., Goenrich, M., Schick, M., Hiromoto, T., and Shima, S. (2010). Hydrogenases from methanogenic archaea, nickel, a novel cofactor, and H_2 storage. Annu. Rev. Biochem. *79*, 507–536.

Tichi, M.A., and Tabita, F.R. (2001). Interactive control of *Rhodobacter capsulatus* redox-balancing system during phototrophic metabolism. J. Bacteriol. *183*, 6344–6354.

Timmins, M., Thomas-Hall, S.R., Darling, A., Zhang, E., Hankamer, B., Marx, U.C., and Schenk, P.M. (2009). Phylogenetic and molecular analysis of hydrogen-producing green algae. J. Exp. Bot. *60*, 1691–1702.

Torzillo, G., Scomaa, A., Faralonia, C., Enaa, A., and Johanningmeierb, U. (2009). Increased hydrogen photoproduction by means of a sulfur-deprived *Chlamydomonas reinhardtii* D1 protein mutant. Int. J. Hydrogen Energy *34*, 4529–4536.

Tripathi, S.A., Olson, D.G., Argyros, D.A., Miller, B.B., Barrett, T.F., Murphy, D.M., McCool, J.D., Warner, A.K., Rajgarhia, V.B., Lynd, L.R., *et al.* (2010). Development of pyrF-based genetic system for targeted gene deletion in *Clostridium thermocellum* and creation of a pta mutant. Appl. Environ. Microbiol. *76*, 6591–6599.

Tye, J.W., Hall, M.B., and Darensbourg, M.Y. (2005). Better than platinum? Fuel cells energized by enzymes. Proc. Natl. Acad. Sci. U.S.A. *102*, 16911–16912.

Van Niel, E.W.J., Budde, M.A.W., De Haas, G.G., Vander Wal, F.J., Claassen, P.A.M., and Stams, A.J.M. (2002). Distinctive properties of high hydrogen producing extreme thermophiles, *Caldicellulosiruptor saccharolyticus* and *Thermotoga elfii*. Int. J. Hydrogen Energy *27*, 1391–1398.

Vasilyeva, L., Miyake, M., Khatipov, E., Wakayama, T., Sekine, M., Hara, M., Nakada, E., Asada, Y., and Miyake, J. (1999). Enhanced hydrogen production by a mutant of *Rhodobacter sphaeroides* having an altered light-harvesting system. J. Biosci. Bioeng. *87*, 619–624.

Veit, A., Akhtar, M.K., Mizutani, T., and Jones, P.R. (2008). Constructing and testing the thermodynamic limits of synthetic NAD(P)H: H_2-pathways. Microbial Biotechnol. *1*, 382–394.

Vignais, P.M., and Billoud, B. (2007). Occurrence, classification, and biological function of hydrogenases: an overview. Chem. Rev. *107*, 4206–4272.

Vignais, P.M., Colbeau, A., Willison, J.C., and Jouanneau, Y. (1985). Hydrogenase, nitrogenase, and hydrogen metabolism in the photosynthetic bacteria. Adv. Microbiol. Physiol. *26*, 155–234.

Waks, Z., and Silver, P.A. (2009). Engineering a synthetic dual-organism system for hydrogen production. Appl. Environ. Microbiol. 75, 1867–1875.

Wang, D., Zhang, Y.P., Pohlmann, E.L., Li, J.L., and Roberts, G.P. (2011). The poor growth of *Rhodospirillum rubrum* mutants lacking RuBisCo is due to the accumulation of ribulose-1,5-bisphosphate. J Bacteriol. *193*, 3293–3303.

Wang, R., Zong, W., Qian, C., Wei, Y., Yu, R., and Zhou, Z. (2011). Isolation of *Clostridium perfringens* strain W11 and optimization of its biohydrogen production by genetic modification. Int. J. Hydrogen Energy *36*, 12159–12167.

Wells, M.A., Mercer, J., Mott, R.A., Pereira-Medrano, A.G., Burja, A.M., Radianingtyas, H., and Wright, P.C. (2011). Engineering a non-native hydrogen production pathway into *Escherichia coli* via a cyanobacterial NiFe hydrogenase. Metabol. Eng. *13*, 445–453.

Werken, H.J.G.V.D., Verhaart, M.R.A., Van Fossen, A.L., Willquist, K., Lewis, D.L., Nichols, J.D. *et al.* (2008). Hydrogenomics of the extremely thermophilic bacterium *Caldicellulosiruptor saccharolyticus*. Appl. Environ. Microbiol. 74, 6720–6729.

Wexler, M., Sargent, F., Jack, R.L., Stanley, N.R., Bogsch, E.G., Robinson, C., Berks, B.C., and Palmer, T. (2000). TatD Is a cytoplasmic protein with DNase activity. J. Biol. Chem. *275*, 16717–16722.

Weyman, P.D., Vargas, W.A., Chuang, R.Y., Chang, Y., Smith, H., and Xu, Q. (2011). Heterologous expression of *Alteromonas macleodii* and *Thiocapsa roseopersicina* NiFe hydrogenases in *Escherichia coli*. Microbiology *157*, 1363–1374.

Willquist, K., Zeidan, A.A., and Niel, E.W.V. (2010). Physiological characteristics of the extreme thermophile Caldicellulosiruptor saccharolyticus: an efficient hydrogen cell factory. Microb. Cell Fact. *9*, 89.

Yoshida, A., Nishimura, T., Kawaguchi, H., Inui, M., and Yukawa, H. (2006a). Enhanced hydrogen production from glucose using ldh- and frd-inactivated *Escherichia coli* strains. Appl. Microbiol. Biotechnol. *73*, 67–72.

Yoshida, A., Nishimura, T., Kawaguchi, H., Inui, M., and Yukawa, H. (2006b). Enhanced hydrogen production from formic acid by formate hydrogen lyase over expressing *Escherichia coli strains*. Appl. Environ. Microbiol.72, 1716–1716.

Yoshida, A., Nishimura, T., Kawaguchi, H., Inui, M., and Yukawa, H. (2007). Efficient induction of formate hydrogen lyase of aerobically grown *Escherichia coli* in a three-step biohydrogen production process. Appl. Microbiol. Biotechnol. 74, 754–760.

Yu, R., Wang, R., Bi, T., Sun, W., and Zhou, Z. (2013). Blocking the butyrate-formation pathway impairs hydrogen production in *Clostridium perfringens*. Acta Biochim. Biophys. Sin. 45, 408–415.

Zhang, S., and Bryant, D.A. (2011). The tricarboxylic acid cycle in cyanobacteria. Science 334, 1551–1553.

Zhao, H.X., Ma, K., Lu, Y., Zhang, C., Wang, L.Y., and Xing, X.H. (2009). Cloning and knockout of formate hydrogen lyase and H2-uptake hydrogenase genes in *Enterobacter aerogenes* for enhanced hydrogen production. Int. J. Hydrogen Energy *34*, 186–194..

Zheng, H., Zhang, C., Lu, Y., Jiang, P.X., and Xing, X.H. (2012). Alteration of anaerobic metabolism in *Escherichia coli* for enhanced hydrogen production by heterologous expression of hydrogenase genes originating from *Synechocystis* sp. Biochem. Eng. J. *60*, 81–86.

Zorin, B., Lu, Y., Sizova, I., and Hegemann, P. (2008). Nuclear gene targeting in Chlamydomonas as exemplified by disruption of the *PHOT* gene. Gene *432*, 91–96.

Biogas-producing Microbes and Biomolecules

2

Kornél L. Kovács, Norbert Ács, Tamás Böjti, Etelka Kovács, Orsolya Strang, Roland Wirth and Zoltán Bagi

Abstract
Biogas, essentially a mixture of methane (CH_4) and carbon dioxide (CO_2), can be used to generate electricity and heat or as bio-methane (bio-CH_4). Bio-CH_4 can be utilized in all applications in which natural gas is used today. Biogas is generated by a complex microbial community in nature and in man-made biogas reactors. An understanding of the interactions between the members of such communities and the build-up of the networks of microbial complex systems is extremely important. Culture-dependent methods cannot supply sufficient information about the life of microbial community. However, since the birth of molecular biology, more accurate taxonomic and physiological relationships between the individual microbes and within their communities can be studied and better understood. The methodology developed for molecular identification and physiological studies and recent results are reviewed. Strategies to generate biogas from unconventional substrates such as proteins, lignocelluloses and algae are discussed.

2.1 Introduction

Throughout the history of mankind, the benefits of microbial metabolism have frequently been utilized, for example in food production and preservation, though a long time had to elapse before humans realized that microbes are responsible for these actions. As the relevant microbiological techniques evolved, it became clear not only that microbes are present in enormous numbers in every habitat, but also that they play significant roles in various aspects of the ecosystem on Earth. It is now well known that the biosphere is dominated by microorganisms (around $4–6 \times 10^{30}$ of them thrive relative to the 7 10^9 human beings). They outnumber all animal and plant cells together by a factor of two or three orders of magnitude. Microbes are highly diverse and account for around 60% of the planet's biomass (Singh *et al.*, 2009). They are key players in the construction and decomposition of organic matter, and their decisive role in the recycling of essential elements such as carbon, nitrogen, phosphorus and sulfur is undisputed. Obviously, an understanding of the interactions between the members of such communities and the build-up of the networks of microbial complex systems is extremely important. In consequence of the technical difficulties involved, culture-dependent methods cannot supply sufficient information about the invisible life of microorganisms. However, since the birth of molecular biology, more accurate taxonomic and physiological relationships between the individual microbes and within their communities can be studied and better understood.

The burning of fossil fuels accounts for some 60% of the total greenhouse gas emission worldwide (IEA, 2012). Potential alternatives for energy generation are therefore currently intensively studied. Ideal energy carriers should be renewable and carbon-neutral or even carbon-negative. Biogas is one of the most important potential options. Biogas can be produced from a wide range of solid or liquid organic substances. On average, 80% of the substrate biomass in industrial biogas plants today is maize (Sildermann, 2012) but the production of biogas from maize is becoming less accepted, because of the cultivation costs and on a global scale because of the increasing competition with food production.

The biogas sector is growing worldwide. Thousands of household biogas facilities provide energy in rural areas in several Asian and a few African countries. In other regions, larger, industrial-size biogas power plants are operated. Biogas, essentially a mixture of methane (CH_4) and carbon dioxide (CO_2), can be used to generate electricity and heat in combined heat and power type end-uses or (following more rigorous purification) as bio-CH_4. Bio-CH_4 can be utilized in all applications in which natural gas is used today because the two are indistinguishable chemically or in any other properties. The only significant difference between the two as energy carriers is that natural gas is a fossil fuel that will become depleted in the foreseeable future, whereas bio-CH_4 is renewable and can be generated from biomass continuously. The large-scale use of bio-CH_4 involves a major part in regenerative and storable energy supplies. Typical air emission reductions, when bio-CH_4 is used instead of diesel fuel as vehicle fuel, are from 60% to 85% for nitrogen oxides, from 60% to 80% for particulates and from 10% to 70% for carbon monoxide (Bordelanne *et al.*, 2011; Murphy *et al.*, 2013).

2.2 Biogas microbiology

The decomposition of organic materials by a microbial community under anaerobic conditions leads to biogas formation (Eder and Schulz, 2005). The great variety of diverse microbes that participate in the microbial food chain gradually degrade the complex molecules to a mixture of CH_4 and CO_2 (Bayer *et al.*, 2004; Cirne *et al.*, 2007; Darke *et al.*, 2002; Shin and Youn, 2005). The actions of the various microbes, involving members of the Eubacteria and Archaea are coordinated by environmental and internal factors. The composition of this microbial community depends on various factors, such as the substrate ingredients, temperature, pH, mixing, etc. The active microorganisms have to work together, and both the nutritional and the environmental requirements of a large number of microorganisms therefore have to be satisfied for the biogas process to function as a whole. A clear understanding of the organization and behaviour of this multifarious community is crucial for optimization of the performance and attainment of the stable operation of the system. Some of the general features of the microbiological events will be summarized here; for more detailed descriptions, consultation with several excellent reviews is recommended (Angelidaki and Sanders, 2004; Gerardi, 2003; Klocke *et al.*, 2007; Schnürer *et al.*, 1999; Sousa *et al.*, 2009; Zakrzewski *et al.*, 2012).

2.2.1 Hydrolysis of large molecules

In the first step, sugars, fats and proteins are converted into smaller organic compounds such as amino acids, simple sugars, fatty acids, and some alcohols (Fig. 2.1). This process is very important because large organic molecules are simply too large to be devoured directly and used by microorganisms as a substrate/food source. To accomplish biodegradation, certain microorganisms secrete different types of extracellular enzymes, which hydrolyse some of the bonds holding together the large substrate molecules, in this way producing smaller building blocks that the microorganism can take into the cell and use as sources of energy and nutrition. The rate of decomposition during the hydrolysis stage depends greatly on the nature of the substrate. The transformation of cellulose and hemicellulose generally takes place more slowly than the decomposition of proteins or fats.

Hydrolysis of proteins

Proteins are found in high concentrations in, for example, meat-derived substrates and chicken or pig manure. Amino acids are the primary end-products of the hydrolysis. The decomposition of glycoproteins produces not only amino acids but also various carbohydrates. Among the proteolytic organisms in the biogas process are the genera *Clostridium*, *Peptostreptococcus* and *Bifidbacterium* (Nisman, 1954; Ramsay and Pullammanappallil, 2001).

Hydrolysis of lipids

Fats consist of glycerol and different fatty acids, which are held together by ester bonds and decomposed upon hydrolysis (McInerney, 1988). Examples of materials that are high in fat

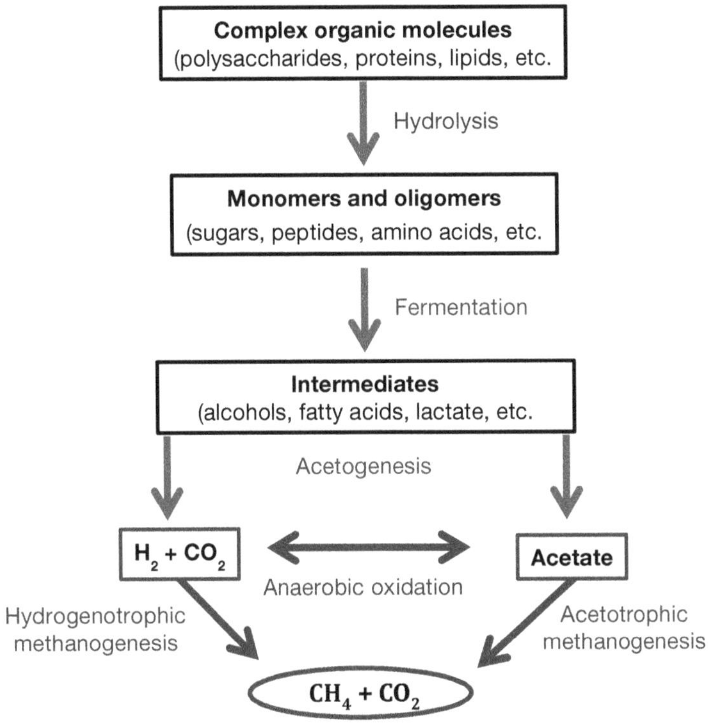

Figure 2.1 Stepwise decomposition of organic matter into biogas. A colour version of this figure is available in the plate section at the back of the book.

are slaughterhouse waste and grease-separation sludge. Most of the known lipases are produced by aerobic or facultative aerobic microorganisms. Strict anaerobes that secrete lipases include the genus *Clostridium* (Gupta *et al.*, 2004; Petersen and Daniel, 2006). Polyhydroxy-alkanoates, the carbon and energy storage material in most prokaryotes, are also linked together by ester bonds and can therefore be degraded by esterases, lipases and other hydrolytic enzymes.

Hydrolysis of polysaccharides

Cellulose, hemicellulose and starch are important components of plant material and are found in fruits, grains, vegetables and root crops. The polysaccharide glycogen functions as a sugar reserve, primarily in animals. Pectin is common in fruits, its structure varying in the different fruits with the degree of ripeness. Polysaccharides may be linear (e.g. cellulose) or branched chains of sugars (e.g. hemicellulose, starch, glycogen or pectin). The hydrolysis of cellulose results in the formation of cellobiose and glucose. Starch and glycogen are both cleaved into glucose units, and several different sugars are formed from the hemicellulose and pectin. Organisms that are active in biogas production during the hydrolysis of polysaccharides include various bacterial groups, e.g. the genera *Bacillus, Clostridium* and *Synergista* (Cirne *et al.*, 2007; Doi, 2008). Some of these organisms have several different enzymes combined into entities known as cellulosomes, situated on the cell wall of the organism. Besides enzymes, the cellulosomes also contain proteins that have the ability to bind to cellulose (Ding *et al.*, 2008). Binding to the substrate makes decomposition more efficient because the enzymes can then work directly 'on-site'.

2.2.2 Fermentation

A wide variety of different microorganisms are active during this stage because the diversity of the molecules to be decomposed is highest in this phase (Colberg, 1988; McInerney, 1988). Many of the organisms that carry out fermentation are related to those that effect hydrolysis during the first stage, but other active genera include, such as *Enterobacterium*, *Clostridium* and *Bacteroides*. During fermentation, the products from the previous hydrolysis stage are used as substrates by a number of microorganisms. Sugars, amino acids, alcohols, etc. can be used as substrates by fermenting microorganisms (Madigan and Martinko, 2006; Thauer *et al.*, 1977). Through various fermentation reactions, the products of hydrolysis are converted mainly into volatile organic acids (acetic acid, propionic acid and butyric acid), or non-volatile acids (succinic acid, lactic acid, long-chain fatty acids, etc.), alcohols, ammonia (NH_3), CO_2 and hydrogen (H_2). Exactly which compounds are formed depends on the substrate and the environmental conditions, as well as on what microorganisms are present. A typical feature of the acids that are formed is that the charged form (without protons) is in equilibrium with the uncharged form. In a biogas process at $pH > 7$, acids are mainly in the negatively charged form (anion). At this stage, they tend to form salts with different metal ions, such as sodium and potassium ions.

Fermentation products from the same compound may differ, depending on the degrading microorganisms. For the organism that produces them, fermentation products are waste products that are excreted and are of no further use. However, they serve as a substrate for other microorganisms in the biogas-producing microbial food chain, including other fermenting microorganisms, which further decompose them.

The role of hydrogen

Some of the products formed during the fermentation stage are further broken down by anaerobic oxidation. This is a very important step in the biogas process and requires close cooperation between the microorganisms that carry out oxidation and some of the CH_4-producing organisms that are active in the actual formation of CH_4. The situation in two different groups of organisms have to work together is strongly regulated by the actual concentration of H_2. During anaerobic oxidation, protons are used as final electron acceptors and H_2 is produced. For thermodynamic reasons, H_2 will be formed only effectively if its concentration is constantly kept at a very low level. If the H_2 formed is not removed continuously, anaerobic oxidation will stop because the microorganisms will then no longer receive enough energy for growth (Schink, 1997; 2002; Schnürer *et al.*, 1999). Hydrogenotrophic methanogens remove H_2 from the system by reducing CO_2 with it to form CH_4 (Bagi *et al.*, 2007; Conrad, 1985; Schmidt and Ahring, 1993; Wolin, 1975). In biological systems other than the biogas process, there are alternative H_2-consuming organisms that can drive anaerobic oxidation, such as sulfate-reducing or nitrate-reducing microorganisms. This cooperation between microorganisms and the transfer of H_2 directly between microbial species is called 'Inter-species Hydrogen Transfer' (IHT) (Schink, 1997; 2002). The mechanism of IHT is still not fully understood; evidence has been presented for direct electron exchange between the partner microbes in syntrophic relationship (Summers *et al.*, 2010).

Not all H_2 producing microorganisms depend on a partner organism and IHT (Dolfing, 1988; Gottschalk, 1986). Certain fermentative organisms produce H_2 in the absence of a H_2-consuming organism, but at much lower concentrations (Ivanova *et al.*, 2009; Kovacs, 2001). Many syntrophs that form H_2 can also use alternative breakdown pathways in the absence of a H_2-consuming partner, which does not lead to H_2 formation. *Syntrophomonas*, *Syntrophus*, *Clostridium* and *Syntrobacter* are examples of genera that can perform different anaerobic oxidations in syntrophy with an organism that uses H_2 (McInerney *et al.*, 2008; Schnürer *et al.*, 1999; Schnürer and Nordberg, 2008).

An alternative pathway for acetate conversion has been described (Hattori, 2008; Schnürer and Nordberg, 2008). Factors that are considered to influence the development of this path in a biogas process are the contents of NH_3 and acetate, and the types of active methanogens. The retention time in the biogas process has also proven to be significant, as has with temperature. Through this pathway, biogas is not generated directly from acetate by acetotrophic CH_4 production (i.e. acetate splitting); instead, acetate is first converted by non-CH_4-producing bacteria into H_2 and CO_2. These products are then used by hydrogenotrophic methanogens to form biogas.

This cooperation between two different groups of organisms is called syntrophic acetate oxidation. For the conversion of acetate to H_2 and CO_2 to take place, the H_2 partial pressure must be kept low, which is taken care of by the CH_4 producers. This pathway of acetate utilization is slower than that of the acetotrophic (acetate splitting) CH_4 producers, which results in the slower breakdown of organic matter and biogas production when the syntrophic acetate oxidation route is used.

2.2.3 Methanogenesis

In this stage, CH_4 and CO_2 (biogas) are formed by various CH_4-producing microorganisms, called methanogens. The most important substrates for these organisms are H_2, CO_2, and acetate, which are formed during the previous stages of the anaerobic degradation of the organic material. However, other substrates such as methylamines, some alcohols, and formate can also be used for the production of biogas (Liu and Whitman, 2008). One of the most abundant groups of CH_4 producers is that of the acetotrophic methanogens, which use acetate as substrate. In their metabolism, acetate is cleaved into CH_4 and CO_2 and the chemical energy stored in the carbon–carbon bond is utilized by the microbes. Acetate is the predominant product of the previous fermentation steps (Gerardi, 2003). The hydrogenotrophs are another important group of methanogens thriving in biogas reactors, for which the primary substrates for the formation of CH_4 are H_2 and CO_2 (Ács *et al.*, 2013; Schlüter *et al.*, 2008; Wirth *et al.*, 2012). There are only two known groups of methanogens that break down acetate, *Methanosaeta* and *Methanosarcina*, whereas there are many different groups of methanogens that use H_2, including *Methanobacterium*, *Methanococcus*, *Methanogenium* and *Methanobrevibacter* (Garcia *et al.*, 2000; Gerardi, 2003; Liu and Withman, 2008). *Methanosarcina* grows faster, but finds it difficult to use acetate at low concentrations, when *Methanosaeta* has an advantage. As CH_4 producers generally grow very slowly, this is often the rate-limiting stage of the biogas process (Gerardi, 2003; Liu and Withman, 2008). Too short a retention time (less than 12 days) increases the risk that these organisms will be washed out from the biogas reactor.

Methanogens differ from the other organisms in the biogas process because they are not common bacteria. Instead, methanogens are part of the group of organisms called archaea (Garcia *et al.*, 2000; Gerardi, 2003). Archaea possess unique structural and functional properties, e.g. the composition of membrane lipids, cell wall, coenzymes and DNA replication, which affect their role in the microbial community. Methanogens are often the first to be influenced by various disturbances such as pH changes, or the presence of toxic compounds such as the ions of heavy metals or organic pollutants (Y. Chen *et al.*, 2008; Liu and Withman, 2008). Since these organisms are also of great importance for the function of anaerobic oxidation, inhibition or intensification of methanogenic activity can seriously affect the entire process (Bagi *et al.*, 2007; Y. Chen *et al.*, 2008; Herbel *et al.*, 2010).

2.3 Biomethane

The utilization of fossil fuels on a global scale is limited by the availability of the fossil resources and by the harmful environmental effects of their excessive exploitation. The production of renewable energy carriers is therefore currently receiving increasing attention worldwide. Biogas is a promising candidate as the technology of its production may combine the treatment of various organic wastes with the generation of an energy carrier suitable for the most versatile applications (Angelidaki and Ellegaard, 2003; Daniels, 1992; Santosh *et al.*, 2004; Weiland, 2003). Biogas can be converted into heat and/or electricity, and its purified derivative, bio-CH_4, is suitable for every function for which fossil natural gas is used today. A rapidly increasing number of biogas plants produce bio-CH_4, the upgraded form of biogas, a large proportion of which is fed into the natural gas grid.

2.3.1 Biogas components

The content of biogas depends on various factors, such as the production process or the raw material used for anaerobic digestion (AD). Biogas from sewage digesters usually contains 55–65% CH_4, 35–45% CO_2 and less than 1% N_2. Biogas produced from municipal organic waste comprises 50–70% CH_4, 30–50% CO_2 and less than 1% N_2, while in landfills gas the CH_4 content is lower (45–55%), as is that of CO_2 (30–40%), whereas there is a much higher content of N_2 (5–15%). In addition to the main components, biogas also contains hydrogen sulfide (H_2S) and other reduced sulfur compounds, siloxanes, and aromatic and halogenated organic compounds. In the presence of water the reduced sulfur compounds can cause corrosion to compressors, gas storage tanks and engines, and these compounds must therefore be removed before the biogas can be utilized as an energy carrier (Ahring, 2003; Al Seadi *et al.*, 2008; Eder and Schulz, 2005; Gerardi, 2003). A common task in the various biogas technologies is the elimination of all undesired components before utilization. H_2O is relatively easy to remove, but CO_2 is more challenging as many of its properties are rather similar to those of CH_4, the target compound. Methods to achieve these goals will be summarized in the following sections.

2.3.2 Upgrading with physical methods

Removal of water

When it leaves the digester the biogas is saturated with water vapour, this water may condense in gas pipelines and cause corrosion. H_2O can be removed by cooling, compression, absorption or adsorption. Cooling is simply achieved by passing the gas pipeline through the soil and incorporating a condensate trap. H_2O can also be removed by means of technologies such as absorption in glycol solution or the use of hygroscopic salts (Ryckebosch *et al.*, 2011).

Water scrubbing of biogas

This is the most common method of purifying biogas to remove CO_2. The raw biogas from the digester is compressed and fed into a scrubber vessel, where a passing water stream absorbs the gaseous contaminants, leaving near pure CH_4. CO_2 is more soluble than CH_4 in water. The water leaving the absorption column is transferred to a flash tank where the dissolved gas, which contains some CH_4 but mainly CO_2 and other contaminants, is released and transferred back to the raw gas inlet. Finally, after removal of contaminants the biogas is dried on desiccant in the drier columns and exits the system as >98% pure methane (Krischan *et al.*, 2012; Tippayawong and Thanompongchart, 2010).

Pressure swing adsorption

Pressure swing adsorption (PSA) is a technique which involves the use of four adsorber vessels, a compressor and an H_2S removal step. The four adsorber vessels contain adsorbent material (carbon molecular sieve) which adsorbs N_2, O_2, H_2O, CO_2 and H_2S. Each vessel operates in an alternating cycle of adsorption, regeneration and pressure build-up. During the adsorption phase, biogas enters at the bottom of one of the adsorption columns. CO_2 is then adsorbed by the adsorption material. The gas leaving at the top of the vessel contains >97% CH_4. Another adsorption column is then switched to the adsorption mode in order to maintain continuous operation. Regeneration of the adsorption material is achieved by a stepwise depressurization of the column to atmospheric pressure and finally to near vacuum conditions. During regeneration, significant amounts of CH_4 can also be recovered which were trapped in the voids of the adsorption material. The advantages of PSA technology are the high enrichment of CH_4 (>97%) and the low level of emissions and waste (Alonso-Vicario *et al.*, 2010).

Cryogenic separation

The cryogenic separation of biogas is based on the fact that CO_2, H_2S and all other biogas contaminants liquefy at different temperature-pressure domains, which makes it easy to separate them from CH_4 (Ryckebosch *et al.*, 2011). However, this separation process operates at low temperatures, near −100°C, and at high pressures, almost 40 bar. The main advantages of cryogenic separation are both the large quantities and the high purity of the upgraded biogas. The yield of this separation is 99% for CH_4.

Membrane technology

Membranes for biogas upgrading are made of materials that are permeable to CO_2, H_2O and NH_3. H_2S and O_2 permeate through the membrane to some extent, while N_2 and CH_4 pass to only a very low extent. Membranes are usually in the form of hollow fibres bundled together (Kárászová *et al.*, 2012; Makaruk *et al.*, 2010; Molino *et al.*, 2013; Scholz *et al.*, 2012).

2.3.3 Upgrading with chemical methods

Iron-based systems

Ferrous materials in the form of natural soils or certain iron ores are often employed to remove H_2S (Zhang and Tong, 2006). The ferrous material is placed in a closed, gas-tight container. The gas to be purified flows upward through the ferrous absorbing agent and leaves the container at the top, freed from H_2S. The absorbing material must contain iron in the form of oxides, hydrated oxides or hydroxides.

2.3.4 Upgrading with biological methods

Biological sulfide oxidation

H_2S can be oxidized by microorganisms of the genera *Thiobacillus* and *Sulfolobus* (Oprime *et al.*, 2001). The degradation requires O_2 and therefore a small amount of air (or pure O_2 if levels of N_2 are to be minimized) is added to the gas storage facility for biological desulfurization to take place. The degradation can occur inside the digester and can be facilitated by immobilizing the microorganisms occurring naturally in the digestate (Al Seadi *et al.*, 2008).

Biofilter

The biofilter is a fixed bed bioreactor where the microorganisms used in the degradation process are immobilized in a packing medium. The contaminated gases pass through the porous medium. The contaminants pass into the biofilm surrounding the packing material and microbial colonies, where degradation occurs. Two types of biofiltration are open designed biofilters and enclosed design biofilters. The open design generally requires a large area, and is usually installed outside, with only an ascending gas flow. The closed biofilter has a more restricted volume and can have either an ascending or a descending inlet gas flow (Burgess *et al.*, 2001). The nutrient solution and moisture are added non-continuously. Positive aspects of biofilters include low operating and capital costs and their self-containment. Drawbacks include the deterioration of the packing material during time and being not very suited for high concentrations of contaminants. Moreover, the moisture and pH are difficult to control, and clogging and pressure drops may occur as time passes (Burgess *et al.*, 2001).

Biotrickling filters

A biotrickling filter is similar to a biofilter except that the nutrient solution flow and the wetting of the filter material are continuous. The gas mixture flows through a fixed bed, while the bed is continuously kept wet with an aqueous solution supplemented with the nutrients needed for the microbial system. It has been reported that the use of a co-current versus a counter-current flow system makes no difference to the degradation performance (Burgess *et al.*, 2001).

Bioscrubbers

In general a bioscrubber consists of an absorption tower and a bioreactor. The gas and liquid flows run counter-current to one another within the column (Lam and Lee, 2011). The absorption column can include packaging, but it is not definitely required. Packaging can assist in increasing the surface area of contact between the contaminant and the aqueous phase. The microorganisms are suspended in an aqueous growth medium in the bioreactor. Since the bioscrubber separates the absorption and the biological processes, both processes can be stabilized and optimized separately (Burgess *et al.*, 2001). High pressure drops across the bed and installation space may be minimized.

2.4 Molecular methods for the study and control of biogas production

Molecular identification is a tool commonly used to assess microbial communities from various environments (Ivey and Phister, 2011; Schlüter *et al.*, 2008; Singh *et al.*, 2009).

2.4.1 Culture-dependent methods

By definition, this approach involves methods that require cultivation of the selected strain in pure culture. This is often difficult because the vast majority (~99%) of microbes cannot grow under defined laboratory conditions. Nevertheless, many important bacteria and archaea have been identified through the use of classical microbiological methods. Enrichment cultures are also employed, in which the composition of the growth medium is designed to mimic the specific ecological niche of the microbes (Enthceva *et al.*, 2001). New techniques involve the use of laser tweezers and flow cytometry, which are effective, but require sophisticated and costly instrumentation (Vartoukian *et al.*, 2010).

2.4.2 Culture-independent methods

The specificity of the detection is increased by using DNA-based approaches, and the representation of the microbes in the sample can be revealed much more precisely. These methods are usually divided into two major groups. In one, only a partial section of the community is interrogated (for example, a characteristic gene sequence), while in the second strategy the full-blown community is examined.

Regardless of the scheme selected, the initial steps are sample preparation and DNA extraction. A systematic bias can be introduced by choosing the wrong DNA extraction protocol, which changes the apparent microbial community structure and will eventually lead to incorrect results. The optimal DNA extraction protocol involves a mixture of physical and chemical cell disruption methods (Martin-Laurent *et al.*, 2001).

Partial community analysis approaches

The most widely used techniques target a portion of the genetic information of the whole community. The basic idea of these molecular biological tools is that a certain part of the total genomic context is selected and investigated thoroughly. This can be done with signature genes that are suitable for identification of the microbes. The commonly used primary target gene is the gene coding for the 16S rRNA. There are two major features which make this gene preferred. The size of the gene is just about right for molecular analysis, and it has both conserved and variable sequence regions (Drancourt *et al.*, 2000; Woese *et al.*, 1990). Universal primers that can bind to the conservative parts of the gene can easily designed, while the hypervariable regions are suitable for distinction of the various microorganisms. A huge database of sequence data has already been amassed. In addition, a number of other markers have been studied, including the genes coding for methyl coenzyme-m reductase (*mcrA*), the RNA polymerase beta subunit (*rpoB*), the gyrase beta subunit (*gyrB*) or recombinase A (*recA*). These approaches generally involve a PCR (polymerase chain reaction) step in order to amplify the selected region from the genomic DNA pool (Case *et al.*, 2007).

Clone library analysis

Since the original sample contains a mixture of DNA of various origins, the PCR product will also contain various amplified DNA fragments, which are more or less of the same size, but with different base sequences. Conventional agarose gel electrophoresis cannot always resolve these fragments, and hence alternative methods are needed. A relatively simple solution is to clone the PCR-amplified gene fragments into circular cloning vectors. After transformation into *Escherichia coli*, the different inserts can be separated, as every bacterium can pick up one vector at a time, and the screening of sufficient individual colonies will therefore result in a whole library of clones, which all carry distinct inserts. The origin of any specific insert is then traced back by DNA sequencing of the cloned fragments. Unfortunately, the number of clones needed for a full coverage of a complex sample is enormous. More than 40,000 clones represent merely about 50% of the richness of a soil sample (Dunbar *et al.*, 2002). The operational costs and labour time and efforts necessary to handle such huge clone libraries do not allow the general application of this approach. Further, with the new generation sequencing platforms, a dataset at least one magnitude larger can be collected for only a fraction of the price. The diversity of activated sludge communities was investigated via this method by utilizing several sampling sites and screening at least 100 individual clones per library. The results corroborated that the microbiological diversity of the analysed samples is much higher than expected by using conventional plating methods (Blackall *et al.*, 1998).

Quantitative PCR

Real-time PCR, or Q-PCR, is based upon the intercalation of fluorescent dyes such as SYBR Green, or fluorescent probes (e.g. TaqMan). The accumulation of fluorescent dye is measured in every cycle, and the initial DNA concentration can be calculated from the fluorescence intensity.

Quantitative determination of genes coding for important physiological groups of proteins (such as NH_3 oxidizers, CH_4 oxidizers or sulfate reducers) can be achieved, and gene expression studies can be performed through coupling with reverse transcription (RT). The methanogenic population has been investigated through real-time PCR, with targeting of the *mcrA* gene and determination of the gene copy number in order to correlate it with the CH_4 producing activity (Ács *et al.*, 2013; Luton *et al.*, 2002). TaqMan probes have also been used to detect up to nine different phylogenetic groups in Q-PCR assays. The results demonstrated the highly different community structures of the various samples (Steinberg and Regan, 2009).

Fluorescence *in situ* hybridization

Fluorescence *in situ* hybridization (FISH) is used to study the phylogenetic diversity with the help of whole-cell hybridization and fluorescently labelled oligonucleotide probes. A rapid identification can be achieved with this method. Its drawback is that the oligonucleotide probes can be designed only on the basis of known gene sequences, and a search for new strains in a given sample by this method is therefore not possible. Nevertheless, it provides useful information about the presence/absence or a shift in the number of cells. A specific oligonucleotide probe (Arc915) has been successfully implemented for the quantification of methanogens by using the FISH method. The results revealed the high level of specificity and reproducibility of the comparative estimation of the cell number (Stabnikova *et al.*, 2006).

Fingerprinting methods

Genetic fingerprinting methods investigate a portion of a certain set of DNA. PCR amplification of the desired amplicons is inevitable in order to determine the microbial content of the examined sample. In general, fingerprinting methods are rapid, and allow the investigation of multiple samples simultaneously. The most commonly used techniques will be introduced in the sections below.

Denaturing gradient gel electrophoresis (DGGE) or temperature gradient gel electrophoresis (TGGE)

The PCR products of genomic DNA templates extracted from an environmental sample are separated by means of polyacrylamide gel electrophoresis (PAGE), containing a linear gradient of a denaturing agent (temperature in the case of TGGE; urea or formamide in the case of DGGE. The method is based on the fact that, although the PCR products have similar lengths, their sequences are not identical, which results in distinct melting temperatures (T_m). Besides the universal primer used for taxonomic determination, these methods require a GC clamp on the 5′ end of the forward primer, which is a relatively short (30–50 nucleotides) sequence consisting of only G or C nucleotides. This clamp prevents the complete melting of the PCR product at its T_m. If the experiment is carried out correctly, a pattern can be visualized after staining. The relevant bands can subsequently be excised, reamplified and sequenced in order to identify the corresponding microbe. A drawback of this method is that only a limited number of bands (~15–20) can be separated at a time on one gel. An additional difficulty stems from the fact that different DNA fragments can have similar melting points and, due to the sequence heterogeneity among multiple rRNA operons in one strain, multiple bands can be produced in the gel. A further drawback is that limited sequence information (~500 bp) can be achieved, which makes the identification cumbersome (Muyzer and Smalla, 1998). The overall diversity of archaea has been determined using the DGGE method when dairy wastewater was fermented. Numerous physical and chemical parameters were monitored during the digestion. It emerged, that at 35°C microwave radiation favours the growth of *Methanosarcina barkeri*, while at 55°C the predominant microbe is *Methanosarcina thermophila*, replacing *Methanoculleus palmolei* (Zielinska *et al.*, 2013).

Single-strand conformation polymorphism (SSCP)

This method uses single-stranded DNA as a target for the identification. The basic principle of SSCP is that single-stranded DNA fragments with different sequences have unique secondary structures after reannealing. The technique is applied mainly for environmental samples. PCR products are denatured after amplification, while the separation of the single-stranded conformation is carried out on a non-denaturant PAGE or in a capillary gel electrophoretic system. Thanks to the nature of the method capillary gel electrophoresis provides significantly higher resolution. Samples with medium complexity can be taxonomically resolved with this method, though SSCP is primarily used for the identification of different pure cultures. A major limitation is the high degree of reannealing of the DNA strands, but this can be overcome by the use of phosphorylated primers during the PCR step. Closely related species have been distinguished through the use of this molecular biological tool (Hiss *et al.*, 1994).

Random amplified polymorphic DNA (RAPD)

The RAPD method is popular among the molecular ecological tools, because of its simplicity and low operating costs. This technique also uses PCR amplification, but, unlike the regular PCR reactions, the primers are usually random hexamers, which anneal randomly to the template DNA. The typically low annealing temperature (<35°C) is essential for the generation of the taxonomically useful pattern. The result of RAPD is a set of PCR amplicons with various lengths, which can be separated on agarose gel or PAGE. On visualization, a specific pattern will emerge, that is characteristic of the examined microbial community. The drawback of this method is the difficulty of producing consistent parallel patterns from the same sample. Unfortunately, many parameters can affect the PCR (e.g. $MgCl_2$ concentration), resulting in differences in the fragment pattern. With this method, important population genetic parameters (such as gene and genotype frequencies, within- and between-population heterozygotes, degree of inbreeding and population subdivision, and degree of individual relatedness) have been determined successfully (Lynch and Milligan, 1994).

Amplified ribosomal DNA restriction analysis (ARDRA)

ARDRA involves the direct examination of the sequence variations that can occur in the 16S rRNA genes with the use of restriction endonucleases. The amplified sequence of the gene in question must be digested with enzymes, e.g. AluI or RsaI, that recognize a 4-bp sequence, in order to generate the maximum number of fragments. On visualization, this method results in a pattern which can give information about alterations in the community. Unfortunately, because of the huge amount of fragments produced, direct conclusions cannot be drawn concerning the species that comprise the microbial community. A major limitation of this method is that many ARDRA patterns are too complex to resolve on a conventional agarose gel or PAGE. Nevertheless, ARDRA has been applied to assess the complexity of certain microbial communities. For example, the usefulness of this method was demonstrated by distinguishing the type strains of the genus *Bacillus* from *Peanibacillus* or the genus *Alcaligenes* from *Bordatella* (Heyndrickx *et al.*, 1996).

Terminal restriction fragment length polymorphism (T-RFLP)

T-RFLP is very similar to ARDRA, since the main target gene is the one coding for 16S rRNA. Restriction digestion of the PCR product is carried out with restriction enzymes having 4-bp recognition sequences. The simultaneous use of two enzymes is recommended. The method simplifies the pattern of the microbial community by using a 5′ fluorescently labelled forward primer during the PCR step. Because of this alteration, only those fragments remain visible which are closest to the 5′ end of the PCR product when the sample is separated by capillary gel electrophoresis with a fluorescence detector.

Theoretically, every terminal restriction fragment (TR-F) detected in the pattern will correspond to one strain in the complex environmental sample, while its abundance is related to the area of the peak in question. The restriction pattern itself can provide basic information about the composition of the environmental community, and it is also possible to identify the corresponding strains if a clone library is produced in parallel. After selection and screening of the positive clones, the clones relating to the unique restriction peaks can be identified by cutting the inserts with the same restriction enzyme(s). These patterns are then compared with the restriction patterns, allowing the identification of the predominant groups in the sample. The clones carrying a fragment are sequenced for strain identification. Web-based T-RFLP analyses and BLAST tools are available to help the evaluation of the data. The rapid detection, high throughput and relatively low operational costs make T-RFLP one of the most favoured molecular biological tools in microbial taxonomy. This technique has some drawbacks, the most disturbing phenomenon being that different bacterial species can produce TR-Fs of the same size. This difficulty can be effectively alleviated by using two restriction enzymes. T-RFLP has been applied for the analysis of numerous microbial communities and environmental samples, including studies of biogas-producing microbial communities (Ács *et al.*, 2013; Collins *et al.*, 2003; Klocke et al, 2007; Liu *et al.*, 1997; Osborne *et al.*, 2000).

Length heterogeneity PCR (LH-PCR)

LH-PCR is similar to T-RFLP analysis, the main difference being the lack of the restriction digestion step. The method takes advantage of the unique structure of the 16S rRNA gene. Owing to the hypervariable regions, the size of the gene differs in the different strains. If the product of the PCR amplification is separated with the high-resolution capillary gel system, detectable differences are observed in a complex sample. For example, several environmental samples have been investigated through the use of the LH-PCR method, which demonstrated significant variations in the composition of the microbial communities (Ritchie *et al.*, 2000).

Ribosomal intergenic spacer analysis (RISA)

RISA examines the sequence length of the intergenic spacer region (ISR) between the small (16S) and large (23S) subunits of the rRNA gene. The ISR is known to display significant heterogeneity in the various species. The primers used in this method are designed with regard to the conserved regions of the genes flanking the DNA sequence that separates them from each other. RISA generates a specific profile that can be evaluated in order to obtain taxonomic information on the sample. The automated version of this method, ARISA, uses fluorescent primers with a laser detector to enhance the accuracy of length detection. In this way, multiple samples can be investigated simultaneously, which demands care as RISA is known to overestimate microbial richness and diversity. Nevertheless, the method is a rapid and effective one that can be used in conjunction with other techniques to assess the microbial diversity of a given sample (Fisher and Triplett, 1999).

DNA microarrays

The DNA microarray technology has the highest throughput of the methods discussed in this section. It uses known molecular probes fixed on a microarray chip. The PCR products can be directly hybridized to these probes in order to detect differences or similarities between samples. Since the PCR products are fluorescently labelled, the positive signal can be detected via confocal laser scanning microscopy. This allows rapid and precise detection of a signal, which is proportional to the abundance of the target microorganism. Two major forms of microarray technology are distinguished. The first type is the 16S rRNA gene microarray (or PhyloChip), while the second type focuses on the functional genes (functional gene array: FGA) (DeSantis *et al.*, 2007). The former comprises about 3×10^4 probes of 16S rRNA genes, covering all 121 prokaryotic orders. Theoretically, this method allows the simultaneous detection of 8741 bacterial and archaeal taxa (He *et al.*, 2007). The latter can be used to shed light on the functional complexity of a certain sample. FGA can also reveal the metabolic potential of an environmental sample, since it contains over 2.4×10^4 probes from a wide range of known metabolic genes. The two major limitations of the microarray technologies are cross-hybridization and the fact that the detection of novel strains is impossible due to the pre-produced DNA probes. Despite these disadvantages, the robust method is commonly used when rapid examination of highly diverse samples is needed. A specifically designed microarray technology has also been developed for the investigation of methanogenic communities: ANAEROCHIP, which uses more than 100 probes to detect most of the relevant species involved in CH_4 production. ANAEROCHIP has demonstrated its suitability for the analysis of methanogens in sludge samples (Daniel, 2005; Franke-Whittle *et al.*, 2009).

Whole-community analysis

The methods discussed in this section are applied to investigate the entire genetic information of any given sample. Perhaps the most important feature that distinguishes whole-community analysis from the previously introduced methods is the limited use of PCR amplification. PCR is known to introduce a certain degree of bias, which may lead to inaccuracy in the taxonomical determination. These problems can be eliminated when the whole intact genetic material of the sample is tested and a more complete picture of the microbial community inhabiting a certain environment is obtained. Another problem with the partial community approaches is that in many cases the 16S rRNA gene cannot provide sufficient resolution to differentiate strains at a species level. Whole-genome methods are more comprehensive, making them more suitable for the analysis of complex samples.

DNA–DNA hybridization (DDH) kinetics

This approach offers genome-wide comparisons between organisms. Although the method was originally developed for the analysis of pure cultures, in some cases it additionally furnishes valuable information about the richness of an environmental sample. In practice, this means that the whole genomic DNA is extracted and the DNA is then melted. Incubation under appropriate conditions results in rehybridization or re-association of the single-stranded DNA. The rate of this step is an indicator of the genomic complexity or the diversity present in the sample (Goris *et al.*, 2007). It has been shown that DDH can yield helpful information about closely-related species. When Goris *et al.* (2007) examined 28 strains belonging in six distinct groups of bacteria, which had ≥ 94% 16S rRNA gene sequence identity, the results revealed a close relationship between DDH and ANI (average nucleotide identity), which led to the conclusion that ANI can replace DDH in certain cases when the whole-genome sequence is known.

Guanine-plus-cytosine content fractionation

A genomic DNA mixture can be fractionated through density-gradient centrifugation, when the difference in the guanine-plus-cytosine (G+C) content of the DNA results in a separation pattern after the centrifugation step. Since the various prokaryotic groups differ significantly in G+C content, taxonomic identification can be carried out after the fractionation. Another molecular technique (such as DGGE or ARDRA) is usually employed for the secondary identification of the separated DNA. To carry out an experiment of this kind, a relatively large amount of pure genomic DNA (about 50 μg), and a long time (4 days) are required. The G+C method in combination with radioactive tracers, or with blotting and hybridization to various probes, can be useful for the further investigation of the environmental samples (Tiedje *et al.*, 1999).

Whole-microbial-genome sequencing

In this approach, the DNA of a complex environmental sample is first digested into small fragments. These fragments are then cloned into a suitable host in plasmids. The plasmids are sequenced after extraction from the host bacterium, which is *E. coli* in most cases. This approach requires a large shotgun clone library with the desired insert size (~2 kb), and at least a 6-fold coverage of the whole environmental DNA material. The data are usually collected by bidirectional Sanger sequencing of the clones, and the sequences are then aligned and assembled into contigs by using bioinformatics tools. In order to build a comprehensive database, a considerable amount of sequencing should be carried out, which increases the costs of such investigations. With the discovery of the next-generation sequencing methods this approach has lost ground, because it cannot compete with the efficacy of the next-generation DNA sequencing techniques. In their classical work, Venter *et al.* (2004) (who had the financial and technical support to carry out such a large-scale project) collected genomic DNA from the Sargasso Sea and, by sequencing more than one billion base pairs, identified at least 1800 genomic species, 148 of them previously unknown. Among the sequence data, over 1.2×10^6 previously undiscovered genes were represented, which demonstrated the lack of our knowledge about our environment (not only in the deep sea).

Next-generation sequencing (NGS)

The automated Sanger method is now regarded as first-generation DNA sequencing technology, while later methods are usually referred to as next-generation sequencing (NGS), and the most recent developments lay claim to be the third-generation sequencing (TGS) platforms. The widespread application of these technologies has fundamentally changed the approach to both general and specific problems in basic, applied and clinical research. These methods have opened up new avenues in the field of microbial ecology referred to as metagenomics. Metagenomics is defined as the direct genetic analysis of genomes in an environmental sample. A total genomic DNA extract is prepared from the samples and is characterized by whole-genome sequencing. The resulting DNA sequences, individual reads or assembled sequence contigs are assigned to individual taxonomic groups or known genome sequences by sophisticated bioinformatic tools. For example, there are a number of tools that provide an overview of the species composition of metagenomic samples via direct nucleotide sequence (Brady and Salzberg, 2009; Huson *et al.*, 2007) or conserved protein domain-coding sequence comparisons (Finn *et al.*, 2010), identification of 16S rRNA sequences within the sample (Schloss *et al.*, 2009), or oligonucleotide frequencies (The reference genome group, 2009). Other common types of whole genomic sequencing metagenomic analyses, include determination of the functional composition of a microbial community based on the assignment of protein-coding open reading frames to functional categories, such as protein domain families (Finn *et al.*, 2010) or gene ontologies (The reference genome group, 2009). Sequence data must be evaluated via relative abundances rather than absolute presences/absences. With the advent of NGS platforms, genomics is changing rapidly, a trend that is expected to continue. The major advance offered by NGS and TGS is the ability to produce an enormous volume of data at relatively low cost. This feature expands the realm of experimentation beyond merely determining the order of nucleotide bases. As sequencing costs continue to decrease, it may be anticipated that more applications will be developed. The current costs of storing, processing and analysing large sequence data sets already exceed the cost of generating them. The ability to sequence the whole genome of many related organisms has allowed large-scale comparative and evolutionary studies that were unimaginable only a few years ago.

In 2003, the adenovirus genome was sequenced by using what was then called NGS technology, 454 pyrosequencing (Margulies *et al.*, 2005). This platform uses an array-based approach that allows the simultaneous detection of millions of sequencing reactions, thereby significantly increasing the number of reads per sequencing reaction. Subsequent NGS technologies include a number of methods that are grouped broadly as template preparation sequencing and imaging. The unique combination of specific protocols distinguishes one technology from another and determines the type of data produced from each platform. These differences in data output present challenges when comparing platforms based on data quality and cost.

Sequencing by ligation

While all other currently available sequencing technologies depend on the nucleotide specificity and the accuracy of DNA polymerases during the sequencing process, sequencing by oligonucleotide ligation and detection (SOLiD) platform developed by Applied Biosystems Life Technologies, uses DNA ligase to synthesize a complementary copy of immobilized DNA fragment templates on an array (Durfee *et al.*, 2008; Srivatsan *et al.*, 2008). Following emulsion PCR (similar to 454), DNA fragments are hybridized to bead-bound primers. A set of four fluorescently labelled dibase hexamer probes compete for ligation to the sequencing primer. The specificity of the dibase probe is achieved by interrogating every first and second base in each ligation reaction. Multiple cycles of ligation, detection and cleavage are performed, with the number of cycles determining the eventual read length. The template is then reset with a primer complementary to the n-1 position for a second round of ligation cycles. Five rounds of primer reset are completed for each sequence tag. Through the primer reset process, each base is interrogated in two independent ligation reactions by two different primers, increasing the sequencing accuracy of this technology. A total of 60–100 Gbp is generated per run with an accuracy of 99.94%.

Between NGS and TGS

One technology that sits between the NGS and TGS categories is the Ion Torrent semiconductor sequencer (Rothberg *et al.*, 2011). The Ion Torrent sequencing instrument employs a state-of-the-art semiconductor technology, originally developed by DNA Electronics Ltd, by sensing the release of hydrogen ions as part of the base incorporation process, and creates a high-density array of micro-machined wells that carry out sequencing by synthesis. This process eliminates the need for light, scanning and cameras to monitor the sequencing process, thereby simplifying the overall process, dramatically accelerating the time to result, reducing the overall footprint of the instrument, and lowering the costs to make DNA sequencing more generally accessible to all. However, this technology is still a 'wash-and-scan' system, like all current NGS technologies, requiring PCR amplification of the DNA template in each well, and also termination events typically halting sequencing after each nucleotide incorporation, in order to monitor in succession the incorporation of each of the four bases across all DNA templates. As a result of this technology, the overall read length is limited to that of current NGS systems. The semiconductor sequencers are capable of generating 100–200-bp-long reads (similarly to 454) with an accuracy of 98–99%, but the generated sequence reads are lower in number.

Third generation sequencing (TGS)

TGS technologies use single molecule sequencing and can be grouped into three different categories: (1) sequencing by synthesis technologies in which single molecules of DNA polymerase are observed as they synthesize a single molecule of DNA; (2) nanopore-sequencing technologies in which single molecules of DNA are threaded through a nanopore and individual bases are detected as they pass through the nanopore; and (3) direct imaging of individual DNA molecules using advanced microscopy techniques. Each of these technologies provides novel approaches to sequencing DNA and has advantages and disadvantages with respect to specific applications. These technologies are at varying stages of development. However, if the full potential of these technologies is realized within a few years, whole genome sequencing will probably be fast and inexpensive enough to sequence genomes as needed for any application.

TGS by synthesis

The Helicos Genetic Analysis Platform was the first commercially available sequencing instrument to carry out single molecule sequencing (Bowers *et al.*, 2009; Harris *et al.*, 2008; Lipson *et al.*, 2009; Tessler *et al.*, 2009). The Helicos sequencing platform works by imaging individual DNA molecules affixed to a planar surface as they are extended, using a defined primer and a modified polymerase together with proprietary fluorescently labelled nucleotide analogues, referred to as virtual terminator nucleotides, in which the dye is attached to the nucleotide via a chemically cleavable group that allows stepwise sequencing. As termination of the reaction is still required in this process (similarly as in NGS), the time needed to sequence a single nucleotide is long, and the read lengths realized are ~32 nucleotides. However, given the single molecule sequencing nature of this technology, no PCR is required for sequencing, a significant advantage over NGS technologies. The sample preparation in this approach involves fragmenting genomic

DNA into smaller pieces, adding a 3′ poly(A) tail to the fragments, labelling and blocking by terminal transferase. These templates are then captured onto a surface with covalently bound 5′ dT (50) oligonucleotides via hybridization. The surface is next imaged by using charge-coupled device (CCD) sensors, where those templates that have been appropriately captured are identified and then tracked for sequencing by synthesis. The subsequent steps resemble the 'wash-and-scan' strategy of NGS in which a labelled nucleotide and polymerase mixture are flooded onto the system and incubated for a period of time, and the surface is then washed to remove the synthesis mixture and scanned to detect the fluorescent label. The dye–nucleotide linker is subsequently cleaved to release the dye, and this process is repeated in cycles. A drawback is the relatively high raw read error rates (>5%), although the highly parallel nature produces plentiful coverage and the consensus read accuracy (>99%) is satisfactory. Unlike NGS, the hundreds of millions of sequencing reactions can be carried out asynchronously, a hallmark of TGS. Further, given the individual monitoring of templates, the enzymatic incorporation step does not need to be driven to completion, which serves to reduce the overall error rate.

The single-molecule real-time sequencing approach developed by Pacific Biosciences is the first TGS approach in which a single molecule of DNA polymerase is directly observed as it synthesizes a strand of DNA, leveraging the speed and processivity of this enzyme to address many of the shortcomings of synthesis (Eid *et al.*, 2009; Levene *et al.*, 2003). Given that a single DNA polymerase molecule is of the order of 10 nm in diameter, two important obstacles needed to be overcome to enable direct observation of DNA synthesis as it occurs in real time. One was confining the enzyme to an observation volume that was small enough to achieve the signal-to-noise ratio needed to accurately call bases as they were incorporated into the template of interest. The second was labelling the nucleotides to be incorporated in the synthesis process in such a way that the dye–nucleotide linker is cleaved after completion of the incorporation process so that a natural strand of DNA remains for continued synthesis and so that multiple dyes are not held at a time. The problem of observing a DNA polymerase working in real time, detecting the incorporation of a single nucleotide taken from a large pool of potential nucleotides during DNA synthesis, was solved by using zero-mode waveguide (ZMW) technology (Levene *et al.*, 2003). A ZMW is a hole, tens of nanometres in diameter, fabricated in a 100-nm-thick metal film deposited on a glass substrate. The small size of the ZMW prevents visible laser light, which has a wavelength of ~600 nm, from passing entirely through the ZMW. Rather than passing through, the light exponentially decays as it enters the ZMW. Therefore, if laser illumination is shone up through the glass into the ZMW, only the bottom 30 nm of the ZMW becomes illuminated. Within each ZMW, a single DNA polymerase molecule is anchored to the bottom glass surface, using the biotin/streptavidin interaction. Nucleotides, each type labelled with a different coloured fluorophore, are then flooded above an array of ZMWs at the required concentration. The sequencing instrument detects this as a flash whose colour corresponds to the base identity. Following incorporation, the signal immediately returns to baseline and the process repeats, with the DNA polymerase continuing to incorporate multiple bases per second. Thus, the ZMW has the ability to detect a single incorporation event against the background of fluorescently labelled nucleotides at biologically relevant concentrations. Each ZMW is capable of containing a DNA polymerase loaded with a different strand of DNA sample. As a result, the array enables the potential detection of ~7.5×10^4 single molecule sequencing reactions in parallel. At present, approximately a third of the ZMWs of a given array are active for a given run because the DNA polymerase and DNA template to be sequenced are delivered to ZMWs via a random diffusion process (Schadt *et al.*, 2010).

Nanopore sequencing technologies

Oxford Nanopore offers a commercial system for DNA sequencing based on three biological molecules that have been engineered to work as a system (Clarke *et al.*, 2009; Howorka *et al.*, 2001; Stoddart *et al.*, 2009). The biological nanopore is constructed from a modified α-haemolysin pore that has an exonuclease attached on the normally extracellular face of the pore. A synthetic cyclodextrin sensor is also covalently attached, but to the inside surface of the nanopore. This assembly is positioned in a synthetic lipid bilayer so that when DNA is loaded onto its exonuclease-containing face and a voltage is applied across the bilayer by changing the concentration of salt, the exonuclease can cleave off individual nucleotides. Once they are cleaved, the individual nucleotides are detected through their characteristic disruption of the ionic current flowing through the pore. Reliable throughput at high multiplex may be difficult to achieve with this system on the use of natural lipid bilayers, but synthetic membranes and solid-state nanopores may help to overcome this problem. One significant challenge in nanopore-based sequencing lies in the need for the simultaneous monitoring of a large number of nanopores. The first parallel readout of any nanopore-based method has recently been demonstrated through the use of optical multipore detection (McNally *et al.*, 2000). In this approach, the contrast between the four bases is first increased off-line through a biochemical process that converts each base in the DNA into a specific, ordered pair of concatenated oligonucleotides. Subsequently, two different fluorescently labelled molecular beacons are hybridized to the converted DNA. The beacons are then sequentially unzipped from the DNA molecules as they are translocated through a nanopore. Each unzipping event unquenches a new fluorophore, resulting in a series of dual- colour fluorescence pulses that are detected by a high-speed CCD camera with a conventional total internal reflection fluorescence microscopy setup. The unzipping process is slowed down by adjusting the voltage governing DNA translocation through the nanopore to a speed compatible with single-molecule optical detection.

Direct imaging of DNA using electron microscopy

Halcyon Molecular is pioneering a single molecule sequencing approach using transmission electron microscopy (TEM) for the direct imaging and chemical detection of atoms that would uniquely identify the nucleotides comprising a DNA template. The approach reliably detects atoms on a planar surface, using annular dark-field imaging in aberration-corrected scanning TEM (Krivanek *et al.*, 2010). A number of supporting technologies required to carry out TEM-based DNA sequencing should also be developed, such as the use of functionalized needles to attach stretched molecules of DNA to a substrate for the direct imaging procedure.

Biogas-producing microbial community studies with metagenomics approach

In studies based on NGS methods, designed to improve the understanding of microbial communities in biogas-producing anaerobic digestors, the 454-pyrosequencing technique (454 GS FLX and Titanium) was first employed (Jaenicke *et al.*, 2011a; Krause *et al.*, 2008; Kröber *et al.*, 2009; McHugh *et al.*, 2003; Schlüter *et al.*, 2008). The substrates fed into the reactors included animal manure and green plant biomass (maize or green rye silage). The Titanium provides somewhat longer reads and increased throughput relative to GS FLX (see section 'Sequencing by synthesis' above). The SOLiD sequencing method produces short individual reads (50 nucleotides) in a significantly higher number than does pyrosequencing (see section 'Sequencing by ligation' above). In spite of the distinct sequencing strategy, the analysis of the data at the species level revealed a strong correlation between the findings of the 454-pyrosequencing and SOLiD NGS technology platforms (Wirth *et al.*, 2012).

In all cases, the class clostridia constituted the most widespread group of microbes in the biogas fermenters. The Clostridia are noted for their highly effective cellulose degradation potential (Jaenicke *et al.*, 2011b), and are therefore essential in the breakdown of lignocellulosic substrates in the biogas process. It was also noted that hydrogenases have been found in large quantity among the redox enzymes in the biogas-producing community. Thus, the Clostridia may contribute to the widening of at least two bottlenecks in the biogas process, through the hydrolysis of large polymeric substrates and the *in situ* production of H_2, an important substrate for the hydrogenotrophic methanogens.

At the level of resolution of the abundances

of individual strains, the most frequently occurring species likewise displayed a good correlation. Strains noted for their highly efficient polysaccharide degradation capabilities, such as *Clostridium thermocellum*, *C. cellulolyticum* and *Caldicellulosiruptor saccharolyticus*, were found to be the most abundant, regardless of the sequencing method used for their identification.

Similarly to the Bacteria, the members of the Archaea domain demonstrated a markedly comparable community structure. The Methanomicrobiales were indicated to constitute the majority of the Archaea in this environment by the sequencing with the 454 GS FLX, 454 Titanium and SOLiD platforms alike. Within this taxon, the predominant genus is Methanoculleus, and the most abundant species is *M. marisnigri*. The DNA-based community structure analysis of anaerobic degradation samples has therefore demonstrated the importance of hydrogenotrophic methanogens in biogas formation (Wirth *et al.*, 2012).

Postgenomic approaches

Methods will next be discussed which are aimed not at the DNA sequences of a certain microbial community, but at other markers also carrying taxonomic information. These approaches avoid the potential inaccuracy of the PCR step, and protein and RNA markers can reveal functional relationships inherent in the biogas-producing community. The goal is not merely to identify the microbes in the environmental sample, but additionally to shed light on the metabolic pathways that play a key role in the function of the microbial community.

Microbial lipid analysis

Fatty acids are present in a relatively constant proportion in the cell biomass. They function as a signature for identification, and in this way major taxonomic groups can be recognized. This procedure is carried out by saponification of the fatty acids, creating fatty acid mono esters, which are than analysed by gas chromatography. The emerging pattern is compared with a reference database, to identify the corresponding microbial signatures (Ritchie *et al.*, 2000). This strategy can be used either to identify unknown microbes in pure cultures, or to acquire information on the whole microbial community in a given sample. General methods have been optimized in order to extract the lipids in significant quantities from the sample to be examined. The lipid fraction is usually separated, by a one-phase chloroform/methanol/water extraction. This is followed by alkaline methanolysis of the polar fraction containing the phospholipids, and the fatty acids can be recovered as methyl esters. The bacterial and archaeal community can be examined separately, because the phospholipid fatty acids are characteristic for the bacteria domain, while the phospholipid diether lipids are derived from the (methanogenic) archaea. This method has been used to assess the changes in the microbial community of a laboratory-scale biogas fermentor upon overloading with glucose Sundh *et al.*, 2003). The bacteria inhabiting the digester responded very quickly to the feeding condition. Their cell density almost doubled briefly after the sugar was added into the fermentor. The methanogen population was apparently not affected by the glucose addition.

RNA sequencing (transcriptomics)

The first question in studying any complex microbial community is 'What is there?', but at least as important is the next question, 'What do they do?'. DNA-based approaches offer only tentative information about the metabolic activity of the microbes. The presence of the DNA of a specific strain can be detected regardless of the functional state of the cells, which may introduce considerable ambiguity in DNA-based physiological conclusions. For a full understanding of the events, functional markers should be studied. RNA is a suitable candidate since mRNA, the transcript of genetic information, reflects the biological activity of the community. A key step in these methods is the purification of the RNA fraction because of the high risk of RNase contamination and degradation of the RNA molecules. First the total RNA content is purified and this is followed by separation of the various RNA fractions, as only the mRNA yields information concerning the active metabolic reactions in the cells. When the metatranscriptome is studied by sequencing, only the mRNA is used in order to avoid the generation of useless 'junk' reads. The translation of the purified mRNA to cDNA cannot be omitted, because of the unstable nature of RNA. Once the cDNA population has been produced, any of the high-throughput sequencing methods discussed above can be applied. Enormous sequence data sets are created, which must be dealt with by means of specialized bioinformatic tools to generate contigs and to identify metabolically active genes. The identification of these genes supplies important information relating to the metabolic activity of the community at the time of RNA sampling. The analysis of multiple samples taken at various times, or under various conditions, reveals the metabolic complexity and capability of the examined microbial community (Giannoukos *et al.*, 2012; Wang *et al.*, 2009; Zakrzewski *et al.*, 2012).

Proteomics (metaproteomics)

Conventional proteomics approaches must start with the extraction of the total protein content of the sample. The separation of the individual proteins is the following task. The most commonly used method is two dimensional gel electrophoresis (2D-PAGE) or capillary chromatography. 2D-PAGE employs partition of the molecules, making use of two distinct chemical properties. The first is their isoelectric point, and the method is called isoelectric focusing. The isoelectric point is the pH value at which the cumulative positive and negative charges are equal, and therefore the molecule does not display net charge. Since the charged amino acid side-chains are a function of the protein sequence, the isoelectric point is a distinctive property of the protein. The protein mixture is loaded on a matrix containing a linear pH gradient. In an electric field generated by direct current, the proteins migrate until they reach the pH value corresponding to their isoelectric point and stop there because they no longer interact with the electric field. Complex protein samples may contain several proteins which have similar isoelectric points. These are resolved in the second dimension, which is routinely SDS-PAGE. SDS-PAGE separates the proteins according to their size. The result of 2D-PAGE is a protein map in the PAGE, where ideally each spot represents a unique protein. Computer programs are available to analyse the spot intensities and to highlight the differences in comparative studies. The protein spots of interest are removed and partially digested with proteases (trypsin is generally used) for further investigation. The size distribution of the resulting peptide fragments is determined by using matrix-assisted laser desorption/ionization – time of flight (MALDI-TOF) mass spectrometry (Gorg *et al.*, 2004). This protocol can be employed in dealing with multiple microorganisms from a complex sample. A limitation of this method stems from the resolution capacity of 2D-PAGE. The number of individual protein spots is usually in the range of a few hundred, but there can be several thousand individual proteins in a complex sample (Hanreich *et al.*, 2012; Heyer *et al.*, 2013; Wilmes and Bond, 2004).

2.5 Biogas from unconventional substrates

2.5.1 Low C/N substrates

One of the most important process parameters controlling efficient biogas production is the substrate composition. Its commonly used measure is the carbon over nitrogen (C/N) content of the material fed into the reactor. It is a rule of thumb that C/N should be between 20 and 30 for an anaerobic digestor to operate most successfully and efficiently (Al Seadi *et al.*, 2008; Eder and Schulz, 2005; Gerardi, 2003). Substrates that have C/N values either below or above the advised range are not recommended for biogas production. In the European Union (EU), more than 8000 biogas plants are currently operating, but none of them process primarily protein-rich waste ($C/N < 20$), despite huge amounts of such materials being generated continuously, with an immense biogas potential. Together with manure from animal production (mainly cattle and pig farms), decentralized co-digestion plants have the greatest potential for biogas production, but the use of sludge and food industry waste and household waste likewise offers enormous opportunities (van Foreest, 2012). These pollutants mount up in vast quantities and their disposal by composting or incineration is energy-consuming and costly. Anaerobic digestion is an excellent supplement and alternative to composting and has proved to be preferentially suited for waste materials with high water content (Braun *et al.*, 2003).

The roughly 10^6 tons of protein-rich waste produced annually by pig, cattle and bird breeding worldwide contains manure, blood and feathers (Gendebien *et al.*, 2001), a biomass type classified as hazardous waste in several countries. The poultry sector has been the most dynamic meat sector during the last decade, showing the greatest growth of all meat sectors as reflected in world consumption. The world's broiler chicken production is growing faster than any other form of meat production. Poultry meat represents about 33% of global meat production: in 2007, some 2.7×10^8 tons of meat were produced globally, of which 0.9×10^8 tons were poultry meat (FAO Statistics Division 2009; FAO Investment Centre Division and FAO's Rural Infrastructure and Agro-Industries Division, 2010). 30.9% of the world's poultry meat is supplied by the EU (Grabkowsky and Windhorst, 2009). The EU egg and poultry sector contributes 6.3% to the overall EU agricultural output, accounting for some €2.33×10^{10} annually (Caspari *et al.*, 2010). Chicken guts, heads, feathers, blood and wastewater that remain from the processing have to be treated before being disposed of or further utilized, e.g. as fertilizer (FAO Investment Centre Division and FAO's Rural Infrastructure and Agro-Industries Division, 2010). This means 5×10^5 tons of protein-rich waste per year in the EU alone. Many other parts of the world face similar problems. In eastern Asia, the shortage of supply of pork meat already observed in 2007 and resulted in a gradual shift in consumption from pork to poultry. In the Russian Federation and Latin America, income growth significantly promoted the consumption of meat in general and poultry meat in particular. The demand for poultry meat was even more important in the Middle East, where competition by pork is almost non-existent (FAO Investment Centre Division and FAO's Rural Infrastructure and Agro-Industries Division, 2010). Chicken manure has a high nitrogen content, which is in two main forms: uric acid and undigested proteins, representing 70% and 30% of the total nitrogen, respectively. Anaerobic degradation (AD) of these components in chicken manure can result in the production of inhibitory concentrations of unionized NH_3 and NH_4^+ ions (Y. Chen *et al.*, 2008a; Salminen and Rintala, 2002).

Protein also accounts for a major part of the organic load in a dairy wastewater stream (Ramsay and Pullammanappallil, 2001). A low rate of anaerobic digestion (AD), together with inhibition problems caused by certain components, have been identified as responsible for the restricted activity of biogas reactors operated with dairy effluent (Braun *et al.*, 2003; Perle *et al.*, 1995). Other branches of the food industry, such as abattoirs, and the processing of whey, cheese, casein and certain vegetables, typically produce wastewater containing relatively high concentrations of protein (Ramsay and Pullammanappallil, 2001). Such forms of protein-rich biomass would be extremely valuable as biogas substrates if the basic questions concerning their AD could be solved. Besides the advantages of the biogas produced, the environmentally harmless fermentation residue is an excellent fertilizer.

Proteins are composed of α-amino acids linked together by peptide bonds into polypeptide chains, which are hydrolysed by proteases upon decomposition. An amino acid is characterized by an amino group ($-NH_2$) and a carboxylic acid group (-COOH) linked to a carbon, known as the α-carbon. The degradation products include short- or branched-chain organic acids, NH_3, CO_2, H_2S and H_2. Amino acids are metabolized through two main routes: pairs of amino acids can be decomposed through the Stickland reaction; and single amino acids can be degraded in the presence of H_2-utilizing bacteria (Nisman, 1954; Ramsay and Pullammanappallil, 2001). This pathway involves two amino acids: the first undergoes oxidative deamination, and the second acts as an electron acceptor through reductive deamination. This redox reaction results in two keto acids that are employed to synthesize ATP via substrate-level phosphorylation (de Vladar, 2012). The Stickland reaction is the simplest way to degrade amino acids and provides the cell with 0.5 mole of ATP per mole of amino acid transformed (Andreesen *et al.*, 1989). Stickland reactions are usually faster than uncoupled amino acid fermentation (Ramsay and Pullammanappallil, 2001).

Ammonia toxicity

Being a microbial process, CH_4 formation is influenced by many environmental factors, e.g. the NH_3 concentration (van Velsen, 1979). In anaerobic digestion processes, NH_3 is produced through the degradation of the nitrogenous matter present in the feedstock, mostly in the form of proteins (Kayhanian, 1999; Kotsyurbenko *et al.*, 2004). The NH_3 concentration has often been regarded as the main reason for digester failures and for the inability to treat certain substrates anaerobically (Braun *et al.*, 1981; Schmid and Lipper, 1969). The optimal carbon–nitrogen–phosphorus (C/N/P) ratio for a high methane yield is around 100:3:1 (Conrad *et al.*,1989; Rajeshwari *et al.*, 2000). The most important nutrients are nitrogen and phosphorus (Rajeshwari *et al.* 2000). If the C/N ratio is high there is a risk of nutrient deficiency and a low buffering capacity will result in a more sensitive process (Nyns, 1986). A decrease in the CH_4 production rate results in an accumulation of organic acids and causes a decrease in pH and a poor buffering capacity (van Velsen, 1979).

The quantity of NH_3 generated from AD of an organic substrate can be estimated (Nielsen and Angelidaki, 2008). Inhibitory thresholds for NH_3 have been reported in a number of studies, but the determined limiting concentrations vary significantly (Angelidaki and Ahring, 1993; Banks and Wang, 1999; Hansen *et al.*, 1996; Kroeker *et al.*, 1979). Free NH_3 is the main cause of inhibition since it is membrane-permeable (de Baere *et al.*, 1984; Kroeker *et al.*, 1979) and it causes a proton imbalance and/or a potassium deficiency (Gallert *et al.*, 1998; Sprott *et al.*, 1984; Sprott and Patel 1986). The ammonium ion (NH_4^+) is less toxic (Hejnfelt and Angelidaki, 2009). At least two possible mechanisms of NH_3 toxicity have been postulated: (i) NH_3 itself could directly inhibit the activity of cytosolic enzymes, or (ii) NH_4^+ accumulated inside cells might be toxic via its effects on the intracellular pH (Kleiner, 1993; Sprott *et al.*, 1984) or the concentrations of other cations such as potassium (Sprott and Patel, 1986). Proteinaceous substrates are easily degradable and yield high biogas yields. However, the rapid degradation of proteins leads to elevated NH_3 concentrations, which can cause a rapid decrease of biogas production. The molar ratio of NH_4^+/NH_3 decreases and, in correlation with this decline the biogas production ceases abruptly as the pH increases. A rise in pH from 7.0–7.35 (NH_4^+/NH_3 ≈ 200–400) to 8.5–9.0 (NH_4^+/NH_3 ≈ 5–15) was accompanied by a loss of CH_4 production to only 15–20% of the original level (Ahring, 2003).

NH_3 is essential for anaerobic microorganisms; concentrations between 0.05 and 0.2 g/l are beneficial (Bhattacharya and Parkin, 1989). According to van Velsen (1979) NH_3-nitrogen concentration between 0.2 and 1.5 g/l has no effect on the CH_4-forming microbes. Koster and Lettinga (1984) indicated that under mesophilic conditions the maximum methanogenic activity was unaffected at a total NH_3-nitrogen concentration of 0.68 g/l. However, a concentration between 1.5 and 3 g/l was apparently inhibitory at pH > 7.4 (McCarty and McKinney, 1961; van Velsen, 1979). Concentrations of 4 g/l or more inhibited thermophilic digestion of cattle manure (Angelidaki and Ahring, 1993).

The main operational parameters affecting the NH_3 concentration are as follows:

- the organic and inorganic nitrogen in the substrate, i.e. the C/N ratio;
- the reactor loading (Luste and Luostarinen, 2010);
- the buffering capacity and resulting pH;
- the temperature of the reactor (Straka *et al.*, 2007).

There are difficulties associated with reporting NH_3 inhibition based on total NH_3-N rather than free NH_3, as the NH_3-N inhibitory concentrations reported from different studies are not comparable unless these additional conditions are also specified.

Treatment of ammonia toxicity

Ammonia removal by sulfuric acid

Sulfuric acid (H_2SO_4) reacts with NH_3 to form ammonium sulfate ($(NH_4)_2SO_4$). Use of H_2SO_4 would increase the operational costs. In exchange, $(NH_4)_2SO_4$ can be utilized as a fertilizer (Lei *et al.*, 2007) and accelerate the recycling of nitrogen because farms can handle the slurry or solid fertilizer more easily than the residue after wet AD (Abouelenien *et al.*, 2010).

Ammonia stripping and removal

Stripping of the liquid phase is an efficient way to avoid the accumulation of NH_3 during biogas fermentation. Chicken manure diluted with water could be treated anaerobically either in a semi-solid form, i.e. 10–11.5% total solid content, or in a wet form (0.5–3% total solids; Bujoczek *et al.*, 2000). Other endeavours applied to reduce the inhibitory effect of NH_3 have involved physical–chemical or biological removal of NH_3 from methanogenic sludge (Abouelenien *et al.*, 2010). In the former solutions, NH_3 is adsorbed onto zeolite or clay (Milán *et al.*, 2001). A promising biological procedure involves the anaerobic NH_3 oxidation (Anammox) process and nitrification/denitrification (Dong and Tollner, 2003). In anaerobic NH_3 oxidization gaseous N_2 is formed when nitrite (NO_2^-) is utilized as the electron acceptor (Jetten *et al.*, 1998). Partial nitrification (Surmacz-Gorska *et al.*, 1997) occurs and NO_2^- accumulates when the second step of the nitrification is inhibited. The first alternative was based on Anammox and the liquid which had a high NH_4^+/NH_3 ratio was removed from the bottom of the reaction vessel. About half of the NH_4^+ content in this liquid was converted to NO_2^- and the liquid was circulated back into the anaerobic digester. Recirculation also served to mix the reactor content uniformly. In another solution, denitrification and AD steps were combined. The NH_4^+ in the leachate was completely nitrified to nitrate (NO_3^-) followed by conversion of NO_3^- to N_2 gas. Denitrification took place in the anaerobic digester where the organic intermediates of the AD products were used as electron donors. A difficulty in applying these concepts originates from the very slow growth of Anammox microorganisms, which therefore cannot compete with the denitrifying organisms for NO_2^- in the anaerobic digester.

NH_3 stripping has been employed for the removal of NH_3 in dry anaerobic fermentation of chicken manure (Abouelenien *et al.*, 2009) and for stripping AD effluent (Lei *et al.*, 2007) or poultry litter leachate (Gangagni Rao *et al.*, 2008). The discharge can be utilized safely as fertilizer. The NH_3 stripping technique can achieve removal of about 40% of the organic carbon and about 80% of the organic nitrogen without significant nitrate pollution of ground water (Abouelenien *et al.*, 2009).

Struvite production

This technology removes both nitrogen and phosphorus from agro-industrial and livestock wastewaters. Magnesium ammonium phosphate hexahydrate ($MgNH_4PO_4 \times 6H_2O$, MAP) is a promising product for nutrient recovery from wastes with high nitrogen and phosphorus contents and has the general name struvite (Demeestere *et al.*, 2001). Struvite precipitates spontaneously from some wastewaters. It has a low solubility constant (pK_s) between 12.6 and 13.2 at 25°C, and can therefore readily be transformed to its insoluble form. In this case, separation from the aqueous phase is simple. Numerous waste streams, including municipal wastewater, tend to be rich in NH_4^+-N, but deficient in Mg^{2+}, and supplementation of the Mg^{2+} is required. Struvite precipitation from anaerobic digestion effluents has great advantage due to the predominance of NH_4^+-N and PO_4^{3-}-P, so that addition of the necessary chemicals can be minimized (Marti *et al.*, 2008; Ohlinger *et al.*, 2000). MAP precipitation, combined with AD, offers important advantages in terms of renewable energy production in the form of CH_4 gas and the recovery of MAP sludge as a valuable slow-release fertilizer for agricultural use (Yetilmezsoy *et al.*, 2009).

NH_4^+ and CO_2 removal by algae

The high nitrogen concentration in livestock liquid manure causes eutrophication, besides the odour nuisance, if proper and sufficient treatment is not applied. After AD the levels of nutrients, such as NH_3 or NO_3^- are only partially reduced as the microorganisms preferentially consume organic nitrogen (Cheng and Liu, 2002). Microalgae, such as cyanobacteria and green algae, incorporate both inorganic NH_4^+, NO_3^- and phosphate into their biomass in addition to CO_2 during photosynthesis (Yoshihara *et al.*, 1996). A microalgal system can therefore be employed as an alternative secondary or post-secondary process to remove excess nutrients from wastewater, liquid manure or AD effluents, particularly when a high level of inorganic nitrogen remains after the organic matter degradation (Nunez *et al.*, 2001; Przytocka-Jusiak, 1984; Sriram and Seenivasan, 2012). Moreover, microalgae can produce a valuable biomass which can be used as biofuel feedstock for biodiesel, H_2 or biogas production or as an animal feed additive (Chisti, 2007; Melis and Happe, 2001). Application of microalgae as a substrate for biogas production will be discussed in detail in section 'Co-digestion'.

2.5.2 Co-digestion

The digestibility of nitrogen-rich wastes can be improved by mixing them with substrates of high carbon content, thereby improving the C/N ratio (Ahring *et al.*, 1992; Kaparaju *et al.*, 2002; Macias-Corral *et al.*, 2008). Co-digestion of various wastes is an attractive approach for perfecting the efficiency of biotransformation because this treatment can use the bacterial diversity in those wastes to optimize the digestion process (Wang *et al.*, 2012). Many successful co-fermentation experiments, using different substrates, have achieved a considerable increase in CH_4 productivity relative to separate digestion of the individual substrate components (Xiaojiao *et al.*, 2012). Co-digestion has important benefits, including the balancing of the C/N ratio in the co-substrate mixture and also the macro and micronutrients, pH, inhibitors/toxic compounds and dry matter (Banks and Humphreys, 1998; Campos at al., 1999; Hartmann and Ahring, 2005). Co-digestion of dairy manure, chicken manure and wheat straw gave a better result than separate AD treatment of the materials. It is noteworthy that the mixture displayed a higher synergetic effect than for single manure mixtures with wheat straw. C/N ratios of 25:1 and 30:1 led to better digestion behaviour because of the stable pH and low concentrations of free NH_3 and total NH_4^+-N. Maximum CH_4 production was achieved on co-digestion with a dairy manure/chicken manure ratio of 40.3:59.7 and a C/N ratio of 27.2:1, determined by response surface methodology (Xiaojiao *et al.*, 2012).

Co-digestion of chicken manure with other types of livestock manure such as pig waste (Magbanua *et al.*, 2001), cattle manure (Güngör-Demirci and Demirer, 2004) and anaerobically digested sludge (Bujoczek *et al.*, 2000) also improved the biogas productivity. Co-digestion with a carbon-rich substrate, e.g. vegetable processing waste, similarly improved the AD of poultry litter, due to the increased C/N ratio. The addition of vegetable processing waste (25%) resulted in increases in biogas and CH_4 production. However, more than 25% of this substrate gave rise to inhibition of the biogas production, probably due to the pH decrease and an excessive volatile fatty acids accumulation (Molinuevo *et al.*, 2008).

Co-digestion of chicken and cattle manure has been carried out in a biogas plant under mesophilic conditions, with maize silage added as co-substrate (Cigdem and Izzet, 2013). Both the daily bio-CH_4 and the total net energy production increased about 1.2-fold. The elevated CH_4 yields demonstrated that co-digestion with maize silage as co-substrate is viable.

2.5.3 Acclimation to low C/N

Microbial adaptation to higher NH_3 concentrations and cation antagonism effects (Y. Chen *et al.*, 2008; Lapp *et al.*, 1975) could contribute to the broad range of NH_3 inhibitory thresholds reported in the literature. Higher tolerances can be achieved by acclimation and NH_3-adapted anaerobic consortia are inhibited at 0.7–1.1 g NH_3-N/l (Angelidaki and Ahring, 1993; Hansen *et al.*, 1996). The acclimation of biogas-forming microbial communities to a wide variety of potentially inhibitory substances has been reported (Y. Chen *et al.*, 2008). Hashimoto (1986) observed that NH_3 inhibition started at about 2.5 g/l and 4 g/l for unacclimated and acclimated thermophilic methanogens, respectively. In the undisturbed natural microbial community, the inhibitory NH_3 level can be as low as 0.08–0.10 g NH_4^+-N/l (Braun *et al.*, 1981; de Baere *et al.*, 1984). Nevertheless, Parkin *et al.* (1983) reported that levels as high as 8–9 g/l of total NH_4^+-N could be tolerated with no significant decrease in methane production after acclimation. A tolerance of up to 3–4 g NH_4^+-N/l for an adapted process has also been reported (Angelidaki and Ahring, 1993). Control of the pH may reduce NH_3 toxicity by maintaining NH_3 at low levels (Kadam and Boone, 1996).

Edström *et al.* (2003) used blood, stomach and intestinal contents and food waste in co-fermentation with animal manure. Feedstock mixtures containing 8–15% of animal waste products could be co-digested under stable conditions at a total concentration of 4.5–5.0 g NH_4^+-N/l. Successful operation of anaerobic filters was also achieved at 6.0–7.8 g N/l after an adaptation period (de Baere *et al.*, 1984; Parkin *et al.*, 1983).

It is generally accepted that, of all the microorganisms involved in anaerobic digestion, the methanogens are the least tolerant to environment inhibitors and the most likely to cease growth due to NH_3 inhibition (Kayhanian, 1994; Kotsyurbenko *et al.*, 2004; McMahon *et al.*, 2001). Elevated NH_3 levels (0.8–6.9 g NH_4^+-N/l) cause changes in microbial communities; the shift from aceticlastic methanogenesis to syntrophic acetate oxidation is a consequence of the effect of the inhibition by NH_3 (Schnürer and Nordberg, 2008; Westerholm *et al.*, 2012). The relative number of some aceticlastic methanogens decreased in the NH_3-stressed reactor, whereas the abundance of bacteria increased. Tests of NH_3 toxicity on the acetate- and H_2-utilizing populations revealed a higher sensitivity of the aceticlastic than of the hydrogenotrophic methanogens; the specific growth rate for the aceticlastic methanogens was halved at NH_3 concentrations of 3.5 g N/l, relative to 7 g N/l for the hydrogenotrophic methanogens (Angelidaki and Ahring, 1993). The NH_3 tolerance of some syntrophic acetate-oxidizing bacteria, such as *Clostridium ultunense*, *Syntrophaceticus schinkii* and *Tepidanaerobacter acetatoxidans*, apparently bestowed a competitive advantage on the community (Westerholm *et al.*, 2012). Koster and Lettinga (1988) studied the changes in the microbial activity of the acidogenic and methanogenic population in granular sludge as the NH_3 concentration was increased in the range 4.051–5.734 g N/l (Koster, 1986). They demonstrated that the methanogenic population lost more than 50% of its activity, whereas the acidogenic populations were hardly affected. An earlier experiment indicated that H_2-utilizing methanogens were more susceptible than acetoclastic methanogens to NH_3; moreover, H_2 accumulation subsequently caused a build-up of propionate, which acted as an inhibitor of the acetoclastic methanogens (Wiegant and Zeeman, 1986). It was concluded that it was possible to produce biogas effectively at a concentration exceeding 4–5 g NH_4^+-N/l after an initial adaptation phase (Y. Chen *et al.*, 2008; Nielsen and Angelidaki, 2008). A lower biogas yield and/or CH_4 yield was observed, however, in the case of elevated NH_3 loads (Borja *et al.*, 1996; Koster and Lettinga, 1988).

In contrast, more recent studies indicate that acetoclastic methanogens are more sensitive than hydrogenotrophic ones to NH_3 toxicity at both mesophilic and thermophilic temperatures (Angelidaki and Ahring, 1993; Robbins *et al.*, 1989; Schnürer and Nordberg, 2008). Koster and Koomen (1988) studied NH_3 inhibition of hydrogenotrophic methanogens by using non-adapted sludge. The hydrogenotrophic population could grow well at high NH_3 concentrations (6.3 g N/l). An adaptation period was found to be absolutely necessary for acetoclastic methanogens, which again supports the finding that hydrogenotrophic methanogens are less susceptible to NH_3 toxicity.

A common feature of most NH_3 adaptation attempts was that the substrate with high nitrogen content was fed together with a substantial amount of carbon-rich materials in order to approach the recommended C/N/P ratio. A small number of toxicity studies have been carried out to examine the different inhibitory concentrations of NH_3 toward pure cultures of hydrogenotrophic and acetoclastic methanogens. Sprott *et al.* (1984) used pure cultures to study the NH_3 tolerance in four methanogen strains (*Methanospirillum hungatei*, *Methanosarcina barkeri*, *Methanobacterium thermoautotrophicum* and *Mb. formicicum*) commonly isolated from sludge digesters, which grow on H_2 and CO_2. *Ms. hungatei* was apparently the most sensitive of the four tested strains to NH_3, it could tolerate NH_3 concentrations up to 4.2 g N/l, while the other three strains were resistant to concentrations above 10 g N/l after acclimation. In similar pure culture studies, conducted on four strains of thermophilic hydrogenotrophic methanogens, population growth could still be observed for some strains even at an NH_3 concentration of 9 g N/l (Hendriksen and Ahring, 1991). Other experiments revealed that members of the

Methanosarcinaceae and Methanosaetaceae were strongly inhibited by NH_3 (Hajarnis and Ranade, 1994; Sprott and Patel, 1986).

The NH_3 concentrations tolerated by methanogens in those investigations were significantly higher than in other studies in which mixed cultures were used (Poggi-Varaldo *et al.*, 1997; Pohland and Ghosh, 1971) corroborating that hydrogenotrophic methanogens have higher tolerance toward NH_3 toxicity. Archaea and bacteria grew similarly in the presence of 0.8–2.3 g NH_4^+-N/l. At NH_4^+ concentrations of 4.4 g/l and 8.6 g/l, archaea were quasi-absent or absent, respectively. In AD, the main competitors of the syntrophic acetate-oxidizing bacteria are the aceticlastic methanogens, including Methanosarcina and Methanosaeta (Hattori, 2008).

2.5.4 High C/N substrates

The abundance of lignocelluloses available worldwide, together with the high carbohydrates content of these materials, makes lignocelluloses a potentially appealing feedstock for biofuel production. An additional advantage of the digestion of these substances is that they do not compete with land usage for feed or food production. Plants produce the largest amount of biomass on Earth, harvesting solar energy and utilizing it in photosynthesis to assemble organic molecules from inorganic ones. The primary composite materials are lignocelluloses, which have been estimated to account for approximately 50% of the biomass in the world, and to have a yearly production of some 2×10^{11} tons per year. These materials are carbohydrate-rich and can therefore be used for AD. Lignocelluloses consist of highly ordered cellulose fibres held together by hemicellulose and lignin (Lynd *et al.*, 2002).

Cellulose is an unbranched polymer chain composed of glucose units linked by β-1,4-glycosidic bonds (Brett, 2000). The degree of polymerization can be up to 1.4×10^4 glucose units, where each glucose unit is rotated 180° relative to the adjoining unit. The cellulose chains form tight aggregates in the form of three-dimensional microfibrils. Each microfibril consists of 30–36 parallel cellulose chains, these being stabilized by hydrogen bonds and van der Waals linkages. This makes the configuration tightly packed and stable. The half-life of crystalline cellulose at neutral pH is 1×10^8 years (Wilson, 2008).

Hemicelluloses are a group of branched polysaccharides, composed of both pentoses (e.g. xylose and arabinose) and hexoses (e.g. glucose, mannose and galactose). The compositions of hemicelluloses vary from plant to plant and from cell to cell. Lignocelluloses consist of approximately 20–35% hemicelluloses, the degree of polymerization often being around 200.

Next to cellulose and hemicellulose, lignin is the third most abundant substance in nature. Lignins are complex, amorphous, branched polymers constructed of different phenylpropane units (Van Dyk and Pletschke, 2012). The monomeric composition varies. There are interlinkages between lignin and hemicellulose and also between lignin and cellulose that keep the lignocellulosic structure together. As a result of these strong linkages, lignin is extremely resistant to biodegradation (Triolo *et al.*, 2012). One of the greatest challenges to the use of lignocelluloses as a feedstock for biogas production is to open up and alter the lignin structure. A high lignin content in the biomass has been associated with a lower CH_4 yield (Jiminez *et al.*, 1990).

Numerous types of lignocellulose-containing biomasses have been tested for biogas production (e.g. Bagi *et al.*, 2007; Herbel *et al.*, 2010; Hutnan *et al.*, 2000; Kalia *et al.*, 2000; Menardo and Balsari, 2012; Rani and Nand, 2004; Triolo *et al.*, 2012; Zhong *et al.*, 2011, 2012). Unfortunately, the results are usually difficult to compare, partly because a more or less standardized procedure is rare (e.g. VDI-Handbuch Energietechnik, 2006) and the data are not expressed in consistent units of volatile solids (VS), or synonyms such as volatile suspended solids (VSS), organic dry matter (oDM) or organic total solids (oTS).

Pretreatment

Various pretreatment methods have been developed to open up the structure and break down lignin to a certain degree, resulting in a more easily digested feedstock for anaerobic microorganisms. The pretreatment of lignocellulosic biomass includes biological, mechanical and chemical methods or their combinations. The choice of the optimal pretreatment process depends on the feedstock, enzymes and organisms to be applied, the economic assessment and the environmental impact. Each pretreatment has its advantages and disadvantages.

In many ways, the pretreatment of lignocelluloses prior to AD has the same objectives as in the case of ethanol production from the same substrate, except that the pretreatment can be less extensive since the anaerobic microorganisms are, to some extent, themselves able to degrade both crystalline cellulose structures and hemicelluloses (Kaparaju *et al.*, 2009; Petersson *et al.*, 2007). From a broad perspective, the requirements of the pretreatment are (1) to improve the accessibility of the enzymes to cellulose and hemicelluloses, (2) to avoid degradation or loss of the carbohydrates, (3) to avoid the production of potential inhibitors, (4) to be cost and energy efficient, and (5) to have as low an impact as possible on the environment (Menon and Rao, 2012; Taherzadeh and Karimi, 2008).

Disintegrating pretreatment

Size reduction

The lignocellulosic biomass requires preliminary mechanical processing for size reduction. This results in a larger specific surface area of the biomass and in a decreased degree of polymerization. The various pretreatments require the size reduction of the collected feedstock to different extents. With some, more easily degraded lignocellulosic-rich material, size reduction or milling can serve as the only necessary pretreatment. The size reduction of rice straw has been shown to increase the production of CH_4 although the best results were obtained with a combination of milling and other pretreatment methods (Zhang and Zhang, 1999). The milling of different municipal solid waste fractions proved to enhance CH_4 yields by 5–25% (Delgenés *et al.*, 2002; Palmowski and Muller, 1999). The drawback of extensive milling, however, is the high energy consumption and therefore the high cost of the method, which makes it inappropriate in many cases (Johnson and Elander, 2008). A somewhat contradictory finding emerged from batch AD experiments of sunflower oil cake within the particle size range of 0.355–2.0 mm. The largest size fraction (1.4–2.0 mm) produced the highest CH_4 yield, 213±8 ml CH_4/g VS, smaller fractions yielding only 182–190±4 ml CH_4/g VS added (De la Rubia *et al.*, 2011).

Ultrasonication

The application of ultrasound is a mechanical method that disintegrates and destroys the biomass, e.g. sludge particle from wastewater treatment. The efficiency of ultrasound pretreatment is affected by the frequency, the time, the energy level and the characteristics of the sludge. The treatment acts by disrupting the microbial cell structure and releasing the internal cellular material from the cells. When used in full-scale sewage sludge plants, ultrasonication led to an increase of 50% in biogas production (Yeow and Peng, 2012). However, a combined alkaline and ultrasound pretreatment of thickened pulp mill waste activated sludge did not improve the accumulated CH_4 yield, though the initial digestion rate was improved substantially (Park *et al.*, 2012).

Thermal pretreatment

The most important processes in this group include: steam explosion, catalysed (SO_2 or CO_2) steam explosion, NH_3 fibre explosion (AFEX), liquid hot water and microwave-chemical pretreatment (Menon and Rao, 2012).

Steam explosion

The biomass is treated with high-pressure saturated steam, and the pressure is then suddenly reduced, which causes the materials to undergo an explosive decompression. This high-temperature process leads to hemicellulose degradation and lignin transformation, thereby improving lignocellulose hydrolysis. A combination of thermal and chemical pretreatments usually gives a better result than either individual pretreatment (Menon and Rao, 2012).

Ammonia fibre explosion (AFEX) and ammonia recycle percolation (ARP)

AFEX is a physicochemical pretreatment process in which lignocellulosic biomass is exposed to liquid NH_3 at high temperature and pressure for a period of time, and the pressure is then suddenly reduced. The AFEX process is similar to that of steam explosion. Because of the structural disruption, AFEX increases the accessible surface area and improves the wettability of the treated biomass. An advantage of AFEX is that the NH_3 used during the process can be recovered and reused (Zuo *et al.*, 2012). In ARP, aqueous NH_3 passes through the biomass at elevated temperatures. The NH_3 is then separated and recycled. Under these conditions, aqueous NH_3 reacts primarily with lignin, causing its depolymerization and cleaving lignin–carbohydrate linkages.

Microwave-chemical pretreatment
The combination of microwave and chemical pretreatment has been found to be more effective than conventional heating and chemical pretreatment; it accelerates the reactions during the pretreatment process (Menon and Rao, 2012). Cheng and Liu (2010) developed a microwave-assisted alkaline pretreatment method for the enhancement of biogas production from herbal-extraction process residues.

Liquid hot water pretreatment
In liquid hot water pretreatment, pressure is utilized to maintain water in the liquid state at elevated temperatures. The biomass undergoes high-temperature cooking under these conditions. Application of this pretreatment led to an increased biogas yield from winter rye straw (Petersson *et al.*, 2007). In the experiments of Kaparaju *et al.* (2009) biogas energy output from pretreated wheat straw was approximately 10% higher than that of untreated wheat straw.

Chemical pretreatments
These technologies were originally developed and have been extensively used in the paper industry for delignification of cellulosic materials to produce high-quality paper products. Chemical pretreatment is the most extensively studied approach among pretreatment categories. The various techniques include acid, alkali, organic acid, pH-controlled liquid hot water and ionic liquids treatments.

Acid pretreatment
Acid pretreatment involves the use of concentrated or diluted acids to disrupt the rigid structure of the lignocellulosic material. Most commonly used is dilute sulfuric acid (H_2SO_4), though traces of H_2SO_4 may remain in the treated material, resulting in an elevated of H_2S content of the produced biogas, as an undesired by-product of this pretreatment procedure. Other acids have also been studied, such as hydrochloric acid (HCl), phosphoric acid (H_3PO_4) and nitric acid (HNO_3). Acid pretreatment (removal of hemicellulose) followed by alkali pretreatment (removal of lignin) results in relatively pure cellulose.

Alkaline pretreatment
In these methods, bases such as sodium, potassium, calcium or ammonium hydroxide are used. Alkali metal hydroxides degrade the ester and glycosidic bonds, resulting in alterations of the lignin structure, cellulose swelling, partial decomposition of the cellulose, and partial solvation of the hemicellulose. A neutralizing step to remove the lignin and inhibitors (salts, phenolic acids, furfural and aldehydes) is required before the subsequent enzymatic hydrolysis. Alkaline pretreatment is more effective on hardwood and agricultural residues than on softwood. Solid-state anaerobic digestion of 5.0% NaOH-pretreated corn stover produced 37% more biogas than the untreated material (Zhu *et al.*, 2010). Following the alkaline pretreatment of softwood spruce and hardwood birch at different temperatures and subsequent enzymatic hydrolysis, the AD of the pretreated materials resulted in 0.31 CH_4/g volatile solids (VS) compared with 0.231 CH_4/g VS for untreated birch, and 0.21 l CH_4/g VS as compared with 0.031 CH_4/g VS for untreated spruce (Mohsenzadeh *et al.*, 2012). Similar positive results have been reported from the use of various lignocellulosic feedstocks under various conditions (He *et al.*, 2008; Himmelsbach *et al.*, 2010; Monlau *et al.*, 2012; Salehian *et al.*, 2013; Xie *et al.*, 2011; Zheng *et al.*, 2009).

Oxidative pretreatment
An oxidizing compound such as hydrogen peroxide or peracetic acid is added to the biomass, and in the wet oxidation method O_2 is also delivered into the pretreatment reactor at temperatures up to 200°C and pressures up to 1.5 MPa (Johnson and Elander, 2008). It has been demonstrated that

addition of hydrogen peroxide was most effective at pH > 10; no significant delignification was found below this pH (Gould, 1984).

Green solvents (ionic liquids)

A new class of solvents has recently emerged: ionic liquids/green solvents. These solvents, which are often fluid at room temperature, consist entirely of ionic species. In general, ionic liquids comprise of a salt in which one or both of the ions are large, and the cation has a low degree of symmetry. Ionic liquids come in two main categories: simple salts (consisting of a single anion and cation) and binary ionic liquids (salts where equilibrium is involved). 3-N-methylmorpholine-N-oxide (NMMO) can dissolve cellulose and efficiently decrease its crystalline structure. Moreover, it is an environmentally friendly solvent, more than 98% of it can be recovered, it needs no chemical derivatization and toxic waste pollutants are not produced. When Teghammar *et al.* (2012) pretreated softwood spruce (chips and milled), rice straw and triticale straw with NMMO prior to AD, the pretreatment improved the CH_4 yields from these lignocellulosic materials by a remarkable 400–1200%. Almost complete digestion of high-crystalline cellulose was achieved by pretreatment with NMMO for 15 days (Jeihanipour *et al.*, 2005). Following the pretreatment of oil palm empty fruit bunch with NMMO, Purwandari *et al.* (2013) observed a 48–167% increase in biogas productivity was relative to the untreated material.

Biological pretreatment

Biological pretreatments have mainly been investigated with lignin-degrading fungi. A group of basidiomycetes called white-rot fungi are known to initiate the degradation of lignin, leaving most of the hemicellulose and cellulose unchanged. The most widely studied white-rot organism is *Phanerochaeta chrysosporium*, one of the holobasidimomycetes (Ulmer *et al.*, 1983). *Trichoderma reesei* and its mutants, likewise effective lignocellulose degrading fungi, are also employed for the commercial production of hemicellulases and cellulases. Many fungi excrete ligninolytic enzymes, such as lignin peroxidase, manganese peroxidase and laccase (Isroi *et al.*, 2011). In the routine pretreatment procedure, the biomass is inoculated at room temperature with the fungi and is left for several weeks. The advantages of these biological pretreatment methods include low energy consumption and the freedom from environmentally harmful chemicals. There are drawbacks, however, such as the partial degradation of the cellulose and hemicelluloses together with the lignin during the long incubation (Bruni *et al.*, 2010). Unfortunately, in most of the experimental systems enriched mixtures of fungi or fungi and bacteria were used that were poorly identified (Yuan *et al.*, 2011; Zhang *et al.*, 2011; Zhong *et al.*, 2011). The mechanism of the apparently efficient pretreatment is therefore impossible to decipher and the experiments generally cannot be repeated. Among the few exceptions, a mesophilic lignocellulolytic microbial consortium was established by successive inoculation (Yan *et al.*, 2012). The consortium was subjected to diversity analysis by ARDRA (see section 'Amplified ribosomal DNA restriction analysis (ARDRA)' above) of the 16S rDNA clone library. The most abundant taxonomic groups represented in the clone library were Firmicutes (5.96%), Bacteroidetes (40.0%), Deferribacteres (8.94%), Proteobacteria (16.17%), Lentisphaerae (2.13%), Fibrobacteraceae (1.7%) and uncultured bacteria (25.1%).

When the leaves of *Prunus serrulata* were pretreated with lignolytic *Pleurotus pulmonarius*, the CH_4 production increased 5-fold (Takacova *et al.*, 2012). Isolated strains of anaerobic fungi have been added directly into the AD process (Prochazka *et al.*, 2012) and the increase in biogas production has attained 22% (*Anaeromyces* sp.). The biogas yield increased throughout the semi-continuous experiments, although the concentration of the fungal biomass plunged very rapidly.

Inhibitors

Substances formed during the pretreatment of lignocellulosic feedstock inhibit the enzymatic hydrolysis and the subsequent fermentation steps. Ingredients that inhibit the microorganisms include phenolic compounds and other aromatics, aliphatic acids, furan aldehydes, inorganic ions and alcohols or other fermentation products (Jing *et al.*, 2009; Jonsson *et al.*, 2013).

Aromatic compounds

A variety of different phenolic compounds and other aromatics are formed during the degradation of lignocellulose. Carboxylic acids contribute to a shift toward an acidic pH. The effects of phenolics and other aromatic compounds which may inhibit both microbial growth and product yield are very variable and can be related to specific functional groups. Enzymes such as laccases and peroxidases oxidize phenols to radicals, which are then coupled to larger molecules and become less toxic to fermenting microbes (Jonsson *et al.*, 2013).

Aliphatic acids

Lignocellulose hydrolysis products contain acetic acid, formic acid and laevulinic acid. Acetic acid is formed primarily by the decomposition of hemicellulose, while formic acid and laevulinic acid are released in acid-catalysed degradation products from polysaccharides. Formic acid is a breakdown product of furfural and 5-hydroxymethylfurfural (HMF), while laevulinic acid is formed by the degradation of HMF. The concentration of aliphatic acids varies strongly depending on the feedstock and the severity of the pretreatment. Feedstock with highly acetylated xylan content, typically agricultural residues and hardwood, yields more aliphatic acids than does softwood (de Souza, 2013).

Furan aldehydes

The decomposition of pentose and hexose sugars lead to the furan aldehydes, HMF and furfural, which are commonly found among lignocellulose hydrolysis products (Jonsson *et al.*, 2013).

Inorganic compounds

Inorganic ions present among lignocellulose hydrolysis products originate from the feedstock, from the chemicals added during the pretreatment, conditioning or hydrolysis, and possibly from process equipment (de Souza, 2013).

Cellulolytic co-cultures

Cellulolytic organisms can be found in diverse environments in nature. A detailed molecular and microbiological analysis of these habitats can lead to more efficient mapping of the species and consortia. The process of AD of rice straw in paddy soil was dominated by the genus *Clostridium* within the domain Bacteria. Other taxonomic groups such as Proteobacteria and the genus *Bacillus* were present, but not very abundantly (Weber *et al.*, 2001).

Sequence analysis of the 16S rRNA gene has been utilized to study the bacterial diversity of a forest soil and of the same soil enriched with cellulolytic bacteria (Talia *et al.*, 2012). The investigation revealed high bacterial diversity in the native soil sample, with the detection of at least 10 phyla, among them Actinobacteria, Proteobacteria and Acidobacteria accounting for more than 76% of all sequences. In the enriched samples, members of Proteobacteria were the most frequently represented. Cellulolytic taxa were additionally classified, such as *Acidothermus, Micromonospora, Streptomyces, Paenibacillus* and *Pseudomonas*, which indicates that this ecosystem could be an attractive source of novel enzymes for cellulose degradation.

It is noteworthy that cellulolytic organisms in co-cultures with methanogens may achieve more effective cellulose degradation. The effects of syntrophic co-culturing on the decomposition of a cellulosic material in a co-culture of *Clostridium clariflavum* strain CL-1 and the hydrogenotrophic methanogen *Methanothermobacter thermautotrophicus* strain DH under thermophilic conditions were investigated and compared with the efficacy of the monoculture of strain CL-1 (Sasaki *et al.*, 2012). The degradation efficiency and cell density of strain CL-1 were 2.9- and 2.7-fold higher in the co-culture than in the monoculture, respectively after 60 h of incubation. The decomposition product analyses confirmed H_2 scavenging by the hydrogenotrophic methanogen strain DH as the responsible partner, to shift the metabolic pathway to the favourable acetate production in the co-culture. Other studies disclosed that the presence of the methanogens *Methanobacterium arboriphilus, Methanobacterium bryantii* or *Methanobrevibacter smithii* boosted the level of cellulose fermentation by 5 to 10% in co-cultures of several genera of anaerobic fungi (Marvin-Sikkema *et al.*, 1990). *Neocallimastix* sp. strain L2 grown together with methanogens also improved the rate of cellulose fermentation relative to the pure cultures of the fungus (Marvin-Sikkema *et al.*, 1990). The presence of methanogens brought about a shift in the fermentation product balance to more acetate and less lactate, succinate and ethanol. Similar observations were made by Mountfort *et al.* (1982) in the fermentation of cellulose by a rumen anaerobic fungus in the presence of *Methanobrevibacter* sp. strain RAl and *Methanosarcina barkeri* strain 227. During the time course of cellulose fermentation by the triculture, H_2 was not detected ($< 1 \times 10^{-5}$ atm or < 0.001 kPa) and only acetate exhibited a transitory accumulation, which was higher than the total acetate yield in the fungus monoculture. The maximum rate of cellulose degradation in the triculture (3 mg/ml per day) was higher than previously reported for bacterial co-cultures and within 16 days the degradation was complete. The triculture was used successfully in the production of biogas from the plant fibrous material sisal (fibre from the leaves of *Agave sisalona* L.) and barley straw.

2.6 Future trends: algae

Several directions for further research and development activities in the foreseeable future have been discussed and pointed out in the previous sections, together with the introduction of the state-of-the-art. As a largely unexplored area, the use of algal biomass for biogas production will be considered here, and the coupling of other emerging fields of algal biotechnology with biogas production will be emphasized. Clearly, there are huge energetic and biorefinery opportunities in the conversion of solar energy by photosynthetic microorganisms. Hence, the interest in photosynthetic living creatures is growing worldwide. Some of the photosynthetic organisms are used for food and feed production. Energy production routes that do not compete with food and feed production should be preferred. Algae are a large and diverse group of simple, typically autotrophic organisms, ranging from unicellular to multicellular forms. Macroalgae are algae which form a multicellular thallus, i.e. seaweeds (Anderson, 2005; Richmond, 2004). Microalgae, as the name suggests, are small, microscopic algae (5–50 μm) and oxygenic photosyntetic bacteria, such as cyanobacteria, which grow profusely in both salt water or fresh water (Richmond, 2004). Algal biomass production has several advantages over conventional, field-grown energy corps. Microalgae are able to double their biomass within a day so that they produce many times more biomass and lipids per hectare than any kind of terrestrial plant. A very small land area is needed to cultivate algae and it can be arable or marginal land, which further decreases the competition for agricultural land usage and the production price. Macroalgal plantations could easily be sea-based where cultivation takes place in the water column above the seabed. Algae are not demanding creatures, usually not requiring herbicides or pesticides or special feed additives. They grow in an aquatic medium, but use less water than terrestrial crops (Rodolfi *et al.*, 2009). Biogas from algae is high in CH_4 content (≈60–70%) and this is an additional advantage. Moreover, CO_2 is sequestered by these organisms so that the CO_2 released during biogas combustion is in fact recycled through photosynthesis. The contribution to the net greenhouse gas emission can therefore be zero if biogas is produced from algae, as CO_2 is consumed during the photosynthetic growth of their next generation. Algae producing bio-H_2 or biodiesel or other valuable products before biogas generation are ideal for various biorefinery applications (Lakaniemi *et al.*, 2011; Sialve *et al.*, 2009). Among the disadvantages of algal biomass production and exploitation are the relatively high cost of sophisticated photobioreactors and the considerable energy requirements of the cultivation and harvesting of the algae (Uduman *et al.*, 2010). Although practically none of these considerations are applicable if algae are used for biogas production, such energy generation systems should be carefully designed and constructed.

2.6.1 Macroalgae for biogas production

The first experiments on bioenergy production from macroalgae were carried out as part of the US Ocean Food and Energy Farm Project initiated in 1973 and lasting for about 10 years (North, 1987; Wilcox, 1977). Within the framework of this project, an ocean farm was constructed and *Macrocysits* species were cultivated. Farming this seaweed presented several technical challenges, but finally the biogas production from macroalgal biomass furnished remarkably positive results (Kelly and Dworjanyn, 2008). This and other results clearly demonstrated the significant advantages of macroalgae over other sources of biomass for biofuels (Brenam and Owende, 2010). Although the environmental impact of large-scale macroalgae production on sea biodiversity is currently unknown, it does not seem to be a significant danger. Macroalgal plantations, like other marine environments such as corals and seagrasses, can become biodiversity hot spots, providing a sheltered habitat for a wide range of organisms, including fish and birds (Christine *et al.*, 1998).

Macroalgal production

At presently more than 100 species of macroalgae are used for food, the production of unique chemicals, in medicines or as fertilizers (Ortiz *et al.*, 2006). Several macroalgae are under large-scale cultivation in eastern Asia, although domestication of the macroalgae began only in the 20th century. The world's largest producer of cultivated seaweed is China; macroalgae are mostly grown on long-line systems and the hatchery-produced seedlings are transplanted to sea. In 2010 the global production of the large brown *Laminaria japonica* (also known as haidai), alone was 5.14×10^6 tons with a value of 3×10^9 USD. It is grown primarily for food, but also for iodine and alginates (Duarte *et al.*, 2005; Ortiz *et al.*, 2006). The selective breeding of Laminaria species resulted in faster growth and a higher iodine and a lower water content relative to the natural population (Field *et al.*, 2008; Wu and Pang, 2006). Adjacent to salmon cages in Scottish sea lochs, native *Saccharina latissima* macroalgae are farmed producing 200 tons wet weight/ha/year ($\approx$20 tons dry weight per year). Macroalgal crops contribute to the fuel supply, although the biomass has to be collected from very large areas. Seaweed cultivation on a large scale will require the development of more mechanized technologies for planting and harvesting, and more specialized vessels and work boats are needed than are currently employed (Duarte *et al.*, 2009).

Anaerobic digestion of seaweed

Macroalgae contain less cellulose and lignin than the terrestrial plant biomass crops and therefore undergo more complete hydrolysis during AD. Tests for anaerobic fermentation and biogas production furnished encouraging results (Adams *et al.*, 2011; Cecchi *et al.*, 1996; Chynoweth *et al.*, 1987; Matsui *et al.*, 2006). Seaweeds have low nitrogen content, and can therefore be co-digested with other, N-rich substrates such as waste food or agricultural slurries (Cecchi *et al.*,1996). Biogas yields depend on a wide range of parameters, such as inoculum, digester system configuration and feed stock composition, and hence the fermentation conditions should be selected carefully for maximum energy output.

The potential of CH_4 production from macroalgae has been tested in a commercial scale four-stage anaerobic digester (Nkemka and Murto, 2010). The CH_4 yield was about twice the production from average maize fermentation (Shilton and Guieysse, 2010), which suggests that at a regional level large-scale macroalgal culture for biofuels offers real promise for coastal communities. There are several possibilities for the further optimization of biogas yields, such as co-digestion or manipulation of the microbial composition of the inoculum.

2.6.2 Biogas production from microalgae

Microalgae have received considerable attention as alternative feedstock for biofuel production and other exploitations (Becker, 1994). In view of the enormous biodiversity of microalgae and the recent developments in genetic engineering, this group of organisms is one of the most promising sources for a new generation of biofuels. As photosynthetic organisms, microalgae contain chlorophyll and carotenes that could be used in the food and cosmetic industries (Spolaore *et al.*, 2006). Some strains produce surplus lipids and polysaccharides, which can be extracted or converted into biofuels. Microalgae were employed for bio-H_2 production (Hankamer *et al.*, 2007; Melis *et al.*, 2000; Melis and Happe, 2001), they provide excellent lipids for biodiesel applications (Hu *et al.*, 2008; Scott *et al.*, 2010) and the residual biomass can be utilized for bio-CH_4 generation (Vergara-Fernandez *et al.*, 2008). The bottlenecks are due to the relatively high costs of biomass production and downstream processing.

Cultivation of microalgae

An effective system should meet the following criteria: (1) optimal gas–liquid transfer, (2) effective illumination area, (3) easy to operate, (4) low capital and production costs, (5) low contamination level, (6) minimal land area requirement and (7) high biomass production yield (Xu *et al.*, 2009).

Although closed systems are currently favoured (Singh and Gu, 2010), the lower costs of construction and maintenance of open ponds are attractive for biogas production, where the microbiological purity of the culture is not critical.

Open ponds

Open ponds have many different shapes and forms, each with certain advantages and disadvantages. Raceway ponds, circular ponds, shallow large ponds and partially closed ponds are the preferred arrangements on an industrial scale. The system usually includes a paddle wheel to avoid sedimentation, and CO_2 is delivered at the bottom of the vessel from the atmosphere or from a CO_2 gas tank (Benemann *et al.*, 1980; Chisty, 2008; Lam and Lee, 2012; Stephenson *et al.*, 2010). Owing to the lack of the sophisticated process control involved with open systems, the local climate determines biomass productivity. Additional growth-limiting parameters are the light intensity, water evaporation, pH and dissolved O_2 and CO_2 concentrations. Contamination by a predator organism or fast-growing bacteria is another issue of concern (Blanco *et al.*, 2007; Hase *et al.*, 2000). *Dunaliella salina* is one of the most successful species that have been used for the production of carotenoids because the high salt tolerance of this alga protects the culture in an open pond system (Lee, 2001).

Closed systems

There are numerous designs of closed photobioreactors, such as air-lift tubular, flat plate, vertical column, etc. The main advantage of farming microalgae in closed photobioreactors is sterility and the maintenance of optimal growth conditions, but the costs of operation are high (Chisti, 2007).

Tubular reactors are considered to be more suitable for outdoor cultivation. The tubes can be arranged in various configurations, depending on the specific needs (Harun *et al.*, 2010; Ugwu *et al.*, 2008). The vertical design allows greater mass transfer and a decrease of energy consumption, while the horizontal reactor is easier to scale up, although it usually requires a large area (Hu *et al.*, 1996; Molina Grima *et al.*, 2003). An oscillatory flow reactor consists of equally spaced plate baffles (Stephenson *et al.*, 2010); short retention time is attained due to the superior mixing intensity.

Dewatering of microalga cultures

The first step in the downstream procedure is the separation of the microalgae from the bulk water. The harvesting of microalgae is a challenging process because of the small size of the particles (5–50 μm) suspended in water. Several methods have been applied.

Flocculation

Microalgae carry a negative charge, which prevents them from self-aggregation. The surface charge can be neutralized by the addition of chemicals and this brings about the coagulation of the algae. The frequently used flocculants include $FeCl_3$ (ferric chloride), $Al_2 (SO_4)_3$ (aluminium sulfate), and $Fe_2 (SO_4)_3$ (ferric sulfate). Flocculation has been tested successfully on microalgae (L. Chen *et al.*, 2008; Knuckey *et al.*, 2006; Molina Grima *et al.*, 2003). Other flocculants are polyelectrolytes, which are cationic polymers. Polymeric flocculants have the advantage of linking cells together chemically. Polymeric flocculation has also been investigated (Divakaran and Pillai, 2002; Tenney *et al.*, 1969). In an innovative solution, magnetite is blended into the growing algal culture. The algae enclose these iron oxide particles, which are only a few micrometres in size. The resulting mixture can then be easily extracted using a magnet (http://www.siemens.com/innovation/en/news/2009/using-magnets-to-harvest-algae-for-biogas-plants.htm).

Centrifugation
Centrifugation is definitely the most straightforward method for harvesting algal cells, but is also the most expensive. During centrifugation, the shear forces occurring in the course of spinning can disrupt the cells, and this can be a problem if the aim is to use the intact algae for a subsequent downstream procedure.

Filtration
This is perhaps the most competitive of the harvesting options. There are many different forms and devices for filtration, such as microfiltration, ultrafiltration, pressure filtration, dead end filtration, tangential flow filtration, and vacuum filtration (Danaquah *et al.*, 2009). Depending on the method applied the extensive running cost could be prohibitive (Ugwu *et al.*, 2008).

Anaerobic fermentation of microalgae

Algae have a number of advantages relative to higher plants (Stephens *et al.*, 2010). The typical algal biomass contains proteins, carbohydrates, lipids and other valuable components, e.g. pigment, anti-oxidants, fatty acids, vitamins, etc. The protein and carbohydrate contents in some strains reach up to 50% of its dry weight (Becker, 1994; Singh and Gu, 2010). In an early study (Golueke *et al.*, 1956), a comparison was made between sewage sludge and green algae (*Scenedesmus* sp. and *Chlorella* sp.). Following such pioneering experiments, only a few investigations have dealt with the anaerobic fermentation of microalgae (Hernández and Córdoba, 1993; Hughes *et al.*, 2012; Keenan, 1977) until recently. When various fresh and salt water alga strains were compared under mesophilic conditions (Mussgnug *et al.*, 2010), the biogas potential proved to depend strongly on the species and on the cell disruption method (Mussgnug *et al.*, 2010). Drying of the algal biomass resulted in a general decrease in biogas production by around 20%. It was a noteworthy feature that the CH_4 content of the biogas from the microalgae was 7–13% higher than that from maize silage (Mussgnug *et al.*, 2010).

Various research strategies on CO_2 sequestration have been proposed, including physical, chemical and biological methods. The most efficient biological method is the conversion of CO_2 to biomass by algae (Converti *et al.*, 2009; Holm-Nielsen *et al.*, 2009; Kao et al, 2012). Microalgae fix CO_2 10–50 times faster than do terrestrial plants (Wang *et al.*, 2008).

Biogas from alga-biorefinery

The industrial-scale cultivation of microalgae is still in its infancy and the algal biomass is therefore expensive at present. Two strategic approaches can be envisaged to overcome the problem of the biomass production costs. In one option, unsophisticated, simple cultivation technology is selected, e.g. an open pond raceway, natural lakes or the sea. These cultures, with relatively low productivity, may become infected yielding biomass that is not suitable for the extraction of high-value products such as biodiesel or complex chemicals for the food, cosmetics or pharmaceutical industries. Nevertheless, this biomass is superb material for biogas production, where otherwise mixed organic substrates are used. The second option to compensate for the cultivation costs is one or other of the various biorefinery concepts, where algae are used to produce valuable substances. For these applications, advanced and sterile photobioreactors should be built in order to meet the necessary quality standards. Numerous approaches (Cherubini, 2010) have been proposed and many have been tested on a small industrial size. Unfortunately, utilization of the wet biomass left over after the extracting platform of specific chemicals is not considered as important part of most of these concepts. The generation of valuable bio-CH_4 is recommended to be included in these schemes as a highly efficient disposal method with the associated energy or biofuel and fertilizer production.

Coupled biodiesel and biogas production has already been investigated in some detail (Sialve *et al.*, 2009), and it was confirmed that anaerobic digestion is a key process that can solve the treatment of microalgal biodiesel residues in an economically and energetically feasible way. Some potential bottlenecks have also been identified. The high cellular protein content results in NH_3 accumulation, which can lead to toxicity in the biogas reactor, and the presence of sodium can also affect the digester performance, especially when marine species are used.

Chlamydomonas reinhardtii is in the spotlight because of the smart physiological switch to produce H_2 from its own storage materials amassed during photosynthesis (Melis *et al.*, 2000; Melis and Happe, 2001). Its biomass is also an outstanding candidate for biogas production. Harvesting the gaseous H_2 product does not require energy-consuming downstream processing (Mussgnug *et al.*, 2010). The biogas yield increased to a notable 123% relative to the fresh algal biomass when the biomass after the H_2 production cycle was used. The accumulation of storage compounds with high biogas potential, i.e. starch and lipids, is facilitated in response to the induction of the H_2 production cycle (Timmins *et al.*, 2009). Biogenic H_2 and CH_4 production was examined using *Chlorella vulgaris* (a fresh water microalga) and *Dunaliella tertiolectca* (a highly salt-tolerant alga). Low levels of H_2 were produced by anaerobic enrichment cultures, but the H_2 was subsequently consumed even in the presence of 2-bromoethanesulfonic acid, an inhibitor of methanogens (Lakaniemi *et al.*, 2011).

Preparation of alga for biogas

The only task prior to adding microalgae to a biogas digester is to make the algal cells available for the degrading bacteria. The cell walls of algae are biodegradable, but the process is time-consuming. Some species, e.g. *C. vulgaris* or *C. reinhardtii*, have relatively thin cell walls but others have developed thick ones (e.g. *Scenedesmus*). There are several methods for algal cell disruption, including ultrasonication, bead beating, microwaves (at 100°C), osmotic shock and autoclaving, with diverse results (Parabakaran and Ravindran, 2011). Heat treatment of a *Scenedesmus* sp. biomass at 90°C for 1 h increased the CH_4 production about 3-fold (Gonzales-Fernåndez *et al.*, 2013). Despite the enhanced CH_4 production, however, the anaerobic biodegradability of the *Scenedesmus* biomass remained low (32%). Sonication for 15 min also significantly increased the concentration of dissolved carbohydrates from microalgae (Jeon *et al.*, 2013), indicating that various cell disruption methods should be employed.

2.7 Conclusions

It is clear from the vast scientific and practical endeavours that biogas production from biomass is an important technological possibility among the renewables. Unfortunately, it is not valued everywhere as its capabilities deserve. The relative simplicity of the basic technology, stable operation on both the small and the large scale, a wide selection of available substrate biomass feedstock, and versatile utilization of the product biogas elevate this technology to premium ranking among the energetic routes of biomass conversion.

Less developed countries should particularly favour this route because it can be implemented at low cost, and local materials are used for both construction and as input source even in rural areas. In the regions where more advanced technologies are available, biogas is primarily employed as a means through which to control environmentally troublesome waste-disposal problems.

This is the only renewable biotechnology that is able to couple the biotechnological remedying of organic waste with the production of a useful energy carrier, bio-CH_4. Novel approaches should therefore be applied to meet the requirements of densely populated industry-based societies. Although the majority of the large-scale, industrial biogas facilities produce 'green' electricity today, alternative utilization of the CH_4-rich gas should be considered. Purification to bio-CH_4 offers the possibility of using this energy carrier of biological origin in all areas of everyday life, where fossil and increasingly more expensive natural gas is consumed today. The technologies required for the efficient purification of biogas to furnish bio-CH_4 are available and waiting for widespread utilization.

Energy generation from biomass along the biogas route is still more costly than mining the dwindling natural gas or oil reserves, but the economic gap is shrinking and ongoing research should aim to make biogas technologies more feasible and competitive than they are today. Apart from minor polishing in the methodology, there seem to be two significant areas for rapid progress. First, the control and biotechnological engineering of the composition of the biogas-producing microbial communities should be implemented by using the rapidly progressing 'omics' approaches, which allow an understanding of the delicate and complex behaviour of these multifarious communities. These methods offer a new avenue for the advance of biogas technology that was not available until recently. The second and almost inexhaustible territory is the introduction of a wide range of novel organic substrates into biogas digesters. Materials and waste streams rich in protein and lignocellulose, should be utilized for biogas production more efficiently, and algae stand out as highly promising biogas substrates, which do not compete with food production and can be cultivated and utilized resourcefully.

Acknowledgements

This work was supported by EU projects HUSRB/1002/214/041 IPA, HURO/1001/193/2.2.2. CBC, and IEE/10/235 SI2.591589 GreenGasGrids. Funding from TÁMOP 4.2.4.A/2-11-1-2012-0001 and TÁMOP-4.2.2/B-10/1-2010-0012 are gratefully acknowledged.

Web resources

Biogas purification

Water scrubbing technology
http://www.showtimesdaily.com/fleetsfuels/cow-product-cow-power

PSA technology
http://students.chem.tue.nl/ifp24/techn_psa.htm

Cryogenic technology
http://students.chem.tue.nl/ifp24/techn_cryo.htm

Ferrous purification technology
http://www.eisenbau-heilbronn.de/desulfurization.html

Molecular techniques

Quantitative PCR
http://www.promega.com/~/media/images/resources/figures/8400-8499/8477ta.jpg?la=en

TaqMan probe method
http://upload.wikimedia.org/wikipedia/en/0/07/Taqman.png

FISH method
http://www.nature.com/nrmicro/journal/v6/n5/images/nrmicro1888-f1.jpg

DGGE method
http://upload.wikimedia.org/wikipedia/en/thumb/6/68/Step-by-step_procedure_of_using_DGGE_analysis_in_microbiology.pdf/page1-800px-Step-by-step_procedure_of_using_DGGE_analysis_in_microbiology.pdf.jpg

T-RFLP method
http://upload.wikimedia.org/wikipedia/en/thumb/c/c5/Step-by-step_procedure_of_using_T-RFLP_analysis_in_microbiology.pdf/page1-776px-Step-by-step_procedure_of_using_T-RFLP_analysis_in_microbiology.pdf.jpg

Anaerochip
http://ars.els-cdn.com/content/image/1-s2.0-S0167701209003017-gr2.jpg

Shotgun sequencing
http://www.nature.com/nature/journal/v406/n6797/images/406799aa.2.jpg

Next generation sequencing
http://www.oxbridgebiotech.com/review/research-and-policy/whats-so-special-about-next-generation-sequencing/
http://www.nature.com/nrg/journal/v11/n1/fig_tab/nrg2626_F3.html

Third-generation sequencing
http://www.nature.com/nbt/journal/v28/n5/fig_tab/nbt0510-426_F1.html
http://www.nature.com/msb/journal/v9/n1/full/msb201261.html

Proteomics
http://www.nature.com/nrm/journal/v5/n3/box/nrm1336_BX2.html

Microbial lipid analysis
http://www.microbialid.com/images/pages/fatty_acid_materials_303.jpg

References

Abouelenien, F., Fujiwara, W., Namba, Y., Kosseva, M., Nishio, N., and Nakashimada, Y. (2010). Improved methane fermentation of chicken manure via ammonia removal by biogas recycle. Bioresour. Technol. *101*, 6368–6373.

Ács, N., Kovács, E., Wirth, R., Bagi, Z., Strang, O., Herbel, Zs., Rákhely, G., and Kovács, K.L. (2013). Changes in the archaeal microbial community when the biogas fermenters are fed with protein-rich substrates. Biores. Technol. *131*, 121–127.

Adams, J.M.M., Toop, T.A., Donninson, I.S., and Gallagher, J.A. (2011). Seasonal variation in *Laminaria digitata* and its impact on biochemical conversion routes to biofuels. Bioresour. Technol. *102*, 9976–9984.

Ahring, B.K. (ed.) (2003). Biomethanation I. Springer-Verlag, Berlin.

Ahring, B.K., Angelidaki, I., and Johansen, K. (1992). Anaerobic treatment of manure together with industrial waste. Water Sci. Technol. *30*, 241–249.

Alonso-Vicario, A., Ochoa-Gómez, J.R., Gil-Río, S., Gómez-Jiménez-Aberasturi, O., Ramírez-López, C.A., Torrecilla-Soria, J., and Domínguez, A. (2010). Purification and upgrading of biogas by pressure swing adsorption on synthetic and natural zeolites. Micropor. Mesopor. Mater. *134*, 100–107.

Al Seadi, T., Rutz, D., Prassl, H., Köttner, M., Finsterwalder, T., Volk, S., and Janssen, R. (2008). Biogas Handbook. Univ. Southern Denmark, Esbjerg.

Anderson, R.A. (2005). Algal Culturing Technologies. Elsevier Academic Press, Oxford.

Andreesen, J.R., Bahl, H., and Gottschalk, G. (1989). Introduction to the Physiology and Biochemistry of the Genus Clostridium. Minton NP & Clarke DC, New York.

Angelidaki, I., and Ahring, B.K. (1993). Thermophilic anaerobic digestion of livestock waste: the effect of ammonia. Appl. Microbiol. Biotechnol. *38*, 560–564.

Angelidaki, I., and Ellegaard, L. (2003). Codigestion of manure and organic wastes in centralized biogas plants: status and future trends. Appl. Biochem. Biotechnol. *109*, 95–105.

Angelidaki, I., and Sanders, W. (2004). Assessment of the anaerobic biodegradability of macropollutants. Rev. Environ. Sci. Biotechnol. 3, 117–129.

de Baere, L.A., Devocht, M., van Assche, P., and Verstraete, W. (1984). Influence of high NaCl and NH_4Cl salt levels on methanogenic associations. Water Res. *18*, 543–548.

Bagi, Z., Ács, N., Bálint, B., Horváth, L., Dobó, K., Perei, K.R., Rákhely, G., and Kovács, K.L. (2007). Biotechnological intensification of biogas production. Appl. Microbiol. Biotechnol. *76*, 473–482.

Banks, C.J., and Humphreys, P.N. (1998). The anaerobic treatment of a ligno-cellulosic substrate offering little natural pH buffering capacity. Water Sci. Technol. *38*, 29–35.

Banks, C.J., and Wang, Z. (1999). Development of a two phase anaerobic digester for the treatment of mixed abattoir wastes. Water Sci. Technol. *40/1*, 69–76.

Bayer, E.A., Belaich, J.P., Shoham, Y., and Lamed, R. (2004). The cellulosomes: multienzyme machines for degradation of plant cell wall polysaccharides. Annu. Rev. Microbiol. *58*, 521–554.

Becker, E.W. (1994). Microalgae: Biotechnology and Microbiology. Cambridge, Cambridge University Press.

Benemann, J., Kopman, B., Weissman, D., Eisenberg, D., and Goebel, R. (1980). Development of microalgae harvesting and high rate pond technologies in California. Algal Biomass 457.

Bentley, D.R., Balasubramanian, S., Swerdlow, H.P., Smith, G.P., Milton, J., Brown, C.G., Hall, K.P., Evers, D.J., Barnes, C.L., Bignell, H.R., *et al.* (2008). Accurate whole human genome sequencing using reversible terminator chemistry. Nature *456*, 53–59.

Bhattacharya, S.K., and Parkin, G.F. (1989). The effect of ammonia on methane fermentation processes. J. Water Pollution Control Fed. *61*, 55–59.

Blackall, L.L., Burrell, P.C., Gwilliam, H., Bradford, D., Bond, P.L., and Hugenholtz, P. (1998). The use of 16S rDNA clone libraries to describe the microbial diversity of activated sludge communities. Water Sci. Technol. *37*, 451–454.

Blanco, A.M., Moreno, J., Del Campo, J.A., Rivas, J., and Guerrero, M.G. (2007). Outdoor cultivation of lutein-rich cells of *Murellopsis sp.* in open ponds. Appl. Microbiol. Biotechnol. 73, 1259–1266.

Bordelanne, O., Montero, M., Bravin, F., Prieur-Vernat, A., Oliveti-Selmi, O., Pierre, H., Papadopoulo, M., and Muller, T. (2011). Biomethane CNG hybrid: A reduction by more than 80% of the greenhouse gases emissions compared to gasoline. J. Nat. Gas. Sci. Eng. *3*, 617–624.

Borja, R., Sanchez, E., and Duran, M.M. (1996). Effect of the clay mineral zeolite on ammonia inhibition of anaerobic thermophilic reactors treating cattle manure. J. Environ. Sci. Health *31*, 479–500.

Bowers, J., Mitchell, J., Beer, E., Buzby, P.R., Causey, M., Efcavitch, J.W., Jarosz, M., Krzymanska-Olejnik, E., Kung, L., and Lipson, D. (2009). Virtual terminator nucleotides for next-generation DNA sequencing. Nature Methods *6*, 593–595.

Brady, A., and Salzberg, S.L. (2009), Metagenomic phylogenetic classification with interpolated Markov models. Nature Methods *6*, 673–676.

Braun, R., Huber, P., and Meyrath, J. (1981). Ammonia toxicity in liquid piggery manure digestion. Biotechnol. Lett. 3, 159–164.

Braun, R., Brachtl, E., and Grasmug, M. (2003). Codigestion of proteinaceous industrial waste. Appl. Biochem. Biotechnol. *109*, 139–153.

Brenam, L., and Owende, P. (2010). Biofuels from microalgae – a review of technologies for production, processing, and extraction of biofuels and co-products. Renew. Sustain. Energy Rev. *14*, 557–577.

Brett, C.T. (2000). Cellulose microfibrils in plants: biosynthesis, deposition, and integration into the cell wall. Int. Rev. Cytol. *199*, 161–199.

Bruni, E., Jensen, A.P., and Angelidaki, I. (2010). Comparative study of mechanical, hydrothermal, chemical and enzymatic treatments of digested biofibers to improve biogas production. Bioresour. Technol. *101*, 8713–8717.

Bujoczek, G., Oleszkiewicz, J.A., Sparling, R.R., and Cenkowski, S. (2000). High solid anaerobic digestion of chicken manure. J. Agric. Eng. Res. *76*, 51–60.

Burgess, J.E., Parsons, S.A., and Stuetz, R.M. (2001). Developments in odour control and waste gas treatment biotechnology: a review. Biotechnol. Adv. *19*, 35–63.

Campos, E., Palatsi, J., and Flotats, X. (1999). Co-digestion of pig slurry and organic wastes from food industry. In: J. Mata-Álvarez, A. Tilche, and F. Cecchi eds. Proceedings of the Second Symposium on Anaerobic Digestion of Solid Waste, Barcelona, Spain, pp. 192–195.

Case, R.J., Boucher, Y., Dahllof, I., Holmstrom, C., Doolittle, F.W., and Kjelleberg, S. (2007). Use of 16s rRNA and *rpob* genes as molecular markers for microbial ecology studies. Appl. Environ. Microbiol. 73, 278–288.

Caspari, C., Oliver, E., Nganga, J., Ricci, M., van Horne, P., and Magdelaine, P. (2010). The poultry and egg sectors: Evaluation of the current market situation and future prospects. Availabe online: http://www.europarl.europa.eu/studies.

Cecchi, F., Pavan, P., and Mata-Alvarez, J. (1996). Anaerobic co-digestion of sewage sludge: application to the macroalgae from the Venice lagoon. Resour. Conserv. Recycl. *17*, 57–66.

Chen, L., Li, P., Liu, Z., and Jiao, Q. (2008). The released polysaccharide of the cyanobacterium *Aphanothece halophytica* inhibits flocculation of the alga with ferric chloride. J. Appl. Phycol. *6*, 327–331.

Chen, Y., Cheng, J.J., and Creamer, K.S. (2008). Inhibition of anaerobic digestion process: a review. Bioresour. Technol. *99*, 4044–4064.

Cheng, J., and Liu, B. (2002). Swine wastewater treatment in anaerobic digesters with floating medium. Trans. Amer. Soc. Agric. Biol. Eng. *45*, 799–805.

Cheng, X., and Liu, C. (2010). Enhanced biogas production from herbal-extraction process residues by microwave-assisted alkaline pretreatment. J. Chem. Technol. Biotechnol. *85*, 127–131.

Cherubini, F. (2010). The biorefinery concept: Using biomass instead of oil for producing energy and chemicals. Energy Conversion and Management 51, 1412–1421.

Chisti, Y. (2007). Biodiesel from microalgae. Biotechnol. Adv. *25*, 294–306.

Chisti, Y. (2008). Biodiesel from microalgae beats bioethanol. Trends Biotechnol. *25*, 126–131.

Christine, H., Fredriksen, S., and Rinde, E. (1998). Regrowth of kelp and colonization of epiphyte and fauna community after kelp trawling at the coast of Norway. Hydrobiologia *375–376*, 49–58.

Chynoweth, D.P., Fannin, K.F., and Srivastava, V.J. (1987). Biological gasification of marine algae. In: K.T. Bird and P.H. Benson, eds. Biomass Conversion Process for Renewable Resources. Developments in Aquaculture and Fisheries Science. Elsevier, Amsterdam, pp. 287–303.

Cigdem, Y., and Izzet, O. (2013). Effect of maize silage addition on biomethane recovery from mesophilic co-digestion of chicken and cattle manure to suppress ammonia inhibition. Energy Conversion and Management *71*, 92–100.

Cirne, D.G., Lehtomäki, A., Björnsson, L., and Blackhall, L.L. (2007). Hydrolysis and microbial community analysis in two-stage anaerobic digestion of energy crops. J. Appl. Microbiol. *103*, 516–527.

Clarke, J., Wu, H.C., Jayasinghe, L., Patel, A., Reid, S., and Bayley, H. (2009). Continuous base identification for single-molecule nanopore DNA sequencing. Nature Nanotechnol. *4*, 265–270.

Colberg, P.J. (1988). Anaerobic microbial degradation of cellulose, lignin, oligolignols and monoaromatic lignin derivatives. In: J.B. Zehnder, ed. Biology of Anaerobic Microorganisms.. J. Wiley and Sons Inc., pp. 333–372.

Collins, G., Woods, A., McHugh, S., Carton, M.W., and O'Flaherty, V. (2003). Microbial community structure and methanogenic activity during start-up of psychrophilic anaerobic digesters treating synthetic industrial wastewaters FEMS Microbiol. Ecol., *46*, 159–170.

Conrad, R., Phelps, T.J., and Zeikus, J.G. (1985). Gas metabolism evidence in support of juxtaposition of hydrogen-producing and methanogenic bacteria in sewage sludge and lake sediments. Appl. Environ. Microbiol. *50*, 595–601.

Conrad, R., Bak, F., Seitz, H.J., Thebrath, B., Mayer, H.P., and Schütz, H. (1989). Hydrogen turnover by psychrotrophic homoacetogenic and mesophilic methanogenic bacteria in anoxic paddy soil and lake sediment. FEMS Microbiol. Lett. *62*, 285–293.

Converti, A., Oliveira, R.P.S., Torres, B.R., Lodi, A., and Zilli, M. (2009). Biogas production and valorization by means of a two-step biological process. Biosour. Technol. *100*, 5771–5776.

Danaquah, M., Ang, L., Uduman, N., Moheimani, N., and Forde, G. (2009). Dewatering of microalgal culture for biodiesel production: exploring polymer flocculation and tangential flow filtration. J. Chem. Technol. Biotechnol. *84*, 1078–1083.

Daniel, R. (2005). The metagenomics of soil. Nat. Rev. Microbiol. *3*, 470–478.

Daniels, L. (1992). Biotechnological potential of methanogens. Biochem. Soc. Symp. *58*, 181–193.

Darke, H.L., Kusel, K., and Matthies, C. (2002). Ecological consequences of the phylogenetic and physiological diversities of acetogens. Antonie Van Leeuwenhook *81*, 203–213.

De la Rubia, M.A., Fernández-Cegrí, V., Raposo, F., and Borja, R. (2011). Influence of particle size and chemical composition on the performance and kinetics of anaerobic digestion process of sunflower oil cake in batch mode. Biochem. Eng. J. *58–59*, 162–167.

DeSantis, T.Z., Brodie, E.L., Moberg, J.P., Zubieta, I.X., Piceno, Y.M., and Andersen, G.L. (2007). High-density universal 16S rRNA microarray analysis reveals broader diversity than typical clone library when sampling the environment. Microb. Ecol. 53, 371–383.

Delgenés, J.P., Penaud, V., and Moletta, R. (2002). Pretreatments for the enhancement of anaerobic digestion of solid wastes. In: J. Mata-Alvarez, ed. Biomethanization of the Organic Fraction of Municipal Solid Wastes. IWA Publishing, pp. 201–228.

Demeestere, K., Smet, E., Van Langenhove, H., and Galbacs, Z. (2001). Optimization of magnesium ammonium phosphate precipitation and its applicability to the removal of ammonia. Environ. Technol. *22*, 1419–1428.

Ding, S.Y., Xu, Q., Crowley, M., Zeng, Y., Nimlos, M., Lamed, R., Bayer, E.A., and Himmel, M.E. (2008). A biophysical perspective on the cellulosome: new perspectives for biomass conversion. Curr. Opinion Biotechnol. *19*, 218–227.

Divakaran, R., and Pillai, V.N.S. (2002). Flocculation of algae using chitosan. J. Appl. Phycol. *14*, 419–422.

Doi, R.H. (2008). Cellulases of mesophilic microorganisms. Annu. N.Y. Acad. Sci. *1125*, 267–279.

Dolfing, J. (1988). Acetogenic dehydrogenations. In: J.B. Zehnder, ed. Biology of Anaerobic Microorganisms. J. Wiley and Sons Inc., pp. 417–468.

Dong, X., and Tollner, E.W. (2003). Evaluation of Anammox and denitrification during anaerobic digestion of poultry manure. Bioresour. Technol. *86*, 139–145.

Drancourt, M., Bollet, C., Carlioz, A., Martelin, R., Gayral, J.-P., and Raoult, D. (2000). 16S ribosomal DNA sequence analysis of a large collection of environmental and clinical unidentifiable bacterial isolates. J. Clin. Microbiol. *38*, 3623–3630.

Duarte, C.M., Middelburg, J.J., and Caraco, N. (2005). Major role of marine vegetation on the oceanic carbon cycle. Biogeosciences *2*, 1–8.

Duarte, C.M., Holmer, M., Olsen, Y., Soto, D., Marba, N., Guiu, J., Black, K., and Karakassis, I. (2009). Will the oceans help feed humanity? Bioscience *59*, 967–976.

Dunbar, J., Barns, S.M., Ticknor, L.O., and Kuske, C.R. (2002). Empirical and theoretical bacterial diversity in four Arizona soils. Appl. Environ. Microbiol. *68*, 3035–3045.

Durfee, T., Nelson, R., Baldwin, S., Plunkett, G., Burland, V., Mau, B., Petrosino, J.F., Qin, X., Muzny, D.M., Ayele, M., *et al.* (2008). The complete genome sequence of *Escherichia coli* DH10B: Insights into the biology of a laboratory workhorse. J. Bacteriol. *190*, 2597–2606.

Eder, B., und Schulz, H. (2005). Biogas praxis: Grundlagen. Planung, Anlagenbau, Beispiele, Wirtschaftlichkeit. Ökobuch Verlang GmbH. Freiburg.

Edström, M., Nordberg, A., and Thyelius, L. (2003). Anaerobic treatment of animal byproducts from slaughterhouses at laboratory and pilot scale. Appl. Biochem. Biotech. -Part A Enzyme Eng. and Biotech. *109*, 127–138.

Eid, J., Fehr, A., Gray, J., Luong, K., Lyle, J., Otto, G., Peluso, P., Rank, D., Baybayan, P., Bettman, B., *et al.* (2009). Real-time DNA sequencing from single polymerase molecules. Science *323*, 133–138.

Entcheva, P., Liebl, W., Johann, A., Hartsch, T., and Streit, W.R. (2001). Direct cloning from enrichment cultures, a reliable strategy for isolation of complete operons and genes from microbial consortia. Appl. Environ. Microbiol. *67*, 89–99.

FAO Investment Centre Division and FAO's Rural Infrastructure and Agro-Industries Division (2010). https://www.responsibleagroinvestment.org/sites/responsibleagroinvestment.org/files/FAO_Agbiz%20handbook_Poultry_Meat.pdf

FAO Statistics Division 2009 FAO Statistical Yearbook. (2009). http://www.fao.org/docrep/014/am079m/pdf/am079m00a.pdf

Field, C.B., Campbell, J.E., and Lobell, D.B. (2008). Biomass energy: the scale of the potential resource. Trends. Ecol. Evol. *23*, 65–72.

Finn, R.D., Mistry, J., Tate, J., Coggill, P., Heger, A., Pollington, J.E., Gavin, O.L., Gunasekaran, P., Ceric, G., Forslund, K., Holm, L., Sonnhammer, E.L., Eddy, S.R., and Bateman, A. (2010). The Pfam protein families database. Nucleic Acids Res. *38*, D211–D222.

Fisher, M.M., and Triplett, E.W. (1999). Automated approach for ribosomal intergenic spacer analysis of microbial diversity and its application to freshwater bacterial communities. Appl. Environ. Microbiol. *65*, 4630–4636.

Franke-Whittle, I.H., Goberna, M., Pfister, V., and Insam, H. (2009). Design and testing of the ANAEROCHIP microarray for investigation of methanogenic communities. J. Microbiol. Methods *79*, 279–288.

Gallert, C., Bauer, S., and Winter, J. (1998). Effect of ammonia on the anaerobic degradation of protein by a mesophilic and thermophilic biowaste population. Appl. Microbiol. Biotechnol. *50*, 495–501.

Gangagni Rao, A., Sasi Kanth Reddy, T., Surya Prakash, S., Vanajakshi, J., Joseph, J., Jetty, A., Rajashekhara Reddy, A., and Sarma, P.N. (2008). Biomethanation of poultry litter leachate in UASB reactor coupled with ammonia stripper for enhancement of overall performance. Bioresour. Technol. *99*, 8679–8684.

Garcia, J.-L., Patel, B.K.C., and Olliver, B. (2000). Taxonomic, phylogenetic and ecological diversity of methanogenic archaea. Anaerobe *6*, 105–226.

Gendebien, A., Ferguson, R., Brink, J., Horth, H., Sullivan, M., Davis, R., Brunet, H., Dalimier, F., Landrea, B., Krack, D., Perot, J., and Orsi, C. (2001). Survey of Wastes Spread on Land. EU DG Energy. Final report study contract B4-3040/99/110194/MAR/E3

Gerardi, M.H. (2003). The Microbiology of Anaerobic Digesters. John Wiley and Sons, Inc.

Giannoukos, G., Ciulla, D.M., Huang, K., Haas, B.J., Izard, J., Levin, J.Z., Livny, J., Earl, A.M., Gevers, D., Ward, D.V., Nusbaum, C., Birren, B.W., and Gnirke, A. (2012). Efficient and robust RNA-seq process for cultured bacteria and complex community transcriptomes. Genome Biol. *13*, R23.

Golueke, C.G., Oswald, W.J., and Gotaas, H.B. (1956). Anaerobic digestion of algae. Appl. Microbiol. *5*, 47–55.

Gonzalez-Fernández, C., Sialve, B., Berne,t N., and Steyer, J.P. (2013). Effect of organic loading rate on anaerobic digestion of thermally pretreated *Scenedesmus* sp. biomass. Bioresour. Technol. *129*, 219–223.

Gorg, A., Weiss, W., and Dunn, M.J. (2004). Current two-dimensional electrophoresis technology for proteomics. Proteomics *4*, 3665–3685.

Goris, J., Konstantinidis, K.T., Klappenbach, J.A., Coenye, T., Vandamme, P., and Tiedje, J.M. (2007). DNA-DNA hybridization values and their relationship to whole-genome sequence similarities. Int. J. Syst. Evol. Microbiol. *57*, 81–91.

Gottschalk, G. (1986). Bacterial Metabolism. (Springer Verlag New York Inc.)

Gould, J.M. (1984). Alkaline peroxide delignification of agricultural residues to enhance enzymatic saccharification. Biotechnol. Bioeng. *26*, 46–52.

Grabkowsky, B., and Windhorst, H.-W. (2009). Shifts and patterns in global poultry meat trade. World Poultry *25*, 09.

Güngör-Demirci, G., and Demirer, G.N. (2004). Effect of initial COD concentration, nutrient addition, temperature and microbial acclimation on anaerobic treatability of broiler and cattle manure. Bioresour. Technol. 93, 109–117.

Gupta, R., Gupta, N., and Rathi, P. (2004). Bacterial lipases: an overview of production, purification and biochemical properties. Appl. Microbiol. Biotechnol. *64*, 763–781.

Hajarnis, S.R., and Ranade, D.R. (1994). Inhibition of methanogens by n- and iso-volatile fatty acids. World J. Microbiol. Biotechnol. *10*, 350–351.

Hankamer, B., Lehr, F., Rupprecht, J., Mussgnug, J.H., Posten, C., and Kruse, O. (2007). Photosyntetic biomass and H_2 production by green algae: from bioengineering to bioreactor scale up. Physiol. Plant. *131*, 10–21.

Hanreich, A., Heyer, R., Benndorf, D., Rapp, E., Pioch, M., Reichl, U., and Klocke, M. (2012). Determination of the metabolically active part of a thermophilic microbial community producing biogas from agricultural biomass by metaproteome analysis. Can. J. Microbiol. *58*, 917–922.

Hansen, H.K., Angelidaki, I., and Ahring, B.K. (1996). Anaerobic digestion of swine manure: inhibition by ammonia. Water Res. *32 (1)*, 5–12.

Harris, T.D., Buzby, P.R., Babcock, H., Beer, E., Bowers, J., Braslavsky, I., Causey, M., Colonell, J., Dimeo, J., and Efcavitch, J.W. (2008). Single-molecule DNA sequencing of a viral genome. Science *320*, 106–109.

Hartmann, H., and Ahring, B.K. (2005). Anaerobic digestion of the organic fraction of municipal solid waste: influence of co-digestion with manure. Water Res. *39*, 1543–1552.

Harun, R., Singh, M., Forde, G.M., and Danquah, M.K. (2010). Bioprocess engineering of microalgae to produce a variety of consumer products. Renew. Sust. Energy Rev. *14*, 1037–1047.

Hase, R., Oikawa, H., Sasao, C., Morita, M., and Watanabe, Y. (2000). Photosynthetic production of microalgal biomass in a raceway system under greenhouse conditions in Sendai City. J. Biosci. Bioeng. *89*, 157–163.

Hashimoto, A.G. (1986). Ammonia inhibition of methanogenesis from cattle wastes. Agr. Wastes, *17*, 241–261.

Hattori, S. (2008). Syntrophic acetate oxidizing microbes in methanogenic environments. Microbes Environments 23, 118–127.

He, Y., Pang, Y., Liu, Y., Li, X., and Wang, K. (2008). Physicochemical characterization of rice straw pretreated with sodium hydroxide in the solid state for enhancing biogas production. Energy Fuels *22*, 2775–2781.

He, Z., Gentry, T.J., Schadt, C.W., Wu, L., Liebich, J., Chong, S.C., Huang, Z., Wu, W., Gu, B., Jardine, P., Criddle, C., and Zhou, J. (2007). GeoChip: a comprehensive microarray for investigating biogeochemical, ecological and environmental processes. ISME J. *1*, 67–77.

Hejnfelt, A., and Angelidaki, I. (2009). Anaerobic digestion of slaughterhouse by-products. Biomass and Bioenergy 33, 1046–1054.

Hendriksen, H.V., and Ahring, B.K. (1991). Effects of ammonia on growth and morphology of thermophilic hydrogen-oxidizing methanogenic bacteria. FEMS Microbiol. Lett. *85*, 241–245.

Herbel, Zs., Rákhely, G., Bagi, Z., Ivanova, G., Ács, N., Kovács, E., and Kovács, K.L. (2010). Exploitation of the extremely thermophilic *Caldicellulosiruptor saccharolyticus* in hydrogen and biogas production from biomasses. Env. Technol. *31*, 1017–1024.

Hernández, E.P.S., and Córdoba, L.T. (1993). Anaerobic digestion of *Chlorella vulgaris* for energy production. Resour. Conservation Recycling *9*, 127–132.

Heyer, R., Kohrs, F., Benndorf, D., Rapp, E., Kausmann, R., Heiermann, M., Klocke, M., and Reichl, U. (2013). Metaproteome analysis of the microbial communities in agricultural biogas plants. New Biotechnol. http://dx.doi.org/10.1016/j.nbt.2013.01.002

Heyndrickx, M., Vauterin, L., Vandamme, P., Kersters, K., and DeVos, P. (1996). Applicability of combined amplified ribosomal DNA restriction analysis (ARDRA) patterns in bacterial phylogeny and taxonomy. J. Microbiol. Methods *26*, 247–259.

Himmelsbach, J.N., Raman, D.R., Anex, R.P., Burns, R.T., and Faulhabcr, C.R. (2010). Effect of ammonia soaking pretreatment and enzyme addition on biochemical methane potential of switchgrass. Trans. Asabe *53*, 1921–1927.

Hiss, R.H., Norris, D.E., Dietrich, C.H., Whitcombe, R.F., West, D.F., Bosio, C.F., Kambhampati, S., Piesman, J., Antolin, M.F., and Black, W.C.I.V. (1994). Molecular taxonomy using single-strand conformation polymorphism (SSCP) analysis of mitochondrial ribosomal DNA genes. Insect Mol. Biol. *3*, 171–182.

Holm-Nielsen, J.B., Al Seadi, T., and Oleskowicz-Popiel, P. (2009). The future of anaerobic digestion and biogas utilization. Bioresour. Technol. *100*, 5478–5487.

Howorka, S., Cheley, S., and Bayley, H. (2001). Sequence-specific detection of individual DNA strands using engineered nanopores. Nat. Biotechnol. *19*, 636–639.

Hu, Q., Gueterman, H., and Richmond, A. (1996). A flat inclined modular photobioreactor for outdoor mass cultivation of photoautotrophs. Biotechnol. Bioeng. *51*, 51–60.

Hu, Q., Sommerfeld, M., Jarvis, E., Ghirardi, M., Posewitz, M., Seibert, M., and Darzins, A. (2008). Microalgal triacylglycerols as feedstocks for biofuel production: perspectives and advances. Plant J. *54*, 621–689.

Hughes, A.D., Kelly, M.S., Black, K.D., and Stanley, M.S. (2012). Biogas from microalgae: is it time to revisit the idea? Biotechnol. Biofuels *5*, 86.

Huson, D.H., Auch, A.F., Qi, J., and Schuster, C.S. (2007). MEGAN analysis of metagenomic data. Genome. Res. *17*, 377–386.

Hutnan, M., Drtil, M., and Mrafkova, L. (2000). Anaerobic biodegradation of sugar beet pulp. Biodegradation *11*, 203–211.

IEA (International Energy Agency). (2012). The global value of coal. www.iea.org/publications/insights/global_value_of_coal.pdf

Isroi, Millati, R., Syamsiah, S., Niklasson, C., Cahyanto, M.N., Lundquist, K., and Taherzadeh, M.J. (2011). Biological pretreatment of lignocelluloses with white-rot fungi and its application: A review. Bioresources *6*, 5224–5259.

Ivanova, G., Rákhely, G., and Kovács, K.L. (2009). Thermophilic biohydrogen production from energy plants by *Caldicellulosyruptor saccharolyticus* and comparison with related studies. Int. J. Hydrogen Energy *34*, 3659–3670.

Ivey, M.L., and Phister. T.G. (2011). Detection and identification of microorganisms in wine: a review of molecular techniques. J. Ind. Microbiol. Biotechnol. *38*, 1619–1634.

Jaenicke, S., Ander, C., Bekel, T., Bisdorf, R., Dröge, M., Gartemann, K.-H., Jünemann, S., Kaiser, O., Krause, L., Tille, F., Zakrzewski, F., Pühler, A., Schlüter, A., and Goesmann, A. (2011a). Comparative and joint analysis of two metagenomic datasets from a biogas fermenter obtained by 454-pyrosequencing. PLoS One *6*, e14519.

Jaenicke, S., Zakrzewski, M., Jünemann, S., Pühler, A., Groesmann, A., and Schlüter, A. (2011b). Analysis of the metagenome from biogas-producing microbial community by means of bioinformatics methods. In: FJ. Bruijn ed. Handbook of Molecular Microbial Ecology, vol. II. Metagenomics in Different Habitats. Wiley-Blackwell, Hoboken, NJ, pp. 403–414.

Jeihanipour, A., Karimi, K., and Taherzadeh, M.J. (2010). Enhancement of ethanol and biogas production from high-crystalline cellulose by different modes of NMO pretreatment. Biotechnol. Bioeng. *105*, 469–476.

Jeon, B.-H., Choi, J.-A., Kim, H.-C., Hwang, J.-H., Abou-Shanab, R.A., Dempsey, B.A., Regan, J.M., and Kim, J.R. (2013). Ultrasonic digestion of microalgal biomass and consequent of biocassebility/bioavailability in microbial fermentation. Biotechnol. Biofuels *6*, 37.

Jetten, M.S.M., Strous, M., van de Pas-Schoonen, K.T., Schalk, J., van Dongen, U.G.J.M., van de Graaf, A.A., Logemann, S., Muyzer, G., van Loosdrecht, M.C.M., and Kuenen, J.G. (1998). The anaerobic oxidation of ammonium. FEMS Microbiol. Rev. *22*, 421–437.

Jimenez, S., Cartagena, M.C., and Arce, A. (1990). Influence of lignin on the methanization of lignocellulosic wastes. Biomass, *21*, 43–54.

Jing, X., Zhang, X., and Bao, J. (2009). Inhibition performance of lignocellulose degradation products on industrial cellulase enzymes during cellulose hydrolysis. Appl. Biochem. Biotechnol. *159*, 696–707.

Johnson, D.K., and Elander, R.T. (2008). Pretreatments for enhanced digestability of feedstocks. In: M. E. Himmel, ed. Biomass Recalcitrance. Blackwell Publishing, pp. 436–453.

Jonsson, L.J., Alriksson, B., and Nilvebrant, N.O. (2013). Bioconversion of lignocellulose: inhibitors and detoxification. Biotechnol. Biofuels *6*, 16.

Kadam, P.C., and Boone, D.R. (1996). Influence of pH on ammonia accumulation and toxicity in halophilic, methylotrophic methanogens. Abstracts of the General Meeting of the American Society for Microbiology *96*, 304.

Kalia, V., Sonakya, V., and Raizada, N. (2000). Anaerobic digestion of banana stem waste. Bioresour. Technol. *73*, 191–193.

Kao, C.-Y., Chiu, S.-Y., Huang, T.-T., Dai, L., Wang, G.-H., Tseng, C.-P., Chen, C.-H., and Lin, C.-S. (2012). A mutant strain of microalga Chlorella sp. for the carbon dioxide capture from biogas. Biomass Bioenergy *36*, 132–140.

Kaparaju, P., Luostarinen, S., Kalmari, J., and Rintala, J. (2002). Co-digestion of energy crops and industrial confectionery by-products with cow manure: batch-scale and farm-scale evaluation. Water Sci. Technol. *45*, 275–280.

Kaparaju, P., Serrano, M., Thomsen, A.B., Kongjan, P., and Angelidaki, I. (2009). Bioethanol, biohydrogen and biogas production from wheat straw in a biorefinery concept. Bioresour. Technol. *100*, 2562–2568.

Kárászová, M., Vejrazka, J., Vesely, V., Friess, K., Randová, A., Hejtmánek, V., Brabec, L., and Izák, P. (2012). A water-swollen thin film composite membrane for effective upgrading of raw biogas by methane. Sep. Purif. Technol. *89*, 212–216.

Kayhanian, M. (1994). Performance of a high-solids anaerobic digestion process under various ammonia concentrations. J. Chem. Tech. Biotechnol. *59*, 349–352.

Kayhanian, M. (1999). Ammonia Inhibition in high-solids biogasification: an overview and practical solutions. Environ. Technol. *20*, 355–365.

Keenan J.D. (1977). Bioconversion of solar energy to methane. Energy *2*, 365.

Kelly M., and Dworjanyn, S. (2008). The potential of marine biomass for anaerobic biogas production. In The Potential of Marine Biomass for Anaerobic Biogas Production. The Crown Estate.

Kleiner, D. (1993). NH_4^+ transport systems. In: E. Bakker, ed. Alkali Cation Transport Systems in Prokaryotes. CRC Press.

Klocke, M., Mähnert, P., Mundt, K., Souidi, K., and Linke, B. (2007). Microbial community analysis of a biogas-producing completely stirred tank reactor fed continuously with fodder beet silage as monosubstrate. Syst. Appl. Microbiol. *30*, 139–151.

Knuckey, R.M., Brown, M.R., Robert, R., and Frampton, D.M.F. (2006). Production of microalgal concentrates by flocculation and their assessment as aquaculture feeds. Aquacultural Eng. 35, 300–313.

Koster, I.W. (1986). Characteristics of the pH-influenced adaptation of methanogenic sludge to ammonium toxicity. J. Chem. Technol. Biotechnol. *36*, 445–455.

Koster, I.W., and Koomen, E. (1988). Ammonia inhibition of the maximum growth rate of the hydrogenotrophic methanogens at various pH-levels and temperatures. Appl. Microbiol. Biotechnol. *28*, 500–505.

Koster, I.W., and Lettinga, G. (1984). The influence of ammonium-nitrogen on the specific activity of pelletized methanogenic sludge. Agric. Wastes *9*, 205–216.

Koster, I.W., and Lettinga, G. (1988). Anaerobic digestion at extreme ammonia concentrations. Biol. Wastes *25*, 51–59.

Kotsyurbenko, O.R., Chin, K.-J., Glagolev, M.V., Stubner, S., Simankova, M.V., Nozhevnikova, A.N., and Conrad, R. (2004). Acetoclastic and hydrogenotrophic methane production and methanogenic populations in an acidic West-Siberian peat bog. Environ. Microbiol. *6*, 1159–1173.

Kovács, K.L. (2001). Utilization of hydrogen metabolism in biotechnological applications. In: R. Cammack, R.L. Robson, and M. Frey, eds. Hydrogen as a Fuel: Learning from Nature. Chapter 9.1. Taylor & Francis, pp. 181–189.

Krause, L., Diaz, N.N., Edwards, R.A., Gartemann, K.-H., Krömeke, H., Neuwger, H., Pühler, A., Runte, K.J., Schlüter, A., Stoye, J., Szczepanowski, R., and Tauch, A. (2008). Taxonomic composition and gene content of a methane-producing microbial community isolated from a biogas reactor. J. Biotechnol. *136*, 91–101.

Krischan, J., Makaruk, A., and Harasek, M. (2012). Design and scale-up of an oxidative scrubbing process for selective removal of hydrogen sufide from biogas. J. Hazard. Mater. *215–216*, 49–56.

Krivanek, O.L., Chisholm, M.F., Nicolosi, V., Pennycook, T.J., Corbin, G.J., Dellby, N., Murfitt, M.F., Own, C.S., Szilagyi, Z.S., Oxley, M.P., Pantelides, S.T., and Pennycook, S.J. (2010). Atom-by-atom structural and chemical analysis by annular dark-field electron microscopy. Nature *464*, 571–574.

Kröber, M., Bekel, T., Diaz, N.N., Goesmann, A., and Sebastian, J. (2009). Phylogenetic characterization of a biogas plant microbial community integrating clone library 16S-rDNA sequences and metagenome sequence data obtained by 454-pyrosequencing. J. Biotechnol. *142*, 38–49.

Kroeker, E.J., Schulte, D.D., Sparling, A.B., and Lapp, H.M. (1979). Anaerobic treatment process stability. J. Water Pollut. Control Fed. *51*, 718–727.

Lakaniemi, A.-M., Hulatt, C.J., Thomas, D.N., Tuovinen, O.H., and Puhakka, J.A. (2011). Biogenic hydrogen and methane production from *Chlorella vulgaris* and *Dunaliella tertiolecta* biomass. Biotechnol. Biofuels *4*, 34.

Lam, M.K., and Lee, K.T. (2011). Renewable and sustainable bioenergies production from palm oil mill effluent (POME): Win-win strategies toward better environmental protection. Biotechnol. Adv. *29*, 124–141.

Lam, M.K., and Lee, K.T. (2012). Microalgae biofuels: a critical review of issues, problems and the way forward. Biotechnol. Adv. *30*, 673–690.

Lapp, H.M., Schulte, D.D., Kroeker, E.J., Sparling, A.B., and Topnik, B.H. (1975). Start-up of pilot scale swine manure digesters for methane production. Managing livestock wastes. ASAE Publ. Proc. *275*, 234–237.

Leamon, J.H. (2003). A massively parallel PicoTiterPlate based platform for discrete picoliter-scale polymerase chain reactions. Electrophoresis *24*, 3769–3777.

Lee, Y.K. (2001). Microalgal mass culture systems and methods: their limitation and potential. J. Appl. Phycol. *13*, 307–315.

Lei, X., Sugiura, N., Feng, C., and Maekawa, T. (2007). Pretreatment of anaerobic digestion effluent with ammonia stripping and biogas purification. J. Hazard. Mater. *145*, 391–397.

Levene, M.J., Korlach, J., Turner, S.W., Foquet, M., Craighead, H.G., and Webb, W.W. (2003). Zero-mode waveguides for single-molecule analysis at high concentrations. Science *299*, 682–686.

Lipson, D., Raz, T., Kieu, A., Jones, D.R., Giladi, E., Thayer, E., Thompson, J.F., Letovsky, S., Milos, P., and Causey, M. (2009). Quantification of the yeast transcriptome by single-molecule sequencing. Nat. Biotechnol. *27*, 652–658.

Liu, W.T., Marsh, T.L., Cheng, H., and Forney, L.J. (1997). Characterization of microbial diversity by determining terminal restriction fragment length polymorphisms of genes encoding 16S rRNA. Appl. Environ. Microbiol. *63*, 4516–4522.

Liu, Y., and Whitman, W.B. (2008). Metabolic, phylogenetic and ecological diversity of the methanogenic archaea. Ann. N.Y. Acad. Sci. *1125*, 170–189.

Luste, S., and Luostarinen, S. (2010). Anaerobic co-digestion of meat-processing by-products and sewage sludge – effect of hygienization and organic loading rate. Bioresour. Technol. *101*, 2657–2664.

Luton, P.E., Wayne, J.M., Sharp, R.J., and Riley, P.W. (2002). The *mcrA* gene as an alternative to 16S rRNA in the phylogenetic analysis of methanogen populations in landfill. Microbiology *148*, 3521–3530.

Lynch, M., and Milligan, B.G. (1994). Analysis of population genetic structure with RAPD markers, Mol. Ecol. *3*, 91–99.

Lynd, L.R., Weimer, P.J., van Zyl, W.H., and Pretorius, I.S. (2002). Microbial cellulose utilization: fundamentals and biotechnology. Microbiol. Mol. Biol. Rev. *66*, 506–577.

McCarty, P.L., and McKinney, R. (1961). Salt toxicity in anaerobic digestion. J. Water Pollut. Control Fed. *33*, 399–415.

McHugh, S., Carton, M., Mahony, T., and O'Flaherty, V. (2003). Methanogenic population structure in a variety of anaerobic bioreactors. FEMS Microbiol. Lett. *219*, 294–304.

Macias-Corral, M., Samani, Z., Hanson, A., Smith, G., Funk, P., Yu, H., and Longworth, J. (2008). Anaerobic digestion of municipal solid waste and agricultural waste and the effect of co-digestion with dairy cow manure. Bioresour. Technol. *99*, 8288–8293.

McInerney, M.J. (1988). Anaerobic hydrolysis and fermentation of fats and proteins. In Biology of Anaerobic Microorganisms. J.B. Zehnder, ed. (J. Wiley and Sons Inc.), pp. 373–415.

McInerney, M.J., Struchtmeyer, C.G., Sieber, J., Mouttaki, H., Stams, A.J.M., Rohlin, L., and Gunsalus, R.P. (2008). Physiology, ecology, phylogeny, and genomics of microorganisms capable of syntrophic metabolism. Annu. N. Y. Acad. Sci. *1125*, 58–72.

McMahon, K.D., Stroot, P.G., Mackie, R.I., and Raskin, L. (2001). Anaerobic codigestion of municipal solid waste and biosolids under various mixing conditions—II: microbial population dynamics. Water Res. *35*,1817–1827.

McNally, B., Singer, A., Yu, Z., Sun, Y., Weng, Z., and Meller, A. (2000). Optical recognition of converted DNA nucleotides for single-molecule DNA sequencing using nanopore arrays. Nano Lett. *10*, 2237–2244.

Madigan, M.T., and Martinko, J.M. (2006). Brock Biology of Microorganisms, 11th edn. London, Pearson Education.

Magbanua, B.S., Adams, T.T., and Johnston, P. (2001). Anaerobic co-digestion of hog and poultry waste. Bioresour. Technol. *76*, 165–168.

Makaruk, A., Miltner, M., and Harasek, M. (2010). Membrane biogas upgrading processes for the production of natural gas substitute. Sep. Purif. Techol. *74*, 83–92.

Margulies, M., Egholm, M., Altman, W.E., Attiya, S., Bader, J.S., Bemben, L.A., Berka, J., Braverman, M.S., Chen, Y.J., Chen, Z., *et al.* (2005). Genome sequencing in microfabricated high-density picolitre reactors. Nature *437*, 376–380.

Martí, N., Bouzas, A., Seco, A., and Ferrer, J. (2008). Struvite precipitation assessment in anaerobic digestion processes. Chem. Eng. J. *141*, 67–74.

Martin-Laurent, F., Philippot, L., Hallet, S., Chaussod, R., Germon, J.C., Soulas, G., and Catroux, G. (2001). DNA extraction from soils: old bias for new microbial diversity analysis methods. Appl. Environ. Microbiol. *67*, 2354–2359.

Marvin-Sikkema, F.D., Richardson, A.J., Stewart, C.S., Gottschal, J.C., and Prins, R.A. (1990). Influence of hydrogen-consuming bacteria on cellulose degradation by anaerobic fungi. Appl. Environ. Microbiol. *56*, 3793–3797.

Matsui, J.T., Amano, T., Koike, Y., Saiganji, A., and Saito, H. (2006). Methane fermentation of seaweed biomass. In American Institute of Chemical Engineers. (San Francisco)

Melis, A., and Happe, T. (2001). Hydrogen production: green algae as a source of energy. Plant Physiol. *127*, 740–748.

Melis, A., Zhang, L., Forestier, M., Ghirardi, M.L., and Seibert, M. (2000). Sustained photobiological hydrogen gas production upon reversible inactivation of oxygen evolution in the green alga *Chlamydomonas reinhardtii*. Plant Physiol. *122*, 127–136.

Menardo, S., and Balsari, P. (2012). An analysis of the energy potential of anaerobic digestion of agricultural by-products and organic waste. BioEnergy Res. *5*, 759–767.

Menon, V., and Rao, M. (2012). Trends in bioconversion of lignocellulose: biofuels, platform chemicals and biorefinery concept. Progr. Energy Combust. Sci. *38*, 522–550.

Milán, Z., Sánchez, E., Weiland, P., Borja, R., Martin, A., and Ilangovan, K. (2001). Influence of different natural zeolite concentrations on the anaerobic digestion of piggery waste. Bioresour. Technol. *80*, 37–43.

Mohsenzadeh, A., Jeihanipour, A., Karimi, K., and Taherzadeh, M.J. (2012). Alkali pretreatment of softwood spruce and hardwood birch by NaOH/thiourea, NaOH/urea, NaOH/urea/thiourea, and NaOH/PEG to improve ethanol and biogas production. J. Chem. Technol. Biotechnol. *87*, 1209–1214.

Molina Grima, E., Belarbi, E., Acién Fernández, F., Robles Medina, A., and Chisti, Y. (2003). Recovery of microalgal biomass and metabolites: Process options and economics. Biotechnol. Adv. *20*, 491–515.

Molino, A., Migliori, M., Ding, Y., Bikson, B., Giordano, G., and Braccio, G. (2013). Biogas upgrading via membrane process: Modelling of pilot plant scale and the end uses for the grid injection. Fuel. *107*, 585–592.

Molinuevo, B., Garcia, M.C., Leon, M.C., and Acitores, M. (2008). 13th International Conference of the FAO ESCORENA Network on Recycling of Agricultural, Municipal and Industrial Residues in Agriculture. (Castilla and Leon, Spain).

Monlau, F., Barakat, A., Steyer, J.P., and Carrere, H. (2012). Comparison of seven types of thermo-chemical pretreatments on the structural features and anaerobic digestion of sunflower stalks. Bioresour. Technol. *120*, 241–247.

Mountfort, D.O., Asher, R.A., and Bauchop, T. (1982). Fermentation of cellulose to methane and carbon dioxide by a rumen anaerobic fungus in a triculture with *Methanobrevibacter* sp. Strain RA1 and *Methanosarcina barkeri*. Appl. Environ. Microbiol. *44*, 128–134.

Murphy, J.D., Browne, J., Allen, E., and Gallagher, C. (2013). The resource of biomethane, produced via biological, thermal and electrical routes, as a transport biofuel. Renew. Energ. *55*, 474–479.

Mussgnug, J.H., Klassen, V., Schlüter, A., and Kruse, O. (2010). Microalgae as a substrates for fermentative biogas production in a combined biorefinery concept. J. Biotechnol. *150*, 51–56.

Muyzer, G., and Smalla, K. (1998). Application of denaturing gradient gel electrophoresis (DGGE) and temperature gradient gel electrophoresis (TGGE) in microbial ecology. Antonie Van Leeuwenhoek *73*, 127–141.

Nielsen, H.B., and Angelidaki, I. (2008). Strategies for optimizing recovery of the biogas process following ammonia inhibition. Bioresour. Technol. *99*, 7995–8001.

Nisman, B. (1954). The Stickland reaction. Bacteriol. Rev. *18*, 16–42.

Nkemka, V.N., and Murto, M. (2010). Evaluation of biogas production from seaweed in batch tests and in UASB reactors combined with the removal of heavy metals. J. Environ. Manag. *91*, 1573–1579.

North, W.J. (1987). Oceanic farming of Macrocystis, the problems and non-problems. Seaweed cultivation for renewable resources. In: K.T. Bird and P.H. Benson, eds. Seaweed Cultivation for Renewable Resources. Volume 16. Elsevier, Amsterdam, pp. 39–68.

Nunez, V.J., Voltolina, D., Nieves, M., Pina, P., Medina, A., and Guerrero, M. (2001). Nitrogen budget in *Scenedesmus obliquus* cultures with artificial wastewater. Bioresour. Technol. *78*, 161–164.

Nyns, E.J. (1986). Biomethanation processes. In: Schonborn, ed. Microbial Degradations. *W18*, 207–267.

Ohlinger, K.N., Young, T.M., and Schroeder, E.D. (2000). Post digestion struvite precipitation using a fludized bed reactor. J. Environ. Eng. *126*, 361–368.

Oprime, M.E.A.G., Garcia, O., and Cardoso, A.A. (2001). Oxidation of H2S in acid solution by Thiobacillus ferrooxidans and *Thiobacillus thiooxidans*. Process. Biochem. *37*, 111–114.

Ortiz, J., Romero, N., Robert, P., Araya, J., Lopez-Hernandez, J., Bozzo, C., Navarrete, E., Osorio, A., and Rios, A. (2006). Dietary fiber, amino caid, fatty acid and tocopherol contents of the edible seaweeds *Ulva lactuca* and *Durvillaea antarctica*. Food Chem. *99*, 98–104.

Osborn, A.M., Moore, E.R.B., and Timmis, K.N. (2000). An evaluation of terminal-restriction fragment length polymorphisms (TRFLP) analysis for the study of microbial community structure and dynamics. Environ. Microbiol. 2, 39–50.

Palmowski, L., and Muller, J. (1999). Influence of the size reduction of organic waste on their anaerobic digestion. In 2nd International Symposium on Anaerobic Digestion of Solid Waste, Barcelona, pp. 137–144.

Parabakaran P., and Ravindran, A.D. (2011). A comperative study on effective cell disruption methods for lipid extraction from microalgae. Lett. Appl. Microbiol. *53*, 150–154.

Park, N.D., Helle, S.S., and Thring, R.W. (2012). Combined alkaline and ultrasound pretreatment of thickened pulp mill waste activated sludge for improved anaerobic digestion. Biomass Bioenergy *46*, 750–756.

Parkin, G.F., Speece, R.E., Yang, C.H.J., and Kocher, W.M. (1983). Response of methane fermentation systems to industrial toxicants. J. Water Pollut. Control Fed. *55*, 44–53.

Perle, M., Kimchie, S., and Shelef, G. (1995). Some biochemical aspects of the anaerobic degradation of dairy wastewater. Water Res. *29*, 1549–1554.

Petersen, M., and Daniel, R. (2006). Purification and characterization of an extracellular lipase from *Clostridium tetanomorphum*. World J. Microbiol. Biotechnol. *22*, 431–435.

Petersson, A., Thomsen, M.H., Hauggaard-Nielsen, H., and Thomsen, A. (2007). Potential bioethanol and biogas production using lignocellulosic biomass from winter rye, oilseed rape and faba bean. Biomass Bioenergy *31*, 812–819.

Poggi-Varaldo, H.M., Tingley, J., and Oleszkiewicz, J.A. (1991). Inhibition of growth and acetate uptake by ammonia in batch anaerobic digestion. J. Chem. Technol. Biotechnol. *52*, 135–143.

Pohland, F.G., and Ghosh, S. (1971). Developments in anaerobic stabilization of organic wastes– the two-phase concept. Environ. Lett. *1*, 255–266.

Prochazka, J., Mrazek, J., Strosova, L., Fliegerova, K., Zabranska, J., and Dohanyos, M. (2012). Enhanced biogas yield from energy crops with rumen anaerobic fungi. Eng. Life Sci. *12*, 343–351.

Przytocka-Jusiak, M., Duszota, M., Matusiak, K., and Mycielski, R. (1984). Intensive culture of *Chlorella vulgaris* as the second stage of biological purification of nitrogen industry wastewaters. Water Res. *18*, 1–7.

Purwandari, F.A., Sanjaya, A.P., Millati, R., Cahyanto, M.N., Horváth, I.S., Niklasson, C., and Taherzadeh, M.J. (2013). Pretreatment of oil palm empty fruit bunch (OPEFB) by N-methylmorpholine-N-oxide (NMMO) for biogas production: Structural changes and digestion improvement. Bioresour. Technol. *128*, 461–466.

Rajeshwari, K.V., Balakrishnan, M., Kansal, A., Kusum, L., and Kishore, V.V.N. (2000). State of-the-art of anaerobic digestion technology for industrial wastewater treatment. Renew. Sustain. Energy Rev. *4*, 135–156.

Ramsay, I.R., and Pullammanappallil, P.C. (2001). Protein degradation during anaerobic wastewater treatment: deviation of stoichiometry. Biodegradation *12*, 247–257.

Rani, D., and Nand, K. (2004). Ensilage of pineapple processing waste for methane generation. Waste Manag. *24*, 523–528.

Richmond, A. (2004). Handbook of Microalgal Culture: Biotechnology and Applied Phycology. Oxford, Blackwell Science.

Ritchie, N.J., Scutter, M.E., Dick, R.P., and Myrold, D.D. (2000). Use of length heterogeneity PCR and fatty acid methyl ester profiles to characterize microbial communities in soil. Appl. Environ. Microbiol. *66*, 1668–1675.

Robbins, J.E., Gerhardt, S.A., and Kappel, T.J. (1989). Effects of total ammonia on anaerobic digestion and an example of digestor performance from cattle manure-protein mixtures. Biol. Wastes 27, 1–14.

Rodolfi, l., Zitteli, G.C., Bassi, N., Padovani, G., Biondi, N., Bonini, G., and Tredici, M.R. (2009). Microalgae for oil: strain selection, induction of lipid sytesis and outdoor mass cultivation in a low-cost photobioreactor. Bioethanol. Bioeng. *102*, 100–112.

Rothberg, J.M., Hinz, W., Rearick, T.M., Schultz, J., Mileski, W., Davey, M., Leamon, J.H., Johnson, K., Milgrew, M.J., Edwards, M., *et al.* (2011). An integrated semiconductor device enabling non-optical genome sequencing. Nature *475*, 348–352.

Ryckebosch, E., Drouillon, M., and Vervaeren, H. (2011). Techniques for transformation of biogas to biomethane. Biomass. Bioenerg. *35*, 1633–1645.

Salehian, P., Karimi, K., Zilouei, H., and Jeihanipour, A. (2013). Improvement of biogas production from pine wood by alkali pretreatment. Fuel *106*, 484–489.

Salminen, E., and Rintala, J. (2002). Anaerobic digestion of organic solid poultry slaughter house waste – a review. Bioresour. Technol. *83*, 13–26.

Santosh, Y., Sreekrishnan, T.R., Kohli, S., and Rana, V. (2004). Enhancement of biogas production from solid substrates using different techniques a review. Bioresour. Technol. *95*, 1–10.

Sasaki, D., Morita, M., Sasaki, K., Watanabe, A., and Ohmura, N. (2012). Acceleration of cellulose degradation and shift of product via methanogenic co-culture of a cellulolytic bacterium with a hydrogenotrophic methanogen. J. Biosci. Bioeng. *114*, 435–439.

Schadt, E.E., Turner, S., and Kasarskis, A. (2010). A window into third-generation sequencing. Human Mol. Gen. *19*, R227–R240.

Schink, B. (1997). Energetics of syntrophic cooperation in methanogenic degradation. Microbiol. Mol. Biol. Rev. *61*, 262–280.

Schink, B. (2002). Synergistic interactions in the microbial world. Antonie van Leeuwenhoek. *81*, 257–261.

Schloss, P.D., Westcott, S.L., Ryabin, T., Hall, J.R., Hartmann, M., Hollister, E.B., Lesniewski, R.A., Oakley, B.B., Parks, D.H., Robinson, C.J., *et al.* (2009). Introducing mothur: open-source, platform-independent, community-supported software for describing and comparing microbial communities. Appl. Environ. Microbiol. 75, 7537–7541.

Schlüter, A., Bekel, T., Diaz, N.N., Dondrup, M., Eichenlaub, R., Gartemann, K.-H., Krahn, I., Krause, L.B., Krömeke, H., Kruse, O., Mussgnug, J.H., Neuweger, H., Niehaus, K., Pühler, A., Runte, K.J., Szczepanowski, R., Tauch, A., Tilker, A., Viehöver, P., and Goesmann, A. (2008). The metagenome of a biogas-producing microbial community of a production-scale biogas plant fermenter analysed by the 454-pyrosequencing technology. J. Biotechnol., *136*, 77–90.

Schmid, L.A., and Lipper, R.I. (1969). Swine wastes, characterization and anaerobic digestion. Animal Waste Management. Proc. Cornell Agr, Waste Manage, Conf. p. 50.

Schmidt, J.E., and Ahring, B.K. (1993). Effects of hydrogen and formate on the degradation of propionate and butyrate in thermophilic granules from an upflow anaerobic sludge blanket reactor. Appl. Environ. Microbiol. *59*, 2546–2551.

Schnürer, A., and Nordberg, Å. (2008). Ammonia, a selective agent for methane production by syntrophic acetate oxidation at mesophilic temperature. Water Sci. Technol. *57*, 735–740.

Schnürer, A., Zellner, G., and Svenson, B.H. (1999). Mesophilic syntrophic acetate oxidation during methane formation in biogas reactors. FEMS Microbiol. Ecol. *29*, 249–261.

Scholz, M., Melin, T., and Wessling, M. (2012). Transforming biogas into biomethane using membrane technology. Renew. Sust. Energ. Rev. *17*, 199–212.

Scott, S.A., Davey, M.P., Dennis, J.S., Horst, I., Howe, C:J., Lea-Smith, D.J., and Smith, A.G. (2010). Biodiesel from algae: challenges and prospects. Curr. Opinion Biotechnol. *21*, 277–286.

Shilton, A., and Guieysse, B. (2010). Sustainable sunlight to biogas is via marginal organics. Curr. Opin. Biotechnol. *21*, 287–291.

Shin, H.S., and Youn, J.H. (2005). Conversion of food waste into hydrogen by thermophilic acidogenesis. Biodegradation *16*, 33–44.

Sialve, B., Bernet, N., and Bernard, O. (2009). Anaerobic digestion of microalgae as a necessary step to make microalgal biodiesel sustainable. Biotechnol. Adv. *27*, 409–416. doi: 10.1016/j.biotechadv.2009.03.001

Sildermann N. (2012). Maisfeld wird multikulti. Erneuerbare Energien *22*, 78–81.

Singh, B.K., Campbell, C.D., Sorenson, S.J., and Zhou, J. (2009). Soil genomics. Nature Rev. Microbiol. *7*, 756 doi:10.1038/nrmicro2119-c1.

Singh, J., and Gu, S. (2010). Commercialization potential of microalgae for biofuels production. Renew. Sust. Energy Rev. *14*, 2596–2610.

Sousa, D.Z., Smidt, H., Alves. M.M., and Stams, A.J.M. (2009). Ecophysiology of syntrophic communities that degrade saturated and unsaturated long-chain fatty acids. FEMS Microbiol. Ecol. *68*, 257–272.

de Souza, W.R. (2013). Microbial degradation of lignocellulosic biomass. In Sustainable Degradation of Lignocellulosic Biomass – Techniques, Applications and Commercialization. A.K. Chandel and S.S. da Silva, eds. (Intech). DOI: 10.5772/54325

Spolaore, P., Joannis-Cassan, C., Duran, E., and Isambert, A. (2006). Commercial applications of microalgae. J. Biosci. Bioeng. *101*, 87–96.

Sprott, G.D., and Patel, G.B. (1986). Ammonia toxicity in pure cultures of methanogenic bacteria. System. Appl. Microbiol. 7, 358–363.

Sprott, G.D., Shaw, K.M., and Jarrel, K.F. (1984). Ammonia/potassium exchange in methanogenic bacteria. J. Biol. Chem. *259*, 12602–12608.

Sriram, S., and Seenivasan, R. (2012). Microalgae cultivation in wastewater for nutrient removal. J. Algal Biomass Utln. *3*, 9–13.

Srivatsan, A., Han, Y., Peng, J., Tehranchi, A.K., Gibbs, R., Wang, J.D., and Chen, R. (2008). High-precision, whole-genome sequencing of laboratory strains facilitates genetic studies. PLoS Genet. *4*, e1000139.

Stabnikova, O., Liu, X.Y., Wang, J.Y., and Ivanov, V. (2006). Quantification of methanogens by fluorescence *in situ* hybridization with oligonucleotide probe. Appl. Microbiol. Biotechnol. *73*, 696–702.

Steinberg, L.M., and Regan, J.M. (2009). *mcrA*-targeted real-time quantitative PCR method to examine methanogen communities. Appl. Environ. Microbiol. *75*, 4435–4442.

Stephens, E., Ross, I.L., King, Z., Mussung, J.H., Kruse, O., Posten, C., Borowitzka, M.A., and Hankamer, B. (2010). An economic and technical evaluation of microalgal biofuels. Nat. Biotechnol. *28*, 126–128.

Stephenson, A.L., Kazamia, E., Dennis, J.S., Howe, C.J., Scott, S.A., and Smith, A.G. (2010). Life-cycle assessment of potential algal biodiesel production in the United Kingdom: a comparison of raceways and air lift tubular bioreactors. Energy Fuel *24*, 4062–4077.

Stoddart, D., Heron, A.J., Mikhailova, E., Maglia, G., and Bayley, H. (2009). Single-nucleotide discrimination in immobilized DNA oligonucleotides with a biological nanopore. Proc. Natl. Acad. Sci U.S.A. *106*, 7702–7707.

Straka, F., Jenicek, P., Zabranska, J., Dohanyos, M., and Kuncarova, M. (2007). Anaerobic fermentation of biomass and wastes with respect to sulfur and nitrogen contents in treated materials. In Proceedings Sardinia 2007, Eleventh International Waste Management and Landfill Symposium CISA. (Environmental Sanitary Engineering Centre, Italy).

Summers, Z.M., Fogarty, H.E., Leang, C., Franks, A.E., Malvankar, N.S., and Lovley, D.R. (2010). Direct exchange of electrons within aggregates of an evolved syntrophiccoculture of anaerobic bacteria. Science *330*, 1413–1415.

Sundh, I., Carlsson, H., Nordberg, A., Hansson, M., and Mathisen, B. (2003). Effects of glucose overloading on microbial community structure and biogas production in a laboratory-scale anaerobic digester. Bioresour. Technol. *89*, 237–243.

Surmacz-Gorska, J., Cichon, A., and Miksch, K. (1997). Nitrogen removal from wastewater with high ammonia nitrogen concentration via shorter nitrification and denitrification. Water Sci. Technol. *36*, 73–78.

Taherzadeh, M.J., and Karimi, K. (2008).Pretreatment of lignocellulosic wastes to improve ethanol and biogas production: a review. Int. J. Mol. Sci. *9*, 1621–1651.

Takacova, A., Mackulak, T., Smolinska, M., Hutnan, M., and Olejnikova, P. (2012). Influence of selected biowaste materials pretreatment on their anaerobic digestion. Chem. Papers *66*, 129–137.

Talia, P., Sede, S.M., Campos, E., Rorig, M., Principi, D., Tosto, D., Hopp, H.E., Grasso, D., and Cataldi, A. (2012). Biodiversity characterization of cellulolytic bacteria present on native Chaco soil by comparison of ribosomal RNA genes. Res. Microbiol. *163*, 221–232.

Teghammar, A., Karimi, K., Sárvári Horváth, I., and Taherzadeh, M.J. (2012). Enhanced biogas production from rice straw, triticale straw and softwood spruce by NMMO pretreatment. Biomass Bioenergy *36*, 116–120.

Tenney, M.W., Echelberger Jr, W.F., Schussler, R.G., and Pavoni, J.L. (1969). Algal flocculation with synthetic organic polyelectrolytes. Appl. Microbiol. *18*, 965–971.

Tessler, L.A., Reifenberger, J.G., and Mitra, R.D. (2009). Protein quantification in complex mixtures by solid phase single-molecule counting. Anal. Chem. *81*, 7141–7148.

Thauer, R.K., Jungermann, K., and Decker, K. (1977). Energy conservation in chemotrophic anaerobic bacteria. Bacteriol. Rev. *41*, 100–180.

The Reference Genome Group of the Gene Ontology Consortium. (2009). The gene ontology's reference genome project: a unified framework for functional annotation across species. PLoS Comput. Biol. *5*, e1000431. doi:10.1371/journal.pcbi.1000431

Tiedje, J.M., Asuming-Brempong, S., Nüsslein, K., Marsh, T.L., and Flynn, S.J. (1999). Opening the black box of soil microbial diversity. Appl. Soil Ecol. *13*, 109–122.

Timmins, M., Zhou, W.X., Rupprecht, J., Lim, L., Thomas-Hall, S.R., Doebbe, A., Kruse, O., Hankamer, B., Marx, U.C., Smith, S.M., and Schenk, P.M. (2009). The metabolome of *Chlamidomonas reinhardtii* following induction of anaerobic H_2 production by sulfur depletion. J. Biol. Chem. *284*, 23415–23425.

Tippayawong, N., and Thanompongchart, P. (2010). Biogas quality upgrade by simultaneous removal of CO_2 and H_2S in a packed column reactor. Energy 35, 4531–4535.

Triolo, J.M., Pedersen, L., Qu, H., and Sommer, S.G. (2012). Biochemical methane potential and anaerobic biodegradability of non-herbaceous and herbaceous phytomass in biogas production. Bioresour. Technol. *125*, 226–232.

Uduman, N., Qi, J., Danaquah, M.K., Forde, G.M., and Hoadley, A. (2010). Dewatering of microalgal cultures: a major bottleneck to algae-based fuels. J. Renewable Sust. Energy 2, 127001–127015.

Ugwu, C.U., Aoyagi, H., and Uchiyama, H. (2008). Photobioreactors for mass cultivation of algae. Bioresour. Technol. *99*, 4021–4028.

Ulmer, D.C., Leisola, M.S.A., Schmidt, B.H., and Fiechter, A. (1983). Rapid degradation of isolated lignins by *Phanerochaete chrysosporium*. Appl. Environ. Microbiol. *45*, 1795–1801.

Van Dyk, J.S., and Pletschke, B.I. (2012). A review of lignocellulose bioconversion using enzymatic hydrolysis and synergistic cooperation between enzymes – factors affecting enzymes, conversion and synergy. Biotech. Adv. *30*, 1458–1480.

Van Foreest, F. (2012). Perspectives for Biogas in Europe. Oxford Institute for Energy Studies.

Van Velsen, A.F.M. (1979). Adaptation of methanogenic sludge to high ammonia-nitrogen concentrations. Water Res. *13*, 995–999.

Vartoukian, S.R., Palmer, R.M., and Wade, W.G. (2010). Strategies for culture of 'unculturable' bacteria. FEMS Microbiol. Lett. *309*, 1–7.

VDI-Handbuch Energietechnik (2006). Fermentation of organic materials. VDI 4630. (Beuth Verlag GmbH, Berlin)

Venter, J.C., Remington, K., Heidelberg, J.F., Halpern, A.L., Rusch, D., Eisen, J.A., Wu, D., Paulsen, I., Nelson, K.E., Nelson, W., *et al.* (2004). Environmental genome shotgun sequencing of the Sargasso Sea. Science *304*, 66–74.

Vergara-Fernandez, A., Vargas, G., Alarcon, N., and Velasco, A. (2008). Evalution of marine algae as a source of biogas in a two-stage anaerobic reactor system. Biomass Bioenegy *32*, 338–344.

de Vladar, H.P. (2012). Amino acid fermentation at the origin of the genetic code. Biol. Direct. 7, 6. doi: 10.1186/1745-6150-7-6.

Wang, B., Li, Y., Wu, N., and Lan, C.Q. (2008). CO_2 bio-migration using microalgae. Appl. Microbiol. Biotechnol. *79*, 707–718.

Wang, X.J., Yang, G.H., Feng, Y.Z., and Ren, G.X. (2012). Potential for biogas production from anaerobic co-digestion of dairy and chicken manure with corn stalks. Adv. Mat. Res. *347*, 2484–2492.

Wang, Z., Gerstein, M., and Snyder, M. (2009). RNA-seq: a revolutionary tool for transcriptomics. Nature Rev. Genet. *10*, 57–63.

Weber, S., Stubner, S., and Conrad, R. (2001). Bacterial populations colonizing and degrading rice straw in anoxic paddy soil. Appl. Environ. Microbiol. *67*, 1318–1327.

Weiland, P. (2003). Production and energetic use of biogas from energy crops and wastes in Germany. Appl. Biochem. Biotechnol. *109*, 263–274.

Westerholm, M., Hansson, M., and Schnürer, A. (2012). Improved biogas production from whole stillage by co-digestion with cattle manure. Bioresour. Technol. *114*, 314–319.

Wiegant, W.M., and Zeeman, G. (1986). The mechanism of ammonia inhibition in the thermophilic digestion of livestock wastes. Agricultural Wastes *16*, 243–253.

Wilcox, H.A. (1977). The U.S. Navy's Ocean food and energy farm project. In The U.S. Navy's Ocean Food and Energy Farm Project. N.T. Mooney, ed. 83–104. (Naval Ocean Systems Center).

Wilmes, P., and Bond, P.L. (2004). The application of two-dimensional polyacrylamide gel electrophoresis and downstream analyses to a mixed community of prokaryotic microorganisms. Environ. Microbiol. *6*, 911–920.

Wilson, D.B. (2008). Aerobic microbial cellulase systems. In Biomass Recalcitrance, M.E. Himmel, ed. (Blackwell Publishing).

Wirth, R., Kovács, E., Maróti, G., Bagi, Z., Rákhely, G., and Kovács, K.L. (2012). Characterization of a biogas-producing microbial community by short-read next generation DNA sequencing. Biotechnol. Biofuels 5, 41.

Woese, C.R., Kandler, O., and Wheelis. M.L. (1990). Towards a natural system of organisms: proposal for the domains Archae, Bacteria, and Eukarya. Proc. Natl. Acad. Sci. U.S.A. *87*, 4576–4579.

Wolin, M.J. (1975). Interspecies hydrogen transfer between H_2-producing and methane-producing species. In Symposium on Microbial Production and Utilization of Gases. H.G. Schlegel, G. Gottschalk, and N. Pfenning eds. (E. Goltz Göttingen). pp. 141–150.

Wu, C.Y., and Pang, S.J. (2006). World Seaweed Resources – an Authoritative Reference System. (City).

Xiaojiao, W., Gaihe, Y., Yongzhong., F., Guangxin, R., and Xinhui, H. (2012). Optimizing feeding composition and carbon–nitrogen ratios for improved methane yield during anaerobic co-digestion of dairy, chicken manure and wheat straw. Bioresour. Technol. *120*, 78–83.

Xie, S., Frost, J.P., Lawlor, P.G., Wu, G., and Zhan, X. (2011). Effects of thermo-chemical pretreatment of grass silage on methane production by anaerobic digestion. Bioresour. Technol. *102*, 8748–8755.

Xu, L., Weathers, P.J., Xiong, X.R., and Liu, C.Z. (2009). Microalgal bioreactors: Challenges and opportunities. Eng. Life Sci. *9*, 178–189.

Yan, L., Gao, Y., Wang, Y., Liu, Q., Sun, Z., Fu, B., Wen, X., Cui, Z., and Wang, W. (2012). Diversity of a mesophilic lignocellulolytic microbial consortium which is useful for enhancement of biogas production. Bioresour. Technol. *111*, 49–54.

Yeow, S.K., and Peng, W.L. (2012). Application of ultrasound pretreatment for sludge digestion. In: A. Mudhoo (ed.), Biogas Production: Pretreatment Methods in Anaerobic Digestion. Hoboken, NJ, John Wiley & Sons.

Yetilmezsoy, K., Sertyesilisik, B., Kocak, E., and Sapci-Zengin, Z. (2009). Ameliorative effect of different doses of $MgNH_4PO_4.6H_2O$ precipitate recovered from the effluent of UASB treating poultry manure wastewater: growth of *Lolium perenne*. J. Food Agric. Environ. 7, 823–831.

Yoshihara, K.I., Nagase, H., Eguchi, K., Hirata, K., and Miyamoto, K. (1996). Biological elimination of nitric oxide and carbon dioxide from flue gas by marine microalga NOA-113 cultivated in a long tubular photobioreactor. J. Ferment. Bioeng. *82*, 351–354.

Yuan, X., Li, P., Wang, H., Wang, X., Cheng, X., and Cui, Z. (2011). Enhancing the anaerobic digestion of corn stalks using composite microbial pretreatment. J. Microbiol. Biotechnol. *21*, 746–752.

Zakrzewski, M., Goesmann, A., Jaenicke, S., Jünemann, S., Eikmeyer, F., Szczepanowski, R., Al-Soud, W.A., Sørensen, S., Pühler, A., and Schlüter, A. (2012). Profiling of the metabolically active community from a production-scale biogas plant by means of high-throughput metatranscriptome sequencing. J. Biotechnol. *158*, 248–258.

Zhang, J., and Tong, Z. (2006). Study on catalytic wet oxidation of H_2S into sulfur on Fe/Cu catalyst. J. Nat. Gas. Chem. *15*, 63–69.

Zhang, Q., He, J., Tian, M., Mao, Z., Tang,, L., Zhang, J., and Zhang, H. (2011). Enhancement of methane production from cassava residues by biological pretreatment using a constructed microbial consortium. Bioresour. Technol. *102*, 8899–8906.

Zhang, R.H., and Zhang, Z.Q. (1999). Biogasification of rice straw with an anaerobic-phased solids digester system. Bioresour. Technol. *68*, 235–245.

Zheng, M., Li, X., Li, L., Yang, X., and He, Y. (2009). Enhancing anaerobic biogasification of corn stover through wet state NaOH pretreatment. Bioresour. Technol. *100*, 5140–5145.

Zhong, W., Zhang, Z., Luo, Y., Sun, S., Qiao, W., and Xiao, M. (2011). Effect of biological pretreatments in enhancing corn straw biogas production. Bioresour. Technol. *102*, 11177–11182.

Zhong, W., Zhang, Z., Luo, Y., Qiao, W., Xiao, M., and Zhang, M. (2012). Biogas productivity by co-digesting Taihu blue algae with corn straw as an external carbon source. Bioresour. Technol. *114*, 281–286.

Zhu, J., Wan, C., and Li, Y. (2010). Enhanced solid-state anaerobic digestion of corn stover by alkaline pretreatment. Bioresour. Technol. *101*, 7523–7528.

Zielinska, M., Cydzik-Kwiatkowska, A., Zielinski, M., and Debowski, M. (2013). Impact of temperature, microwave radiation and organic loading rate on methanogenic community and biogas production during fermentation of dairy wastewater. Bioresour. Technol. *129*, 308–314.

Zuo, Z., Tian, S., Chen, Z., Li, J., and Yang, X. (2012). Soaking pretreatment of corn stover for bioethanol production followed by anaerobic digestion process. Appl. Biochem. Biotechnol. *167*, 2088–2102.

Engineering Recombinant Organisms for Next-generation Ethanol Production 3

Eugéne van Rensburg, Riaan den Haan, Daniël C. la Grange, Heinrich Volschenk, Willem H. van Zyl and Johann F. Görgens

Abstract

Development of microbes for the production of fuel ethanol continues at a rapid pace, for established first generation processes relying on sugar- and starch-based food crops, and second generation processes for conversion of lignocellulose to ethanol by pretreatment-hydrolysis-fermentation. The production of bio-ethanol is dominated by *Saccharomyces cerevisiae* as fermentative organism, although *Escherichia coli*, *Zymomonas mobilis*, *Klebsiella oxytoca* and *Kluyveromyces marxianus*, among others, were considered for second generation processes. For *S. cerevisiae*, key developments for fermentation of raw and cooked starch, as well as lignocellulose, include consolidated bioprocessing (CBP) where enzyme production for substrate hydrolysis and fermentation of released sugars are performed by a single organism. Further yeast developments required for second generation ethanol include engineering of xylose utilization capability, to increase yields from lignocellulose, and increasing the resistance of yeasts to inhibitory compounds formed during lignocellulose pretreatment prior to hydrolysis and fermentation. The present chapter will review the full scope of organism development for ethanol production, followed by examination of *S. cerevisiae* strain development for xylose utilization, CBP and hardening, as well as metabolic engineering, application of –omics technologies, and synthetic biology as examples of global efforts in organism development for ethanol production.

3.1 Introduction

Significant technological advances influencing all spheres of modern life inspired a quest for alternative and sustainable energy, *in lieu* of finite fossil fuel resources. Core to this changing paradigm lies four key drivers, namely security of energy supply, improved trade balances by limiting oil import expenditure, social development of rural communities through increased agriculture, and mitigation of greenhouse gas (GHG) emissions (Sims *et al.*, 2010). Ethanol derived from biomass (bioethanol) has been identified as a prime candidate for liquid fuel for the transport sector due to established production technology and minimal impact on existing supply chain infrastructure, especially when used as a 'drop in' additive (Zhang and Hu, 2013). Data showed that world bioethanol production increased from 340 PJ to more than 2 045 PJ between 2000 and 2012, which equates to a volumetric increase from 16 to 85.2 Gl ethanol (Anon, 2012). This growth can be attributed to a deliberate effort by governments to incorporate renewable fuel into the value chain, for example the US Renewable Fuel Standard (RFS2) mandating the use of 136 billion litres of bioethanol, incorporating 61 billion litres of cellulosic ethanol by 2022 (Schnepf and Yacobucci, 2013). Uncertainty about instability in many OPEC countries and wild fluctuations in oil prices further prompted many countries, especially in the OECD, to invest in, develop and trade non-petroleum fuels (de Gorter *et al.*, 2013).

Ethanol production from renewable resources can be divided into two main technologies. First generation (1G) ethanol is produced from sugar- or starch-containing crops, where the largest ethanol producers are Brazil using sugarcane and the US using maize. In fact, since 2005, the US became the largest ethanol producer in the world (Heinimö and Junginger, 2009), with exports to Brazil of up to 500 Ml/year, and to the EU and Australia of up to 100 Ml/year (REN21, 2012). China is the second largest exporter of ethanol, mostly to other countries in the Orient (Scott and Junyang, 2009). Second generation (2G) ethanol is produced from lignocellulose, which is regarded as the most abundant carbonaceous material on earth. Sources of lignocellulose can be divided into three main groups, namely (i) agricultural residues (straw from cereals, bagasse from sugarcane, sorghum and triticale and residues from forestry), (ii) the 'green waste' component of municipal solid waste and (iii) dedicated crops, such as grasses, short-rotation forests and energy crops including sorghum and triticale (Sims *et al.*, 2010).

In spite of its abundance, the commercialization of lignocellulosic ethanol faces significant challenges, primarily stemming from its recalcitrance to cellulase (Lynd *et al.*, 2002) and the lack of low-cost technologies to overcome this obstacle (Van Zyl *et al.*, 2007). Biomass-to-liquids (BTL) processes, including gasification combined with synthesis, such as the Fischer–Tropsch process, or fast pyrolysis to liquid followed by chemical upgrading, offers a possible recourse (Anex *et al.*, 2010). Integration of BTL technology into biological processing, by using unconverted, lignin-rich residues of lignocellulose as feedstock, has a promising future (Van Zyl *et al.*, 2011). The gasification of lignocellulose followed by syngas fermentation to ethanol, using organisms such as *Clostridium ljungdahlii*, *Clostridium autoethanogenum*, *Acetobacterium woodii*, *Clostridium carboxidivorans* and *Peptostreptococcus productus* represents a hybrid approach between thermochemical and biological conversion of lignocellulose. Fermentation of syngas, obtained by gasification of lignocellulose or lignin-residues from bioprocessing of lignocellulose, to ethanol by bacteria represents an opportunity to utilize the complete biomass, including the lignin component, at ambient conditions without excessive poisoning of the microbial catalyst (Munasinge and Khanal, 2010).

Much room remains for the development of biological processing of lignocellulose, which could result in this technology becoming increasingly cost competitive, especially given the economic potential of integrating 1G fermentation-distillation processes with similar 2G processes based on lignocellulose pretreatment-hydrolysis-fermentation-distillation. Lignocellulose fermentation can make a sizable contribution to the ethanol yield obtained from 1G energy crops. On average, the sucrose in sugarcane has a yield of 6 900 l ethanol/ha, whereas the bagasse has a potential yield of 3 000 l/ha (Somerville *et al.*, 2010), implying that utilizing the waste for ethanol would increase the total yield by 44%. 2G ethanol production through sugar fermentation also benefits from established 1G process technologies, which should lead to a less steep learning curve. Several studies have demonstrated the technical and economic advantages of integrating of 1G and 2G technologies for sugar fermentation to ethanol (Erdei *et al.*, 2013; Macrelli *et al.*, 2012), in terms of both yield and production cost, by sharing feedstock supply, infrastructure and process utilities/units. Over and above ethanol production, the celluloses and lignin can also serve as a source of several high-value components through a biorefinery concept (Menon and Rao, 2012). Further expansion to aviation fuels is also a reality (Klein-Marcuschamer *et al.*, 2013), where conversion of alcohols to jet fuel components is a leading technology, nearing commercialization. Several commercial plants presently produce cellulosic ethanol, such as the 76 million litres per annum plant by Beta Renewables in Cesentino, Italy using giant reed (*Arundo donax*) and wheat straw, the Abengoa plant in Salamanca, Spain, producing 5.3 million litre per year from wheat and barley straw and the 15 000 ton bioethanol plant of Domsjö Fabriker in Örnsköldsvik, Sweden, using softwood (reviewed by Kudakasseril Kurian *et al.*, 2013).

Sugar, starch and cellulose appears to be the main viable carbon sources for future liquid fuel production using fermentation, provided that oil prices do not decrease below US$70/barrel (Timilsina and Shresta, 2011). Starch serves as an energy reserve in most plants and can constitute up to 90% of the mass of the plant seeds (Preiss,

2009). The nature of starch, the amylase enzymes acting on starch to liberate glucose and gelatinization and liquefaction steps for starch fermentation, is well-described (Van Zyl *et al.*, 2012), and will be extensively reviewed in a chapter sub-section.

Lignocellulosic plant biomass consists of 40–55% cellulose, 25–50% hemicellulose and 10–40% lignin, depending on the plant source (Sun and Cheng, 2002). The main polysaccharide present is water-insoluble crystalline cellulose, which also contains the major fraction of fermentable sugars. The hydrolysis of crystalline cellulose to fermentable sugars is achieved by the co-operative action of three basic types of activity namely (i) endoglucanases (EGs), acting in the amorphous regions of cellulose to release cellodextrins and provide free chain ends; (ii) exoglucanases, including cellodextrinases and cellobiohydrolases (CBHs), which act on crystalline cellulose in a processive manner starting at the free chain ends and releasing mainly cellobiose; and (iii) β-glucosidases (BGLs), which hydrolyse cellobiose and small cello-oligosaccharides to glucose (Lynd *et al.*, 2002; Zhang and Lynd, 2004). However, in nature there is a wide array of hydrolytic and oxidative enzymes active on cellulose en hemi-cellulose (http://www.cazy.org/).

Conversion of biomass to ethanol or other commodities via hydrolysis-fermentation is initiated with a pretreatment process, rendering the polymeric fractions more accessible to enzymatic hydrolysis (Van Zyl *et al.*, 2013). Whereas steam-pretreatment with or without a catalyst is one of the preferred methods for lignocellulose, high temperature processes reaching 85°C (Tester and Morrison, 1990) is mostly used for the gelatinization of starch before liquefaction can occur using thermophilic amylases. For ethanol production from monomeric sugars, four biologically mediated steps convert pre-treated lignocellulose to ethanol. These steps are (i) production of polymer degrading enzymes, (ii) hydrolysis of the polysaccharides, (iii) fermentation of the produced hexose and (iv) pentose sugars present in the liquor (Lynd *et al.*, 2002). Any improvement of this conversion process generally involves combining two or more of these steps. Polysaccharide hydrolysis and sugar fermentation steps are combined in simultaneous saccharification and fermentation (SSF) of hexoses or simultaneous saccharification and co-fermentation (SSCF) of both hexoses and pentoses in the case of lignocellulose, if a suitable fermentative organism is available (Lynd *et al.*, 2002; Lynd *et al.*, 2005). SSF or SSCF avoids the feedback inhibition effect of the sugars produced by the action of the hydrolytic enzymes. However, using a mesophilic ethanologen requires a lower operating temperature to levels that are inevitably suboptimal for enzymatic activity. Furthermore, several by-products from lignocellulose processing could result in sub-optimal fermentation performance (Almeida *et al.*, 2009, Palmqvist and Hahn-Hägerdal, 2000a), requiring the use of targeted metabolic engineering and evolutionary engineering to instil a degree of robustness in the fermenting yeast.

Consolidated bioprocessing (CBP) refers to the merging of all four biologically mediated process steps for lignocellulose hydrolysis and sugar fermentation to ethanol, and entails direct microbial conversion of pretreated lignocellulose or starches to bioethanol in a single reactor, without the requirement for addition of exogenous enzymes. CBP would signify a breakthrough for low-cost biomass processing due to the economic benefits of process integration and avoiding the high costs of enzymes that make the biological conversion route less attractive (Anex *et al.*, 2010; Hamelinck *et al.*, 2005; Van Zyl *et al.*, 2011). Process performance in terms of yield and productivity can be further enhanced through integration of 1G and 2G processes, combined with robust ethanologenic organisms, predominantly *S. cerevisiae*, harbouring the required genetic material for pentose co-fermentation.

The purpose of this chapter is to provide a reference source on recent developments in the field of biomass to ethanol processing, which should serve as a guiding text for future research. Several excellent reviews are available dealing with different aspects of ethanol production, but here we aimed to provide a comprehensive body of information addressing biomass to ethanol with a key emphasis on integration on the genomic, as well as the process level. Following a short section of different microbial technologies developed for 1G and 2G ethanol production, the discussion will primarily focus on *S. cerevisiae*, the most commonly used ethanol producer in fermentation. A comprehensive discussion will deal with the core principles of introducing xylose fermenting ability into the yeast and key results will be summarized. This work pertains specifically to lignocellulose fermentation where xylose presents a substantial underutilized carbon component in the case of fermentations using wild-type or commercial yeast. The discussion will then progress to aspects related to yeast tolerance to various inhibitors found in lignocellulosic hydrolysates. This is a natural progression for the discussion, given the involvement of several co-factors in by-product detoxification that are normally associated with xylose utilization. Next, the discussion will turn to the CBP application of soluble and insoluble (raw, granular or uncooked) starch where recent advances in recombinant organisms to utilize starch without prior gelatinization will be accentuated. The CBP theme will continue to lignocellulose fermentation, focusing on the different phenotypes, genetic configurations and interactions associated with the expression of recombinant cellulases from fungi and bacteria in *S. cerevisiae*. The presented information as outlined above will then be integrated into a short discussion on the application to processes, the challenges associated with biomass processing, and an illustration of the benefits of 1G and 2G integration from a techno-economic perspective. The chapter will end of with a brief summary highlighting the requirements for and future prospects of ideal ethanol-producing organisms, as well as the challenges that commercialization might hold.

3.2 Overview of all microbial technologies for first- (1G) and second-generation (2G) ethanol production

For 2G ethanol production, a wide range of microbes have been considered, based on inherent qualities that might make them desirable. New developments in the combination of lignocellulose gasification with syngas fermentation, using organisms such as *Clostridium ljungdahlii*, *Clostridium autoethanogenum*, *Acetobacterium woodii*, *Clostridium carboxidivorans* and *Peptostreptococcus productus* for ethanol production, have gained commercial interest and are under development, where such fermentation requires efficient conversion of CO and H_2 as the major components of syngas to ethanol (Henstra *et al.*, 2007). Mesophilic microorganisms can convert syngas to short-chain fatty acids and alcohols, while no thermophiles with similar efficiencies are known. Strain development is focused on selection methods, rather than genetic engineering, where few tools are available, in particular to improve ethanol yields (Munasinghe and Khanal, 2010). Mass transfer limitations between syngas and liquid fermentation medium also limit the efficiency of microbial conversions (Munasinghe and Khanal, 2010; Ungerman and Heindel, 2007). The presence of nitric oxide in syngas from lignocellulose may also inhibit microbial activity (Ahmed and Lewis, 2007).

In contrast to the gasification–fermentation route, the saccharification of biomass for the release of sugars for fermentation is the most prominent route being developed for ethanol production from lignocellulose. Organisms with promising ethanol yields and titres generally do not produce the enzyme systems to hydrolyse the sugar polymers in lignocellulose (Lynd *et al.*, 2005). Conversely, (hemi-) cellulose degraders generally produce inadequate amounts of a desired product. Three different approaches have been followed to develop organisms for 2G biofuels using the saccharification and fermentation route: (i) engineering product forming ability into organisms that can degrade biomass efficiently, (ii) engineering cellulolytic ability into organisms that have desired product producing attributes and (iii) engineering cellulolytic and product formation abilities into organisms with other positive attributes (Van Zyl *et al.*, 2013).

Cellulolytic bacteria such as *Clostridium thermocellum* and filamentous fungi such as *Trichoderma reesei* and *Fusarium oxysporum* have been engineered to produce ethanol as main fermentation product and to tolerate higher levels of ethanol (Xu *et al.*, 2009). However, significant improvements in all aspects are still required before these organisms can be considered as industrial ethanologens. The Gram-negative fermenting bacterium *Z. mobilis* and the yeast *S. cerevisiae* are both used to produce ethanol at industrial scale and possess excellent ethanol production rates, titres and tolerance. However, both organisms have limited substrate ranges and lack the ability to utilize polymeric carbohydrate substrates as carbon sources (La Grange *et al.*, 2010). D-Xylose utilization and some degree of cellulase production have been engineered into both organisms. While complete conversion of crystalline celluloses by recombinant strains has not yet been achieved, *S. cerevisiae* strains were engineered that significantly decreased the quantity of exogenous enzymes that had to be added to the fermentation broth for SSF of pretreated hardwood (McBride *et al.*, 2010). *Escherichia coli* possess remarkable metabolic engineering capacity, as well as the ability to produce mainly ethanol, together with other biofuels or biofuel precursors as fermentation products. Even the ability to secrete cellulases was engineered into the organism, albeit at relatively low production levels (La Grange *et al.*, 2010; Bräu and Sahm 1986; Bokinsky *et al.*, 2011). The yeast, *K. marxianus* has the ability to utilize a wider range of sugars compared to *S. cerevisiae* and strains were isolated that can grow at temperatures in excess of 50°C (reviewed by Fonseca *et al.*, 2008). A cellulase-expressing *K. marxianus* strain that could convert amorphous cellulose to ethanol was created (Yanase *et al.*, 2010). Other interesting organisms that are being investigated for 2G bioethanol production include the thermophilic bacteria *Geobacillus* spp. and *Thermoanaerobacterium saccharolyticum* (Van Zyl *et al.*, 2013).

There are thus many organisms that are being considered for 2G bioethanol production through CBP. However, perhaps the most developed among these is the yeast *S. cerevisiae*. This level of development, coupled with other advantages, including the generally regarded as safe (GRAS) status, the existence of industrial fermentation technology using this organism and genetic flexibility, implies that this organism is the most likely candidate to be the first ethanologen used in 2G bioethanol processes. For this reason the remainder of this chapter will focus on the various aspects of development required to use *S. cerevisiae* for bioethanol production, with a specific emphasis on 2G ethanol and co-fermentation of carbon sources.

3.3 Xylose fermentation by *Saccharomyces cerevisiae*

Xylose is the second most abundant carbohydrate in nature, after glucose. Given this abundance, microorganisms capable of fermenting xylose to ethanol at high yields and rates are critical to the development and success of cost-effective large scale technologies for the conversion of lignocellulosic biomass to ethanol. A large number of wild-type organisms capable of fermenting xylose were described, including bacteria (Dien *et al.*, 2003), yeast (Martiniano *et al.*, 2013; Weigert *et al.*, 1988) and filamentous fungi (Okamoto *et al.*, 2012; Wohlbach *et al.*, 2013).

Fungi and bacteria generally employ two different strategies for the metabolism of xylose (Fig. 3.1). In most fungi, D-xylose is reduced to D-xylitol by a D-xylose reductase (XR) (EC 1.1.1.21) and subsequently oxidized to D-xylulose by D-xylitol dehydrogenase (XDH) (EC 1.1.1.9) (Chandel *et al.*, 2011). Some D-xylose reductases have an absolute requirement for NADPH as co-factor, whereas others were also shown to use NADH as co-factor (Hahn-Hägerdal *et al.*, 2001). D-Xylitol dehydrogenases almost exclusively uses NAD^+ as electron acceptor. Conversely, D-xylose can also be converted directly to D-xylulose by a xylose isomerase (XI) (EC 5.3.1.5) without the need for any co-factor, although this pathway is predominantly found in bacteria and usually proceeds at a slower rate than the XR/XDH pathway (Hahn-Hägerdal *et al.*, 2007). Finally, before entering the pentose phosphate pathway (PPP), D-xylulose is phosphorylated by xylulokinase (EC 2.7.1.17).

Bacteria generally produce a mixture of acids (lactic acid, acetic acid, propionic and butyric acid) during fermentation of carbohydrates (Papoutsakis and Meyer, 1985; Williams *et al.*, 2013; Zhu and Yang, 2004). On the other hand, genetic engineering of bacteria, such as *E. coli*, *K. oxytoca*, and *T. saccharolyticum* to selectively produce ethanol from substrates like xylose has been very successful. *E. coli* is capable of metabolizing all the major sugars present in plant biomass (Alterthum and Ingram, 1989), but only half of the pyruvate produced during growth is converted to ethanol (Ingram *et al.*, 1997). However, through high level expression of the *Z. mobilis* pyruvate

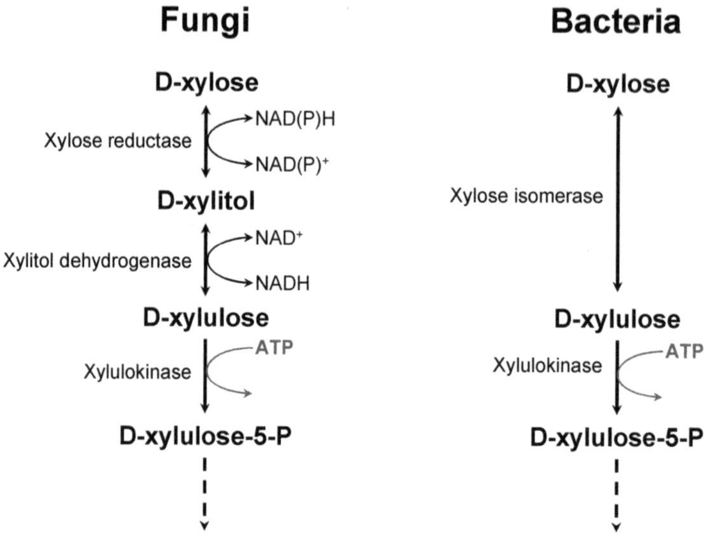

Figure 3.1 Metabolic pathways for the utilization of xylose by fungi and bacteria. A colour version of this figure is available in the plate section at the back of the book.

decarboxylase gene in *E. coli*, ethanol production levels could be increased to levels similar to that observed in *Z. mobilis* (Table 3.1) (Bräu and Sahm, 1986; Ingram *et al.*, 1987, 1991; Ingram and Conway, 1988; Ohta *et al.*, 1991a). These recombinant strains are genetically stable, produce high concentrations of ethanol and are more efficient than pentose-fermenting yeast (Table 3.1). More recent studies focused on improving the ethanol tolerance of recombinant *E. coli* strains (Qureshi *et al.*, 2006, 2012).

Similar to *E. coli*, *K. oxytoca* can also metabolize all major sugars in plant biomass. It also has the ability to metabolize cellobiose, cellotriose, xylobiose and xylotriose (Zhou and Ingram, 1999). *K. oxytoca* has a number of metabolic pathways, producing a range of products, including acetate, lactate, succinate and butanediol (Ohta *et al.*, 1991b). However, through genetic engineering, product formation during growth on soluble sugars could be altered to enable *K. oxytoca* to produce ethanol at 95% of the maximum theoretical yield (Wood and Ingram, 1992). *K. oxytoca* has the additional advantage of being able to ferment glucose and xylose at the same rate, significantly shortening fermentation times (Ohta *et al.*, 1991b).

T. saccharolyticum can grow at temperatures as high as 65°C (Shaw *et al.*, 2008b), allowing for processes with substantially improved kinetics. Furthermore, *T. saccharolyticum* does not suffer from the same level of carbon catabolite repression as *E. coli* and is capable of utilizing glucose and xylose at essentially the same rate (Shaw *et al.*, 2008a). *T. saccharolyticum* is capable of metabolizing all the sugars present in biomass, producing organic acid in addition to ethanol. With the development of genetic systems for engineering *T. saccharolyticum* (Desai *et al.*, 2004; Shaw *et al.*, 2010; Tyurin *et al.*, 2005) it became possible to create the necessary gene knock-out mutant strains to direct product formation almost exclusively to ethanol (Shaw *et al.*, 2008b). Recently, Tsakralides *et al.* (2012) began characterizing carbon catabolite repression in *T. saccharolyticum*. The authors produced HPr mutants with derepressed carbon catabolite repression. These mutants displayed significantly improved simultaneous fermentation of glucose, xylose, galactose, arabinose and mannose.

Unlike the above-mentioned three organisms, wild type strains of *Zymomonas mobilis* produce ethanol at high rates and yields (da Silveria dos Santos *et al.*, 2009; Joachimsthal *et al.*, 1998), but it can only utilize glucose and fructose as carbon source. Zhang *et al.* (1997) patented a *Z. mobilis* strain capable of fermenting both xylose and arabinose. This required the introduction of seven

Table 3.1 Ethanol production from xylose

Organisms	Ethanol yield (g ethanol/g xylose)	Ethanol productivity		Temperature	Reference
		g/l/h	g/g DW/h		
S. cerevisiae[a]	0.48	3.3		30°C	Cavka *et al.* (2011), Pagliardini *et al.* (2013)
E. coli FBR5	0.44	1.0		35°C	Martin *et al.* (2006)
E. coli KO11	0.47	0.88	2.24	35°C	Kim *et al.* (2007), Tao *et al.* (2001)
K. oxytoca	0.48	2.0		30°C	Ohta *et al.* (1991b)
T. saccharolyticum ALK2	0.46	2.2		55°C	Shaw *et al.* (2008b)
Z. mobilis	0.48	0.57			Lawford and Rousseau (1999), Zaldivar *et al.* (2001)
S. stipitis	0.48	0.9		30°C	Jeffries *et al.* (1996), Skoog and Hahn-Hägerdal (1990)
H. polymorpha		0.054		48°C	Dmytruk *et al.* (2013)
K. marxianus SUB-80-S	0.28	0.1		42°C	Margaritis and Bajpai (1982)
K. marxianus YZB014[b]	0.19	0.111		42°C	Zhang *et al.* (2013)
K. marxianus YRL005[b,c]	0.38	0.069		42°C	Wang *et al.* (2013b)
S. cerevisiae TMB 3400 (XR/XDH)	0.33		0.43	30°C	Karhumaa *et al.* (2007)
S. cerevisiae TMB 3066 (XI)	0.43		0.34	30°C	Karhumaa *et al.* (2007)
S. cerevisiae Scf2	0.37	0.52		30°C	Zhang and Geng (2012)
S. cerevisiae			0.67		Demeke *et al.* (2013a)

[a]*S. cerevisiae* on glucose as benchmark.
[b]Express the *S. stipitis* xylose reductase and xylitol dehydrogenase.
[c]Express a codon-optimized *Orpinomyces* xylose isomerase.

different genes encoding xylose isomerase, xylulokinase, L-arabinose isomerase, L-ribulokinase, L-ribulose-5-phosphate 4-epimerase, transaldolase, and transketolase. Although co-fermentation of glucose, xylose and arabinose was possible, there was a preferential order of utilization: glucose first, then xylose and lastly arabinose (Mohagheghi *et al.*, 2002).

The yeast *Scheffersomyces stipitis*, formerly *Pichia stipitis*, was isolated from decaying wood and the larvae of wood-feeding insects and is one of the best natural xylose fermenting microorganisms known (Jeffries *et al.*, 2007). It can convert most of the major sugars in plant biomass to ethanol. Whereas this yeast cannot ferment arabinose, it is still able to utilize this carbon source for cell biomass production. Compared to *S. cerevisiae*, *S. stipitis* is not an especially good ethanol producer, but it is able to metabolize acetic acid (Agbogbo and Coward-Kelly, 2008). It is also capable of reducing the furan ring in furfural and hydroxymethylfurfural (HMF), enabling the yeast to remediate some of the most common inhibitors in pretreated biomass (Almeida *et al.*, 2008).

Chen *et al.* (2011) identified XR as the rate limiting step during D-xylose fermentation. Through directed evolution these researchers increased enzyme activity 5-fold. This led to a 36% increase in the volumetric production and a 100% increase in the specific production rate of ethanol on D-xylose. Effective co-factor recycling between XR and XDH is another area considered to be key in the fermentation of D-xylose by *S. stipitis*. The different co-factor requirements of these two enzymes cause an intracellular redox imbalance under anaerobic conditions, leading to accumulation of D-xylitol and low ethanol yields (Hahn-Hägerdal *et al.*, 2013; Skoog and Hahn-Hägerdal, 1990). Efforts to change the co-factor requirements of these enzymes have been made. In 2005, Watanabe *et al.* (2005) managed to completely reverse the co-factor specificity of XDH from NAD^+ to $NADP^+$. The group subsequently also succeeded in creating a strictly NADPH-dependent xylose reductase in order improve co-factor recycling in *S. stipitis* (Khattab *et al.*, 2011).

Systems biology approaches for *S. stipitis* based on metabolic reconstruction and flux analysis lead to the development of genome-scale metabolic models (GSMMs) which hold the potential to engineer other promising hosts through the application of comparative flux analysis. For example, by analysing the metabolic pathways of *S. stipitis*, known for its ability to ferment xylose without any concomitant xylitol production, the reconstructed GSMMs can inform rational metabolic engineering strategies of *S. cerevisiae* for enhanced pentose to ethanol fermentation (Chung *et al.*, 2013).

The thermotolerant methylotrophic yeast *Hansenula polymorpha* is capable of fermenting glucose, cellobiose and xylose (Dmytruk *et al.*, 2013; Ishchuk *et al.*, 2008; Ryabova *et al.*, 2003). Whereas this yeast grows optimally at 37°C, it can tolerate temperatures of up to 48°C. This high tolerance to temperature provides for significantly improved enzyme kinetics in a SSCF process, since a 10°C increase in temperature approximately doubles the reaction rate of the cellulase enzymes (Ibrahim and El-diwany, 2007). Furthermore, its ability to resist heavy metals and oxidative stress makes *H. polymorpha* a good candidate for the production of ethanol from lignocellulosic materials.

H. polymorpha can naturally ferment xylose to ethanol. However, a considerable amount of ethanol is reutilized, resulting in a substantial decrease in the ethanol yield. Using UV mutagenesis, Ishchuk *et al.* (2008) created a mutant with an impaired ability to reutilize ethanol as carbon source. High level expression of the *Kluyveromyces lactis* pyruvate decarboxylase in this mutant lead to a 2.2-fold increase in ethanol production from D-xylose. Changing the co-factor requirements of the *H. polymorpha* XR to favour NADH, resulted in a 7.3-fold increase in the ethanol productivity on xylose and a significant reduction in D-xylitol production (Table 3.1) (Dmytruk *et al.*, 2013). Therefore, it is clear that *H. polymorpha* has several attributes that make it a good candidate for lignocellulosic ethanol production. However, the current ethanol productivity levels of this yeast remains comparatively low compared with other D-xylose-fermenting strains. Transcriptome analysis revealed that some genes involved in glycolysis and ethanol fermentation are repressed during growth on D-xylose (refer to discussion in next section). It is thus likely that future studies will focus on increasing the expression of these repressed genes (O.C. Kim *et al.*, 2013).

The thermo-tolerant yeast, *K. marxianus* can utilize a wide range of sugars, including xylose (Kuhad *et al.*, 2011; Zhang *et al.*, 2013), where several strains of *K. marxianus* can grow at temperatures as high as 45–52°C. In fact, at these high temperatures the fermentation efficiency of this yeast is as efficient as that of *S. cerevisiae* at 30°C (Nonklang *et al.*, 2008). However, similar to most organisms that convert D-xylose to D-xylulose by means of two oxido-reductases (XR/XDH), *K. marxianus* also suffers from a redox imbalance under oxygen-limited conditions. Margaritis and Bajpai (1982) evaluated eight strains of *K. marxianus* for ethanol production on D-xylose. Their best strain designated SUB-80-S produced 0.28 g ethanol/g D-xylose at a rate of 0.1 g/l/h. In an attempt to increase ethanol production, two independent groups replaced the native *K. marxianus* XR and XDH with *S. stipitis* enzymes (Goshima *et al.*, 2013; Zhang *et al.*, 2013). In both cases this modification led to a significant increase in ethanol production compared to the parental strains. Zhang *et al.* (2013) measured an increase from 0.06 up to 0.19 g ethanol/g D-xylose with this strategy. A mutant strain of *S. stipitis* harbouring XR with a preference for NADH as co-factor was also evaluated. Unfortunately, expression of the mutant gene was too low to support growth and fermentation of D-xylose. In another study by Wang *et al.* (2013b) the XR and XDH were replaced by a codon-optimized xylose isomerase (XI) from an *Orpinomyces* sp. This strain fermented D-xylose to ethanol at good yields without the formation of D-xylitol (Table 3.1).

The yeast *S. cerevisiae* has been used for centuries for the production of ethanol from hexose sugars. Among the several attributes that makes it a good candidate for industrial production of ethanol, this species is one of the best ethanol producing organisms, at volumetric rates as high as 3.3 g/l/h, using glucose as sole carbon source. However, *S. cerevisiae* is generally considered to be unable to ferment xylose. Van Zyl *et al.* (1989) described four *S. cerevisiae* strains capable of consuming xylose under aerobic conditions at different efficiencies. The degree of utilization depended on the presence of other substrates or yeast extract. The highest level of xylose utilization of 69% was observed in presence of D-ribose. Using [^{14}C]-xylose in the presence of D-ribose under aerobic conditions, the authors illustrated that most of the radioactive labelled carbon was distributed to D-xylitol (91%), CO_2 (2.6%) and biomass (1.7%). However, although a small amount of ethanol was produced under anaerobic conditions, none of the labelled carbon was incorporated into this metabolic end-product, even in the presence of D-ribose.

By overexpressing the endogenous non-specific aldose reductase (*GRE3*) and the XDH (*YLR070c*, *ScXYL2*) Toivari *et al.* (2004) could illustrate growth on D-xylose in aerobic cultures. They also found that XDH activity significantly increased during growth on D-xylose, although this was not the case with the XR. The recombinant strains converted 55% of the consumed D-xylose to D-xylitol when expressing the endogenous enzymes, which was slightly greater than a recombinant strain with the XDH from *S. stipitis*. Expression of the *S. stipitis* XR (*XYL1*) and XDH (*XYL2*) resulted in a strain that grew more rapidly on D-xylose than the strains over-expressing endogenous *S. cerevisiae* genes (Kotter and Ciriacy, 1993; Tantirungki *et al.*, 1993). These strains accumulated less D-xylitol and also produced ethanol. Further improvements were made by the over-expression of the *S. cerevisiae* xylulokinase (XK) (Ho *et al.*, 1998; Jin *et al.*, 2003; Toivari *et al.*, 2001) and later also the *S. stipitis* version of this gene (*XYL3*). However, it soon became evident that too high XK activity has a negative influence on ethanol production (Jin *et al.*, 2002). Optimization of the *S. stipitis* genes in *S. cerevisiae* has been the focus of much research. Eliasson *et al.* (2001) used a kinetic model to calculate the optimal ratio of XR/XDH/XK to be 1:≥10:≥4. The authors proved this ratio experimentally with strain TBM3004. This strain produced 0.35 g ethanol/g D-xylose and D-xylitol production was reduced to 0.19 g/g D-xylose.

With the XR/XDH-pathway, the availability of NAD^+ for XDH is limited under anaerobic conditions, because of the preference of the *S. stipitis* XR for NADPH as co-factor (Walfridsson *et al.*, 1997). By changing the co-factor requirements of the XR, D-xylitol production was reduced from 0.19 to 0.12 g D-xylitol/g D-xylose, which led to an increase of 0.35 to 0.39 g ethanol/g D-xylose (Bengtsson *et al.*, 2009; Runquist *et al.*, 2010a). Many of the problems associated with the XR/XDH-pathway could theoretically be solved by expressing a XI in *S. cerevisiae*, since this would enable the one-step conversion of D-xylose to D-xylulose, without the need for co-factors. For many years, scientists attempted to express XIs from different sources in *S. cerevisiae*, with very limited success (Amore *et al.*, 1989; Gardonyi and Hahn-Hägerdal, 2003; Sarthy *et al.*, 1987; Walfridsson *et al.*, 2013). Eventually an XI with sufficient activity to support growth on D-xylose was obtained from the anaerobic fungus *Piromyces* sp. E2 (Kuyper *et al.*, 2003). The *S. stipitis* XK (XYL3) was required to enable rapid fermentation of D-xylose under anaerobic conditions. To eliminate the production of D-xylitol, the *S. cerevisiae* non-specific aldose reductase (*GRE3*) was deleted, which was a critical necessity, since the *Piromyces* XI is inhibited in the presence of D-xylitol (Brat and Boles, 2013). Recently the *Clostridium phytofermentans* XI was successfully expressed in *S. cerevisiae* (Brat *et al.*, 2009). The kinetic properties of this enzyme is very similar to that of the *Piromyces* XI, but it is significantly less sensitive to D-xylitol inhibition (Brat *et al.*, 2009; Brat and Boles, 2013).

Karhumaa *et al.* (2007) compared the XR/XDH or XI-pathways in recombinant *S. cerevisiae* to settle the question on which of these pathways are superior in terms of ethanol production from xylose. The final ethanol concentration with the XR/XDH strain was 13.3 g/l, whereas 7.3 g/l ethanol was produced by the strain containing the XI in anaerobic batch cultures. Furthermore, the rate of D-xylose consumption and specific ethanol productivity were significantly higher in the XR/

XDH strain. However, the XI strain produced significantly less D-xylitol, 0.04 g D-xylitol/g D-xylose compared to the XR/XDH strain with 0.22 g D-xylitol/g D-xylose. The XI strain therefore had a superior ethanol yield of 0.43 g/g D-xylose vs. 0.33 g/g in the case of the XR/XDH strain (Table 3.1). A recent metabolomic study concluded that low D-xylose uptake by the XI-strain resulted in carbon starvation (Bergdahl *et al.*, 2012). This was not the case with the XR/XDH-strain, as it was able to obtain sufficient energy form D-xylose.

With the improvements made in the rate at which D-xylose is utilized, D-xylose uptake became the rate-limiting step. *S. cerevisiae* consumes D-xylose via high-affinity hexose transporters (Gardonyi *et al.*, 2003). These generally have a two orders of magnitude lower affinity for D-xylose than for hexose sugars (Lee *et al.*, 2002). Expression of heterologous transport proteins improved D-xylose uptake at low D-xylose concentrations of 0 to 15 g/l (Du *et al.*, 2010; Gardonyi *et al.*, 2003; Lee *et al.*, 2002; Runquist *et al.*, 2010b; Saloheimo *et al.*, 2007). Studies confirmed that catabolite repression regulate native *S. cerevisiae* transporters (Runquist *et al.*, 2010b). D-Xylose transport increased significantly during cultivation on D-xylose in the presence of low concentrations of D-glucose. To enable simultaneous metabolism of D-glucose and D-xylose, a D-xylose transporter able to function at high D-glucose concentrations is essential.

D-Xylose fermentation is complex and increasing the expression of three genes is not the complete story. Wahlbom *et al.* (2003) used chemical mutagenesis and selected for improved growth on D-xylose and CO_2 production. The ethanol yield in the best mutant strain TBM 3400 increased from 0.21 to 0.25 g ethanol/g D-xylose under oxygen-limited conditions. Ethanol yield under strictly anaerobic condition were lower at 0.18 g ethanol/g D-xylose.

It took approximately 20 years to create a good D-xylose-fermenting *S. cerevisiae* strain using rational metabolic engineering. These developments involved constitutive expression of multiple genes followed by mutagenesis and post evolutionary engineering (Gong *et al.*, 2009). Whole genome engineering enables simultaneous changes at different positions throughout the genome and does not require prior genome sequence information. Using a modified whole genome shuffling technique Zhang and Geng (2012) constructed a recombinant *S. cerevisiae* strain with enhanced D-xylose-fermenting ability. The technique involved recursive recombination of the *S. stipitis* genome with that of *S. cerevisiae* through direct genome isolation and transformation, followed by a carefully-designed screening strategy. The best strain obtained with this method, *S. cerevisiae* ScF2, displayed increased D-xylose consumption and ethanol production when compared with the parental strains. It could also produce ethanol from D-xylose at a higher rate than *S. stipitis* (Table 3.1). Furthermore, in a recent study, B. Kim *et al.* (2013) employed elements of synthetic and systems biology to create a random DNA assembler platform that created random pathways of xylose utilization in *S. cerevisiae*. This powerful strategy led to the ability to select for the best randomly assembled combination enzymes required for improve xylose utilization.

Another powerful technique for the construction of *S. cerevisiae* strains for the production of lignocellulosic ethanol involves the combination of yeast strains with desired traits. Demeke *et al.* (2013a) screened 580 yeast strains for high inhibitor tolerance. Through sporulation, segregants were obtained with even higher inhibitor tolerance. One of these segregants was crossed with a D-xylose fermenting strain, which also exhibited good ethanol producing ability and the segregants were screened for inhibitor tolerance and efficient D-xylose fermentation. The best strains showed good D-xylose utilization rates and were able to ferment undetoxified acid pretreated spruce, producing 23% more ethanol than one of the parental strains.

3.4 Hardening of *S. cerevisiae* against inhibitors formed during lignocellulose pretreatment

Thermo-chemical pretreatment of lignocellulose has become accepted as the norm for 2G ethanol production due to its ease of execution and ability to accommodate throughput of large volumes of feedstock. However, the pretreatment method in itself remains a delicate balance between enhancing the digestibility of the material to maximize sugar yields from cellulose, while limiting the degradation of the hexose and pentose sugars destined for fermentation. By-products from pretreatment are traditionally divided into three main categories (Klinke *et al.*, 2004; Luo *et al.*, 2002; Palmqvist *et al.*, 1999), namely the furan derivatives, weak acids and phenolics. Furaldhehydes are derived from sugar degradation, whereas lignin is mostly a source of aldehydes, phenolics and weak acids, although acetic acid is also derived from hemicellulose (Almeida *et al.*, 2007; Palmqvist and Hahn-Hägerdal, 2000b; Van Maris *et al.*, 2006). In spite of extensive research towards optimizing pretreatment to maximize sugar recovery, a degree of by-product formation will always occur that would affect downstream hydrolytic and fermentative performance.

During the fermentation of lignocellulose, ethanologenic microorganisms are subjected to a range of hostile conditions, including high levels of acidity, osmotic stress from elevated sugar levels, limitations in nitrogen and ethanol toxicity (Casey *et al.*, 2013; Zhao and Bai, 2009). In addition to its predominance in terms of sugar conversion efficiency, *S. cerevisiae* possess an innate resistance to many of these conditions, especially ethanol tolerance (Breisha 2010; Gibson *et al.*, 2007), which resulted in this microorganism being extensively used and most studied for fuel ethanol production using 1G and 2G technologies. However, in spite of having naturally robust characteristics (Olsson and Hahn-Hägerdal, 1993), many of the toxic components arising from lignocellulose pretreatment, even when present in low concentrations, could negatively affect fermentative performance, although responses are varied depending on the type of cultivation and the manner in which the culture was exposed to the inhibitors. For the purposes of this section, fermentative performance will be defined as the ethanol yield, expressed as either biomass produced from sugar consumed (g/g) or as a percentage of the theoretical maximum (%), the volumetric productivity (g/l/h) and the maximum specific growth rate of the culture ($\mu_{max}\, h^{-1}$). Whereas a high ethanol yield enjoys preference above growth rate from a process point of view, the latter is an important indicator of the physiological 'fitness' of the cells.

Excellent reviews in which growth inhibition was quantified are available (Almeida *et al.*, 2009; Palmqvist and Hahn-Hägerdal, 2000a; Klinke *et al.*, 2004), where key studies will be highlighted in the present text. Taherzadeh *et al.* (2000a) found that addition of 4 g/l HMF in the mid-exponential growth phase of cultures of *S. cerevisiae* CBS 8066 resulted in a marked decrease in the maximum specific growth rate, as evident from the biomass concentration as well as the carbon evolution rate, where the latter decreased by 32% (μ_{max} values were not mentioned). However, the ATP yield increased by 70%, which coincided with a slight increase in the ethanol yield to 0.42 g/g, but an almost 30% decrease in the biomass yield to 59 mg/g on 50 g/l glucose. On the other hand, Martín and Jönsson (2003), who compared 11 strains of *S. cerevisiae* with two *Zygosaccharomyces* strains, observed a decrease in the ethanol yield on glucose of between 10% and 56% for the two best-performing *S. cerevisiae* strains CEN.PK and #2SSL3 when exposed to an undiluted inhibitor cocktail. Similarly, the volumetric rate of ethanol production decreased by 84% and 82% for the two respective strains. In a study by Palmqvist *et al.* (1999) using commercial bakers' yeast, acetic acid together with furfural was shown to have negative interactions on the growth rate, biomass and ethanol yield as well as the volumetric rate of ethanol production. In the presence of 10 g/l acetic acid and 1.5 g/l furfural, the μ_{max} decreased by 72% to $0.07 h^{-1}$, whereas the volumetric ethanol productivity decreased from 0.7 to 0.1 g/l/h. Biomass and ethanol yields respectively decreased by 47% and 51%, to final values of 0.027 and 0.19 g/g. The proportionally low biomass yield was attributable to cultivations under fermentative conditions, i.e. minimal exposure to atmospheric oxygen.

Several strategies can be followed to minimize or remove the inhibitory by-products, including detoxification (see below). However, since a detoxification approach is often unpractical from a process cost point of view, much attention focused on the engineering of microbial strains, most notably *S. cerevisiae*, for enhanced inhibitor tolerance. Most of these techniques revolved around metabolic and evolutionary engineering, within context of the central carbon metabolism related to glucose and xylose utilization. Although biotransformation of by-products were described for a variety of microorganisms, including *K. marxianus*, *P. stipitis*, *Zygosaccharomyces* spp., *Candida shehatae* (Almeida *et al.*, 2007; Delgenes *et al.*, 1996; Martín and Jönsson, 2003; Oliva *et al.*, 2003; Olsson and Hahn-Hägerdal, 1996) and the bacteria *Z. mobilis* and *E. coli* (Boopathy *et al.*, 1993; Franden *et al.*, 2013; Klinke *et al.*, 2004; Ranatunga *et al.*, 1997; Zaldivar *et al.*, 2001; and reviewed by Almeida *et al.*, 2009), by far the most developmental work to introduce tolerance into the fermenting organism focused on *S. cerevisiae*. Therefore, in addition to a concise background on inhibitors, this section of the chapter will focus on short- and long-term adaptation strategies that were followed to introduce enhanced tolerance into this ethanologenic yeast.

3.4.1 Detoxification and fermentation strategies

Overcoming inhibitor toxicity after thermochemical treatment requires a multifaceted approach, incorporating the selection of feedstock types and varieties most amenable to pretreatment-hydrolysis (Palmqvist and Hahn-Hägerdal, 2000a; Wyman *et al.*, 1992), optimization of hydrolysis and adapting and/or engineering the microorganism for enhanced tolerance to inhibitors from pretreatment, to eventually ensure maximal ethanol productivity and yield. Meticulous control of the fermentation process, primarily using fed-batch culture to minimize high level exposure to inhibitors, were recommended (Almeida *et al.*, 2007; Rudolf *et al.*, 2005; Taherzadeh *et al.*, 2000b; Wierckx *et al.*, 2011). In this way, the overall substrate loading could be increased, while maintaining inhibitors at subtoxic levels during the course of the fermentation, in addition to avoiding viscosity and mass transfer limitations. This method has the added advantage that hydrolytic enzyme loadings as well as yeast inoculum size can be decreased (reviewed by Olofsson *et al.*, 2008).

A common strategy for inhibitor removal is by washing the substrate (Weiss *et al.*, 2013), whereas detoxification, or substrate conditioning, is another possible process strategy (Alriksson *et al.*, 2011; Cantarella *et al.*, 2004; Millati *et al.*, 2002; Miyafuji *et al.*, 2003; Mussatto and Roberto, 2004; Pienkos and Zhang, 2009). However, such procedures usually imply additional process complexity and cost. Typical detoxification procedures include physical (evaporation and washing) and chemical (solvent extraction, charcoal absorption, ion exchange resins and over-liming) procedures, although biological options such as treatment with laccase-producing fungi were also considered where inhibitor molecules are either inactivated or metabolized (see Jönsson *et al.*, 2013, for a comprehensive review). However, biological treatment has the disadvantage of decreased sugar concentrations and increased process time, resulting in a decreased overall process productivity. Therefore, the use of a fermenting organism with the ability to carry out detoxification combined with fermentation, i.e. without additional detoxification steps, in an optimized fermentation process using high solids loading would be most favourable from a process and cost point of view (Karimi *et al.*, 2008; Öhgren *et al.*, 2006; Rudolf *et al.*, 2005), especially if such an organism is engineered with CBP capability (Van Zyl *et al.*, 2011).

3.4.2 Strategies to enhance tolerance

Adaptation of the culture to inhibitors can be divided into two main categories, namely short-term and long-term adaptation (Almeida *et al.*, 2009). Short-term adaptation is typically encountered when the fermenting microorganism is exposed to the inhibitors before inoculation of the fermentation broth, whether for SSF or simultaneous hydrolysis and fermentation (SHF; Lynd *et al.*, 2005) and this exposure results in the temporary upregulation of enzymes associated with inhibitor reduction or detoxification. However, as the ensuing discussion will show, such adaptation is highly complex and caution should be exercised to generalize the adaptive response.

Alkasrawi *et al.* (2006) clearly illustrated that short-term adaptation effectively enhances the ethanol yield in SSF culture of steam-pretreated spruce, where the inoculum for SSF was grown in pretreatment liquor. In SSF culture at a solids loading of 8%, a commercially-available packaged strain pre-adapted to hydrolysate liquor produced an ethanol yield of 92% of the theoretical maximum, which was 53% greater than when the strains were not pre-adapted before SSF. Fermentation of the hydrolysate further confirmed the beneficial effect of induced tolerance where the pre-adapted commercial strain exhibited a volumetric productivity (calculated from the data as the highest ethanol concentration divided by fermentation time) of 3.29 g/l/h in a hydrolysate concentration corresponding to 10% water-insoluble solids (WIS), that was 77% greater than that of the un-adapted strain. An improvement in the volumetric rate of especially HMF conversion was also evident, where the adapted strain converted HMF at a rate of approximately 0.3 g/l/h, which was more than double that of the un-adapted strain.

Short-term adaptation can further be extrapolated to fed-batch culture where prolonged exposure to sub-lethal levels of inhibitors would result in overall improved performance in terms of ethanol productivity and yield, as opposed to batch SSF culture where the yeast would be exposed to the full extent of the toxicity of the liquor upon inoculation. Using fed-batch addition of hydrolysate to cultures of *S. cerevisiae* strains TMB3000 (isolated from SSL liquor) and TMB3400, Modig *et al.* (2008) observed respective specific rates of ethanol production of 0.61 and 0.35 g/g/h, that were respectively 15% and 35% greater than in batch culture after pulse additions of hydrolysate during the exponential growth phase. Throughout the fed-batch cultivations, no furfural accumulation was observed, although no clear trend in the specific rates of HMF consumption was evident. *In vitro* assays indicated an increase NADPH-dependent reductase activity in these and other strains tested in the same study.

Although preconditioning of the yeast culture is occasionally mentioned (for example Rudolf *et al.*, 2008), studies where the effect of yeast preconditioning was evaluated on SSF performance (such as that by Alkasrawi *et al.*, 2006) are relatively rare. Furthermore, limited information was available with regards to the degree by which the strains retained the induced level of tolerance after the culture was grown in the absence of the hydrolysate. In a study by Wright *et al.* (2011), cultures subjected to evolutionary engineering for tolerance to acetic acid lost their ability to grow on xylose at an acetic acid concentration of 5 g/l after storage and subsequent cultivation in the absence of acetic acid. Conversely, Aarnio *et al.* (1991) isolated acetic acid-resistant bakers' yeast strain from a turbidostat that retained resistance to up to 0.5% (w/v) acetate on agar plates. The plates were inoculated using culture that was subjected to 15 sequential rounds of subculture without selective pressure in shake flasks. The molecular mechanism for the resistance was not identified in this study. No similar studies were found where the same investigations were done using either furaldehydes or hydrolysate.

Long-term adaptation, on the other hand, requires a more comprehensive intervention and can also be divided into two main categories, namely targeted engineering and evolutionary engineering (Sauer, 2001). Targeted engineering typically involves overexpression of key enzymes and requires a thorough knowledge of the metabolic pathways involved. These manipulations are often conducted in conjunction with transcriptomics, metabolomics and metabolic flux analysis. Whereas metabolomics would reveal the presence and concentrations of metabolites, disturbances in the homeostasis might radically influence the flux of metabolites through different pathways. Random mutagenesis followed by strong selective pressure presents an alternative strategy, where prior knowledge of the metabolic pathways is not necessarily required (Demeke *et al.*, 2013b). Ideally, recombinant technology or random mutagenesis should be developed in conjunction with evolutionary engineering, where the latter is discussed in a following section.

3.4.3 Enhancing resistance to furaldehydes

The influence of the furaldehydes on yeast performance during fermentation of lignocellulosic substrates is by far the most comprehensively studied in the literature, due to their prominence in hydrolysate toxicity (Taherzadeh *et al.*, 1997) and proven synergism (Liu *et al.*, 2004; Palmqvist *et al.*, 1999). The two respective furaldehydes formed from the degradation of hexoses (glucose, mannose) and pentoses (xylose, arabinose) are

HMF and furfural (Larsson *et al.*, 1999; Palmqvist and Hahn-Hägerdal, 2000b; Parawira and Tekere, 2011), where the toxicity of these compounds can be attributed to the aldehyde group rather than the furan ring *per se* (Liu, 2011). The concentration of these aldehydes vary according to feedstock and pretreatment method and could range between 2 and 6 g/l for HMF and up to 2 g/l for furfural (comprehensively reviewed by Almeida *et al.*, 2009; Wierckx *et al.*, 2011). The method of toxicity of these molecules reside in the generation of reactive oxygen species that cause damage to DNA, proteins and cellular membranes, and may induce apoptosis (Laluce *et al.*, 2012), in addition to inhibition of key metabolic enzymes (see below). Over and above a decrease in culture productivity, a prolonged lag phase is a distinguishing characteristic of furaldehyde toxicity. Compared to a control, Liu *et al.* (2004) observed lag phases of up to 8 and 4 h for cultures of *S. cerevisiae* strain ATCC 211239 exposed to furfural and HMF at concentrations of 10 and 30 mM, respectively. This lag phase can be attributed to the time required during which the aldehyde groups were reduced to 2,5-bis-hydroxymethylfuran from HMF and furfuryl alcohol or furoic acid from furfural (Taherzadeh *et al.*, 1999, 2000a; Taherzadeh and Karimi, 2011) and cellular damage was repaired (Allen *et al.*, 2010). To date, limited data are available where the maintenance energy requirement was quantified of the cells during this damage repair process as a result of furaldehyde toxicity (Helle *et al.*, 2003).

The Embden–Meyerhof–Parnas (EMP) pathway and PPP should be considered jointly when examining the effect of especially aldehyde inhibitors on the microbial metabolism (Liu, 2011) to identify targets to introduce enhanced tolerance. These pathways are interconnected through the requirement of the cell to maintain a redox balance, where enzymes are either up-regulated or repressed, with the apparently express purpose of increasing the levels of NAD(P)H. In turn, these reducing equivalents facilitate the reduction of furaldhehydes (Liu, 2011; Palmqvist and Hahn-Hägerdal, 2000a). The general effect that inhibitors seem to elicit is repression of glycolysis and up-regulation of enzymes in the PPP. Liu and Moon (2009) demonstrated that reprogrammed metabolic pathways in cultures of the tolerant *S. cerevisiae* strain NRRL Y-50049 exhibited up-regulation of *HXK1*, *HXK2* and *GLK1* expression, from which hexokinases and glucokinases are produced for phosphorylation of glucose in the first step in the EMP pathway. On the other hand, *PGI1* (phosphoglucose isomerase transforming glucose-6-phosphate to fructose-6-phosphate) became repressed and in addition, Modig *et al.* (2002) reported repression of several central glycolytic pathway enzymes, such as alcohol dehydrogenase (ADH), pyruvate dehydrogenase (PDH) and aldehyde dehydrogenase (ALD4). Simultaneously, Liu and Moon (2009) found that *ZWF1* (glucose-6-phosphate dehydrogenase), *SOL3* (lactonase), *GND1* and *GND2* (6-phosphogluconate dehydrogenase), all associated with the PPP, were up-regulated. This work was corroborated by the findings of Gorsich *et al.* (2006), who observed a high degree of sensitivity to HMF by *ZWF1* and *GND1* deletion mutants.

The involvement of the PPP can be attributed to the requirement for NADPH, of which this pathway is a major source (Chu and Lee, 2007; Jeppson *et al.*, 2002; Van Maris *et al.*, 2007), for aldehyde reduction by several alcohol and aldehyde dehydrogenases (Almeida *et al.*, 2007). Usually, this co-factor is mostly associated with amino acid and nucleotide synthesis, whereas NADH is generally required for ATP production during respiration, as well as for the reduction of acetaldehyde to ethanol by ADH1 under fermentative conditions. Control of the expression of the various oxidoreductases (see below) is highly complex, and was found to be primarily regulated by the transcription factor YAP1 (see Liu, 2011, for a comprehensive review). Kim and Hahn (2013) elucidated the underlying mechanism of oxidative stress attributable to the furaldehydes and the involvement of Yap1 transcription factor, a major regulator of the oxidative stress response. It was found that reactive oxygen species accumulate in the cell due to depletion of cellular glutathione, resulting in the activation of this transcription factor, which forms part of the hydrogen peroxide-independent pathway (Alriksson *et al.*, 2010). The main reason why this factor is activated was that HMF and furfural acted as thiol-reactive electrophiles (Kim and Hahn, 2013). Other transcription factor genes that might serve as candidates for targeted engineering include *YAP5*, *YAP6*, *PDR1*, *PDR3*, *RPN4* and *HSF1* (Liu, 2011).

Up-regulation of furaldehyde reductase enzymes could also be beneficial to strains engineered for xylose utilization, especially via xylose reductase (XR) and xylitol dehydrogenase (XDH). Wahlbom and Hahn-Hägerdal (2002) illustrated decreased xylitol and increased ethanol production after pulse additions with acetoin and furfural. Pulse additions with acetoin resulted in a more than two-fold increase in the ethanol yield to reach 1.35 mol ethanol/mol xylose (corresponding to 0.31 g/g). Since a similar response was observed with furfural, it was concluded that decreases in xylitol production during consumption of xylose could be attributed to increased NAD^+ production from furfural reduction and thus a greater level of oxidative power for xylitol conversion by XDH in *S. cerevisiae* strain TMB3001. In turn, decreases in xylitol resulted in an increase in ethanol production. However, the *S. stipitis* XR used in this strain showed a co-factor preference for NADPH (Verduyn *et al.*, 1985) and although either NADH or NADPH can be used for furaldehyde reduction, strain TMB3400 demonstrated an elevated degree of NADPH-dependent activity with furfural, compared to a control strain (Almeida *et al.*, 2008). These findings again supported the intimate involvement of the PPP in furaldehyde reduction.

Several aldehyde-reducing enzymes that could serve as suitable targets for targeted engineering were shown to utilize both NADH and NADPH in the reduction of the furaldehydes, including enzymes coded by the genes *ADH6* utilizing NAD(P)H. Using strain TMB3400 transformed with the xylose reductase/xylitol dehydrogenase (XR/XDH) pathway for xylose utilization, Almeida *et al.* (2008) observed a specific ethanol productivity of 0.35 g/g/h after a pulse addition of spruce hydrolysate. This productivity was 67% greater than that of the parental *S. cerevisiae* strain USM21 (Modig *et al.*, 2008). Although strain USM21 did not have xylose-utilizing capabilities, this improved performance by strain TMB 3400 could not be explained to the low xylose concentration in the medium, which implied the involvement of the XR in furaldehyde reduction. The latter was proved in *in vitro* reactions of crude cell extracts of *S. cerevisiae* strain TMB 3260, which was also derived from strain USM21, but not subjected to random mutagenesis and engineered to overproduce *S. stipitis* XR. Compared to a control strain, the overproducer exhibited 3-fold improved specific activity towards HMF, using NADPH as co-factor. Using a tolerant *S. cerevisiae* strain NRRL Y-50049, obtained through directed evolution, Liu *et al.* (2008) recorded ethanol concentrations of almost 10 g/l at HMF and furfural concentrations of 12 mM each after 24 h, whereas little glucose consumption or ethanol production was observed for the control strain Y-12632. Overexpression of ADH6 and ADH7 resulted in strong cell growth (reported as optical density) in the presence of 40 mM HMF, after a lag phase of approximately 100 h. Similar results for ADH6 were reported by Petersson *et al.* (2006). Other instances where oxidoreductases were reported to improve tolerance to furaldehydes included NADH-utilizing enzymes from *ADH7*, *ALD4* and *GRE3* (Liu *et al.*, 2008), *ADH1* (Almeida *et al.*, 2008); *GRE2* (Moon and Liu, 2012) and *Y62* and *Y76* (Liu, 2011). All reductases reduced both HMF and furfural, except the latter two that were only found to be involved in furfural reduction. Interestingly, Larroy *et al.* (2002) showed that the reductive activity of alcohol dehydrogenases from the genes *ADH6* and *ADH7* were up to 100-fold more efficient compared to their oxidative activity.

3.4.4 Enhancing resistance towards weak acids

Acetic, formic and laevulinic acids are the most abundant weak acids in lignocellulose hydrolysates (Almeida *et al.*, 2007), although tannic and terpene acids can be derived from raw material extractives (Chandell *et al.*, 2013). Acetic acid is primarily formed from the de-acetylation of the hemicelluloses fraction (Almeida *et al.*, 2007), over and above that formed by the fermenting or contaminating organisms. Whereas laevulinic acid is formed from degraded HMF, formic acid originates from the degradation of HMF and furfural (Jönsson *et al.*, 2013; Ulbricht *et al.*, 1984). The high degree of acetylation of hardwood and agricultural residue typically yield weak acid concentrations in excess of 100 mM (Jönsson *et al.*, 2013), whereas concentrations lower than 100 mM were reported for softwoods (Jönsson *et al.*, 2013; Larsson *et al.*, 1999). The mode of toxicity of acetic and formic acids is well described and relates to acidification of the cytoplasm (Palqvist and Hahn-Hägerdal, 2000b). In the case of acetic acid, the undissociated acid permeates freely across the cell membrane and dissociates in the cytoplasm, resulting in uncoupling of the energy metabolism (Guldfeldt and Arneborg, 1998; Pampulha and Loureiro-Dias, 1989; Russel, 1992; Taherzadeh *et al.*, 1997). The small molecular size of formic acid facilitates its cellular penetration, whereas ease of permeation of laevulinic acid could be attributed to its high degree of hydrophobicity (Larsson *et al.*, 1999). Over and above ATP expenditure to pump protons from the cell, acid toxicity was shown to lead to increased turgor pressure, oxidative stress, protein aggregation, lipid peroxidation and the inhibition of membrane trafficking via perturbation of plasma/vacuolar membranes (Chandel *et al.*, 2013), and ultimately cell death via the target of rapamycin (TOR) pathway (Mira *et al.*, 2010a). Typically, acetic acid will result in a decreased growth rate and ethanol yield (Helle *et al.*, 2003; Palmqvist and Hahn-Hägerdal, 2000a). Acetic acid toxicity is of specific relevance to yeast used in lignocellulose fermentations since this acid was shown to affect xylose consumption (Bellissimi *et al.*, 2009; Hasunuma *et al.*, 2011).

Activation of the transcriptional activator Haa1p was identified as a key response in *S. cerevisiae* during weak acid adaptation (Fernandes *et al.*, 2005; Mira *et al.*, 2010a) and the overexpression of *HAA1* resulted in a strain with enhanced acetic acid tolerance as evident from cultures spotted on agar plates (Tanaka *et al.*, 2012). Production of the histone deacetylase complex from SAP30 and protein kinase from HRK1, regulating transporting activity, was also found to be up-regulated under acid stress (Mira *et al.*, 2010b; Zhang *et al.*, 2011). Several multidrug resistance transporter genes were reported to enhance acid tolerance by this yeast, including *TPO1* that infers resistance to 2,4-D artesunic and mycophenolic acids (Alenquer *et al.*, 2006; Mira *et al.*, 2010a), and *TPO2* and *TPO3* enhancing resistance to benzoic, propanoic and acetic acid (Fernandes *et al.*, 2005). Formic acid resistance was reported in strains over-expressing the *FDH1* (formate dehydrogenase) gene (Geertman *et al.*, 2006). Deletion of *FPS1*, encoding the aquaglyceroporin through which acetic acid enters the cell, was investigated since this porin is degraded during acid stress response (Mollapour *et al.*, 2009; Zhang *et al.*, 2011). Results from deletion studies revealed a 31.6% increase in acetic acid tolerance as well as improved xylose fermentation (Zhang *et al.*, 2011; Wei *et al.*, 2013). Similar to the furaldehydes, acetic acid was also found to cause inhibition of non-oxidative enzymes in the PPP. Accumulation of ribulose-5-phosphate and sedoheptulose-7-phosphate led Hasunuma *et al.* (2011) to overexpress transaldolase and transketolase in a xylose-utilizing recombinant strain of *S. cerevisiae* harbouring the XR/XDH system. Compared to a control strain, the ethanol yield on xylose by strain MT8-1X/TAL improved by more than six-fold to 0.328 g/g in the presence of 60 mM acetic acid. Although a strain with the *TKL* gene (MT8–1X/TKL) resulted in a slightly improved xylose consumption rate and ethanol concentration, the latter strain was less resistant to acetic acid.

From the discourse thus far two key themes emerged, namely that the PPP is strongly impacted by both furaldehydes and acetic acid, suggesting that this pathway should enjoy core focus, especially in light of the requirement for co-fermentation of xylose. Secondly, to enhance acid resistance the focus should rest on controlling cell membrane permeability through porins, proton pumps and lipid composition.

3.4.5 Enhancing resistance towards phenolic compounds

Compared to the furaldehydes, relatively little information is available on inhibitors arising from lignin, mostly due to the complexity of the lignin structure and the wide variety of lignin degradation products, including aldehydes and aromatic molecules. The toxicity of the phenolic compounds is attributed to their ability to compromise membrane integrity, which in turn compromises selectivity and enzyme anchoring (Heipieper *et al.*, 1994; Weber and De Bont, 1996). Ma and Liu (2010) identified transcriptional factors under control of *YAP1* conferring resistance to coniferaldehyde and HMF, whereas

PAD1 encoding phenylacrylic acid decarboxylase resulted in increased conversion of ferulic and cinnamic acid in *S. cerevisiae* (Larsson *et al.*, 2001). The products from *PAD1* and *FDC1* (ferulic acid decarboxylase) was shown to act synergistically in the conversion of phenylacrylic acids (ferulic acid, *p*-coumaric acid, cinnamic acid) to corresponding vinyl derivatives (Mukai *et al.*, 2010). Heterologous production of LCC2 laccase from the white rot fungus *Trametes versicolor* in *S. cerevisiae* was also shown to increase resistance to phenolic inhibitors in lignocellulose hydrolysates. The use of laccase was also proposed as a general biological detoxification of lignocellulose hydrolysates (Jönsson *et al.*, 2013).

3.4.6 Evolutionary engineering

The key purpose of evolutionary engineering is to design and implement cultivation strategies to select for strains with desirable traits (Wright *et al.*, 2011; Sauer, 2001). Two approaches can be followed for long-term evolutionary engineering, namely repeated subculture of batch-grown cultures and continuous culture (reviewed by Parawira and Tekere, 2011). In this way, the condition, such as the concentration of the primary growth limiting nutrient or an inhibitor, imposes constant selective pressure on the culture, allowing for selection of strains more proficient at adapting to these conditions. Evolution is usually evident from increases in growth rate and biomass yield. Furthermore, evolutionary engineering can also be used to enhance the ability of cells to simultaneously utilize different carbon sources, as reported by Wisselink *et al.* (2009) and Sanchez *et al.* (2010). Comprehensive evolutionary engineering is based on the combination of two actions, namely (i) exposure of cells to random mutagenesis or metabolic engineering and (ii) long-term cultivation in the presence of inhibitors for the purpose of reprogramming metabolic pathways. Furthermore, these techniques can be applied in a rational engineering approach (Nevoigt, 2008) where the adaptation in the cells is assessed by also taking into account the changes that occur in the genotype, instilling a holistic approach to the evolutionary engineering effort. Such an approach was aptly illustrated by Almario *et al.* (2013) using visualizing evolution in real-time (VERT), whereby molecular mechanisms of adaptive mutants was uncovered using transcriptomics during evolutionary engineering on hydrolysate.

Whereas evolutionary engineering to enhance resistance to furaldehyes was investigated in several instances (see below), comprehensive engineering approaches combined with metabolic engineering for enhanced organic acid tolerance are only recently emerging (Parawira and Tekere 2011; Van Maris *et al.*, 2006; Wright *et al.*, 2011). Furthermore, many studies focused on the use of synthetic inhibitor cocktails, whereas the use of hydrolysate would in fact be a more effective approach to account for inhibitor synergistic effects (Ding *et al.*, 2011). However, contamination, the volume of hydrolysate required for continuous culture on laboratory scale and keeping inhibitor concentrations consistent (variation in content and evaporation) remains challenging.

Wahlbom *et al.* (2003) developed a furaldehyde-resistant *S. cerevisiae* strain TMB3400 through successful combination of mutagenesis and long-term adaptation to increasing concentrations of HMF and furfural over 300 generations in batch culture. This resulted in enhanced furan aldehyde-resistant strains with decreased lag phase of 19 h, compared to a lag phase of 90 h of the parental strain in the presence of 17 mM furfural (Heer and Sauer, 2008). Using continuous culture at a dilution rate of $0.1 h^{-1}$ for 353 h and increasing concentrations of hydrolysate from sugarcane bagasse, Martín *et al.* (2007) demonstrated a more than two-fold increase in the ethanol yield and specific ethanol productivity of *S. cerevisiae* TMB3001 in batch culture when compared to the untreated parental strain. Using over 100 sequential subcultures in a synthetic medium supplemented with 30 mM HMF and furfural or 60 mM furfural and 30 mM HMF, Liu *et al.* (2005) demonstrated up to 100% conversion of HMF to furan di-methanol (FDM) by adapted strains of *S. cerevisiae* 307-12H60 and 307-12H120 and *S. stipitis* 307-10H60. Other instances where evolutionary engineering was carried out using long-term exposure to inhibitors can be found in Amartey and Jeffries (1996), Keller *et al.* (1998) and Koppram *et al.* (2012).

3.5 CBP application to soluble and insoluble (raw, uncooked) starch fermentation

Starch is after cellulose the most abundant glucan polymer in plants (Kennedy *et al.*, 1987). Starch granules are found in the seeds, fruits, leaves, tubers and bulbs of plants as food reserve for periods of dormancy, germination and growth (Preiss, 2009). Starchy crops (maize, grains, potatoes, rice, etc.) constitute 50 to 70% of the energy in human and most domestic animal diets. In contrast to the recalcitrant nature of cellulose, starch is readily degradable by α-1,4 and α-1,6-debranching hydrolases, which are widely distributed in the animal, microbial and plant kingdoms (Solomon, 1978; Van der Maarel *et al.*, 2002).

In the search for renewable alternatives to fossil liquid fuels, the sustainable production of biofuels received much attention in recent years. In the USA, commercial production of bioethanol from maize starch (maize contains on average 62% starch) through liquefaction, hydrolysis and fermentation is a well-established and mature technology that yielded just below 50 BL in 2010, superseding ethanol production from sugarcane in Brazil (Anon., 2011; Gray *et al.*, 2006). However, the energy costs for this process is high due to the requirement for cooking and hydrolysis to gelatinize and liquefy starch (Gray *et al.*, 2006). In fact, the energy required for heating in the conventional process is about 10 to 20% of the fuel value of the ethanol product (Robertson *et al.*, 2006), or 6 to 13% of the total energy consumption in a typical maize-to-ethanol plant (Meredith, 2003). The use of commercial enzymes, as well as caustic soda, lime and sulfuric acid for pH control all add to the process cost.

More cost-efficient utilization of starch substrates for ethanol production could be achieved by either finding ways whereby heating and/or enzyme costs can be decreased or through the biorefinery concept optimized for maximal energy efficiency and where the full array of carbon sources (including lignocellulose) and products are exploited to ensure optimum utilization of biomass (Gallezot, 2011; Hatti-Kaul, 2010). To enhance the energy efficiency of the starch to ethanol process, the focus of the remaining section will rest on employing raw starch degrading enzymes (RSDEs). These enzymes would allow savings in energy costs where the process temperature of saccharification and fermentation are matched using SSF, or where starch hydrolysis and subsequent sugar utilization occurs a single CBP step, which could render the process even more cost-effective. The concept of CBP application in starch fermentation to ethanol is already demonstrated by the commercially available yeast, TransFerm® by Lallemand and Mascoma Corp. (http://www.ethanoltech.com/transferm/), which secretes glucoamylase enzymes required for starch saccharification after liquefaction. Application of this yeast in existing processes for ethanol production from starch may substantially reduce the amount of exogenous glucoamylase required for the saccharification process. Similar application of CBP to the conversion of granular (raw, uncooked) starches in an integrated hydrolysis-fermentation process is under development, and will be considered here.

3.5.1 Starch substrates for bioprocessing

Commercial sources of starch include seeds of maize, wheat, barley, oats, rice and sorghum, tubers and roots of potato, arrow root, and cassava and piths of the sago palm. The starch content differs between these sources, e.g. maize kernel contains on average 72% starch (Eckhoff and Watson, 2009), triticale on average 62.4 to 70.9% (Burešová *et al.*, 2010) and cassava roots with 60 to 70% water typically having a starch content of 15 to 33% (Breuniger *et al.*, 2009).

Starch contains two types of polysaccharides in varying concentrations, namely amylose (about 25%) and amylopectin (about 75%) (French, 1973). Native amylose forms a colloidal dispersion in hot water, but amylopectin remains completely water-insoluble. Amylose is a linear molecule consisting of about 1000 α-1,4-linked D-glucopyranose units (Tester *et al.*, 2004). The amylose chain forms a spiral due to the angle of the α-1,4 linkages, with hydrogen atoms on the inside of the helix. The inside of the spiralled molecule is hydrophobic and allows amylose to form complexes with fatty acids, alcohols and iodine (Whistler and BeMiller, 1997). Amylopectin consists of α-1,4-D-glucopyranose chains with α-1,6 branching points found at 10 to 12 glucose unit intervals (Stevnebø *et al.*, 2006; Tester *et al.*, 2004). Amylopectin constitutes the bulk structure, where intertwined amylose can be removed from the structure by hot water leaching (Oates, 1997). The ratio of amylose and amylopectin can vary considerably between plants, e.g. from 18% to 21% amylose in potato starch and up to 33 to 36% amylose in peas (Buléon *et al.*, 1998). In cereals, the amylose content also varies, for example 27% in maize (Shibanuma *et al.*, 1994) and 24% in triticale (Burešová *et al.*, 2010).

Traditionally, acids were used to hydrolyse starch into simpler carbohydrates called dextrins, whereas enzymes were preferred for this function since the 1970s due to milder process conditions that are less corrosive and deliver higher sugars yields (Robertson *et al.*, 2006). Starch granules contain both ordered, semi-crystalline regions and amorphous regions, which are less ordered and more susceptible to attack by enzymes or acids. The glucan molecules in unmodified (raw) starch are arranged in dense, concentric semi-crystalline rings, which renders raw starch insoluble in cold water and resistant to enzymatic attack (Daniel *et al.*, 2000). Additional heating (typically 60 to 80°C, depending on the type of starch) is therefore needed to gelatinize the raw starch in water prior to enzymatic degradation (Mitsuiki *et al.*, 2005). As the amylose content of the starch rises, the level of crystallinity also increases with a consequent increase in the gelatinization temperature (Ellis *et al.*, 1998; Tester *et al.*, 2004). Gelatinization causes breakage of the hydrogen bonds, which results in an uptake of water and swelling of the molecules, leading to the disruption of the semi-crystalline structure and leaching of the amylose molecules. However, this change in the properties of the slurry results in high viscosity (Daniel *et al.*, 2000), that can make it difficult to stir and pump, and on equipment specifications and maintenance.

3.5.2 Enzymatic degradation of starch

Efficient starch hydrolysis requires the activities of both α-1,4 and α-1,6-debranching hydrolases, which include endo-amylases, exo-amylases, debranching enzymes and transferases. Isoamylases and α-amylases (EC 3.2.1.1) are endo-type enzymes (degrades the internal bonds of the starch) responsible for the liquefaction of starch. The isoamylases and pullulanases (EC 3.2.1.41) cleave the α-1,6-bonds within the starch chain, converting amylopectin into amylose and oligosaccharides of various lengths (25 to 30 glucose residues) and pullulan into maltotriose (Gomes *et al.*, 2003). The β-amylases (EC 3.2.1.3) are exo-type enzymes that hydrolyse the oligosaccharides from the non-reducing ends to maltose residues, which are hydrolysed by maltases into two D-glucose residues.

The two most prominent enzymes required for cooked and raw starch conversion are α-amylases and glucoamylases (EC 3.2.1.3). α-Amylases hydrolyse the internal α-1,4-bonds of amylose and amylopectin at random, producing maltodextrins with a length of 10 to 20 glucose residues, as well as maltose and free glucose. Microbial α-amylases are not able to hydrolyse 1,6-linked units and a number of α-limit dextrins containing at least one 1,6-linkage are usually generated (Kennedy *et al.*, 1987), hence the required synergy between α-amylases, isoamylases and pullulanases. Glucoamylases hydrolyse the terminal 1,4-linked α-D-glucopyranosyl residues successively from non-reducing ends of the starch chains (Kuriowa *et al.*, 2005; Miao *et al.*, 2009; Saha *et al.*, 1989) to release β-D-glucose in an exo-fashion (Belshaw and Williamson, 1993). Most forms of the enzyme can rapidly hydrolyse 1,6-α-D-glucosidic bonds when the next bond in the sequence is 1,4-linked (Fierobe *et al.*, 1996), but the specific activity towards the 1,6-linkage is only 0.2% of that for the 1,4-linkage (Fierobe *et al.*, 1996; Frandsen *et al.*, 1995; Sierks and Svensson, 1994).

Approximately 10% of amylolytic enzymes have the ability to hydrolyse linkages in raw or unmodified starch (Machovic *et al.*, 2005) and generally contain a carbohydrate binding module, called the starch-binding domain (SBD) (Coutinho and Reilly, 1997). The SBD brings the catalytic site of the enzyme in closer proximity with the substrate, thereby enhancing the rate of the catalytic activity (Cornett *et al.*, 2003). Only a few enzymes were identified to hydrolyse raw starch without the presence of a specialized binding domain (Robert *et al.*, 2005; Søgaard *et al.*, 1993).

When attacking the raw starch granule, α-amylases hydrolyse the interior linkages of the starch molecule on the granule surface, supplying the glucoamylases with non-reducing ends as substrate. This hydrolysis results in the formation of small holes in starch granules, which allows the α-amylase entry into the interior of the starch molecule (Buléon *et al.*, 1998). Glucoamylases hydrolyse the shorter dextrins from the non-reducing ends of the starch molecule resulting in the exclusive formation of glucose. Consequently, the combined action of the RSDE (α-amylases and glucoamylases) results in the complete degradation of raw starch into glucose units (Sun *et al.*, 2010).

3.5.3 Natural starch-utilizing microorganisms and their enzymes

Many fungi and yeast possesses the ability to utilize starch as carbon source. Although fungi in general secrete a vast array of enzymes into the substrate on which they grow, few have been identified with the ability to grow well on raw starch. *Mucor, Rhizopus, Rhizomucor, Lipomycetes, Penicillium, Aspergillus* and *Fusarium* species express raw starch α–amylases or glucoamylases, or both in most cases (Mukerjea *et al.*, 2006; Yoon *et al.*, 2007). Of the limited number of yeasts that naturally secrete both α-amylase and glucoamylase enzymes, *Lipomyces kononenkoae* and *Saccharomycopsis fibuligera* are known for their ability to degrade raw starch. However, these yeasts have poor fermentation capabilities and low ethanol tolerance and are thus not ideal for ethanol production (Eksteen *et al.*, 2003a; Eksteen *et al.*, 2003b). The *yeast S. cerevisiae* remains the preferred microorganism for ethanol production and has been used in most fermented beverages (Jespersen 2003; Lodolo *et al.*, 2008). Numerous α-amylase and glucoamylase genes from different sources have been expressed *in S. cerevisiae*, including RSDEs, and will be discussed in the following section.

3.5.4 Engineering yeast for starch bioconversion

A variety of amylase genes were expressed in *S. cerevisiae* for starch conversion over the past three decades. Early reports focused on the conversion of soluble starch and applications in beer fermentation, but most recent studies looked at the conversion of raw starch to bioethanol. For more examples of amylase genes expressed in *S. cerevisiae*, also consult Pretorius (1997) and Sun *et al.* (2010). As the synergistic interaction of both α-amylases and glucoamylases are required for efficient conversion of starch, many groups focused on the co-expression of different α-amylases and glucoamylase or fusion proteins in *S. cerevisiae*. Two general strategies were generally followed, namely the secretion of amylases into the fermentation broth and tethering of enzymes on the cell surface.

Co-expression of secreted enzymes showed good conversion of soluble starch to ethanol, but fermentation often took several days as the enzymes had to be produced *de novo* (Altintas *et al.*, 2003; Eksteen *et al.*, 2003a; Ghang *et al.*, 2007; Kim *et al.*, 2010; Shibuya *et al.*, 1992; Steyn and Pretorius, 1991). The use of tethered enzymes expressed on the cell surface has been eloquently demonstrated by different groups over the past two decades (Kondo *et al.*, 2002; Ma *et al.*, 2000; Murai *et al.*, 1997, 1999). The advantage of this approach is that cells loaded with tethered enzymes are inoculated at high cell densities, shortening fermentation time. However, the interaction of tethered enzymes with the substrate is limited to the immediate vicinity of the cell surface and Liao *et al.* (2010) demonstrated that yeast strains producing secreted enzymes gave higher ethanol yields than yeast strains with tethered enzymes. In some cases, flocculation was also found to interfere with starch hydrolysis employing yeast with tethered enzymes (Khaw *et al.*, 2006b).

Recent studies demonstrated that expression of heterologous genes in yeast improves with increasing cell ploidy. In a study by Yamada *et al.* (2009a), multiple copies of the *Streptococcus bovis* α-amylase and *Rhizopus oryzae* glucoamylase genes were integrated into the *URA3* and *HIS3* loci of separate haploid laboratory strains, with surface display of the glucoamylase and secretion of α-amylase. A diploid strain generated by mating the haploid strains displayed improved amylase activity and substantially better fermentation performance, yielding 89% conversion of 100 g/l raw maize starch to produce 39 g/l of ethanol after 84 h. In a follow-up study, Yamada *et al.* (2010a) performed polyploidization of haploid strains expressing amylases through cell fusion. Diploids generated by mating of haploid strains were used to generate a tetraploid strain through protoplast fusion. Whereas diploidization resulted in an 85% improvement in amylase production, polyploidy resulted in a further 88% increase on glucose and raw starch. Diploid and tetraploid strains exhibited similar growth and ethanol production on glucose, but substantially better than that observed for haploid strains.

3.5.5 Commercial production of bioethanol from starchy feedstocks

The process generally used in industry for the conversion of maize starch to ethanol is conducted in a five-step sequence (Fig. 3.2A). The maize starch is milled (mostly through dry milling), the starch slurry (at about 30% by weight) is adjusted to pH 6.0 and first heated to above 100°C to allow gelatinization and then kept for at least 30 min at 80–90°C in the presence of a thermostable bacterial α-amylase to facilitate liquefaction (hydrolysis to starch oligosaccharides). The mixture is cooled to 60°C, adjusted to pH 4.5 and fungal glucoamylase is added to ensure saccharification to primarily glucose and maltose residues. The saccharification step is often combined with the fermentation step (simultaneous saccharification and fermentation, SSF) during which industrial yeast strains readily ferment the available sugars to ethanol levels of about 16 to 20% (Bothast and Schlicher, 2005; Puligundla *et al.*, 2011). The excess energy demand required for the heating of the slurry during the liquefaction and saccharification steps has been calculated to be between 10% and 20% of the fuel value of ethanol produced (Robertson *et al.*, 2006). Furthermore, thermal liquefaction and solubilization of the starch slurry can increase the viscosity with up to 20-fold, which makes it difficult to pump or stir.

The conversion of starch to ethanol can,

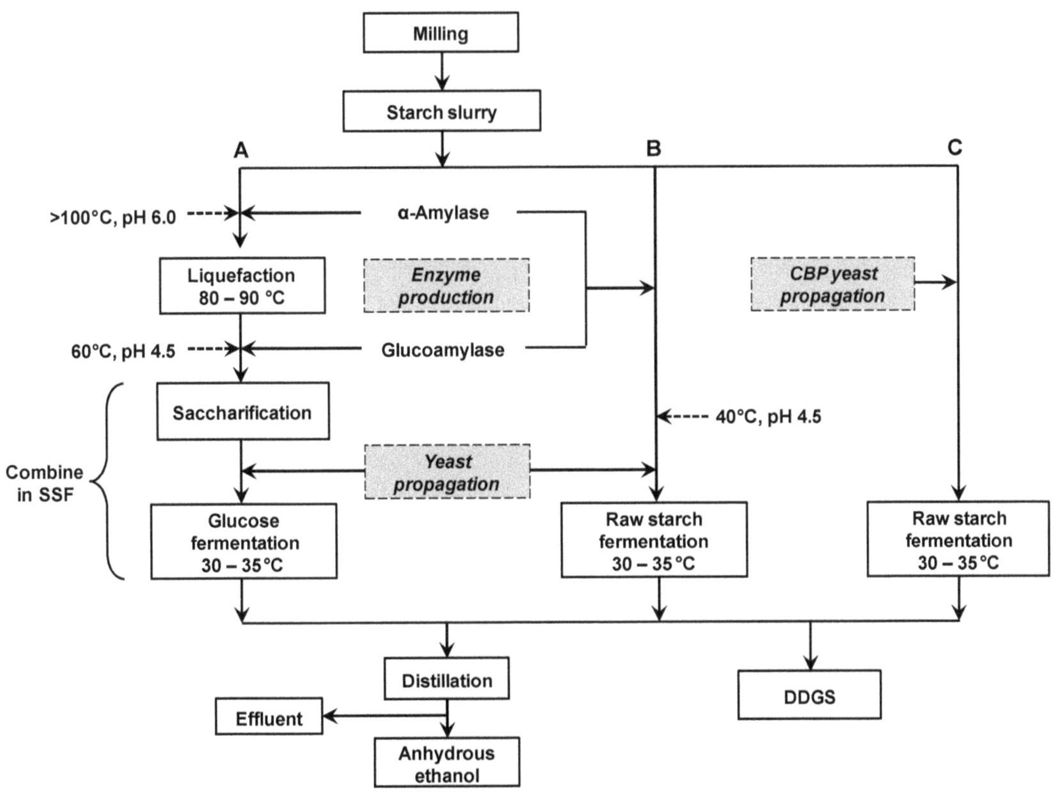

Figure 3.2 Process configurations for starch hydrolysis and fermentation.

therefore, benefit from the design of a more energy-efficient, one-step processes, with SHF of raw, uncooked starch into ethanol (Fig. 3.2B), with reduced energy consumption (Gray *et al.*, 2006; Robertson *et al.*, 2006; Reddy *et al.*, 2009; Rudolf *et al.*, 2009). To carry out the hydrolysis and fermentation of raw, uncooked starch in a single step requires that the hydrolysis temperature of the starch granules be similar to that of the fermentation step, i.e. below the temperature at which starch gelatinization occurs (Lynd *et al.*, 1999; Devantier *et al.*, 2005; Robertson *et al.*, 2006). The direct conversion of raw, granular starch into ethanol on industrial scale has been shown to achieve the same or better ethanol yields and productivity than the conventional cooking process, based on the use of RSDEs (Matsumoto *et al.*, 1982). The present day commercial production of cocktails of RSDEs has enabled the development of low temperature processes for starch hydrolysis-fermentation. Genencor (Dupont-Danisco) recently released the Stargen 002 RSDE cocktail, as a replacement of the previous Stargen 001 version, whereas POET implemented a similar cocktail developed by Novozymes, referred to as POET BPX technology (http://www.poet.com/advantage). The use of these RSDEs proved to be a viable alternative to the conventional liquefaction and saccharification enzymes in a dry-grind maize process, for industrial ethanol production. However, higher enzyme loadings, especially α-amylase, are required for raw starch conversion relative to soluble starch hydrolysis, which partly negate the energy savings by using lower temperatures (Robertson *et al.*, 2006). The next challenge in cost-effective conversion of raw starch requires the expression of starch-hydrolysing enzymes in a fermenting yeast strain to enable one-step CBP (Fig. 3.2C). A CBP process for raw starch conversion to ethanol can save on the excess energy demand in heating of the starch slurry, pumping or stirring of the slurry as well as on the cost of the exogenously-added enzyme cocktail. However, CBP of raw starch would require recombinant *S. cerevisiae* strains to produce sufficient quantities of raw starch degrading enzymes to ensure hydrolysis at a high substrate loading and at moderate temperatures (Robertson *et al.*, 2006). This has become the primary focus of several research groups in recent years and good progress towards proof of concept has been made.

Although several amylolytic *S. cerevisiae* strains could show high conversion of raw starch, it was mostly demonstrated at low (1 to 2%) starch loadings (De Moraes *et al.*, 1995; Favaro *et al.*, 2012; Liao *et al.*, 2010, 2012; Wong *et al.*, 2005). Only a few groups concentrated on the conversion of raw starch at higher substrate loadings. An amylolytic polyploid distiller's strain of *S. cerevisiae* was engineered through multiple integration of genes encoding the glucoamylase of *Aspergillus awamori* and α-amylase of *Debaryomyces occidentalis* into delta (δ) sites distributed throughout the yeast genome (Kim *et al.*, 2010, 2001). The recombinant yeast converted 79% of 20% (w/v) raw maize starch to 10.3% (v/v) ethanol after 6 days of fermentation. Although the ethanol yield was impressive, rich medium containing YP (1% yeast extract and 1% Bacto-peptone) with 20% raw starch and a relatively high cell inoculum (a 10-fold dilution of a cell culture entering stationary phase on YP with 2% soluble starch) was used. Delayed conversion of starch could most likely be attributed to the recombinant yeast that had to produce and secrete sufficient enzymes before effective raw starch conversion could commence. Furthermore, higher enzyme loadings are required for raw starch conversion relative to soluble starch hydrolysis (Robertson *et al.*, 2006; Wong *et al.*, 2005).

Considering the challenges of high substrate-loading for raw starch conversion highlighted above, different Japanese groups made a concerted effort in recent years using tethered amylases displayed on the surface of recombinant *S. cerevisiae* strains to produce ethanol from 5 to 20% raw starch using high cell density loadings. By having enzymes immobilized on the cell surface, the enzymes are retained and available for bioconversion when cells are pre-cultured and used at high cell loadings in starch conversion fermentations (Hasunuma and Kondo, 2012). By combining multi-copy expression of efficient raw starch enzymes (the glucoamylase of *R. oryzae* and α-amylase of *S. bovis*) with high cell loading (10 to 100 g/l wet cell mass), they could demonstrate up to 93.3 g/l ethanol production from 200 g/l raw maize starch (Chen *et al.*, 2008;

Khaw *et al.*, 2006a; Kotaka *et al.*, 2008; Shigechi *et al.*, 2004; Yamada *et al.*, 2010a; Yamakawa *et al.*, 2010). The second advantage of tethered enzymes immobilized on the cell surface, is that cells can be recycled. For example, a stable *S. cerevisiae* strain producing the *R. oryzae* glucoamylase and *S. bovis* α-amylase could be re-used in 23 repetitive fermentations of 100 g/l raw starch without loss of either glucoamylase or α-amylase activity (Yamakawa *et al.*, 2012).

An interesting observation with raw starch converting yeast strains is that α-amylase plays a more critical role than glucoamylase in the synergistic amylases activities, being the rate-limiting step in raw starch fermentation. α-Amylase convert raw starch to oligosaccharides and thereby creates opportunities for synergism by generating new ends for exo-acting glucoamylases (Khaw *et al.*, 2006b; Wong *et al.*, 2007; Yamada *et al.*, 2010a). Analysis of raw starch fermenting strains of *S. cerevisiae* confirmed that insufficient levels of α-amylase production often limited raw starch hydrolysis-fermentation by recombinant *S. cerevisiae*.

Engineered strains of *S. cerevisiae* is not yet ready for industrial employment due to sluggish fermentation times and lower ethanol yields (Yamakawa *et al.*, 2010), particular when compared to the conventional process, with high temperature liquefaction followed by saccharification during fermentation (SSF) and the more recent processes based on RSDEs. The combination of high ethanol yields with near-complete raw starch utilization and volumetric productivities in excess of 1 g/l/h during simple batch culture is still out of reach of the best engineered yeast strains. Improved enzyme production under anaerobic conditions and hardening of engineered strains to grow and ferment raw starches with limited nutrient supplements remain major challenges for the future (Yamada *et al.*, 2009b). A further challenge for raw starch CBP will also be to curb microbial contamination without excessive use of antibiotics (Narendranath, 2003; Robertson *et al.*, 2006). Until superior CBP yeast strains have been developed that can overcome these limitations, the dosage of exogenous RSDEs during the early stages of the fermentation might be essential to ensure industrial application in the shorter term.

3.5.6 Future outlook

The utilization of starch as feedstock for bioethanol production has been met by resistance from some countries who fear that a stronger demand for grain and maize starch will result in an increase in food and feed prices. If more cost-effective raw starch conversion technologies can be developed, alternative sources of starch can also be considered, for example micro- and macro-algae, which has received much attention lately as renewable sources of biomass. Apart from oil production, several micro-algae belonging to the genera *Chlorella*, *Dunaliella*, *Chlamydomonas*, *Scedesmus* and *Spirulina* are known to contain > 50% starch and glycogen, which is still an untapped feedstock for the production of bio-products (John *et al.*, 2011). Algal biomass presents the major advantage in that its use does not compete directly with agricultural food production, especially if saltwater-based production systems are used (Radakovits *et al.*, 2010). Cost-effective methods for harvesting micro-algal cells in industrial-scale production systems, overcoming productivity and contamination challenges, still requires development, but starch converting yeast could play a key role in converting this renewable source of biomass into valuable bio-products.

3.6 Conversion of cellulose to ethanol by *S. cerevisiae* in a CBP configuration

In the last 25 years, several authors reported the expression of cellulase encoding genes in the yeast *S. cerevisiae* (reviewed by Den Haan *et al.*, 2013). Some sought to produce cellulases in an organism that would not yield interfering activities to gain insight into their underlying mechanisms, whereas others hoped to enable the yeast to hydrolyse cellulosic substrates. While complete conversion of insoluble cellulosic substrates by heterologous *S. cerevisiae* strains have been difficult to achieve, recombinant strains were created that could grow on and ferment cellobiose, the repeated monomer of cellulose, at roughly the same rate as on glucose (Van Rooyen *et al.*, 2005; McBride *et al.*, 2005; Wilde *et al.*, 2012). Furthermore, given that enzyme cost could constitute a substantial portion of the total process direct cost, cellobiose-fermenting strains would represent a clear cost benefit in SSF conversions in that such functionality forgoes the need for exogenously added BGL enzyme.

A strain expressing *S. fibuligera* β-glucosidase (*bgl1*) had a similar growth rate when grown on cellobiose or glucose, reached similar levels of biomass and obtained a final ethanol yield of 2.3 g/l from 5 g/l cellobiose, compared to the 2.1 g/l from 5.26 g/l glucose (Van Rooyen *et al.*, 2005). A BGL-secreting strain of *Aspergillus aculeatus* efficiently produced ethanol from pretreated maize-cob in an SSF process, obviating the need for exogenous BGL supplementation (Wang *et al.*, 2013a). An alternative strategy to enable cellobiose utilization was to reconstitute the high affinity cellodextrin transport system of the fungus *Neurospora crassa* into *S. cerevisiae* (Galazka *et al.*, 2010). This approach enabled a recombinant strain that also produced an intracellular *β*-glucosidase to grow on cellodextrins up to cellotetraose. Recently, simultaneous directed evolution of both genes in this pathway allowed for increased growth and cellobiose consumption rates (Eriksen *et al.*, 2013) and it was shown that a similarly engineered strain performed well in SSF experiments without supplementation with exogenous BGLs (Lee *et al.*, 2013). Heterologous production of the *Kluyveromyces lactis* lactose permease also facilitated transport of cellobiose into a recombinant *S. cerevisiae* strain (Sadie *et al.*, 2011). A strain co-producing the permease and the *Clostridium stercorarium* cellobiose phosphorylase was able to grow on cellobiose as sole carbohydrate source. Recently a strain was engineered that produced both a xylose isomerase and cellobiose phosphorylase from *Ruminococcus flavefaciens* that enabled fermentation of glucose, xylose, and cellobiose under anaerobic conditions (Aeling *et al.*, 2012).

Several researchers illustrated successful heterologous production of endoglucanases (EG) and cellobiohydrolases (CBH) individually and in combination in *S. cerevisiae* (Van Zyl *et al.*, 2007; Ilmén *et al.*, 2011). Using a strain co-producing a *β*-glucosidase and an exo/endocellulase in an SSF configuration, Cho *et al.* (1999) demonstrated that required loadings of externally added cellulase could be reduced. This work paved the way for engineering strains that could grow on and ferment cellulosic substrates without the need for external enzyme addition. Attempts to engineer these strains have generally followed three different strategies: (i) production of free enzymes secreted into the growth media, (ii) production of enzymes tethered to the cell wall and (iii) production of enzymes associated as mini-cellulosomes.

To produce a basic free enzyme cellulolytic system, analogous to that of cellulolytic filamentous fungi, an *S. cerevisiae* strain co-expressing the *T. reesei* EG1 (*cel7B*) and *S. fibuligera β*-glucosidase (*cel3A*) was constructed (Den Haan *et al.*, 2007). This strain grew on phosphoric acid swollen cellulose (PASC) and converted 10 g/l PASC to ~1 g/l ethanol. A similar strain that produced significantly higher endoglucanase activity showed improved conversion of PASC to ethanol (Jeon *et al.*, 2009). Given that exoglucanase activity is critical for the hydrolysis of crystalline cellulose, high-level secretion of CBHs by strains actively hydrolysing amorphous cellulose should enable the effective conversion of crystalline cellulose to ethanol (Olson *et al.*, 2012). Put differently, if efficient hydrolysis of amorphous cellulose can be demonstrated, such strains would also greatly benefit the conversion of crystalline cellulose, provided the enzymes are excreted at sufficiently high levels. While high-titre production of CBHs in yeast had proven problematic (Van Zyl *et al.*, 2007), the production of relatively high levels of CBH1 and CBH2 in *S. cerevisiae* was recently reported (Ilmén *et al.*, 2011; McBride *et al.*, 2010). Recombinant CBH1 and CBH2 were shown to be secreted to levels of up to 0.3 g/l and 1 g/l, respectively. Using the CBH1 and CBH2 enzymes that could be secreted at the highest levels, a yeast strain was constructed that could convert most of the glucan in paper sludge to ethanol without the need for additional enzymes (McBride *et al.*, 2010). The use of this strain resulted in a ~60% decrease in the enzyme loading usually required for the conversion of crystalline cellulose in pretreated hardwood to ethanol in an SSF configuration. A similar strain expressing three alternative BGL, EG and CBH encoding genes produced ethanol from pretreated corn stover in a single step without the addition of exogenous enzymes (Khramtsov *et al.*, 2011). This strain converted 63% of the available cellulose to 2.6% (v/v) ethanol in 96 hours.

The strategy to develop yeast cell factories for CBP of lignocellulose to bioethanol through cell surface engineering was recently reviewed (Hasunuma and Kondo, 2012). Originally, Fujita *et al.* (2002, 2004) reported co-expression and surface display of cellulases in *S. cerevisiae* where high cell density suspensions of a strain displaying the *T. reesei* EG2 and CBH2, and the *A. aculeatus* BGL converted 10 g/l PASC to approximately 3 g/l ethanol. Using the same cellulases, Yamada *et al.* (2010b) later developed a method to integrate a cocktail of cellulase genes through multi-copy δ-integration, to optimize expression levels. Different cellulase expression cassettes encoding the same three main cellulase activities as before were integrated into *S. cerevisiae* chromosomes in one step, and strains expressing an optimum ratio of these cellulases were selected after growth on media containing PASC as carbon source. Total integrated gene copy numbers of an efficient 'cocktail' δ-integrant strain was roughly half that of a conventional δ-integrant strain, however its PASC degradation activity (64.9 mU/g wet cells) was higher than that of the conventional strain (57.6 mU/g wet cells). These results clearly demonstrated that optimization of the cellulase expression ratio improved PASC degradation activity to a greater extent than simple overexpression of cellulase genes. More recently the PASC degradation and fermentation ability of this *S. cerevisiae* strain was improved by co-expressing the cellulases with the *Aspergillus oryzae* AoelpI expansin-like protein on the cell surface. The co-expressing strain showed 246 mU/g wet cells of PASC degradation activity, which corresponded to 2.9-fold higher activity than that of the parental cellulase-expressing strain. In PASC fermentation experiments this strain produced 3.4 g/l ethanol after 96 h of fermentation, a concentration that was 1.4-fold higher than that achieved by the parental strain. Alternatively, the cocktail δ-integrant strain was improved by also co-producing a *N. crassa* cellodextrin transporter and an intracellular version of the *A. aculeatus* BGL (Yamada *et al.*, 2013a). The cellulase/cellodextrin transporter co-expressing strain produced 4.3 g/l ethanol from PASC, 1.7-fold more during a 72 h fermentation than the parental strain.

Bacteria, such as *Clostridium thermocellum*, produce highly efficient cellulase complexes called cellulosomes on their cell wall surfaces when cultured on cellulosic substrates (Bayer *et al.*, 2008). Multiple catalytic components can be assembled on a scaffoldin subunit through strong non-covalent protein–protein interactions between cohesin modules on the scaffoldin and dockerin modules on the enzymes (Himmel *et al.*, 2010). This highly ordered structure of multiple enzymes in close proximity to the substrate results in a high level of enzyme–substrate–microbe synergy (Fierobe *et al.*, 2002; Lu *et al.*, 2006). Several groups attempted to reconstruct a mini-cellulosome on the *S. cerevisiae* cell surface as a third strategy to enable cellulose conversion by the yeast (Ito *et al.*, 2009; Tsai *et al.*, 2009; Wen *et al.*, 2010). Ito *et al.* (2009) constructed a chimeric scaffoldin on the yeast cell surface that allowed cell surface display of both *T. reesei* EG2 and *A. aculeatus* BGL1. The resulting strain was capable of hydrolysing β-glucan. *S. cerevisiae* strains were also engineered to display a mini-cellulosome consisting of a mini-scaffoldin containing a cellulose binding domain and three cohesin modules, anchored to the cell surface. Three types of cellulases, namely the *T. reesei* EG2 and CBH2 and *A. aculeatus* BGL1, each bearing a C-terminal dockerin were also produced to form a tri-functional multi-enzyme complex (Wen *et al.*, 2010). This strain hydrolysed and fermented PASC to ethanol to a final titre of 1.8 g/l. Tsai *et al.* (2010) engineered yeast strains capable of displaying a scaffoldin carrying three different cohesin domains originating from *C. thermocellum*, *Clostridium cellulolyticum* and *Ruminococcus flavefaciens*. In addition, yeast strains were constructed that secreted one of three corresponding dockerin-attached cellulases namely an EG from *C. thermocellum*, an exoglucanase from *C. cellulolyticum*, or a BGL from *R. flavefaciens*. When using a yeast consortium composed of one strain displaying the mini-scaffoldin and three strains secreting the dockerin-tagged cellulases, the cellulases were docked onto the displayed mini-scaffoldin in a predictable manner. Adjusting the ratio of different populations in the consortium allowed fine-tuning of cellulose hydrolysis and ethanol production and 30% of 10 g/l PASC was solubilized in 73 h.

These results demonstrate that, while significant improvements in enzyme titres and activities are required, cellulolytic *S. cerevisiae* strains can be used as a platform for developing an economically viable biofuel process. The strategies used to enable cellulose hydrolysis each have distinct advantages and disadvantages making the best choice of strategy difficult to ascertain simply from a direct comparison of the results. The free-enzyme strategy is limited only by the quantity of enzymes that can be secreted and not by surface display restrictions as with the other two strategies. However, the enzymes diffuse away from the cell into the bulk media and therefore do not directly benefit the cells from which it was secreted. Such a strategy also makes enzyme recycling virtually impossible. In comparison, enzyme immobilization on the cell surface, either in isolation or as mini-cellulosomes maintains the enzymes, and thus their catalytic products, in close proximity to the cell, which would result in improved degrees of enzyme synergism (Yamada *et al.*, 2013b). Immobilized enzymes can also be recycled in batch applications along with the cells (Matano *et al.*, 2013). However, these systems may suffer from comparatively lower levels of activity and inefficiency of immobilized processive enzymes, leading to lower levels of crystalline cellulose degradation compared to free enzymes systems. Inter-laboratory efforts may shed more light on optimal strategies or combinations of strategies to enable cellulase utilization by recombinant yeast strains.

3.7 Mining microbial diversity for novel enzymes for CBP application to starch and lignocellulose, including genomic and metagenomic and/or transcriptomic libraries as sources of novel enzymes/activities

Although significant improvements in enzyme titres and activities were reported to date in *S. cerevisiae* strains with CBP capabilities, the exploration for more efficient cellulases and hemicellulases is ongoing. Driven by the slim profit margins afforded by 2G ethanol to compete with petroleum fuels, much room remains to improve cellulose and hemicellulose conversion efficiencies within the *S. cerevisiae* CBP configuration. Novel cellulases and hemicellulases better suited for industrial application, for example enzymes with higher catalytic efficiency, production levels, temperature and pH stability and tolerance to end-product inhibition would strongly benefit the commercial implementation of the CBP yeast.

Mining for novel enzymes has moved beyond the reliance on the screening and culturing of microorganisms for specific enzyme activities, into an era when novel genes are being discovered by mining entire metagenomes, metatranscriptomes and metaproteomes of complex microbial communities that are yet to be characterized owing to unculturable nature of most microorganisms. Metagenomic studies involve the direct extraction of genomic DNA from environmental samples followed by either functional (activity-based) high-throughput screening techniques of a metagenome library using suitable cloning vectors and hosts or next generation sequencing (sequence-based screening) of the entire metagenome to discover novel genes using a bioinformatics approach. Since the mid 1990s, metagenomics proved to be a powerful tool to understand microbial populations and to discover novel genes/enzymes relevant to lignocellulose degradation, including cellulases, hemicellulases and unique multifunctional enzymes (Xing *et al.*, 2012). Functional and sequencing-based metagenomics have explored soil (Alvarez *et al.*, 2013), the hindgut of termites (Warnecke *et al.*, 2007; Hess *et al.*, 2011), ruminant animals (Dai *et al.*, 2012), snails (Cardoso *et al.*, 2012), rabbit caecum (Feng *et al.*, 2007), compost (Martins *et al.*, 2013), hot springs (Peacock *et al.*, 2013), anaerobic digesters and biogas reactors environments as sources of novel cellulases and hemicellulases. However, the functional metagenome approach faces a distinct challenge. Functional screening of metagenome libraries are often limited to the use of bacterial hosts such as *E. coli*, *Streptomyces* and *Pseudomonas*, with insufficient or biased expression of foreign genes. As such, this strategy has a strong bias towards prokaryotic genes, as eukaryotic gene promoters do not function in the bacterial host and bacteria have no mechanism for mRNA splicing for the removal of introns often found in eukaryotic genes.

In an attempt to overcome this drawback, metatranscriptomic approaches are deemed less biased against eukaryotic genes. Several research papers on this approach have appeared in the last 7 years (Grant *et al.*, 2006; Bailly *et al.*, 2007, Frias-Lopez *et al.*, 2008, Gilbert *et al.*, 2008, Kellner *et al.*, 2011; Damon *et al.*, 2012). Similarly to metagenomic studies, during the metatranscriptomic approach, RNA is directly extracted from environmental samples, converted to cDNA using mRNA as template, followed by either functional (activity-based) screening of a cDNA library using suitable cloning vectors and hosts, or next generation sequencing (sequence-based screening) of the entire metatranscriptome to discover novel genes using a bioinformatics approach. The advantage of this approach is that only coding DNA sequences without non-coding regions such as introns are obtained and expression of the genes are driven by regulatory sequences supplied in the final expression vector. There are, however, some risks associated with working with RNA, which relates mostly to the instability or ribonuclease-sensitivity of mRNA as well as the short half-life of some mRNA species compared to genomic DNA.

Metagenomics and transcriptomics together with other platforms such proteomics and metabolomics form the basis for systems biology approaches, where all of these platforms are used as an integrated ensemble, combined with computational approaches for analysis and model-based prediction, to gain comprehensive insight into complex phenotypes. Insights gained from systems biology approaches consequently influence how metabolic engineering is applied for example to bypass native feedback inhibition, construct heterologous pathways and optimize and rewire metabolic fluxes. Similarly, metabolic engineering has undergone considerable re-invention, from the classical trial-and-error single gene expression metabolic engineering studies to the application of contemporary synthetic biology approaches striving to manipulate entire genetic circuitries or system-wide metabolic modifications (Kondo *et al.*, 2013). The use of systems and synthetic biology approaches to address improved xylose fermentation and inhibitor tolerance have been described in recent reviews (Kung *et al.*, 2012; Hasunuma and Kondo, 2012; Laluce *et al.*, 2012)

3.8 Process configurations for integration of 1G and 2G processes

The technological, economical and environmental advantages of integrating 2G ethanol production with 1G ethanol production in a single facility have been demonstrated in various reports (Ensinas *et al.*, 2007; Pellegrini and de Oliveira Junior, 2011; Macrelli *et al.*, 2012; Dias *et al.*, 2012). The majority of these integration scenarios consider SHF processes, together with separate 1G and 2G fermentation unit operations. The integration of 1G and 2G fermentation process step requires further development, especially for implementation of SSF and CBP to gain advantages in enzymatic hydrolysis. Such integration of 1G and 2G into combined (single) process steps will require micro-organisms with the capability to simultaneously convert glucose, fructose, sucrose and xylose into ethanol at high yield, and in the presence of inhibitors formed during lignocellulose pretreatment. Micro-organisms suitable for separate hydrolysis and co-fermentation (SHCF) and SSCF processes are therefore required. SSCF for integrated 1G-2G fermentation of wheat grain and straw has been demonstrated (Erdei *et al.*, 2013), as well as SHCF for similar feedstocks (Erdei *et al.*, 2012), demonstrating a key requirement for efficient co-fermentation of all sugars in the presence of inhibitors by fermentative organisms. SSCF capability also forms the baseline for CBP technology, for single-step conversion of pretreated lignocelluloses to ethanol.

3.9 Discussion and conclusions

Much progress was clearly made on the development of bio-ethanol as renewable source of liquid fuel, especially for the transport sector. Whereas 1G ethanol is a mature technology, the development of raw starch fermenting yeast, especially in a CBP configuration, holds much promise in making this technology more cost effective and environmentally friendly (Robertson *et al.*, 2006). A key success factor for the successful commercialization of bioethanol, which was also a key theme throughout this chapter, is successful integration on several levels of the bioethanol value chain.

On a cellular level, enzyme production, whether for raw starch or for lignocellulose hydrolysis, should be integrated into the genome of the ethanologen to bring about a decrease enzyme and process costs. The viability of the Beta Renewables plant in Cescentino, Italy, which is based on an SHF configuration, could probably in part be attributed to an effective business model where the prominent enzyme producer Novozymes holds a 10% stake in this plant (Biofuels Digest, 2013). However, in the broader international arena the requirement for a CBP organism for commercial ethanol production cannot be denied, from a cost as well as from an environmental point of view (Lynd *et al.*, 2002). Simultaneously, xylose-fermenting capability should be integrated into the yeast genome to optimize the ethanol yield from total sugar (Oloffson *et al.*, 2008), given the predominant use of *S. cerevisiae* as ethanol producer, and in turn, the strains should exhibit tolerance to pretreatment liquor inhibitors, which was shown here to be intricately linked to the pentose phosphate pathway for pentose utilization (Almeida *et al.*, 2007; Liu, 2011). Whereas the development of such an organism or organisms is a challenge in itself, as demonstrated in the antecedent sections, the licensing of such organisms could present challenges and high barriers to entry when considering rivalry among ethanol producers and resistance from existing enzyme producers opposing replacement products from disruptive technologies, such as embodied in the CBP approach.

Integration of 1G and 2G technologies would also be a key requirement (Marcelli *et al.*, 2012), where existing 1G infrastructure is expanded to incorporate 2G technology, and the plant can benefit from existing knowledge, key relationships with stakeholders such as suppliers and governments, and from existing supply chains and value chain networks. Such integration would mitigate investor risk where such risk from technology that requires further development is diversified through leveraging on mature technologies, i.e. 1G ethanol from either sucrose or starch. The debate on energy crops that compete with food production will, however, remain, even in cases where non-food crops might compete for arable land with food crops (Huang *et al.*, 2012). The challenge thus remains in the international arena for appropriate land husbandry, taking into account both the positive (decreased inflation from decreased fuel imports, energy security and job creation) and negative (increased food prices from supply/demand kinetics) effects that land use for energy production might have on society and social development.

At an even higher level, integration of the multiple stakeholders, including governments, farmers, crop breeders, ethanol producers and even exiting producers of petroleum fuel is of critical importance. A strong political will is required from the governmental sphere to incentivize ethanol production through mandates, tax exemptions and subsidies (Lamers *et al.*, 2011). Crop breeders should work hand in hand with ethanol producers to breed energy crops more amenable to pretreatment and with improved sugar yield, especially due to the diversity in composition of feedstocks around the world, which prevents a 'one size fits all' solution. Given the differences in the life cycle impacts between petroleum-based and renewable fuel, large oil companies should also participate in renewable energy development, given the increasing difficulty and impact of exploiting fossil deposits. Shale oil mining in the Alberta region in Canada (Woynillowicz *et al.*, 2009) and the controversial use of fracking to extract gas deposits (Mooney, 2011; Jenner and Lamadrid, 2013) is a case in point.

Although several challenges thus remain for the production of bioethanol, much room for advancement remains through a variety of routes, including biological and thermo-chemical avenues. However, buy-in of multiple stakeholders in both the private and public sector is required to bring world-wide commercialization of bioethanol to fruition, where such a source of renewable energy would act in synergism with a variety of other renewable sources, including biodiesel and wind, solar and wave energy.

Web resources

Web resources related to energy, data and energy usage

Learn more about energy and source general ethanol and gasoline data related to the US: http://www.eia.gov/

Information on the fuel ethanol industry, policies, infographics and outllook: http://www.ethanolrfa.org/

Energy analysis tools and biomass data resources provided by the National Renewable Energy Laboratory: http://www.nrel.gov/analysis/models_tools.html and http://www.nrel.gov/biomass/data_resources.html

Rich source of data on a variety of alterntive fuels, including ethanol, diesel and hydrogen: http://www.afdc.energy.gov/ and http://www.greenfuels.org/en.aspx

OECD-FAO Agricultural Outlook 2011–2020: manage raw data using an assortment of filters:http://stats.oecd.org/viewhtml.aspx?QueryId=30104&vh=0000&vf=0&l&il=blank&lang=en

Data related to E85 gasoline in the US, including state-by-state price spread: http://www.e85prices.com/

Information on feedstocks and products for breeders, farmers and ethanol producers: https://www.pioneer.com/home/site/us/products/biofuels

Information on energy crops and crop biotechnology: http://www.isaaa.org/default.asp

General information on natural resources: http://www.wri.org/publication/content/8128

Energy statistics, data and helpful publications: http://www.iea.org/statistics/

Ethanol facts: http://www.iowacorn.org/en/ethanol/ and http://westernfarmpress.com/management/40-interesting-facts-about-ethanol

Calculators

Theoretical ethanol yield calculator: http://www1.eere.energy.gov/bioenergy/ethanol_yield_calculator.html

Bioethanol greenhouse gas calculator using different feedstocks to assist in management of biofuel production: http://www.hgca.com/content.output/2135/2135/Resources/Tools/Bioethanol%20Greenhouse%20Gas%20Calculator.mspx

Use this tool to determine the carbon and petroleum footprint of your fleet: http://greet.es.anl.gov/fleet_footprint_calculator

A handy source of information on fuel economy using E10 and E85 fuels: http://www.fueleconomy.gov/feg/ethanol.shtml

Greenhouse gas (GHG) calculator: http://www.biograce.net/content/ghgcalculationtools/recognisedtool

Web resources related to synthetic biology and 'omics' disciplines

The Synthetic Biology Software Suite (SynBioSS) is a software suite for the generation, storing, retrieval and quantitative simulation of synthetic biological networks: http://synbioss.sourceforge.net/

GeneDesign is a web-based program for the design of synthetic genes. It consists of several modules that automate the tasks associated with the manipulation of synthetic sequences: http://54.235.254.95/gd/

RBSDesigner was developed to computationally design synthetic ribosome binding sites (RBS) to control gene expression levels: http://rbs.kaist.ac.kr/

DNAWorks for automatic oligonucleotide design for PCR-based gene synthesis.
http://helixweb.nih.gov/dnaworks/

BioJADE is a design and simulation tool for synthetic biological systems: http://web.mit.edu/jagoler/www/biojade/

The BioMet ToolBox for computational modeling of complex systems: http://129.16.106.142/

Clotho is for engineering synthetic biological systems and managing the data which is used to create them: http://www.clothocad.org/

CycSim for simulating with constraint-based models of metabolism: http://www.genoscope.cns.fr/cycsim/org.nemostudio.web.gwt.App/App.html

GenoCAD is an open-source computer-assisted-design (CAD) application for synthetic biology: http://www.genocad.org/

Integrated Microbial Genomes (IMG) system for the annotation, analysis and distribution of microbial genome and metagenome datasets sequenced: http://img.jgi.doe.gov/

iPATH2 – Interactive Pathways Explorer: http://pathways.embl.de/

MultiGeneBlast for Combined BLAST searches for operons and gene clusters: http://multigeneblast.sourceforge.net/

OPTIMIZER is an on-line PHP application that optimizes the codon usage of a DNA sequence to increase its expression level: http://genomes.urv.cat/OPTIMIZER/

Registry of Standard Biological Parts: http://parts.igem.org/Main_Page

RetroPath – a retrosynthetic biology approach to metabolic pathway design: http://www.issb.genopole.fr/~faulon/retropath.php

TinkerCell is a Computer-Aided Design software tool for Synthetic Biology: http://www.tinkercell.com/

XCMS Online: a web-based platform to process untargeted metabolomic data: https://xcmsonline.scripps.edu.

Joint BioEnergy Institute Inventory of Composable Parts [JBEI-ICE]: http://code.google.com/p/gd-ice/

The MatchMaker for Automated Selection of Synthetic Biology Parts for Genetic Regulatory Networks: https://synbiotools.bbn.com/

The PLOS ONE Synthetic Biology Collection: www.ploscollections.org/synbio

References

Aarnio, T.H., Suihko, M.L., and Kauppinen, V.S. (1991). Isolation of acetic acid-tolerant baker's yeast variants in a turbidostat. Appl. Biochem. Biotechnol. *27*, 55–63.

Aeling, K.A., Salmon, K.A., Laplaza, J.M., Li, L., Headman, J.R., Hutagalung, A.H., and Picataggio, S. (2012). Co-fermentation of xylose and cellobiose by an engineered *Saccharomyces cerevisiae*. J. Ind. Microbiol. Biotechnol. *39*, 1597–1604.

Agbogbo, F.K., and Coward-Kelly, G. (2008). Cellulosic ethanol production using the naturally occurring xylose-fermenting yeast, *Pichia stipitis*. Biotechnol. Lett. *30*, 1515–1524.

Ahmed, A., and Lewis, R.S. (2007). Fermentation of biomass-generated synthesis gas: Effects of nitric oxide. Biotechnol. Bioeng. *97*, 1080–1086.

Alenquer, M., Tenreiro, S., and Sá-Correia, I. (2006). Adaptive response to the antimalarial drug artesunate in yeast involves Pdr1p/Pdr3p-mediated transcriptional activation of the resistance determinants *TPO1* and *PDR5*. FEMS Yeast Res. *6*, 1130–1139.

Alkasrawi, M., Rudolf, A., Lidén, G., and Zacchi, G. (2006). Influence of strain and cultivation procedure on the performance of simultaneous saccharification and fermentation of steam pretreated spruce. Enz. Microb. Technol. *38*, 279–286.

Allen, S.A., Clark, W., McCaffery, J.M., Cai, Z., Lanctot, A., Slininger, P.J., Liu, Z.L., and Gorsich, S.W. (2010). Furfural induces reactive oxygen species accumulation and cellular damage in *Saccharomyces cerevisiae*. Biotechnol. Biofuels 3, 1–10.

Almario, M.P., Reyes, L.H., and Kao, K.C. (2013). Evolutionary engineering of *Saccharomyces cerevisiae* for enhanced tolerance to hydrolysates of lignocellulosic biomass. Biotechnol. Bioeng. *110*, 2616–2623.

Almeida, J., Modig, T., Petersson, A., Hahn-Hägerdal, B., Lidén, G., and Gorwa-Grauslund, M. (2007). Increased tolerance and conversion of inhibitors in lignocellulosic hydrolysates by *Saccharomyces cerevisiae*. J. Chem. Technol. Biotechnol. *82*, 340–349.

Almeida, J.R.M., Modig, T., Roder, A., Liden, G., and Gorwa-Grauslund, M.-F. (2008). *Pichia stipitis* xylose reductase helps detoxifying lignocellulosic hydrolysate by reducing 5-hydroxymethyl-furfural (HMF). Biotechnol. Biofuels *1*, 12.

Almeida, J., Bertilsson, M., Gorwa-Grauslund, M., Gorsich, S., and Lidén, G. (2009). Metabolic effects of furaldehydes and impacts on biotechnological processes. Appl. Microbiol. Biotechnol. *82*, 625–638.

Alriksson, B., Horváth, I., and Jönsson, L.J. (2010). Overexpression of *Saccharomyces cerevisiae* transcription factor and multidrug resistance genes conveys enhanced resistance to lignocellulose-derived fermentation inhibitors. Process Biochem. *45*, 264–271.

Alriksson, B., Cavka, A., and Jönsson, L. (2011). Improving the fermentability of enzymatic hydrolysates of lignocellulose through chemical *in-situ* detoxification with reducing agents. Bioresour. Technol. *102*, 1254–1263.

Alterthum, F., and Ingram, L.O. (1989). Efficient ethanol production from glucose, lactose, and xylose by recombinant *Escherichia coli*. Appl. Environ. Microbiol. *55*, 1943–1948.

Altintas, M.M., Ülgen, K., Kirdar, B., Önsan, Z.I., and Oliver, S.G. (2003). Optimal substrate feeding policy for fed-batch cultures of *S. cerevisiae* expressing bifunctional fusion protein displaying amylolytic activities. Enz. Microb. Technol. *33*, 262–269.

Alvarez, T.M., Goldbeck, R., dos Santos, C.R., Paixão, D.A., Gonçalves, T.A., Cairo, J.P., Almeida, R.F., Oliveira Pereira, I., Jackson, G., and Cota, J. (2013). Development and biotechnological application of a novel endoxylanase family GH10 identified from sugarcane soil metagenome. PloS One *8*, e70014.

Amartey, S., and Jeffries, T. (1996). An improvement in *Pichia stipitis* fermentation of acid-hydrolysed hemicellulose achieved by overliming (calcium hydroxide treatment) and strain adaptation. World J. Microbiol. Biotechnol. *12*, 281–283.

Amore, R., Wilhelm, M., and Hollenberg, C.P. (1989). The fermentation of xylose – an analysis of the expression of *Bacillus* and *Actinoplanes* xylose isomerase genes in yeast. Appl. Microbiol. Biotechnol. *30*, 351–357.

Anex, R.P., Aden, A., Kazi, F.K., Fortman, J., Swanson, R.M., Wright, M.M., Satrio, J.A., Brown, R.C., Daugaard, D.E., Platon, A., Kothandaraman, G., Hsu, D.D., and Dutta, A. (2010). Techno-economic comparison of biomass-to-transportation fuels via pyrolysis, gasification, and biochemical pathways. Fuel *89*, Suppl. 1, S29–S35.

Anon. (2011). Biofuels. OECD/Food and Agriculture Organization of the United Nations (2011). In OECD-FOA Agricultural Outlook 2011–2020, pp. 77–93.

Anon. (2012). Global Ethanol Production to Reach 85.2 Billion Litres in 2012. Renewable Fuels Association. [On line] http://www.ethanolrfa.org/news/entry/global-ethanol-production-to-reach-85.2-billion-litres-in-2012/

Bailly, J., Fraissinet-Tachet, L., Verner, M.C., Debaud, J.C., Lemaire, M., Wésolowski-Louvel, M., and Marmeisse, R. (2007). Soil eukaryotic functional diversity, a metatranscriptomic approach. ISME J. *1*, 632–642.

Bayer, E.A., Lamed, R., White, B.A., and Flint, H.J. (2008). From cellulosomes to cellulosomics. Chem. Rec. *8*, 364–377.

Bellissimi, E., Van Dijken, J.P., Pronk, J.T., and Van Maris, A.J. (2009). Effects of acetic acid on the kinetics of xylose fermentation by an engineered, xylose-isomerase-based *Saccharomyces cerevisiae* strain. FEMS Yeast Res. *9*, 358–364.

Belshaw, N.J., and Williamson, G. (1993). Specificity of the binding domain of glucoamylase 1. Eur. J. Biochem. *211*, 717–724.

Bengtsson, O., Hahn-Hägerdal, B., and Gorwa-Grauslund, M.-F. (2009). Xylose reductase from *Pichia stipitis* with altered coenzyme preference improves ethanolic xylose fermentation by recombinant *Saccharomyces cerevisiae*. Biotechnol. Biofuels 2, 9.

Bergdahl, B., Heer, D., Sauer, U., Hahn-Hägerdal, B., and van Niel, E. (2012). Dynamic metabolomics differentiates between carbon and energy starvation in recombinant *Saccharomyces cerevisiae* fermenting xylose. Biotechnol. Biofuels 5, 34.

Biofuels Digest (2013). Novozymes becomes preferred enzyme supplier for Beta Renewables' current and future cellulosic biofuel projects; embedded in Proesa production technology. [Online] http://www.biofuelsdigest.com/bdigest/2012/10/29/novozymes-acquires-10-of-beta-renewables-for-115m-companies-to-jointly-market-demonstrate-and-guarantee-cellulosic-biofuel-solutions/

Bokinsky, G., Peralta-Yahya, P.P., George, A., Holmes, B.M., Steen, E.J., Dietrich, J., Soon, L.T., Tullman-Ercek, D., Voigt, C.A., Simmons, B.A., and Keasling, J.D. (2011). Synthesis of three advanced biofuels from ionic liquid-pretreated switchgrass using engineered *Escherichia coli*. Proc. Natl. Acad. Sci. U.S.A. *108*, 19949–19954.

Boopathy, R., Bokang, H., and Daniels, L. (1993). Biotransformation of furfural and 5-hydroxymethyl furfural by enteric bacteria. J. Ind. Microbiol. *11*, 147–150.

Bothast, R.J., and Schlicher, M.A. (2005). Biotechnological processes for conversion of corn into ethanol. Appl. Microbiol. Biotechnol. *67*, 19–25.

Brat, D., and Boles, E. (2013). Isobutanol production from D-xylose by recombinant *Saccharomyces cerevisiae*. FEMS Yeast Res. *13*, 241–244.

Brat, D., Boles, E., and Wiedemann, B. (2009). Functional expression of a bacterial xylose isomerase in *Saccharomyces cerevisiae*. Appl. Environ. Microbiol. 75, 2304–2311.

Bräu, B., and Sahm, H. (1986). Cloning and expression of the structural gene for pyruvate decarboxylase of *Zymomonas mobilis* in *Escherichia coli*. Arch. Microbiol. *144*, 296–301.

Breisha, G.Z. (2010). Production of 16% ethanol from 35% sucrose. Biomass Bioenergy *34*, 1243–1249.

Breuniger, W., Piyachomkwan, K., and Sriroth, K. (2009). Tapioca/cassava starch: production and use. In: R.Whistler and J.BeMiller, eds. Starch: Chemistry and Technology, Oxford, UK: Elsevier, pp. 541–568.

Buléon, A., Colonna, P., Planchot, V., and Ball, S. (1998). Starch granules: Structure and biosynthesis. Int. J. Biol. Macromol. *23*, 85–112.

Burešová, I., Sedlá̀èková, I., Famìra, O., and Lipavský, J. (2010). Effect of growing conditions on starch and protein content in triticale grain and amylose content in starch. Plant Soil Environ. *56*, 99–104.

Cantarella, M., Cantarella, L., Gallifuoco, A., Spera, A., and Alfani, F. (2004). Comparison of different detoxification methods for steam-exploded poplar wood as a substrate for the bioproduction of ethanol in SHF and SSF. Process Biochem. *39*, 1533–1542.

Cardoso, A., Cavalcante, J., Cantão, M., Thompson, C., Flatschart, R., Glogauer, A., Scapin, S., Sade, Y., Beltrão, P., and Gerber, A. (2012). Metagenomic analysis of the microbiota from the crop of an invasive snail reveals a rich reservoir of novel genes. PloS One 7, e48505.

Casey, E., Mosier, N.S., Adamec, J., Stockdale, Z., Ho, N., and Sedlak, M. (2013). Effect of salts on the co-fermentation of glucose and xylose by a genetically engineered strain of *Saccharomyces cerevisiae*. Biotechnol. Biofuels *6*, 83.

Cavka, A., Alriksson, B., Rose, S.H., Van Zyl, W.H., and Jonsson, L.J. (2011). Biorefining of wood: combined production of ethanol and xylanase from waste fiber sludge. J. Ind. Microbiol. Biotechnol. *38*, 891–899.

Chandel, A.K., Chandrasekhar, G., Radhika, K., Ravinder, R., and Ravindra, P. (2011). Bioconversion of pentose sugars into ethanol: A review and future directions. Biotechnol. Mol. Biol. Rev. *6*, 8–20.

Chandel, A.K., da Silva, S.S., and Singh, O.V. (2013). Detoxification of lignocellulose hydrolysates: Biochemical and metabolic engineering toward white biotechnology. BioEnergy Res. *6*, 388–401.

Chen, J.P., Wu, K.W., and Fukuda, H. (2008). Bioethanol production from uncooked raw starch by immobilized surface-engineered yeast cells. Appl. Biochem. Biotechnol. *145*, 59–67.

Chen, S.H., Hwang, D.R., Chen, G.H., Hsu, N.S., Wu, Y.T., Li, T.L., and Wong, C.H. (2011). Engineering transaldolase in *Pichia stipitis* to improve bioethanol production. Biochim. Biophys. Res. Commun. *404*, 634–537.

Cho, K.M., Yoo, Y.J., and Kang, H.S. (1999). d-Integration of endo/exo-glucanase and b-glucosidase genes into the yeast chromosomes for direct conversion of cellulose to ethanol. Enzyme Microb. Technol. *25*, 23–30.

Chu, B.C., and Lee, H. (2007). Genetic improvement of *Saccharomyces cerevisiae* for xylose fermentation. Biotechnol. Adv. *25*, 425–441.

Chung, B.K.-S., Lakshmanan, M., Klement, M., Ching, C.B., and Lee, D.Y. (2013). Metabolic reconstruction and flux analysis of industrial *Pichia* yeasts. Appl. Microbiol. Biotechnol. 97, 1865–1873.

Cornett, C.A., Fang, T.-Y., Reilly, P.J., and Ford, C. (2003). Starch-binding domain shuffling in *Aspergillus niger* glucoamylase. Protein Eng. *16*, 521–529.

Coutinho, P.M., and Reilly, P.J. (1997). Glucoamylase structural, functional, and evolutionary relationships. Proteins 29, 334–347.

Dai, X., Zhu, Y., Luo, Y., Song, L., Liu, D., Liu, L., Chen, F., Wang, M., Li, J., and Zeng, X. (2012). Metagenomic insights into the fibrolytic microbiome in yak rumen. PloS One 7, e40430.

Damon, C., Lehembre, F., Oger-Desfeux, C., Luis, P., Ranger, J., Fraissinet-Tachet, L., and Marmeisse, R. (2012). Metatranscriptomics reveals the diversity of genes expressed by eukaryotes in forest soils. PloS one 7, e28967.

Daniel, J., Whistler, R., and Röper, H. (2000). Starch. Ullmann's Encyclopedia of Industrial Chemistry. John Wiley & Sons, pp. 1–25.

Delgenes, J.P., Moletta, R., and Navarro, J.M. (1996). Effects of lignocellulose degradation products on ethanol fermentations of glucose and xylose by *Saccharomyces cerevisiae, Zymomonas mobilis, Pichia stipitis,* and *Candida shehatae*. Enz. Microb. Technol. *19*, 220–225.

Demeke, M.M., Dumortier, F., Li, Y., Broeckx, T., Foulquie-Moreno, R., and Thevelein, J.M. (2013a). Combining inhibitor tolerance and D-xylose fermentation in industrial *Saccharomyces cerevisiae* for efficient lignocellulose-based bioethanol production. Biotechnol. Biofuels *6*, 120.

Demeke, M.M., Dietz, H., Li, Y., Foulquié-Moreno, M.R., Mutturi, S., Deprez, S., Den Abt, T., Bonini, B.M., Lidén, G., Dumortier, F., *et al.*. (2013b). Development of a D-xylose fermenting and inhibitor tolerant industrial *Saccharomyces cerevisiae* strain with high performance in lignocellulose hydrolysates using metabolic and evolutionary engineering. Biotechnol. Biofuels *6*, 89.

De Moraes, L.M.P., Astolfi-Filho, S., and Oliver, S.G. (1995). Development of yeast strains for the efficient utilisation of starch: evaluation of constructs that express a-amylase and glucoamylase separately or as bifunctional fusion proteins. Appl. Microbiol. Biotechnol. *43*, 1067–1076.

Den Haan, R., Rose, S.H., Lynd, L.R., and Van Zyl, W.H. (2007). Hydrolysis and fermentation of amorphous cellulose by recombinant *Saccharomyces cerevisiae*. Metab. Eng. *9*, 87–94.

Den Haan, R., Kroukamp, H., Mert, M., Bloom, M., Gorgens, J.F., and Van Zyl, W.H. (2013). Engineering *Saccharomyces cerevisiae* for next generation ethanol production. J. Chem. Technol. Biotechnol. *88*, 983–991.

Desai, S.G., Guerinot, M.L., and Lynd, L.R. (2004). Cloning of L-lactate dehydrogenase and elimination of lactic acid production via gene knockout in *Thermoanaerobacterium saccharolyticum* JW/SL-YS485. Appl. Microbiol. Biotechnol. *65*, 600–605.

Devantier, R., Pedersen, S., and Olsson, L. (2005). Characterization of very high gravity ethanol fermentation of corn mash. Effect of glucoamylase dosage, pre-saccharification and yeast strain. Appl. Microbiol. Biotechnol. *68*, 622–629.

Dias, M.O., Junqueira, T.L., Cavalett, O., Cunha, M.P., Jesus, C.D., Rossell, C.E., Maciel Filho, R., and Bonomi, A. (2012). Integrated versus stand-alone second generation ethanol production from sugarcane bagasse and trash. Bioresour. Technol. *103*, 152–161.

Dien, B.S., Cotta, M.A., and Jeffries, T.W. (2003). Bacteria engineered for fuel ethanol production: current status. Appl. Microbiol. Biotechnol. *63*, 258–266.

Ding, M.Z., Wang, X., Yang, Y., and Yuan, Y.J. (2011). Metabolomic study of interactive effects of phenol, furfural, and acetic acid on *Saccharomyces cerevisiae*. OMICS *15*, 647–653.

Dmytruk, O.V., Dmytruk, K.V., Abbas, C.A., Voronovsky, A.Y., and Sibirny, A.A. (2013). Engineering of xylose reductase and overexpression of xylitol dehydrogenase and xylulokinase improves xylose alcoholic fermentation in the thermotolerant yeast *Hansenula polymorpha*. Microb. Cell Fact. *7*, 21.

Du, J., Li, S., and Zhao, H. (2010). Discovery and characterization of novel D-xylose-specific transporters from *Neurospora crassa* and *Pichia stipitis*. Mol. BioSyst. *6*, 2150–2156.

Eckhoff, S., and Watson, S. (2009). Corn and sorghum starches: production. In: R.Whistler and J. BeMiller, eds. Starch: Chemistry and Technology. Oxford, UK: Elsevier, pp. 374–439.

Eksteen, J.M., Van Rensburg, P., Cordero Otero, R.R., and Pretorius, I.S. (2003a). Starch fermentation by recombinant *Saccharomyces cerevisiae* strains expressing the a-amylase and glucoamylase genes from *Lipomyces kononenkoae* and *Saccharomycopsis fibuligera*. Biotechnol. Bioeng. *84*, 639–646.

Eksteen, J.M., Steyn, A.J., Rensburg, P.v., Cordero Otero, R.R., and Pretorius, I.S. (2003b). Cloning and characterization of a second a-amylase gene (*LKA2*) from *Lipomyces kononenkoae* IGC4052B and its expression in *Saccharomyces cerevisiae*. Yeast *20*, 69–78.

Eliasson, A., Hofmeyr, J.H.S., Pedler, S., and Hahn-Hägerdal, B. (2001). The xylose reductase/xylitol dehydrogenase/xylulokinase ratio affects product formation in recombinant xylose-utilising *Saccharomyces cerevisiae*. Enz. Microb. Technol. *29*, 288–297.

Ellis, R.P., Cochrane, M.P., Dale, M.F., Duffus, C.M., Lynn, A., Morrison, I.M., Prentice, R.D., Swanston, J.S., and Tiller, S.A. (1998). Starch production and industrial use. J. Sci. Food Agric. *77*, 289–311.

Ensinas, A.V., Nebra, S.A., Lozano, M.A., and Serra, L.M. (2007). Analysis of process steam demand reduction and electricity generation in sugar and ethanol production from sugarcane. Energy Convers. Manag. *48*, 2978–2987.

Erdei, B., Frankó, B., Galbe, M., and Zacchi, G. (2012). Separate hydrolysis and co-fermentation for improved xylose utilization in integrated ethanol production from wheat meal and wheat straw. Biotechnol. Biofuels *5*, 12.

Erdei, B., Galbe, M., and Zacchi, G. (2013). Simultaneous saccharification and co-fermentation of whole wheat in integrated ethanol production. Biomass Bioenergy *56*, 506–514.

Eriksen, D.T., Hsieh, P.C., Lynn, P., and Zhao, H. (2013). Directed evolution of a cellobiose utilization pathway in *Saccharomyces cerevisiae* by simultaneously engineering multiple proteins. Microb. Cell Fact. *12*, 61.

Favaro, L., Jooste, T., Basaglia, M., Rose, S.H., Saayman, M., Görgens, J.F., Casella, S., and Van Zyl, W.H. (2012). Codon-optimized glucoamylase sGAI of *Aspergillus awamori* improves starch utilization in an industrial yeast. Appl. Microbiol. Biotechnol. 95, 957–968.

Feng, Y., Duan, C.J., Pang, H., Mo, X.C., Wu, C.F., Yu, Y., Hu, Y.L., Wei, J., Tang, J.L., and Feng, J.X. (2007). Cloning and identification of novel cellulase genes from uncultured microorganisms in rabbit cecum and characterization of the expressed cellulases. Appl. Microbiol. Biotechnol. 75, 319–328.

Fernandes, A.R., Mira, N.P., Vargas, R.C., Canelhas, I., and Sá-Correia, I. (2005). *Saccharomyces cerevisiae* adaptation to weak acids involves the transcription factor Haa1p and Haa1p-regulated genes. Biochem. Biophys. Res. Comm. *337*, 95–103.

Fierobe, H.P., Stoffer, B.B., Frandsen, T.P., and Svensson, B. (1996). Mutational modulation of substrate bond-type specificity and thermostability of glucoamylase from *Aspergillus awamori* by replacement with short homologue active site sequences and thiol/disulfide engineering. Biochemistry 35, 8696–8704.

Fierobe, H.P., Bayer, E.A., Tardif, C., Czjzek, M., Mechaly, A., Belaich, A., Lamed, R., Shoham, Y., and Belaich, J.P. (2002). Degradation of cellulose substrates by cellulosome chimeras. Substrate targeting versus proximity of enzyme components. J. Biol. Chem. *277*, 49621–49630.

Fonseca, G.G., Heinzle, E., Wittmann, C., and Gombert, A.K. (2008). The yeast *Kluyveromyces marxianus* and its biotechnological potential. Appl. Microbiol. Biotechnol. *79*, 339–354.

Franden, M., Pilath, H., Monagheghi, A., Pienkos, P., and Zhang, M. (2013). Inhibition of growth of *Zymomonas mobilis* by model compounds found in lignocellulosic hydrolysates. Biotechnol. Biofuels *6*, 99.

Frandsen, T.P., Christensen, T., Stoffer, B., Lehmbeck, J., Dupont, C., Honzatko, R.B., and Svensson, B. (1995). Mutational analysis of the roles in catalysis and substrate recognition of arginines 54 and 305, aspartic acid 309, and tryptophan 317 located at subsites 1 and 2 in glucoamylase from *Aspergillus niger*. Biochemistry *34*, 10162–10169.

French, D. (1973). Chemical and physical properties of starch. J. Anim. Sci. *37*, 1048–1061.

Frias-Lopez, J., Shi, Y., Tyson, G.W., Coleman, M.L., Schuster, S.C., Chisholm, S.W., and DeLong, E.F. (2008). Microbial community gene expression in ocean surface waters. Proc. Natl. Acad. Sci. U.S.A. *105*, 3805–3810.

Fujita, Y., Takahashi, S., Ueda, M., Tanaka, A., Okada, H., Morikawa, Y., Kawaguchi, T., Arai, M., Fukuda, H., and Kondo, A. (2002). Direct and efficient production of ethanol from cellulosic material with a yeast strain displaying cellulolytic enzymes. Appl. Environ. Microbiol. *68*, 5136–5141.

Fujita, Y., Ito, J., Ueda, M., Fukuda, H., and Kondo, A. (2004). Synergistic saccharification, and direct fermentation to ethanol, of amorphous cellulose by use of an engineered yeast strain codisplaying three types of cellulolytic enzyme. Appl. Environ. Microbiol. *70*, 1207–1212.

Galazka, J.M., Tian, C., Beeson, W.T., Martinez, B., Glass, N.L., and Cate, J.H. (2010). Cellodextrin transport in yeast for improved biofuel production. Science *330*, 84–86.

Gallezot, P. (2011). Direct routes from biomass to end-products. Catal. Today *167*, 31–36.

Gardonyi, M., and Hahn-Hägerdal, B. (2003). The *Streptomyces rubiginosus* xylose isomerase is misfolded when expressed in *Saccharomyces cerevisiae*. Enz. Microb. Technol. 32, 252–259.

Gardonyi, M., Jeppsson, H., Liden, G., Gorwa-Grauslund, M.-F., and Hahn-Hägerdal, B. (2003). Control of xylose consumption by xylose transport in recombinant *Saccharomyces cerevisiae*. Biotechnol. Bioeng. *82*, 818–824.

Geertman, J.-M.A., Van Dijken, J.P., and Pronk, J.T. (2006). Engineering NADH metabolism in *Saccharomyces cerevisiae*: Formate as an electron donor for glycerol production by anaerobic, glucose-limited chemostat cultures. FEMS Yeast Res. *6*, 1193–1203.

Ghang, D.M., Yu, L., Lim, M.H., Ko, H.M., Im, S.Y., Lee, H.B., and Bai, S. (2007). Efficient one-step starch utilization by industrial strains of *Saccharomyces cerevisiae* expressing the glucoamylase and a-amylase genes from *Debaryomyces occidentalis*. Biotechnol. Lett. *29*, 1203–1208.

Gibson, B.R., Lawrence, S.J., Leclaire, J.P., Powell, C.D., and Smart, K.A. (2007). Yeast responses to stresses associated with industrial brewery handling. FEMS Microbiol. Rev. *31*, 535–569.

Gilbert, J.A., Field, D., Huang, Y., Edwards, R., Li, W., Gilna, P., and Joint, I. (2008). Detection of large numbers of novel sequences in the metatranscriptomes of complex marine microbial communities. PloS One 3, e3042.

Gomes, I., Gomes, J., and Steiner, W. (2003). Highly thermostable amylase and pullulanase of the extreme thermophilic eubacterium *Rhodothermus marinus*: Production and partial characterization. Bioresour. Technol. *90*, 207–214.

Gong, J., Zheng, H., Wu, Z., Chen, T., and Zhoa, X. (2009). Genome shuffling: Progress and applications for phenotype improvement. Biotechnol. Adv. *27*, 996–1005.

Gorsich, S.W., Dien, B.S., Nichols, N.N., Slininger, P.J., Liu, Z.L., and Skory, C.D. (2006). Tolerance to furfural-induced stress is associated with pentose phosphate pathway genes *ZWF1*, *GND1*, *RPE1*, and *TKL1* in *Saccharomyces cerevisiae*. Appl. Microbiol. Biotechnol. *71*, 339–349.

de Gorter, H., Drabik, D., Just, D.R., and Kliauga, E.M. (2013). The impact of OECD biofuels policies on developing countries. Agric. Econ. *44*, 477–486.

Goshima, T., Negi, K., Tsuji, M., Inoue, H., Hoshino, T., and Matsushika, A. (2013). Ethanol fermentation from xylose by metabolically engineered strains of *Kluyveromyces marxianus*. J. Biosci. Bioeng. *116*, 551–554.

Grant, S., Grant, W.D., Cowan, D.A., Jones, B.E., Ma, Y., Ventosa, A., and Heaphy, S. (2006). Identification of eukaryotic open reading frames in metagenomic cDNA libraries made from environmental samples. Appl. Environ. Microbiol. 72, 135–143.

Gray, K.A., Zhao, L., and Emptage, M. (2006). Bioethanol. Curr. Opin. Chem. Biol. *10*, 141–146.

Guldfeldt, L.U., and Arneborg, N. (1998). Measurement of the effects of acetic acid and extracellular pH on intracellular pH of nonfermenting, individual *Saccharomyces cerevisiae* cells by fluorescence microscopy. Appl. Environ. Microbiol. *64*, 530–534.

Hahn-Hägerdal, B., Wahlbom, C.F., Gardonyi, M., Van Zyl, W.H., Cordero Otero, R.R., and Jonsson, L.J. (2001). Metabolic engineering of *Saccharomyces cerevisiae* for xylose utilization. Adv. Biochem. Eng. Biotechnol. 73, 53–84.

Hahn-Hägerdal, B., Karhumaa, K., Fonseca, C., Spencer-Martins, I., and Gorwa-Grauslund, M.-F. (2007). Towards industrial pentose-fermenting yeast strains. Appl. Microbiol. Biotechnol. 74, 937–953.

Hahn-Hägerdal, B., Jeppsson, H., Skoog, K., and Prior, B.A. (2013). Biochemistry and physiology of xylose fermentation by yeasts. Enzyme Microb. Technol. *16*, 933–943.

Hamelinck, C.N., Van Hooijdonk, G., and Faaij, A.P.C. (2005). Ethanol from lignocellulosic biomass: techno-economic performance in short-, middle- and long-term. Biomass Bioenergy 28, 384–410.

Hasunuma, T., and Kondo, A. (2012). Development of yeast cell factories for consolidated bioprocessing of lignocellulose to bioethanol through cell surface engineering. Biotechnol. Adv. *30*, 1207–1218.

Hasunuma, T., Sanda, T., Yamada, R., Yoshimura, K., Ishii, J., and Kondo, A. (2011). Metabolic pathway engineering based on metabolomics confers acetic and formic acid tolerance to a recombinant xylose-fermenting strain of *Saccharomyces cerevisiae*. Microb. Cell Fact. *10*, 2.

Hatti-Kaul, R. (2010). Biorefineries – a path to sustainability? Crop Sci. *50*, S152–S156.

Heer, D., and Sauer, U. (2008). Identification of furfural as a key toxin in lignocellulosic hydrolysates and evolution of a tolerant yeast strain. Microb. Biotechnol. *1*, 497–506.

Heinimö, J., and Junginger, M. (2009). Production and trading of biomass for energy – an overview of the global status. Biomass Bioenergy *33*, 1310–1320.

Heipieper, H.J., Weber, F.J., Sikkema, J., Keweloh, H., and De Bont, J.A. (1994). Mechanisms of resistance of whole cells to toxic organic solvents. Trends Biotechnol. *12*, 409–415.

Helle, S., Cameron, D., Lam, J., White, B., and Duff, S. (2003). Effect of inhibitory compounds found in biomass hydrolysates on growth and xylose fermentation by a genetically engineered strain of *S. cerevisiae*. Enzyme Microb. Technol. *33*, 786–792.

Henstra, A.M., Sipma, J., Rinzema, A., and Stams, A.J. (2007). Microbiology of synthesis gas fermentation for biofuel production. Curr. Opin. Biotechnol. *18*, 200–206.

Hess, M., Sczyrba, A., Egan, R., Kim, T.W., Chokhawala, H., Schroth, G., Luo, S., Clark, D.S., Chen, F., and Zhang, T. (2011). Metagenomic discovery of biomass-degrading genes and genomes from cow rumen. Science *331*, 463–467.

Himmel, M.E., Xu, Q., Luo, Y., Ding, S.Y., Lamed, R., and Bayer, E.A. (2010). Microbial enzyme systems for biomass conversion: emerging paradigms. Biofuels *1*, 323–341.

Ho, N.W.Y., Chen, Z., and Brainard, A.P. (1998). Genetically engineered *Saccharomyces* yeast capable of effective cofermentation of glucose and xylose. Appl. Environ. Microbiol. *64*, 1852–1859.

Huang, J., Yang, J., Msangi, S., Rozelle, S., and Weersink, A. (2012). Biofuels and the poor: Global impact pathways of biofuels on agricultural markets. Food Policy *37*, 439–451.

Ibrahim, A.S.S., and El-diwany, A. (2007). Isolation and identification of new cellulases producing thermophilic bacteria from an Egyptian hot spring and some properties of the crude enzyme. Aust. J. Basic Appl. Sci. *1*, 473–478.

Ilmén, M., Den Haan, R., Brevnova, E., McBride, J., Wiswall, E., Froehlich, A., Koivula, A., Voutilainen, S.P., Siika-aho, M., la Grange, D.C., *et al.* (2011). High level secretion of cellobiohydrolases by *Saccharomyces cerevisiae*. Biotechnol. Biofuels *4*, 30.

Ingram, L.O., and Conway, T. (1988). Expression of different levels of ethanologenic enzymes from *Zymomonas mobilis* in recombinant strains of *Escherichia coli*. Appl. Environ. Microbiol. *54*, 397–404.

Ingram, L.O., Conway, T., Clark, D.P., Sewell, G.W., and Preston, J.F. (1987). Genetic engineering of ethanol production in *Escherichia coli*. Appl. Environ. Microbiol. *53*, 2420–2425.

Ingram, L.O., Conway, T., and Alterthum, F. (1991) Ethanol production by *Escherichia coli* strains co-expressing Zymomonas PDC and ADH genes. US Patent 5,000,000. 3-19-1991.

Ingram, L.O., Gomez, P.F., Lai, X., Moniruzzaman, M., Wood, B.E., Yomano, L.P., and York, S.W. (1997). Metabolic engineering of bacteria for ethanol production. Biotechnol. Bioeng. *58*, 204–214.

Ishchuk, O.P., Voronovsky, A.Y., Stasyk, O.V., Gayda, G.Z., Gonchar, M.V., Abbas, C.A., and Sibirny, A.A. (2008). Overexpression of pyruvate decarboxylase in the yeast *Hansenula polymorpha* results in increased ethanol yield in high-temperature fermentation of xylose. FEMS Yeast Res. *8*, 1164–1174.

Ito, J., Kosugi, A., Tanaka, T., Kuroda, K., Shibasaki, S., Ogino, C., Ueda, M., Fukuda, H., Doi, R.H., and Kondo, A. (2009). Regulation of the display ratio of enzymes on the *Saccharomyces cerevisiae* cell surface by the immunoglobulin G and cellulosomal enzyme binding domains. Appl. Environ. Microbiol. *75*, 4149–4154.

Jeffries, T.W., Davis, B.P., Dahn, K., and Cho, J.Y. (1996). Genetic engineering of xylose fermentation in yeasts. USDA. 1996. 2-26--2013. [Online] http://www2.biotech.wisc.edu/jeffries/bioprocessing/xoferm/xoferm.html

Jeffries, T.W., Grigoriev, I.V., Grimwood, J., Laplaza, J.M., Aerts, A., Salamov, A., Schmutz, J., Lindquist, E., Dehal, P., Shapiro, H., *et al.* (2007). Genome sequence of the lignocellulose-bioconverting and xylose-fermenting yeast *Pichia stipitis*. Nature Biotechnol. *25*, 319–326.

Jenner, S., and Lamadrid, A.J. (2012). Shale gas vs. coal: Policy implications from environmental impact comparisons of shale gas, conventional gas, and coal on air, water, and land in the United States. Energy Policy *53*, 442–453.

Jeon, E., Hyeon, J.E., Suh, D.J., Suh, Y.W., Kim, S.W., Song, K.H., and Han, S.O. (2009). Production of cellulosic ethanol in *Saccharomyces cerevisiae* heterologous expressing *Clostridium thermocellum* endoglucanase and *Saccharomycopsis fibuligera* beta-glucosidase genes. Mol. Cells *28*, 369–373.

Jeppsson, M., Johansson, B., Hahn-Hägerdal, B., and Gorwa-Grauslund, M.F. (2002). Reduced oxidative pentose phosphate pathway flux in recombinant xylose-utilizing *Saccharomyces cerevisiae* strains improves the ethanol yield from xylose. Appl. Environ. Microbiol. *68*, 1604–1609.

Jespersen, L. (2003). Occurrence and taxonomic characteristics of strains of *Saccharomyces cerevisiae* predominant in African indigenous fermented foods and beverages. FEMS Yeast Res. *3*, 191–200.

Jin, Y.-S., Jones, S., Shi, N.Q., and Jeffries, T.W. (2002). Molecular cloning of XYL3 (D-xylulokinase) from *Pichia stipitis* and characterization of its physiological function. Appl. Environ. Microbiol. *68*, 1232–1239.

Jin, Y.-S., Ni, H., Laplaza, J.M., and Jeffries, T.W. (2003). Optimal growth and ethanol production from xylose by recombinant *Saccharomyces cerevisiae* require moderate D-xylulokinase activity. Appl. Environ. Microbiol. *69*, 495–503.

Joachimsthal, E., Haggett, K.D., Jang, J.H., and Rogers, P.L. (1998). A mutant of *Zymomonas mobilis* ZM4 capable of ethanol production from glucose in the presence of high acetate concentrations. Biotechnol. Lett. *20*, 137–142.

John, R.P., Anisha, G.S., Nampoothiri, K.M., and Pandey, A. (2011). Micro and macroalgal biomass: A renewable source for bioethanol. Bioresour. Technol. *102*, 186–193.

Jönsson, L.J., Alriksson, B., and Nilvebrant, N.-O. (2013). Bioconversion of lignocellulose: inhibitors and detoxification. Biotechnol. Biofuels *6*, 16.

Karhumaa, K., Sanchez, R.G., Hahn-Hägerdal, B., and Gorwa-Grauslund, M.-F. (2007). Comparison of the xylose reductase-xylitol dehydrogenase and the xylose isomerase pathways for xylose fermentation by recombinant *Saccharomyces cerevisiae*. Microb. Cell Fact. *6*, 1–10.

Karimi, K., Edebo, L., and Taherzadeh, M.J. (2008). *Mucor indicus* as a biofilter and fermenting organism in continuous ethanol production from lignocellulosic hydrolyzate. Biochem. Eng. J. *39*, 383–388.

Keller, F.A., Bates, D., Ruiz, R., and Nguyen, Q. (1998). Yeast adaptation on softwood prehydrolysate. Appl. Biochem. Biotechnol. *70*, 137–148.

Kellner, H., Luis, P., Portetelle, D., and Vandenbol, M. (2011). Screening of a soil metatranscriptomic library by functional complementation of *Saccharomyces cerevisiae* mutants. Microbiol. Res. *166*, 360–368.

Kennedy, J., Cabral, J., Sá-Correira, I., and White, C. (1987). Starch biomass: a chemical feedstock for enzyme and fermentation processes. In: T. Gaillard, ed. Starch: Properties and Potential. New York: John Wiley & Sons, pp. 115–148.

Khattab, S.M., Watanabe, S., Saimura, M., and Kodaki, T. (2011). A novel strictly NADPH-dependent *Pichia stipitis* xylose reductase constructed by site-directed mutagenesis. Biochem. Biophys. Res. Comm. *404*, 634–637.

Khaw, T.S., Katakura, Y., Koh, J., Kondo, A., Ueda, M., and Shioya, S. (2006a). Evaluation of performance of different surface-engineered yeast strains for direct ethanol production from raw starch. Appl. Microbiol. Biotechnol. *70*, 573–579.

Khaw, T., Katakura, Y., Ninomiya, K., Bito, Y., Katahira, S., Kondo, A., Ueda, M., and Shioya, S. (2006b). Effect of flocculation on performance of arming yeast in direct ethanol fermentation. Appl. Microbiol. Biotechnol. 73, 60–66.

Khramtsov, N., McDade, L., Amerik, A., Yu, E., Divatia, K., Tikhonov, A., Minto, M., Kabongo-Mubalamate, G., Markovic, Z., Ruiz-Martinez, M., and Henck, S. (2011). Industrial yeast strain engineered to ferment ethanol from lignocellulosic biomass. Bioresour. Technol. *102*, 8310–8313.

Kim, B., Du, J., Eriksen, D.T., and Zhao, H. (2013). Combinatorial design of a highly efficient xylose-utilizing pathway in *Saccharomyces cerevisiae* for the production of cellulosic biofuels. Appl. Environ. Microbiol. *79*, 931–941.

Kim, D., and Hahn, J.S. (2013). Roles of the Yap1 transcription factor and antioxidants in *Saccharomyces cerevisiae*'s tolerance to furfural and 5-hydroxymethylfurfural, which function as thiol-reactive electrophiles generating oxidative stress. Appl. Environ. Microbiol. *79*, 5069–5077.

Kim, J.H., Kim, H.R., Lim, M.H., Ko, H.M., Chin, J.E., Lee, H.B., Kim, I.C., and Bai, S. (2010). Construction of a direct starch-fermenting industrial strain of *Saccharomyces cerevisiae* producing glucoamylase, a-amylase and debranching enzyme. Biotechnol. Lett. *32*, 713–719.

Kim, M.D., Rhee, S.K., and Seo, J.H. (2001). Enhanced production of anticoagulant hirudin in recombinant *Saccharomyces cerevisiae* by chromosomal d-integration. J. Biotechnol. *85*, 41–48.

Kim, O.C., Suwannarangsee, S., Oh, D.B., Kim, S., Seo, J.W., Kim, C.H., Kang, H.A., Kim, J.Y., and Kwon, O. (2013). Transcriptome analysis of xylose metabolism in the thermotolerant methylotrophic yeast *Hansenula polymorpha*. Bioproc. Biosyst. Eng. *36*, 1509–1518.

Kim, Y., Ingram, L.O., and Shanmugam, K.T. (2007). Construction of an *Escherichia coli* K-12 mutant for homoethanologenic fermentation of glucose or xylose without foreign genes. Appl. Environ. Microbiol. *73*, 1766–1771.

Klein-Marcuschamer, D., Turner, C., Allen, M., Gray, P., Dietzgen, R.G., Gresshoff, P.M., Hankamer, B., Heimann, K., Speight, R., and Nielsen, L.K. (2013). Technoeconomic analysis of renewable aviation fuel from microalgae, *Pongamia pinnata*, and sugarcane. Biofuels Bioprod. Bioref. 7, 416–428.

Klinke, H.B., Thomsen, A.B., and Ahring, B.K. (2004). Inhibition of ethanol-producing yeast and bacteria by degradation products produced during pretreatment of biomass. Appl. Microbiol. Biotechnol. *66*, 10–26.

Kondo, A., Shigechi, H., Abe, M., Uyama, K., Matsumoto, T., Takahashi, S., Ueda, M., Tanaka, A., Kishimoto, M., and Fukuda, H. (2002). High-level ethanol production from starch by a flocculent *Saccharomyces cerevisiae* strain displaying cell-surface glucoamylase. Appl. Microbiol. Biotechnol. *58*, 291–296.

Kondo, A., Ishii, J., Hara, K.Y., Hasunuma, T., and Matsuda, F. (2013). Development of microbial cell factories for bio-refinery through synthetic bioengineering. J. Biotechnol. *163*, 204–216.

Koppram, R., Albers, E., and Olsson, L. (2012). Evolutionary engineering strategies to enhance tolerance of xylose utilizing recombinant yeast to inhibitors derived from spruce biomass. Biotechnol. Biofuels *5*, 32.

Kotaka, A., Sahara, H., Hata, Y., Abe, Y., Kondo, A., Kato-Murai, M., Kuroda, K., and Ueda, M. (2008). Efficient and direct fermentation of starch to ethanol by sake yeast strains displaying fungal glucoamylases. Biosci. Biotechnol. Biochem. 72, 1376–1379.

Kotter, P., and Ciriacy, M. (1993). Xylose fermentation by *Saccharomyces cerevisiae*. Appl. Microbiol. Biotechnol. *38*, 776–783.

Kudakasseril Kurian, J., Raveendran Nair, G., Hussain, A., and Vijaya Raghavan, G.S. (2013). Feedstocks, logistics and pretreatment processes for sustainable lignocellulosic biorefineries: A comprehensive review. Renew. Sust. Energy Rev. *25*, 205–219.

Kuhad, R.C., Gupta, R., Pal Khasa, Y.P., Singh, A., and Zhang, Y.-H.P. (2011). Bioethanol production from pentose sugars: Current status and future prospects. Renew. Sust. Energy Rev. *15*, 4950–4962.

Kung, Y., Runguphan, W., and Keasling, J.D. (2012). From fields to fuels: Recent advances in the microbial production of biofuels. ACS Synth. Biol. *1*, 498–513.

Kuroiwa, T., Shoda, H., Ichikawa, S., Sato, S., and Mukataka, S. (2005). Immobilization and stabilization of pullulanase from *Klebsiella pneumoniae* by a multipoint attachment method using activated agar gel supports. Process Biochem. *40*, 2637–2642.

Kuyper, M., Harhangi, H.R., Stave, A.K., Winkler, A.A., Jetten, M.S., de Laat, W.T., den Ridder, J.J., Op den Camp, H.J., Van Dijken, J.P., and Pronk, J.T. (2003). High-level functional expression of a fungal xylose isomerase: the key to efficient ethanolic fermentation of xylose by *Saccharomyces cerevisiae*? FEMS Yeast Res. *4*, 69–78.

La Grange, D.C., Den Haan, R., and Van Zyl, W.H. (2010). Engineering cellulolytic ability into bioprocessing organisms. Appl. Microbiol. Biotechnol. *87*, 1195–1208.

Laluce, C., Schenberg, A.C.G., Gallardo, J.C.M., Coradello, L.F.C., and Pombeiro-Sponchiado, S.R. (2012). Advances and developments in strategies to improve strains of *Saccharomyces cerevisiae* and processes to obtain the lignocellulosic rthanol: a review. Appl. Biochem. Biotechnol. *166*, 1908–1926.

Lamers, P., Hamelinck, C., Junginger, M., and Faaij, A. (2011). International bioenergy trade-A review of past developments in the liquid biofuel market. Renewable and Sustainable Energy Reviews *15*, 2655–2676.

Larroy, C., Pares, X., and Biosca, J.A. (2002). Characterization of a *Saccharomyces cerevisiae* NADP (H)-dependent alcohol dehydrogenase (ADHVII), a member of the cinnamyl alcohol dehydrogenase family. Eur. J. Biochem. *269*, 5738–5745.

Larsson, S., Reimann, A., Nilvebrant, N.O., and Jönsson, L.J. (1999). Comparison of different methods for the detoxification of lignocellulose hydrolyzates of spruce. Appl. Biochem. Biotechnol. 77, 91–103.

Larsson, S., Nilvebrant, N.O., and Jönsson, L. (2001). Effect of overexpression of Saccharomyces cerevisiae Pad1p on the resistance to phenylacrylic acids and lignocellulose hydrolysates under aerobic and oxygen-limited conditions. Appl. Microbiol. Biotechnol. *57*, 167–174.

Lawford, H.G., and Rousseau, J.D. (1999). The effect of glucose on high-level xylose fermentations by recombinant *Zymomonas* in batch and fed-batch fermentations. Appl. Biochem. Biotechnol. 77, 235–249.

Lee, W.H., Nan, H., Kim, H.J., and Jin, Y.S. (2013). Simultaneous saccharification and fermentation by engineered *Saccharomyces cerevisiae* without supplementing extracellular b-glucosidase. J. Biotechnol. *167*, 316–322.

Lee, W.J., Kim, M.D., Ryu, Y.W., Bisson, L.F., and Seo, J.H. (2002). Kinetic studies on glucose and xylose transport in *Saccharomyces cerevisiae*. Appl. Microbiol. Biotechnol. *60*, 186–191.

Liao, B., Hill, G.A., and Roesler, W.J. (2010). Amylolytic activity and fermentative ability of *Saccharomyces cerevisiae* strains that express barley a-amylase. Biochem. Eng. J. *53*, 63–70.

Liao, B., Hill, G.A., and Roesler, W.J. (2012). Stable expression of barley a-amylase in *S. cerevisiae* for conversion of starch into bioethanol. Biochem. Eng. J. *64*, 8–16.

Liu, Z.L. (2011). Molecular mechanisms of yeast tolerance and in situ detoxification of lignocellulose hydrolysates. Appl. Microbiol. Biotechnol. *90*, 809–825.

Liu, Z.L., and Moon, J. (2009). A novel NADPH-dependent aldehyde reductase gene from *Saccharomyces cerevisiae* NRRL Y-12632 involved in the detoxification of aldehyde inhibitors derived from lignocellulosic biomass conversion. Gene *446*, 1–10.

Liu, Z.L., Slininger, P.J., Dien, B.S., Berhow, M.A., Kurtzman, C.P., and Gorsich, S.W. (2004). Adaptive response of yeasts to furfural and 5-hydroxymethylfurfural and new chemical evidence for HMF conversion to 2,5-bis-hydroxymethylfuran. J. Ind. Microbiol Biotechnol *31*, 345–352.

Liu, Z.L., Slininger, P.J., and Gorsich, S.W. (2005). Enhanced biotransformation of furfural and hydroxymethylfurfural by newly developed ethanologenic yeast strains. Appl. Biochem. Biotechnol. *121–124*, 451–460.

Liu, Z.L., Moon, J., Andersh, B.J., Slininger, P.J., and Weber, S. (2008). Multiple gene-mediated NAD (P)H-dependent aldehyde reduction is a mechanism of in situ detoxification of furfural and 5-hydroxymethylfurfural by *Saccharomyces cerevisiae*. Appl. Microbiol. Biotechnol. *81*, 743–753.

Liu, Z.L., Ma, M., and Song, M. (2009). Evolutionarily engineered ethanologenic yeast detoxifies lignocellulosic biomass conversion inhibitors by reprogrammed pathways. Mol. Genet. Genomics *282*, 233–244.

Lodolo, E.J., Kock, J.L., Axcell, B.C., and Brooks, M. (2008). The yeast *Saccharomyces cerevisiae*-the main character in beer brewing. FEMS Yeast Res. *8*, 1018–1036.

Lu, Y., Zhang, Y.H., and Lynd, L.R. (2006). Enzyme-microbe synergy during cellulose hydrolysis by *Clostridium thermocellum*. Proc. Natl. Acad. Sci. U.S.A. *103*, 16165–16169.

Luo, C., Brink, D.L., and Blanch, H.W. (2002). Identification of potential fermentation inhibitors in conversion of hybrid poplar hydrolyzate to ethanol. Biomass Bioenergy *22*, 125–138.

Lynd, L.R., Wyman, C.E., and Gerngross, T.U. (1999). Biocommodity engineering. Biotechnol. Prog. *15*, 777–793.

Lynd, L.R., Weimer, P.J., Van Zyl, W.H., and Pretorius, I.S. (2002). Microbial cellulose utilization: Fundamentals and biotechnology. Microbiol. Mol. Biol. Rev. *66*, 506–577.

Lynd, L.R., Van Zyl, W.H., McBride, J.E., and Laser, M. (2005). Consolidated bioprocessing of cellulosic biomass: an update. Curr. Opin. Biotechnol. *16*, 577–583.

Ma, M., and Liu, Z.L. (2010). Comparative transcriptome profiling analyses during the lag phase uncover *YAP1*, *PDR1*, *PDR3*, *RPN4*, and *HSF1* as key regulatory genes in genomic adaptation to the lignocellulose derived inhibitor HMF for *Saccharomyces cerevisiae*. BMC genomics *11*, 660.

Ma, Y., Lin, L., Chien, H., and Hsu, W. (2000). Efficient utilization of starch by a recombinant strain of *Saccharomyces cerevisiae* producing glucoamylase and isoamylase. Biotechnol. Appl. Biochem. *31*, 55–59.

McBride, J.E., Zietsman, J.J., Van Zyl, W.H., and Lynd, L.R. (2005). Utilization of cellobiose by recombinant b-glucosidase-expressing strains of *Saccharomyces cerevisiae*: characterization and evaluation of the sufficiency of expression. Enzyme Microb. Technol. *37*, 93–101.

McBride, J.E., Brevnova, E., Ghandi, C., Mellon, M., Froehlich, A., Delaault, K., Rajgarhia, V., Flatt, J., Van Zyl, W.H., Den Haan, R., *et al.* (2010). Yeast expressing cellulases for simultaneous saccharification and fermentation using cellulose. PCT/US2009/065571. 5-27-2010.

Machovic, M., Svensson, B., Ann MacGregor, E., and Janecek, Š. (2005). A new clan of CBM families based on bioinformatics of starch-binding domains from families CBM20 and CBM21. FEBS J. *272*, 5497–5513.

Macrelli, S., Mogensen, J., and Zacchi, G. (2012). Techno-economic evaluation of 2nd generation bioethanol production from sugar cane bagasse and leaves integrated with the sugar-based ethanol process. Biotechnol. Biofuels *5*, 1–18.

Margaritis, A., and Bajpai, P. (1982). Direct fermentation of D-xylose to ethanol by *Kluyveromyces marxianus* strains. Appl. Environ. Microbiol. *44*, 1039–1041.

Martín, C., and Jönsson, L.J. (2003). Comparison of the resistance of industrial and laboratory strains of *Saccharomyces* and *Zygosaccharomyces* to lignocellulose-derived fermentation inhibitors. Enzyme Microb. Technol. *32*, 386–395.

Martín, C., Marcet, M., Almazán, O., and Jönsson, L.J. (2007). Adaptation of a recombinant xylose-utilizing *Saccharomyces cerevisiae* strain to a sugarcane bagasse hydrolysate with high content of fermentation inhibitors. Bioresour. Technol. *98*, 1767–1773.

Martin, G.J., Knepper, A., Zhou, B., and Pamment, N.B. (2006). Performance and stability of ethanologenic *Escherichia coli* strain FBR5 during continuous culture on xylose and glucose. J. Ind. Microbiol. Biotechnol. *33*, 834–844.

Martiniano, S.E., Chandel, A.K., Soares, L.C.S.R., Pagnocca, F.C., and da Silva, S.S. (2013). Evaluation of novel xylose-fermenting yeast strains from Brazilian forests for hemicellulosic ethanol production from sugarcane bagasse. Biotechnol. *3*, 345–352.

Martins, L.F., Antunes, L.P., Pascon, R.C., de Oliveira, J.C.F., Digiampietri, L.A., Barbosa, D., Peixoto, B.M., Vallim, M.A., Viana-Niero, C., and Ostroski, E.H. (2013). Metagenomic analysis of a tropical composting operation at the Sao Paulo Zoo Park reveals diversity of biomass degradation functions and organisms. PloS One *8*, e61928.

Matano, Y., Hasunuma, T., and Kondo, A. (2013). Cell recycle batch fermentation of high-solid lignocellulose using a recombinant cellulase-displaying yeast strain for high yield ethanol production in consolidated bioprocessing. Bioresour. Technol. *135*, 403–409.

Matsumoto, N., Fukushi, O., Miyanaga, M., Kakihara, K., Nakajima, E., and Yoshizumi, H. (1982). Industrialization of a noncooking system for alcoholic fermentation from grains. Agric. Biol. Chem. *46*, 1549–1558.

Menon, V., and Rao, M. (2012). Trends in bioconversion of lignocellulose: biofuels, platform chemicals & biorefinery concept. Prog. Energy Combust. Sci. *38*, 522–550.

Meredith, J. (2003). Understanding energy use and energy users in contemporary ethanol plants. In The Alcohol Textbook, 4th Edition, K. Jacques, T. Lyons, and D. Kelsall, eds. (Nottingham: Nottingham University Press), pp. 355–361.

Miao, M., Jiang, B., and Zhang, T. (2009). Effect of pullulanase debranching and recrystallization on structure and digestibility of waxy maize starch. Carbohydr. Polym. *76*, 214–221.

Millati, R., Niklasson, C., and Taherzadeh, M.J. (2002). Effect of pH, time and temperature of overliming on detoxification of dilute-acid hydrolyzates for fermentation by *Saccharomyces cerevisiae*. Process Biochem. *38*, 515–522.

Mira, N.P., Becker, J.D., and Sá-Correia, I. (2010a). Genomic expression program involving the Haa1p-regulon in *Saccharomyces cerevisiae* response to acetic acid. OMICS *14*, 587–601.

Mira, N.P., Teixeira, M.C., and Sá-Correia, I. (2010b). Adaptive response and tolerance to weak acids in *Saccharomyces cerevisiae*: A genome-wide view. OMICS *14*, 525–540.

Mitsuiki, S., Mukae, K., Sakai, M., Goto, M., Hayashida, S., and Furukawa, K. (2005). Comparative characterization of raw starch hydrolyzing a-amylases from various *Bacillus* strains. Enzyme Microb. Technol. *37*, 410–416.

Miyafuji, H., Danner, H., Neureiter, M., Thomasser, C., Bvochora, J., Szolar, O., and Braun, R. (2003). Detoxification of wood hydrolysates with wood charcoal for increasing the fermentability of hydrolysates. Enzyme Microb. Technol. *32*, 396–400.

Modig, T., Lidén, G., and Taherzadeh, M. (2002). Inhibition effects of furfural on alcohol dehydrogenase, aldehyde dehydrogenase and pyruvate dehydrogenase. Biochem. J. *363*, 769–776.

Modig, T., Almeida, J.R., Gorwa-Grauslund, M.F., and Lidén, G. (2008). Variability of the response of *Saccharomyces cerevisiae* strains to lignocellulose hydrolysate. Biotechnol. Bioeng. *100*, 423–429.

Mohagheghi, A., Evans, K., Chou, Y.C., and Zhang, M. (2002). Cofermentation of glucose, xylose and arabinose by genomic DNA-integrated xylose/arabinose fermenting stains of *Zymomonas mobilis* AX101. Appl. Biochem. Biotechnol. *98*, 885–898.

Mollapour, M., Shepherd, A., and Piper, P.W. (2009). Presence of the Fps1p aquaglyceroporin channel is essential for Hog1p activation, but suppresses Slt2

(Mpk1)p activation, with acetic acid stress of yeast. Microbiology *155*, 3304–3311.

Moon, J., and Liu, Z.L. (2012). Engineered NADH-dependent GRE2 from *Saccharomyces cerevisiae* by directed enzyme evolution enhances HMF reduction using additional cofactor NADPH. Enz. Microb. Technol. *50*, 115–120.

Mooney, C. (2011). The truth about fracking. Scientific American *305*, 80–85.

Mukai, N., Masaki, K., Fujii, T., Kawamukai, M., and Iefuji, H. (2010). *PAD1* and *FDC1* are essential for the decarboxylation of phenylacrylic acids in *Saccharomyces cerevisiae*. J. Biosci. Bioeng. *109*, 564–569.

Mukerjea, R., Slocum, G., Mukerjea, R., and Robyt, J.F. (2006). Significant differences in the activities of a-amylases in the absence and presence of polyethylene glycol assayed on eight starches solubilized by two methods. Carbohydr. Res. *341*, 2049–2054.

Munasinghe, P.C., and Khanal, S.K. (2010). Biomass-derived syngas fermentation into biofuels: opportunities and challenges. Bioresour. Technol. *101*, 5013–5022.

Murai, T., Ueda, M., Yamamura, M., Atomi, H., Shibasaki, Y., Kamasawa, N., Osumi, M., Amachi, T., and Tanaka, A. (1997). Construction of a starch-utilizing yeast by cell surface engineering. Appl. Environ. Microbiol. *63*, 1362–1366.

Murai, T., Ueda, M., Shibasaki, Y., Kamasawa, N., Osumi, M., Imanaka, T., and Tanaka, A. (1999). Development of an arming yeast strain for efficient utilization of starch by co-display of sequential amylolytic enzymes on the cell surface. Appl. Microbiol. Biotechnol. *51*, 65–70.

Mussatto, S.I., and Roberto, I.C. (2004). Alternatives for detoxification of diluted-acid lignocellulosic hydrolyzates for use in fermentative processes: A review. Bioresour. Technol. *93*, 1–10.

Narendranath, N. (2003). Bacterial contamination and control in ethanol production. In The Alcohol Textbook, 4th Edition, K. Jacques, T. Lyons, and D. Kelsall, eds. (United Kingdom: Nottingham University Press), pp. 287–298.

Nevoigt, E. (2008). Progress in metabolic engineering of *Saccharomyces cerevisiae*. Microbiol. Mol. Biol. Rev. *72*, 379–412.

Nonklang, S., Abdel-Banat, B.M.A., Cha-aim, K., Moonjai, N., Hoshida, H., Limtong, S., Yamada, M., and Akada, R. (2008). High-Temperature ethanol fermentation and transformation with linear DNA in the thermotolerant yeast *Kluyveromyces marxianus* DMKU3–1042. Appl. Environ. Microbiol. *74*, 7514–7521.

Oates, C.G. (1997). Towards an understanding of starch granule structure and hydrolysis. Trends Food Sci. Technol. *8*, 375–382.

Öhgren, K., Bengtsson, O., Gorwa-Grauslund, M., Galbe, M., Hahn-Hägerdal, B., and Zacchi, G. (2006). Simultaneous saccharification and co-fermentation of glucose and xylose in steam-pretreated corn stover at high fiber content with *Saccharomyces cerevisiae* TMB3400. J. Biotechnol. *126*, 488–498.

Ohta, K., Beall, D.S., Mejia, J.P., Shanmugam, K.T., and Ingram, L.O. (1991a). Genetic improvement of *Escherichia coli* for ethanol production: Chromosomal integration of *Zymomonas mobilis* genes encoding pyruvate decarboxylase and alcohol dehydrogenase. Appl. Environ. Microbiol. *57*, 893–900.

Ohta, K., Mejia, J.P., Shanmugam, K.T., and Ingram, L.O. (1991b). Metabolic engineering of *Klebsiella oxytoca* M5A1 for ethanol production from xylose and glucose. Appl. Environ. Microbiol. *57*, 2810–2815.

Okamoto, K., Kanawaku, R., Masumoto, M., and Yanase, H. (2012). Efficient xylose fermentation by the brown rot fungus *Neolentinus lepideus*. Eur. J. Biochem. *50*, 96–100.

Oliva, J.M., Sáez, F., Ballesteros, I., González, A., Negro, M.J., Manzanares, P., and Ballesteros, M. (2003). Effect of lignocellulosic degradation compounds from steam explosion pretreatment on ethanol fermentation by thermotolerant yeast *Kluyveromyces marxianus*. In Biotechnology for Fuels and Chemicals, Springer, pp. 141–153.

Olofsson, K., Rudolf, A., and Lidén, G. (2008). Designing simultaneous saccharification and fermentation for improved xylose conversion by a recombinant strain of *Saccharomyces cerevisiae*. J. Biotechnol. *134*, 112–120.

Olson, D.G., McBride, J.E., Shaw, A.J., and Lynd, L.R. (2012). Recent progress in consolidated bioprocessing. Curr. Opin. Biotechnol. *23*, 1–11.

Olsson, L., and Hahn-Hägerdal, B. (1996). Fermentation of lignocellulosic hydrolysates for ethanol production. Enzyme Microb. Technol. *18*, 312–331.

Pagliardini, J., Hubmann, G., Alfenore, S., Nevoigt, E., Bideaux, C., and Guillouet, E. (2013). The metabolic costs of improving ethanol yield by reducing glycerol formation capacity under anaerobic conditions in *Saccharomyces cerevisiae*. Microb. Cell Fact. *12*, 29.

Palmqvist, E., and Hahn-Hägerdal, B. (2000a). Fermentation of lignocellulosic hydrolysates. I: Inhibition and detoxification. Bioresour. Technol. *74*, 17–24.

Palmqvist, E., and Hahn-Hägerdal, B. (2000b). Fermentation of lignocellulosic hydrolysates. II: Inhibitors and mechanisms of inhibition. Bioresour. Technol. *74*, 25–33.

Palmqvist, E., Grage, H., Meinander, N.Q., and Hahn-Hägerdal, B. (1999). Main and interaction effects of acetic acid, furfural, and *p*-hydroxybenzoic acid on growth and ethanol productivity of yeasts. Biotechnol. Bioeng. *63*, 46–55.

Pampulha, M.E., and Loureiro-Dias, M.C. (1989). Combined effect of acetic acid, pH and ethanol on intracellular pH of fermenting yeast. Appl. Microbiol. Biotechnol. *31*, 547–550.

Papoutsakis, E.T., and Meyer, C.L. (1985). Fermentation equations for propionic-acid bacteria and production of assorted oxychemicals from various sugars. Biotechnol. Bioeng. *17*, 67–80.

Parawira, W., and Tekere, M. (2011). Biotechnological strategies to overcome inhibitors in lignocellulose hydrolysates for ethanol production: Review. Crit. Rev. Biotechnol. *31*, 20–31.

Peacock, J.P., Cole, J.K., Murugapiran, S.K., Dodsworth, J.A., Fisher, J.C., Moser, D.P., and Hedlund, B.P. (2013). Pyrosequencing reveals high-temperature cellulolytic microbial consortia in Great Boiling Spring after in situ lignocellulose enrichment. PloS one *8*, e59927.

Pellegrini, L.F., and Oliveira Junior, S. (2011). Combined production of sugar, ethanol and electricity: thermoeconomic and environmental analysis and optimization. Energy *36*, 3704–3715.

Petersson, A., Almeida, J.R.M., Modig, T., Karhumaa, K., Hahn-Hägerdal, B., Gorwa-Grauslund, M.F., and Lidén, G. (2006). A 5-hydroxymethyl furfural reducing enzyme encoded by the *Saccharomyces cerevisiae ADH6* gene conveys HMF tolerance. Yeast *23*, 455–464.

Pienkos, P.T., and Zhang, M. (2009). Role of pretreatment and conditioning processes on toxicity of lignocellulosic biomass hydrolysates. Cellulose *16*, 743–762.

Preiss, J. (2009). Starch: Chemistry and technology. In Biochemistry and molecular biology of starch biosynthesis, R.L. Whistler and J. BeMiller, eds. (Oxford: Access Online via Elsevier), pp. 83–147.

Pretorius, I.S. (1997). Utilization of polysaccharides by *Saccharomyces cerevisiae*. In Yeast sugar metabolism, F. Zimmerman and K. Entian, eds. (Lancaster: Technomic Publishing Company), pp. 435–458.

Puligundla, P., Smogrovicova, D., Obulam, V.S.R., and Ko, S. (2011). Very high gravity (VHG) ethanolic brewing and fermentation: a research update. J. Ind. Microbiol. Biotechnol. *38*, 1133–1144.

Qureshi, N., Dien, B.S., Nichols, N.N., Saha, B.C., and Cotta, M.A. (2006). Genetically engineered *Escherichia coli* for ethanol production from xylose: Substrate and product inhibition and kinetic parameters. Food Bioprod. Process. *84*, 114–122.

Qureshi, N., Dien, B.S., Liu, S., Saha, B.C., Cotta, M., Hughes, S., and Hector, R. (2012). Genetically engineered *Escherichia coli* FBR5: Part II. Ethanol production from xylose and simultaneous product recovery. Biotechnol. Prog. *28*, 1179–1185.

Radakovits, R., Jinkerson, R.E., Darzins, A., and Posewitz, M.C. (2010). Genetic engineering of algae for enhanced biofuel production. Eukar. Cell *9*, 486–501.

Ranatunga, T.D., Jervis, J., Helm, R.F., McMillan, J.D., and Hatzis, C. (1997). Identification of inhibitory components toxic toward *Zymomonas mobilis* CP4 (pZB5) xylose fermentation. Appl. Biochem. Biotechnol. *67*, 185–198.

Reddy, L.V.A., Reddy, O.V.S., and Basappa, S.C. (2009). Potentiality of yeasts in the direct conversion of starchy materials to ethanol and its relevance in the New Millennium. In Yeast Biotechnology: Diversity and Applications, T. Satyanarayana and G. Kunze, eds. (Netherlands: Springer), pp. 515–549.

Ren21. Renewables 2012. Global Status Report. Paris: REN21 Secretariat. 2013. 10-8-0130. [Online] http://www.map.ren21.net/GSR/GSR2012_low.pdf

Robert, X., Haser, R., Mori, H., Svensson, B., and Aghajari, N. (2005). Oligosaccharide binding to barley a-amylase 1. J. Biol. Chem. *280*, 32968–32978.

Robertson, G.H., Wong, D.W., Lee, C.C., Wagschal, K., Smith, M.R., and Orts, W.J. (2006). Native or raw starch digestion: a key step in energy efficient biorefining of grain. J. Agric. Food Chem. *54*, 353–365.

Rudolf, A., Alkasrawi, M., Zacchi, G., and Lidén, G. (2005). A comparison between batch and fed-batch simultaneous saccharification and fermentation of steam pretreated spruce. Enz. Microb. Technol. *37*, 195–204.

Rudolf, A., Baudel, H., Zacchi, G., Hahn-Hägerdal, B., and Lidén, G. (2008). Simultaneous saccharification and fermentation of steam-pretreated bagasse using *Saccharomyces cerevisiae* TMB3400 and *Pichia stipitis* CBS6054. Biotechnol. Bioeng. *99*, 783–790.

Rudolf, A., Karhumaa, K., and Hahn-Hägerdal, B. (2009). Ethanol production from traditional and emerging raw materials. In Yeast Biotechnology: Diversity and Applications, T. Satyanarayana and G. Kunze, eds. (Netherlands: Springer), pp. 489–514.

Runquist, D., Hahn-Hägerdal, B., and Bettiga, M. (2010a). Increased ethanol productivity in xylose-utilizing *Saccharomyces cerevisiae* via a randomly mutagenized xylose reductase. Appl. Environ. Microbiol. *76*, 7796–7802.

Runquist, D., Hahn-Hägerdal, B., and Radstrom, P. (2010b). Comparison of heterologous xylose transporters in recombinant *Saccharomyces cerevisiae*. Biotechnol. Biofuels 3, 1–7.

Russell, J.B. (1992). Another explanation for the toxicity of fermentation acids at low pH: anion accumulation versus uncoupling. J. Appl. Microbiol. 73, 363–370.

Ryabova, O.B., Chmil, O.M., and Sibirny, A.A. (2003). Xylose and cellobiose fermentation to ethanol by the thermotolerant methylotrophic yeast *Hansenula polymorpha*. FEMS Yeast Res. *4*, 157–164.

Sadie, C.J., Rose, S.H., Den Haan, R., and Van Zyl, W.H. (2011). Co-expression of a cellobiose phosphorylase and lactose permease enables intracellular cellobiose utilisation by *Saccharomyces cerevisiae*. Appl. Microbiol. Biotechnol. *90*, 1373–1380.

Saha, B.C., Shen, G.J., Srivastava, K.C., LeCureux, L.W., and Zeikus, J.G. (1989). New thermostable a-amylase-like pullulanase from thermophilic *Bacillus* sp. 3183. Enz. Microb. Technol. *11*, 760–764.

Saloheimo, A., Rauta, J., Stasyk, O.V., Sibirny, A.A., Penttilä, M., and Ruohonen, L. (2007). Xylose transport studies with xylose-utilizing *Saccharomyces cerevisiae* strains expressing heterologous and homologous permeases. Appl. Microbiol. Biotechnol. 74, 1041–1052.

Sanchez, R.G., Karhumaa, K., Fonseca, C., Nogué, V.S., Almeida, J., Larsson, C.U., Bengtsson, O., Bettiga, M., Hahn-Hägerdal, B., and Gorwa-Grauslund, M.F. (2010). Research Improved xylose and arabinose utilization by an industrial recombinant *Saccharomyces cerevisiae* strain using evolutionary engineering. Biotechnol. Biofuels 3, 13.

Sarthy, A.V., McConaughy, B.L., Lobo, Z., Sundstrom, J.A., Furlong, C.E., and Hall, B.D. (1987). Expression of the *Escherichia coli* xylose isomerase gene in *Saccharomyces cerevisiae*. Appl. Environ. Microbiol. 53, 1996–2000.

Sauer, U. (2001). Evolutionary engineering of industrially important microbial phenotypes. Adv. Biochem. Eng. Biotechnol. 73, 129–169.

Schnepf, R., and Yacobucci, B. Renewable Fuel Standard (RFS): Overview and issues. Congressional Research

Service. 2013. 10–8–0130. [Online] http://www.fas.org/sgp/crs/misc/R40155.pdf

Scott, R., and Junyang, J. China – People's Republic of: Biofuels annual. USDA.GAIN Report CH12044. 2013. 9-10-0130. [Online] http://gain.fas.usda.gov/Recent%20GAIN%20Publications/Biofuels%20Annual_Beijing_China%20-%20Peoples%20Republic%20of_7–9–2012.pdf

Shaw, A.J., Jenney, F.E., Adams, M.W.W., and Lynd, L.R. (2008a). End-product pathways in the xylose fermenting bacterium, *Thermoanaerobacterium saccharolyticum*. Enz. Microb. Technol. *42*, 453–458.

Shaw, A.J., Podkaminer, K.K., Desai, S.G., Bardsley, J.S., Rogers, S.R., Thorne, P.G., Hogsett, D.A., and Lynd, L.R. (2008b). Metabolic engineering of a thermophilic bacterium to produce ethanol at high yield. Proc. Natl. Acad. Sci. U.S.A. *105*, 13769–13774.

Shaw, A.J., Hogsett, D.A., and Lynd, L.R. (2010). Natural competence in *Thermoanaerobacter* and *Thermoanaerobacterium* species. Appl. Environ. Microbiol. *76*, 4713–4719.

Shibanuma, K., Takeda, Y., Hizukuri, S., and Shibata, S. (1994). Molecular structures of some wheat starches. Carbohydr. Polym. *25*, 111–116.

Shibuya, I., Tamura, G., Shima, H., Ishikawa, T., and Hara, S. (1992). Construction of an á-amylase/glucoamylase fusion gene and its expression in *Saccharomyces cerevisiae*. Biosci. Biotechnol. Biochem. *56*, 884–889.

Shigechi, H., Koh, J., Fujita, Y., Matsumoto, T., Bito, Y., Ueda, M., Satoh, E., Fukuda, H., and Kondo, A. (2004). Direct production of ethanol from raw corn starch via fermentation by use of a novel surface-engineered yeast strain codisplaying glucoamylase and α-amylase. Appl. Environ. Microbiol. *70*, 5037–5040.

Sierks, M.R., and Svensson, B. (1994). Protein engineering of the relative specificity of glucoamylase from *Aspergillus awamori* based on sequence similarities between starch-degrading enzymes. Protein Eng. *7*, 1479–1484.

da Silveria dos Santos, D., Camelo, A.C., Rodrigues, K.C.P., Carlos, L.C., and Pereira, N. (2009). Ethanol production from sugarcane bagasse by *Zymomonas mobilis* using simultaneous saccharification and fermentation (SSF) process. Appl. Biochem. Biotechnol. *161*, 93–105.

Sims, R.E., Mabee, W., Saddler, J.N., and Taylor, M. (2010). An overview of second generation biofuel technologies. Bioresour. Technol. *101*, 1570–1580.

Skoog, K., and Hahn-Hägerdal, B. (1990). Effect of oxygenation on xylose fermentation by *Pichia stipitis*. Appl. Environ. Microbiol. *56*, 3389–3394.

Søgaard, M., Kadziola, A., Haser, R., and Svensson, B. (1993). Site-directed mutagenesis of histidine 93, aspartic acid 180, glutamic acid 205, histidine 290, and aspartic acid 291 at the active site and tryptophan 279 at the raw starch binding site in barley alpha-amylase 1. J. Biol. Chem. *268*, 22480–22484.

Solomon, B. (1978). Starch hydrolysis by immobilized enzymes industrial applications. In Advances in Biochemical Engineering, Volume 10, T. Ghose, A. Fiechter, and N. Blakebrough, eds. (New York: Springer-Verlag), pp. 131–177.

Somerville, C., Youngs, H., Taylor, C., Davis, S.C., and Long, S.P. (2010). Feedstocks for lignocellulosic biofuels. Science (Washington) *329*, 790–792.

Stevnebø, A., Sahlström, S., and Svihus, B. (2006). Starch structure and degree of starch hydrolysis of small and large starch granules from barley varieties with varying amylose content. Anim. Feed Sci. Technol. *130*, 23–38.

Steyn, A.J., and Pretorius, I.S. (1991). Co-expression of a *Saccharomyces diastaticus* glucoamylase-encoding gene and a *Bacillus amyloliquefaciens* α-amylase-encoding gene in *Saccharomyces cerevisiae*. Gene *100*, 85–93.

Sun, H., Zhao, P., Ge, X., Xia, Y., Hao, Z., Liu, J., and Peng, M. (2010). Recent advances in microbial raw starch degrading enzymes. Appl. Biochem. Biotechnol. *160*, 988–1003.

Sun, Y., and Cheng, J. (2002). Hydrolysis of lignocellulosic materials for ethanol production: A review. Bioresource Technol. *83*, 1–11.

Taherzadeh, M.J., and Karimi, K. (2011). Fermentation inhibitors in ethanol processes and different strategies to reduce their effects. In: A. Pandey, C. Larroche, S. Ricke, C. Dussap, and E. Gnansounou, eds. Biofuels: Alternative Feedstocks and Conversion Processes. Oxford: Academic Press, pp. 287–311.

Taherzadeh, M.J., Niklasson, C., and Lidén, G. (1997). Acetic acid-friend or foe in anaerobic batch conversion of glucose to ethanol by *Saccharomyces cerevisiae*? Chem. Eng. Sci. *52*, 2653–2659.

Taherzadeh, M.J., Gustafsson, L., Niklasson, C., and Lidén, G. (1999). Conversion of furfural in aerobic and anaerobic batch fermentation of glucose by *Saccharomyces cerevisiae*. J. Biosci. Bioeng. *87*, 169–174.

Taherzadeh, M.J., Gustafsson, L., Niklasson, C., and Lidén, G. (2000a). Physiological effects of 5-hydroxymethylfurfural on *Saccharomyces cerevisiae*. Appl. Microbiol. Biotechnol. *53*, 701–708.

Taherzadeh, M.J., Niklasson, C., and Lidén, G. (2000b). On-line control of fed-batch fermentation of dilute-acid hydrolyzates. Biotechnol. Bioeng. *69*, 330–338.

Tanaka, K., Ishii, Y., Ogawa, J., and Shima, J. (2012). Enhancement of acetic acid tolerance in *Saccharomyces cerevisiae* by overexpression of the *HAA1* gene, encoding a transcriptional activator. Appl. Environ. Microbiol. *78*, 8161–8163.

Tantirungki, M., Nakashima, N., Seki, T., and Yoshida, T. (1993). Construction of xylose-assimilating *Saccharomyces cerevisiae*. J. Ferm. Bioeng. *75*, 83–88.

Tao, H., Gonzalez, R., Martinez, A., Rodriguez, M., Ingram, L.O., Preston, J.F., and Shanmugam, K.T. (2001). Engineering a homo-ethanol pathway in *Escherichia coli*: Increased glycolytic flux and levels of expression of glycolytic genes during xylose fermentation. J. Bacteriol. *183*, 2979–2988.

Tester, R.F., and Morrison, W.R. (1990). Swelling and gelatinization of cereal starches. I. Effects of amylopectin, amylose, and lipids. Cereal Chem. *67*, 551–557.

Tester, R.F., Karkalas, J., and Qi, X. (2004). Starch-composition, fine structure and architecture. J. Cereal Sci. *39*, 151–165.

Timilsina, G.R., and Shrestha, A. (2011). How much hope should we have for biofuels? Energy *36*, 2055–2069.

Toivari, M.H., Aristidou, A., Ruohonen, L., and Penttila, M. (2001). Conversion of xylose to ethanol by recombinant *Saccharomyces cerevisiae*: importance of xylulokinase (XKS1) and oxygen availability. Metab. Eng. 3, 236–249.

Toivari, M.H., Salusjarvi, L.S., Ruohonen, L., and Penttila, M. (2004). Endogenous xylose pathway in *Saccharomyces cerevisiae*. Appl. Environ. Microbiol. *70*, 3681–3686.

Tsai, S.L., Oh, J., Singh, S., Chen, R., and Chen, W. (2009). Functional assembly of minicellulosomes on the *Saccharomyces cerevisiae* cell surface for cellulose hydrolysis and ethanol production. Appl. Environ. Microbiol. *75*, 6087–6093.

Tsai, S.L., Goyal, G., and Chen, W. (2010). Surface display of a functional minicellulosome by intracellular complementation using a synthetic yeast consortium and its application to cellulose hydrolysis and ethanol production. Appl. Environ. Microbiol. *76*, 7514–7520.

Tsakraklides, V., Shaw, A.J., Miller, B.B., Hogsett, D.A., and Herring, C.D. (2012). Carbon catabolite repression in *Thermoanaerobacterium saccharolyticum*. Biotechnol. Biofuels 5, 85.

Tyurin, M.V., Sullivan, C.R., and Lynd, L.R. (2005). Role of spontaneous current oscillations during high-efficiency electrotransformation of thermophilic anaerobes. Appl. Environ. Microbiol. *71*, 8069–8076.

Ulbricht, R.J., Northup, S.J., and Thomas, J.A. (1984). A review of 5-hydroxymethylfurfural (HMF) in parenteral solutions. Fund. Appl. Toxicol. *4*, 843–853.

Ungerman, A.J., and Heindel, T.J. (2007). Carbon monoxide mass transfer for syngas fermentation in a stirred tank reactor with dual impeller configurations. Biotechnol. Prog. *23*, 613–620.

Van Der Maarel, M.J., Van Der Veen, B., Uitdehaag, J., Leemhuis, H., and Dijkhuizen, L. (2002). Properties and applications of starch-converting enzymes of the a-amylase family. J. Biotechnol. *94*, 137–155.

Van Maris, A.J., Abbott, D.A., Bellissimi, E., van den Brink, J., Kuyper, M., Luttik, M.A., Wisselink, H.W., Scheffers, W.A., Van Dijken, J.P., and Pronk, J.T. (2006). Alcoholic fermentation of carbon sources in biomass hydrolysates by *Saccharomyces cerevisiae*: Current status. Antonie Van Leeuwenhoek *90*, 391–418.

Van Maris, A.J., Winkler, A.A., Kuyper, M., de Laat, W.T., Van Dijken, J.P., and Pronk, J.T. (2007). Development of efficient xylose fermentation in *Saccharomyces cerevisiae*: Xylose isomerase as a key component. Adv. Biochem. Eng. Biotechnol. *108*, 179–204.

Van Rooyen, R., Hahn-Hagerdal, B., La Grange, D.C., and Van Zyl, W.H. (2005). Construction of cellobiose-growing and fermenting *Saccharomyces cerevisiae* strains. J. Biotechnol. *120*, 284–295.

Van Zyl, C., Prior, B.A., Killian, K., and Kock, J.L. (1989). D-xylose utilization by *Saccharomyces cerevisiae*. J. Gen. Microbiol. *135*, 2791–2798.

Van Zyl, W.H., Lynd, L.R., Den Haan, R., and McBride, J.E. (2007). Consolidated bioprocessing for bioethanol production using *Saccharomyces cerevisiae*. Adv. Biochem. Eng. Biotechnol. *108*, 205–235.

Van Zyl, W.H., Chimphango, A.F.A., Den Haan, R., Gorgens, J.F., and Chirwa, P.W.C. (2011). Next-generation cellulosic ethanol technologies and their contribution to a sustainable Africa. Interface Focus *1*, 196–211.

Van Zyl, W.H., Bloom, M., and Viktor, M.J. (2012). Engineering yeasts for raw starch conversion. Appl. Microbiol. Biotechnol. *95*, 1377–1388.

Van Zyl, W.H., Den Haan, R., and la Grange, D.C. (2013). Developing cellulolytic organisms for consolidated bioprocessing of lignocellulosics. In Biofuel Technologies, V.K. Gupta and M.G. Tuohy, eds. (Berlin: Springer), pp. 189–220.

Verduyn, C., Van Kleef, R., Frank, J., Schreuder, H., Van Dijken, J.P., and Scheffers, W.A. (1985). Properties of the NAD (P)H-dependent xylose reductase from the xylose-fermenting yeast *Pichia stipitis*. Biochem. J. *226*, 669–677.

Wahlbom, C., and Hahn-Hägerdal, B. (2002). Furfural, 5-hydroxymethyl furfural, and acetoin act as external electron acceptors during anaerobic fermentation of xylose in recombinant *Saccharomyces cerevisiae*. Biotechnol. Bioeng. *78*, 172–178.

Wahlbom, C.F., Van Zyl, W.H., Jonsson, L.J., Hahn-Hägerdal, B., and Cordero Otero, R.R. (2003). Generation of the improved recombinant xylose-utilizing *Saccharomyces cerevisiae* TM 3400 by random mutagenesis and physiological comparison with *Pichia stipitis* CBS 6054. FEMS Yeast Res. *3*, 319–326.

Walfridsson, M., Anderlund, M., Bao, X., and Hahn-Hägerdal, B. (1997). Expression of different levels of enzymes from the *Pichia stipitis* XYL1 and XYL2 genes in *Saccharomyces cerevisiae* and its effects on product formation during xylose utilisation. Appl. Microbiol. Biotechnol. *48*, 218–224.

Walfridsson, M., Bao, X., Anderlund, M., Lilius, G., Bulow, L., and Hahn-Hägerdal, B. (2013). Ethanolic fermentation of xylose with *Saccharomyces cerevisiae* harboring the *Thermus thermophilus xylA* gene, which expresses an active xylose (glucose) isomerase. Appl. Environ. Microbiol. *62*, 4648–4651.

Wang, G., Liu, C., Hong, J., Ma, Y., Zhang, K., Huang, X., Zou, S., and Zhang, M. (2013a). Comparison of process configurations for ethanol production from acid- and alkali-pretreated corncob by *Saccharomyces cerevisiae* strains with and without b-glucosidase expression. Bioresour. Technol. *142*, 154–161.

Wang, R., Li, L., Zhang, B., Gao, X., Wang, D., and Hong, J. (2013b). Improved xylose fermentation of *Kluyveromyces marxianus* at elevated temperature through construction of a xylose isomerase pathway. J. Ind. Microbiol. Biotechnol. *40*, 841–854.

Warnecke, F., Luginbühl, P., Ivanova, N., Ghassemian, M., Richardson, T.H., Stege, J.T., Cayouette, M., McHardy, A.C., Djordjevic, G., and Aboushadi, N. (2007). Metagenomic and functional analysis of hindgut microbiota of a wood-feeding higher termite. Nature *450*, 560–565.

Watanabe, S., Kodaki, R., and Makino, K. (2005). Complete reversal of coenzyme specificity of xylitol dehydrogenase and increase of thermostability by the introduction of structural zinc. J. Biol. Chem. *280*, 10340–10349.

Weber, F., and De Bont, J. (1996). Adaptation mechanisms of microorganisms to the toxic effects of organic solvents on membranes. Biochim. Biophys. Acta *1286*, 225–245.

Wei, N., Xu, H., Kim, S.R., and Jin, Y.S. (2013). Deletion of *FPS1*, encoding aquaglyceroporin Fps1p, improves xylose fermentation by engineered *Saccharomyces cerevisiae*. Appl. Environ. Microbiol. *79*, 3193–3201.

Weigert, B., Klein, C., Rizzi, M., Lauterbach, C., and Dellweg, H. (1988). Xylose fermentation by yeast. Biotechnol. Lett. *10*, 895–900.

Weiss, N., Börjesson, J., Pedersen, L.S., and Meyer, A.S. (2013). Enzymatic lignocellulose hydrolysis: Improved cellulase productivity by insoluble solids recycling. Biotechnol. Biofuels *6*, 5.

Wen, F., Sun, J., and Zhao, H. (2010). Yeast surface display of trifunctional minicellulosomes for simultaneous saccharification and fermentation of cellulose to ethanol. Appl. Environ. Microbiol. *76*, 1251–1260.

Whistler, R., and BeMiller, J. (1997). Carbohydrate chemistry for food scientists. (St.Paul, Minnesota: American Association of Cereal Chemists, Inc (AACC)).

Wierckx, N., Koopman, F., Ruijssenaars, H.J., and de Winde, J.H. (2011). Microbial degradation of furanic compounds: biochemistry, genetics, and impact. Appl. Microbiol. Biotechnol. *92*, 1095–1105.

Wilde, C., Gold, N., Bawa, N., Tambor, J., Mougharbel, L., Storms, R., and Martin, V. (2012). Expression of a library of fungal b-glucosidases in *Saccharomyces cerevisiae* for the development of a biomass fermenting strain. Appl. Microbiol. Biotechnol. *95*, 647–659.

Williams, K., Zheng, Y., McGarvey, J., Fan, Z., and Zhang, R. (2013). Ethanol and volatile fatty acid Production from lignocellulose by *Clostridium cellulolyticum*. ISRN Biotechnol. *2013*, 1–7.

Wisselink, H.W., Toirkens, M.J., Wu, Q., Pronk, J.T., and Van Maris, A.J. (2009). Novel evolutionary engineering approach for accelerated utilization of glucose, xylose, and arabinose mixtures by engineered *Saccharomyces cerevisiae* strains. Appl. Environ. Microbiol. *75*, 907–914.

Wohlbach, D.J., Kuo, A., Sato, T.K., Potts, K.M., Salamov, A.A., LaButti, K.M., Sun, H., Clum, A., Pangilinan, J.L., Lindquist, E.A., *et al.* (2013). Comparative genomics of xylose-fermenting fungi for enhanced biofuel production. Proc. Natl. Acad. Sci. U.S.A. *108*, 13212–13217.

Wong, D.W., Batt, S.B., Lee, C.C., Wagschal, K., and Robertson, G.H. (2005). Characterization of active *Lentinula edodes* glucoamylase expressed and secreted by *Saccharomyces cerevisiae*. Protein J. *24*, 455–463.

Wong, D.W.S., Robertson, G.H., Lee, C.C., and Wagschal, K. (2007). Synergistic action of recombinant a-amylase and glucoamylase on the hydrolysis of starch granules. Protein J. *26*, 159–164.

Wood, B.E., and Ingram, L.O. (1992). Ethanol production from cellobiose, amorphous cellulose, and crystalline cellulose by recombinant *Klebsiella oxytoca* containing chromosomally integrated *Zymomonas mobilis* genes for ethanol production and plasmids expressing thermostable cellulase genes from *Clostridium thermocellum*. Appl. Environ. Microbiol. *58*, 2103–2110.

Woynillowicz, D., and Severson-Baker, C. (2009). Oil sands fever – the environmental implications of Canada's oil sands rush. Canada: The Pembina Institute for Appropriate Development.

Wright, J., Bellissimi, E., de Hulster, E., Wagner, A., Pronk, J.T., and Van Maris, A.J. (2011). Batch and continuous culture-based selection strategies for acetic acid tolerance in xylose-fermenting *Saccharomyces cerevisiae*. FEMS Yeast Res. *11*, 299–306.

Wyman, C.E., Spindler, D.D., and Grohmann, K. (1992). Simultaneous saccharification and fermentation of several lignocellulosic feedstocks to fuel ethanol. Biomass Bioenergy 3, 301–307.

Xing, M.N., Zhang, X.Z., and Huang, H. (2012). Application of metagenomic techniques in mining enzymes from microbial communities for biofuel synthesis. Biotechnol. Adv. *30*, 920–929.

Xu, Q., Singh, A., and Himmel, M.E. (2009). Perspectives and new directions for the production of bioethanol using consolidated bioprocessing of lignocellulose. Curr. Opin. Biotechnol. *20*, 364–371.

Yamada, R., Bito, Y., Adachi, T., Tanaka, T., Ogino, C., Fukuda, H., and Kondo, A. (2009a). Efficient production of ethanol from raw starch by a mated diploid *Saccharomyces cerevisiae* with integrated a-amylase and glucoamylase genes. Enz. Microb. Technol. *44*, 344–349.

Yamada, R., Tanaka, T., Ogino, C., Fukuda, H., and Kondo, A. (2009b). Efficient and practical ethanol production from high yield rice by amylase expressing *Saccharomyces cerevisiae*. J. Biosci. Bioeng. *108*, S50.

Yamada, R., Tanaka, T., Ogino, C., Fukuda, H., and Kondo, A. (2010a). Novel strategy for yeast construction using delta-integration and cell fusion to efficiently produce ethanol from raw starch. Appl. Microbiol. Biotechnol. *85*, 1491–1498.

Yamada, R., Taniguchi, N., Tanaka, T., Ogino, C., Fukuda, H., and Kondo, A. (2010b). Cocktail d-integration: A novel method to construct cellulolytic enzyme expression ratio-optimized yeast strains. Microb. Cell Fact. *9*, 32.

Yamada, R., Hasunuma, T., and Kondo, A. (2013a). Endowing non-cellulolytic microorganisms with cellulolytic activity aiming for consolidated bioprocessing. Biotechnol. Adv. *31*, 754–763.

Yamada, R., Nakatani, Y., Ogino, C., and Kondo, A. (2013b). Efficient direct ethanol production from cellulose by cellulase- and cellodextrin transporter-co-expressing *Saccharomyces cerevisiae*. AMB. Express 3, 34.

Yamakawa, S.I., Yamada, R., Tanaka, T., Ogino, C., and Kondo, A. (2010). Repeated batch fermentation from raw starch using a maltose transporter and amylase expressing diploid yeast strain. Appl. Microbiol. Biotechnol. *87*, 109–115.

Yamakawa, S.I., Yamada, R., Tanaka, T., Ogino, C., and Kondo, A. (2012). Repeated fermentation from raw starch using *Saccharomyces cerevisiae* displaying both glucoamylase and a-amylase. Eukaryotic Cell *50*, 343–347.

Yanase, S., Hasunuma, T., Yamada, R., Tanaka, T., Ogino, C., Fukuda, H., and Kondo, A. (2010). Direct ethanol production from cellulosic materials at high temperature using the thermotolerant yeast *Kluyveromyces marxianus* displaying cellulolytic enzymes. Appl. Microbiol. Biotechnol. *88*, 381–388.

Yoon, S.H., Bruce Fulton, D., and Robyt, J.F. (2007). Formation of covalent a-linked carbohydrate-enzyme intermediates during the reactions catalyzed by a-amylases. Carbohydr. Res. *342*, 55–64.

Zaldivar, J., Nielsen, J., and Olsson, L. (2001). Fuel ethanol production from lignocellulose: a challenge for metabolic engineering and process integration. Appl. Microbiol. Biotechnol. *56*, 17–34.

Zhang, B., Li, L., Zhang, J., Gao, X., Wang, D., and Hong, J. (2013). Improving ethanol and xylitol fermentation at elevated temperature through substitution of xylose reductase in *Kluyveromyces marxianus*. J. Ind. Microbiol Biotechnol *40*, 305–216.

Zhang, J.G., Liu, X.Y., He, X.P., Guo, X.N., Lu, Y., and Zhang, B.r. (2011). Improvement of acetic acid tolerance and fermentation performance of *Saccharomyces cerevisiae* by disruption of the *FPS1* aquaglyceroporin gene. Biotechnol. Lett. 33, 277–284.

Zhang, L., and Hu, G. (2013). Supply chain design and operational planning models for biomass to drop-in fuel production. Biomass Bioenergy *58*, 238–250.

Zhang, M., Chou, Y.C., Picataggio, S.K., and Finkelstein, M. (1997) Single *Zymomonas mobilis* strain for xylose and arabinose fermentation. US5,843,760. 12-1-1997.

Zhang, W., and Geng, A. (2012). Improved ethanol production by a xylose-fermenting recombinant yeast strain constructed through a modified genome shuffling method. Biotechnol. Biofuels 5, 46.

Zhang, Y.H., and Lynd, L.R. (2004). Toward an aggregated understanding of enzymatic hydrolysis of cellulose: noncomplexed cellulase systems. Biotechnol. Bioeng. *88*, 797–824.

Zhao, X.Q., and Bai, F.W. (2009). Mechanisms of yeast stress tolerance and its manipulation for efficient fuel ethanol production. J. Biotechnol. *144*, 23–30.

Zhou, S., and Ingram, L.O. (1999). Engineering endoglucanase-secreting strains of ethanologenic *Klebsiella oxytoca* P2. J. Ind. Microbiol. Biotechnol. *22*, 600–607.

Zhu, Y., and Yang, S.T. (2004). Effect of pH on metabolic pathway shift in fermentation of xylose by *Clostridium tyrobutyricum*. J. Biotechnol. *110*, 143–157.

Production of Biobutanol, from ABE to Syngas Fermentation

4

Michael Köpke, Ryan E. Hill, Rasmus Ø. Jensen and Peter Dürre

Abstract

Butanol is an important commodity chemical and due to its properties considered to represent an ideal advanced biofuel. In recent years, in addition to petrochemical production of butanol, fermentation as a means of biobutanol production has been revived. This so called ABE fermentation is not a novel process, and has already been employed commercially for most of the first half of the 20th century. While most of the principles are the same, the rise of genetic tools and metabolic engineering allowed optimization of this process. Through synthetic biology, industrial workhorses as *E. coli* and yeast have been successfully modified for production of *n*-butanol, and through a non-fermentative pathway also for production of *iso*-butanol, a process that is currently commercialized. While all these approaches rely on use of sugar or starch, alternative processes have been developed for sustainable production of biobutanol from CO_2 and light with cyanobacteria or by fermenting syngas using acetogens.

4.1 Butanol – commodity chemical and advanced biofuel

Butanol is a four-carbon alcohol (C_4H_9OH) with a molecular mass of 74.12 g/mol that exists in four different isomers *n*-butanol, *sec*-butanol (as two stereoisomers), isobutanol, and *tert*-butanol (Table 4.1).

The straight chain *n*-butanol is the most common form with a global consumption of 4.5–5.5 million metric tons, estimated to be worth approximately US$7–8 billion. It is used as a precursor for the production of acrylate and methacrylate esters, glycol ethers, *n*-butyl acetate, amino resins, and *n*-butylamines being used in solvents (paints, coatings, varnishes, resins, dyes, etc.), plasticizers, coatings, textiles (swelling agent), flotation agents, cleaners and polishers, cosmetics (make up, lipsticks, foundations, nail care, etc.), or in de-icing fluids.

Market size for the other butanol isomers is somewhat smaller, between US $0.5–1 million in total. *sec*-Butanol is used as solvent, but mainly serves as a precursor for methyl ethyl ketone (MEK), an important industrial solvent, and esters that are used in perfumes or artificial flavours. Isobutanol is finding use as chemical intermediate for esters and as solvent (coatings and cleaners), extractant (in production of drugs, antibiotics, hormones, vitamins), in textile production, and as anti-corrosion agent. *tert*-Butanol is considered a solvent and gasoline octane booster and oxygenate. It can be reacted with methanol or ethanol to MTBE (methyl tert-butyl ether), respectively ETBE (ethyl tert-butyl ether). MTBE, however, has been banned in many US states in response to environmental and health concerns.

n-Butanol and isobutanol are also considered as gasoline additive and even as direct gasoline replacements due to similar properties (Table 4.2). Both, *n*-butanol and isobutanol have a higher energy density than ethanol, which is currently used in many countries around the world in blends with gasoline of up to 10% (20–25% in Brazil) in regular cars or 85–100% in modified flex fuel vehicles. The use of pure *n*-butanol as fuel in a regular car was demonstrated in 2005 by David Ramey, who toured the US in his 13-year-old Buick fuelled with 100% *n*-butanol. Still, some

Table 4.1 Properties of different butanol isomers

Isomer	*n*-Butanol	*sec*-Butanol	Isobutanol	*tert*-Butanol
IUPAC name	Butan-1-ol	Butan-2-ol	2-Methylpropan-1-ol	2-Methylpropan-2-ol
Other names	1-Butanol, butyl alcohol	2-Butanol, *sec*-butyl alcohol	2-Methyl-1-propanol, isobutyl alcohol	3-Butanol, *tert*-butanol, 2-methyl-2-propanol, *tert*-butyl alcohol
CAS number	71-36-3	78-92-2	78-83-1	75-65-0
Skeletal formula		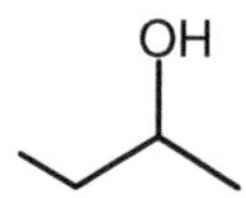	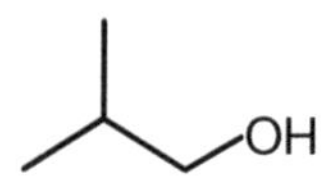	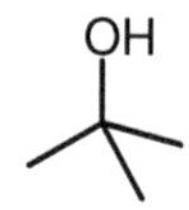
Density (at 20°C)	0.81 g/cm³	0.806 g/cm³	0.802 g/cm³	0.781 g/cm³
Solubility in water (at 20°C)	73 g/l	290 g/l	85 g/l	Miscible
Melting point	−90°C	−115°C	−102°C	25°C
Boiling point	118°C	99°C	108°C	82°C
Flash point	35°C	24°C	28°C	11°C
log *P*[a]	0.839	0.683	0.8	0.584

[a]Log *P*, octanol–water partition coefficient.

Table 4.2 Properties of different butanol isomers

Fuel	Gasoline	Ethanol	*n*-Butanol	Isobutanol
Energy density	32–35 MJ/l	21 MJ/l	29 MJ/l	29 MJ/l
Mileage	100%	61–66%	83–91%	82%
Air–fuel ratio	14.6	9.0	11.2	No data available
Research octane number (RON)	91–99	129	96	113
Motor octane number (MON)	81–89	102	78	94
Vapour pressure (at 20°C)	35–90 hPa	58 hPa	7 hPa	16 hPa
Enthalpy of vaporization	0.36 MJ/kg	0.92 MJ/kg	0.43 MJ/kg	0.58 MJ/kg

scepticism on butanol fuels from car manufacturers remain that needs to be overcome, but several companies have dedicated themselves to commercialize fuel butanol.

Over the last decades, butanol was mainly produced via petrochemical routes. *n*-Butanol is predominantly produced via oxo-synthesis from propylene, *sec*-butanol by hydration of butylene, isobutanol by carbonylation of propylene, and *tert*-butanol is derived from isobutane as a co-product of propylene oxide production or by hydration of isobutylene.

However, biotechnology allows for production of biobutanol independent from petroleum resources and comes at competitive costs in recent years of booming oil prices. Both, *n*-butanol and *sec*-butanol are occurring in nature as fermentation products of microorganisms. While *sec*-butanol is typically only a side product in fermentation of several lactic acid bacteria or yeast strains and thus not the focus of biotechnology, *n*-butanol is the main fermentation product of several clostridia species. Through synthetic biology and metabolic engineering, production of isobutanol by microorganisms became possible too and proved to be an attractive route.

4.2 Classic acetone–butanol–ethanol (ABE) fermentation with solventogenic clostridia

4.2.1 History of the ABE fermentation

Microbial butanol synthesis was first noticed by famous French scientist Louis Pasteur already in the mid-19th century (Pasteur, 1862). At the turn to the 20th century, dedicated efforts were made to isolate butanol-producing organisms in pure cultures by many scientists, amongst them Albert Fitz, Sergei Winogradsky, Martinus Beijerinck, Auguste Fernbach, Moïse Schoen, William Perkins, and Chaim Weizmann (Dürre and Bahl, 1996; Dürre, 2007; Jones and Woods, 1986; Köpke *et al.*, 2011a).

In 1912, the British company Strange and Graham, Ltd. developed a first commercial process for production of *n*-butanol (as a precursor in rubber production for the fast growing car industry at that time) using strains isolated by Fernbach (British patents 15,203 and 15,204). A first plant was built in Rainham and later in King's Lynn. Besides *n*-butanol, acetone and ethanol were also produced on the site, giving the process its name ABE fermentation (for acetone–butanol–ethanol fermentation).

One of the side products, acetone, was suddenly on high demand at the outbreak of the First World War for production of cordite (a smokeless gunpowder). Under the Defence of Realm act, the British government took over the plants from Strange and Graham, Ltd. and switched over to a process developed by Chaim Weizmann (British patent 4845), who had isolated a more efficient strain (later named *Clostridium acetobutylicum*). Production improved from an average of 970 pounds acetone per week to an average of 2340 pounds per week, being a decisive factor to the allied victory. Weizmann refused all honours by the British government, but eventually became the first president of the newly founded state of Israel in 1948 (Dürre and Bahl, 1996; Dürre, 2007; Jones and Woods, 1986; Köpke *et al.*, 2011a).

The ABE fermentation became the second largest biotechnological process behind ethanol fermentation until the 1950s when petrochemical routes outcompeted the process due to cheap and abundant oil becoming available (Dürre and Bahl, 1996; Dürre, 2007; Jones and Woods, 1986; Köpke *et al.*, 2011a). Only in politically isolated countries that were cut off from international supplies such as South Africa (Jones, 2001; Jones and Woods, 1986), the former Soviet Union (Zverlov *et al.*, 2006), or China (Chiao and Sun, 2007), ABE fermentation was carried on at commercial scale until the 1980s or even longer.

4.2.2 Revival of the ABE fermentation and commercial scale production

In recent years, plants in China re-opened and new plants have been built in China, Brazil, and other countries. China leads efforts to re-commercialize the ABE fermentation with an annual solvent production capacity of over 210,000 tons in 2009, which is expected to increase to 1,000,000 tons (Ni and Sun, 2009). Major players are Cathay Industrial Biotech Co. Ltd. (http://www.cathay-biotech.com/en/), who scaled up its *n*-butanol production capacity to 21 million gallons per year (Waltz, 2012), Ji-An Biochemicals Co. Ltd., Guiping Jinyuan Alcohol Industry Co. Ltd., Jinmaoyuan Biochemical Co. Ltd., Lianyungang Lianhua Chemical Product Co. Ltd. and Jiangsu Lianhai Biological Technology Co. Ltd. all operate plants with a solvent capacity of larger than 30,000 metric tons per year (Ni and Sun, 2009).

However, substrates competing with nutrition are no longer used in China for biotechnological processes. Therefore, Songyuan Laihe Chemical is collaborating with Green Biologics GBL (http://www.greenbiologics.com/), a UK/US based company to demonstrate *n*-butanol production on cellulosic feedstocks (Waltz, 2012). Green Biologics also works with other companies to provide strains (Green, 2011) and scale up expertise for ABE fermentation processes (http://www.greenbiologics.com/pr-biobutanolprod.php). In the US, Green Biologics signed a deal with Easy Energy Systems to retrofit their 80,000-l ethanol demonstration plant (http://www.greenbiologics.com/pr-greenbiologicssigns.php), after a merger with Butylfuel Inc. in 2012 (http://www.greenbiologics.com/pr-butylfuel.php).

Also in the US, Cobalt Technologies (http://www.cobalttech.com/) announced successful *n*-butanol production at a fermentation scale greater than 100,000 L (http://www.cobalttech.com/news-item/April%2016,%202013.html). The company is also working with partners on *n*-butanol plants in Brazil (http://www.cobalt-tech.com/media-coverage/October%2030,%202012.html). In Brazil, also a new plant has been built and is operated by HC Sucroquimica, producing 8000 metric tons solvent per annum from sugar cane (Green, 2011).

Other companies working on improvement of the ABE fermentation process are US-based TetraVitae Bioscience, which has been acquired by Eastman Chemicals in 2011 (http://www.eastman.com/company/news_center/2011/pages/eastman_acquires_tetravitae_bioscience_assets.aspx) and France-based Metabolic Explorer (http://www.metabolic-explorer.com/).

4.2.3 ABE fermentation process and strains

The ABE fermentation model organism is *Clostridium acetobutylicum*, which is characterized by its biphasic fermentation (Fig. 4.1). In a first phase (acidogenesis) during exponential growth, mainly acetic and butyric acid are formed and excreted. In a second phase (solventogenesis) these acids are then re-assimilated and converted into solvents *n*-butanol and acetone after a complete shift in metabolism. In addition, small amounts of ethanol, and sometimes lactate and acetoin are formed. Typically, a ratio of 6:3:1

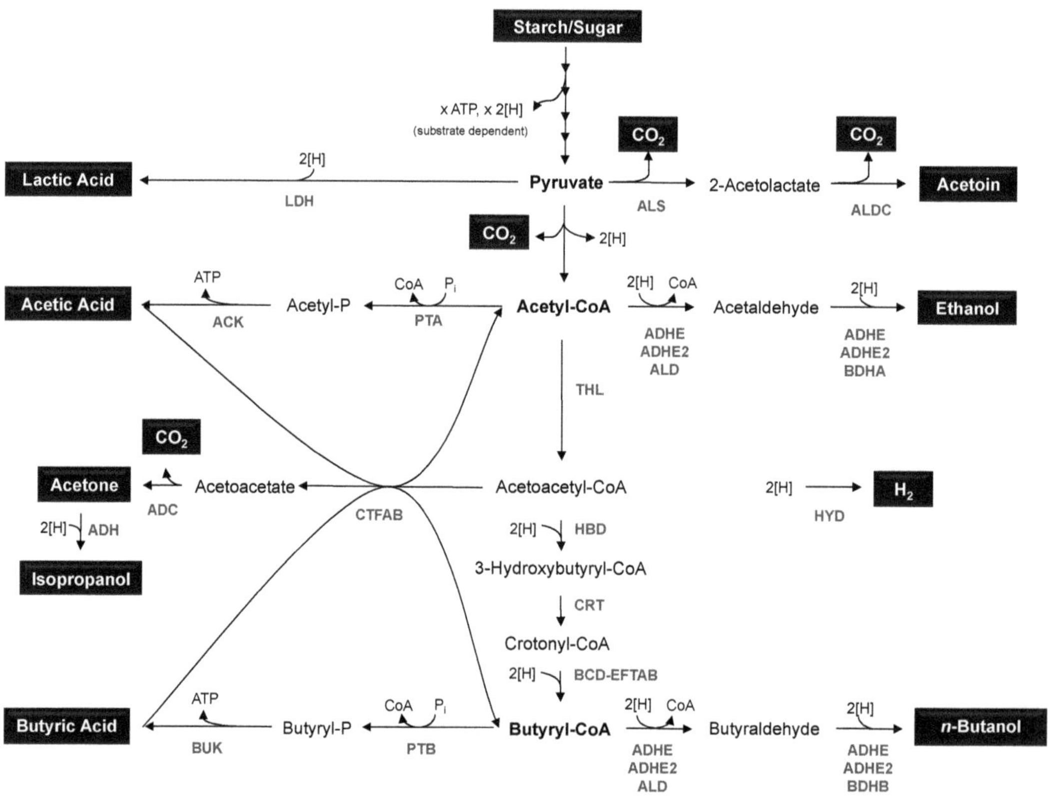

Figure 4.1 Metabolic pathway of solventogenic clostridia. Single reactions shown do not represent stoichiometric fermentation balances. ACK, acetate kinase; ADC, acetoacetate decarboxylase; ADH, primary:secondary alcohol dehydrogenase; ADHE/ADHE2, bifunctional aldehyde/alcohol dehydrogenase; ALD, aldehyde dehydrogenase; ALDC, acetolactate decarboxylase; ALS, acetolactate synthase; BCD, butyryl-CoA dehydrogenase; BDHA/BDHB, alcohol dehydrogenase; BUK, butyrate kinase; CRT, crotonase; ETFAB, electron-transferring flavoproteins; HBD; 3-hydroxybutyryl-CoA dehydrogenase; HYD, hydrogenase; LDH, lactate dehydrogenase; PTA, phosphotransacetylase; PTB, phosphotransbutyrylase.

between butanol–acetone:–ethanol is observed (Dürre, 2005; Jones and Woods, 1986).

The reason for this biphasic metabolism is that the excreted acids pose a serious threat to the organism. Undissociated acids can pass the cell membrane. As the external pH drops and the level of undissociated acids increases (pK_a of acetic acid = 4.75 and pK_a of butyric acid = 4.82), these acids can get back into the cell where they dissociate again into the respective salts and protons due to the higher internal pH. As a consequence, the essential proton gradient across the membrane required for energy conservation and several transport mechanisms is destroyed. To counteract, the cell converts these acids into solvents butanol and acetone. These however, are also toxic to the cell, but allow enough time to initiate sporulation and secure long-term survival (Dürre, 2005; Jones and Woods, 1986). The exact triggers for this metabolic shift are still not completely understood, but it has been shown that several extra- and intracellular parameters as temperature, low pH and high concentration of undissociated acids (Ballongue *et al.*, 1985; Gottwald and Gottschalk, 1985; Hüsemann and Papoutsakis, 1990; Terracciano and Kashket, 1986), levels of butyryl phosphate and butyryl-CoA (Boynton *et al.*, 1994; Gottwald and Gottschalk, 1985; Zhao *et al.*, 2005), limiting phosphate or sulfate concentrations (Bahl *et al.*, 1982; Kanchanatawee and Maddox, 1990), ATP/ADP ratio and NAD(P)H level (Grupe and Gottschalk, 1992) play a role and are interdependent.

Because of this biphasic metabolism, ABE fermentation is typically carried out as a batch process, but semi-continuous processes with series of several linked reactors have been established (Green, 2011; Ni and Sun, 2009; Zverlov *et al.*, 2006). Various sugars and starch can be used as feedstocks, dependent on the strain used. While the original Weizmann strain was best suited for growth on starch, later isolates (now classified as *C. beijerinckii*, *C. saccharobutylicum*, and *C. saccharoperbutylacetonicum*, see below) preferred sugar as a substrate. Usually corn (starch) or molasses from sugar beet and sugar cane (sugar) are used for industrial fermentations (Ezeji *et al.*, 2005).

The first *C. acetobutylicum* strain has been isolated by Chaim Weizmann (DSM 1732, NCIMB 2951), but best characterized is the type strain ATCC824 (DSM792, NRRL B527) (Weyer and Rettger, 1927) for which also a genome sequence is available (Nölling *et al.*, 2001). The genome consists of a 3.94-Mbp chromosome and a 192-kbp megaplasmid, which carries important genes for solventogenesis (see section 4.2.4). Several other strains of *C. acetobutylicum* have been described, one that is of particular interest is a hyper-butanol-producing isolate *C. acetobutylicum* EA 2018 (CCTCC M_94061) that has been obtained through several rounds of mutagenesis (Hu *et al.*, 2011). The genome sequence of this strain has been elucidated too, and between this strain and the type strain of *C. acetobutylicum*, a total of 46 deletion and 26 insertion sites (from which 55 are single indels while 7 indels are of over 100 bp) have been found (Hu *et al.*, 2011). Another sequenced *C. acetobutylicum* strain is DSM 1731 that has 345 SNVs to the type strain (including an 1.7-kbp insertion), and remarkably carrying a second 11 kbp large plasmid (Bao *et al.*, 2011).

Another species of solventogenic clostridia that is commonly used for ABE fermentation is *C. beijerinckii*. The type strain of this species NCIMB 8052 (ATCC 51743) has been sequenced as well. In contrast to *C. acetobutylicum*, it has an almost 50% larger genome of 6.0 Mbp without a megaplasmid, potentially making this strain more stable. While most solventogenic genes are similar to *C. acetobutylicum*, there are a few differences (section 4.2.4). *C. beijerinckii* contains a high number of PTS transporter systems and has a broad substrate spectrum of C5 and C6 sugars. Similar to *C. acetobutylicum*, for *C. beijerinckii* a hyper-butanol-producing mutant strain BA101 (ATCC PTA-1550) has been described (Shi and Blaschek, 2008). Some strains of *C. beijerinckii* as NRRL B-593 (VPI 13437 or DSM 6423) are able to form isopropanol rather than acetone (George *et al.*, 1983).

Besides *C. acetobutylicum* and *C. beijerinckii*, also *C. saccharobutylicum* and *C. saccharoperbutylacetonicum* are considered for ABE fermentation. One should note that several strains have been reclassified over the years. A thorough strain comparison of industrial ABE strains was performed in the lab of David Jones (Jones and Keis, 1995; Keis *et al.*, 1995, 2001; Shaheen *et al.*, 2000).

4.2.4 Pathway, genes, enzymes, and regulation of solventogenic clostridia

ABE fermentation clostridia follow the classic butyric acid fermentation pathway. The entry enzyme is an acetoacetyl-CoA thiolase (ThlA, encoded by gene *thlA*), which condenses two molecules of acetyl-CoA to form acetoacetyl-CoA with the release of a CoA molecule. Acetoacetyl-CoA is reduced to (S)-3-hydroxybutyryl-CoA by a 3-hydroxybutyryl-CoA dehydrogenase (Hbd, encoded by *hbd*) with consumption of NAD(P)H, followed by hydrolysis of the hydroxyl group by the enzyme crotonyl-CoA hydratase (crotonase, Crt, encoded by *crt*) to form crotonyl-CoA. The butyryl-CoA dehydrogenase complex (Bcd-complex, encoded by *bcd-etfAB*) catalyses the hydrogenation/reduction of the double bond in crotonyl-CoA to form butyryl-CoA, which is achieved by coupling the reduction of ferredoxin and oxidation of two NADH molecules in an electron-bifurcating reaction (Li *et al.*, 2008) (Fig. 4.1). While the thiolase gene is monocistronic, all other genes are arranged in one operon, the so-called BCS operon.

Butyryl-CoA is then converted to butyric acid in a two-step process catalysed by phosphotransbutyrylase (Ptb) and butyrate kinase (Buk). Respective genes (*ptb-buk*) are arranged in one operon. Analogous to this, acetic acid is formed from acetyl-CoA by enzymes phosphotransacetylase (Pta, encoded by *pta*) and acetate kinase (Ack, encoded by *ack*). Acetic acid and butyric acid are generated as part of substrate level ATP production (Fig. 4.1).

The acids are then reassimilated by a CoA transferase (CtfAB, encoded by *ctfAB*), transferring the CoA moiety of acetoacetyl-CoA to acetate or butyrate by CoA-transferase and yielding acetoacetate, which is subsequently decarboxylated to acetone by acetoacetate decarboxylase (Adc, encoded by *adc*). Some strains of *C. beijerinckii* also have a primary:secondary alcohol dehydrogenase capable of further reducing acetone to isopropanol in a NADPH-dependent reaction (Ismaiel *et al.*, 1993) (Fig. 4.1).

The final two reactions reduce butyryl-CoA with cleavage of the CoA molecule to butyraldehyde, which is subsequently reduced to butanol. Ethanol synthesis is derived from acetyl-CoA, which is converted to acetaldehyde with the release of CoA and is subsequently reduced to ethanol in a NAD(P)H-dependent two-step reduction. These reactions are catalysed by an aldehyde dehydrogenase (Ald, encoded by *ald*) and alcohol dehydrogenase (Bdh, encoded by *bdhA* and *bdhB*), respectively a bifunctional aldehyde/alcohol dehydrogenase (AdhE and AdhE2, encoded by *adhE* and *adhE2*) (Fig. 4.1).

The genes for solventogenesis are arranged in the so-called *sol* operon that consists of *adhE-ctfA-ctfB* and *adc*. While in *C. acetobutylicum adc* is located in the opposite direction of the other genes, the gene is in frame with the other genes in *C. beijerinckii*. In *C. beijerinckii*, the *adhE* gene is also replaced with an *ald* gene for a monofunctional aldehyde dehydrogenase enzyme Ald (Toth *et al.*, 1999). A major difference between the two species is also that the solventogenic genes in *C. acetobutylicum* are harboured on a megaplasmid (pSOL1), whereas in *C. bejierinckii* the genes are present in the genome.

The master regulator for onset of solventogenesis is the transcription factor Spo0A in its phosphorylated form (Ravagnani *et al.*, 2000; Harris *et al.*, 2002), which also controls sporulation. Unlike *Bacillus*, it is phosphorylated in *C. acetobutylicum* not via a phosphorelay, but instead directly by three different histidine kinases in two different pathways (Steiner *et al.*, 2011). The signals received by the various kinases are still unknown. In addition, several other transcription factors are also involved in regulation. The role of AdcR and AdcS, whose genes are organized in a common operon on the megaplasmid, is not yet clear. A binding site for CodY has been identified upstream of the *sol* operon, indicating that the GTP level inside the cell is important for controlling solventogenesis (Dürre, 2011). Another regulator is CcpA, also important for catabolite repression. It is involved in transcriptional activation of the *sol* operon (Ren *et al.*, 2012). The identified binding sites for Spo0A, CodY, and CcpA upstream of the *sol* operon are shown in Fig. 4.2. However, even more regulatory elements are involved in control of solventogenesis. A small regulatory RNA, SolB, seems to act as an antagonist of Spo0A (Dürre, 2011), while Rex represents a redox sensor and

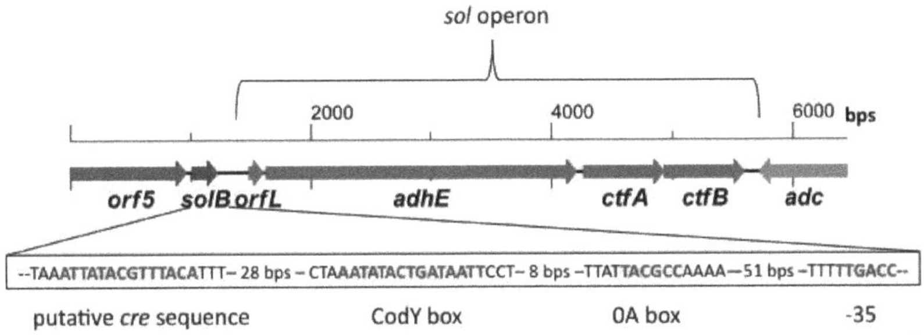

Figure 4.2 Binding sites for regulators upstream of the *sol* operon of *C. acetobutylicum*. Consensus motifs for binding of Spo0A~P (0A), CodY, and CcpA (*cre* sequence) upstream of the *sol* operon promoter are shown (–35 region of this promoter is also indicated). *orf5* encodes a glycosylase/deglycosylase, *solB* a small regulatory RNA. The function of *orfL* is still unknown. A colour version of this figure is available in the plate section at the back of the book.

transcriptional repressor (Wietzke and Bahl, 2012).

Lactate and acetoin are both produced directly from pyruvate; lactate via lactate dehydrogenase (LdhA, encoded by *ldhA*) and acetoin via acetolactate synthase (AlsS, encoded by *alsS*) and acetolactate decarboxylase (encoded by *adcB*). Hydrogen is generated from a hydrogenase for recycling excess reducing equivalents (ferredoxin/NAD(P)H).

4.2.5 Metabolic engineering of solventogenic clostridia to overcome some limitations of the ABE fermentation process

While the ABE fermentation process is a mature process that has been used for almost a century, there are some challenges and limitations to it. Metabolic engineering and systems biology approaches can help to identify and address these challenges, and significant progress has been made to overcome some of the major limitations.

Traditionally, mutagenesis was used to improve strains, but in recent years sophisticated genetic tools for ABE-producing clostridia have become available that allow precise genome modifications, notably two tools developed in the laboratory of Nigel Minton (University of Nottingham), the ClosTron system (a gene disruption tool based on a group II intron) (Heap *et al.*, 2010) and the ACE (Allelic Coupled Exchange) method (that allows cyclic integration of large fragments of DNA into the genome) (Heap *et al.*, 2012). The power of latter method was demonstrated by integration of the whole phage lambda genome of 48.5 kbp into the *pyrE* locus of the *C. acetobutylicum* chromosome in three steps (Heap *et al.*, 2012). Another system has been developed by the laboratory of Terry Papoutsakis (University of Delaware) that allows both markerless chromosomal integration and gene deletions using a *mazF* toxin gene (Al-Hinai *et al.*, 2012), while other methods have been developed as well that allow genome modifications of solventogenic clostridia.

In addition, data from several transcriptomics (Alsaker and Papoutsakis, 2005; Grimmler *et al.*, 2011; Jones *et al.*, 2008; Wang *et al.*, 2012) and proteomics (Han *et al.*, 2013; Janssen *et al.*, 2010; Mao *et al.*, 2010; Sivagnanam *et al.*, 2012) studies as well as genome-scale (McAnulty *et al.*, 2012; Milne *et al.*, 2011; Senger and Papoutsakis, 2008; Wallenius *et al.*, 2013) and kinetic models (Haus *et al.*, 2011; Li *et al.*, 2011; Millat *et al.*, 2013) for both *C. acetobutylicum* and *C. beijerinckii* have become available, guiding strain development efforts.

Most strain engineering is focused on two areas, increasing *n*-butanol yield and titre (which are largely a function of solvent tolerance and product distribution) and optimization of substrate range and utilization.

Solvent tolerance

End-product inhibition is one of the biggest challenges of the ABE fermentation process, as *n*-butanol is toxic to the cells at levels of 1.5–2%. Many of the hyper butanol-producing mutant strains obtained by chemical mutagenesis show a high tolerance and are thus able to produce higher overall *n*-butanol titres. However, one has to be careful as butanol tolerance and production are not necessarily linked to each other and screening for butanol tolerant strains, only in rare cases resulted also in a strain that produces well. Tolerance to butanol is a complex multigenic trait deriving from the action and interactions of a large number of genes and cellular programs (Dunlop *et al.*, 2011).

One promising approach taken was to overexpress chaperons. By overexpressing *groESL* in the type strain of *C. acetobutylicum*, growth was inhibited up to 85% less, resulting in 40% higher solvent titres of 26.7 g/l (17.1 g/l butanol and 8.6 g/l acetone) (Tomas *et al.*, 2003).

To create a more robust strain, also aerotolerance was prolonged by inactivation of *perR* (Hillmann *et al.*, 2008) or enhancing glutathione biosynthesis (Zhu *et al.*, 2011).

Product yield and distribution

A major limitation of the ABE fermentation process is the formation of many byproducts (in particular acetone that cannot be used as fuel) which are pulling carbon away from the main product *n*-butanol. Several attempts have been made to create a homobutanol-producing strain, by knocking out competing pathways.

Elimination of all byproduct forming reactions for acetate, butyrate, acetone, lactate, and hydrogen formation (*Δbuk, Δpta-ack, ΔctfAB, Δldh, ΔhydA*) has been described in *C. acetobutylicum* (WO2010/0086982). While no butanol titres are given, butyrate concentration decreased to a minimal amount of 0.43 g/l. However, one also needs to consider electron flow and energy conservation. A way to minimize acetone production is to reinforce the direct *n*-butanol production route, by simultaneous elimination of acid formation by disruption of acetate and butyrate forming genes (*Δbuk, Δpta-ack*) while overexpressing a (mutated) aldehyde/alcohol dehydrogenase *adhE* at the same time. By this approach a *n*-butanol titre of 18.9 g/l (160% improvement) with a yield of 0.71 mol *n*-butanol/mol glucose (245% improvement) was achieved with the type strain of *C. acetobutylicum* (Jang *et al.*, 2012). Acetone formation was as low as 1.5 g/l (360% decrease). In a fed-batch culture with *in-situ* product removal, 585 g *n*-butanol were produced from 1,861 g glucose (Jang *et al.*, 2012).

Already in 2000, the gene for the butyrate kinase (*Δbuk*) has been inactivated in the type-strain of *C. acetobutylicum*, accounting for a strain with a *n*-butanol titre of 16.7 g/l and a total solvent titre of 23.7 g/l, all with only a single modification. In another promising approach by the same group, inactivation of a postulated *solR* regulatory gene (Nair *et al.*, 1999) and overexpression of the aldehyde/alcohol dehydrogenase *adhE* in the type strain of *C. acetobutylicum* resulted in a *n*-butanol titre of 17.6 g/l and a total solvent titre of 27.9 g/l. It turned out later that this *solR* gene (or *orf5*), which is directly upstream of the *sol* operon (Fig. 4.2), actually encodes an extracellular protein involved in glycosylation-deglycosylation reactions and the repressing effect observed stemmed from an intergenic region between *orf5* and *sol* operon (Thormann and Dürre, 2001; Thormann *et al.*, 2002).

A good overview on additional metabolic engineering approaches to improve butanol yield and titre through reduced by-product formation is given by Gu *et al.* (2011), Lee *et al.* (2008), and Lütke-Eversloh and Bahl (2011).

Several studies also focused on upgrading the byproduct acetone. While acetone cannot be used as fuel, isopropanol has the potential to be used as fuel additive. Few *C. beijerinckii* strains can produce isopropanol naturally, but are not the best *n*-butanol producers. For this reason, the *C. acetobutylicum* type strain (Lee *et al.*, 2011) and also a hyper-butanol-producing mutant strain Rh8 (Dai *et al.*, 2012) have been engineered with the secondary alcohol dehydrogenase enzyme from *C. beijerinckii* to become an IBE (isopropanol–butanol–ethanol) producer.

In another study, byproduct acetoin was converted to 2,3-butanediol with the aim to further convert it into *sec*-butanol (Siemerink *et al.*, 2011).

Substrate utilization (lignocellulose and C_5 sugars)

Feedstock is typically the greatest expense in ABE fermentation. In a conventional plant, corn starch accounts for up to 79% of the overall production cost (Green, 2011). Significant cost reduction can be achieved using cellulosic material and agricultural residues. Solventogenic clostridia can utilize a range of C_6 as glucose and C_5 sugars as xylose and arabinose, but there is a strict catabolite regulation.

C_5 sugar utilization in presence of glucose could be improved by disrupting the *glcG* gene (encoding a putative enzyme II of the glucose PTS system) in the type strain of *C. acetobutylicum*. This modification resulted in a greatly improved xylose and arabinose consumption in the presence of glucose which was still effectively consumed (Xiao *et al.*, 2011). Overexpression of a xylose proton symporter, xylose isomerase, and xylulokinase enhanced *n*-butanol titres from xylose further, achieving a 24% higher solvent titre (16.06 g/l) and a 5% higher yield from xylose (0.28 g/g) (Xiao *et al.*, 2011). Xylose utilization in the type strain of *C. acetobutylicum* could also be improved by expressing a transaldolase gene *talA* from *E. coli* (Gu *et al.*, 2009).

While *C. acetobutylicum* cannot ferment cellulose, it does contain a cellulosome (Sabathé *et al.*, 2002) which, however, is inactive, presumably due to a single inactive component, Cel48A (Tracy, 2012). If this function could be restored, *C. acetobutylicum* could be an ideal platform for a consolidated bioprocess (CBP) that would allow production of *n*-butanol directly from cellulosic material. Recently, the successful expression, secretion, and mini-cellulosome formation of *C. thermocellum*-derived components in *C. acetobutylicum* has been reported, however, without cellulose utilization yet (Kovács *et al.*, 2013).

4.3 Engineering of non-natural butanol producers and synthetic pathways

4.3.1 *n*-Butanol via the CoA-linked pathway of clostridia

A number of other bacteria and yeast have been engineered to produce *n*-butanol using the CoA-dependent pathway from *C. acetobutyliucm* and *C. beijerinckii*, including non-natural butanol producing clostridia (Köpke *et al.*, 2010a; Yu *et al.*, 2011), industrial workhorses as *Escherichia coli* (Atsumi *et al.* 2007; Inui *et al.* 2008; Nielsen *et al.* 2009) or *Saccharomyces cerevisiae* (Steen *et al.*, 2008), and other relevant organisms such as *Pseudomonas putida*, *Bacillus subtilis* (Nielsen *et al.*, 2009), or *Lactobacillus brevis* (Berezina *et al.*, 2010) (Table 4.3). Reconstruction of acetone and isopropanol biosynthesis has also been achieved in *E. coli* using a combination of *C. acetobutylicum* and *C. beijerinckii* genes (Bermejo *et al.*, 1998; Jojima *et al.*, 2008; May *et al.*, 2013).

The study by Atsumi *et al.* (2007) described engineering of a strain of *E. coli* with a minimal *n*-butanol pathway consisting of *hbd*, *crt*, *bcd*, *etfA*, *etfB*, and *adhE2* genes from *C. acetobutylicum* under the control of a synthetic, isopropyl β-D-1-thiogalactopyranoside (IPTG)-inducible promoter P_{LlacO1} (Lutz and Bujard, 1997). Further engineering and optimization of the strain was conducted with the removal of native genes encoding enzymes that diverted acetyl-CoA and NADH towards pathways that synthesize the alternative compounds lactate (*ldhA*), ethanol (*adhE*), succinate (*frdBC*), and acetate (*pta*). A fifth gene *fnr* was removed, which encoded a repressor of the pyruvate dehydrogenase complex (PDHc) under anaerobic conditions. PDHc was theorized to produce greater amounts of NADH over the anaerobic pyruvate formate-lyase (encoded by *pflB*). When grown in a rich media under semi-anaerobic conditions, a maximum of 552 mg/l after 24 h was obtained.

A similar study by Inui *et al.* (2008) engineered an *E. coli* strain capable of producing *n*-butanol. This strain utilized the same pathway as Atsumi *et al.* (2007), but was under the control of an IPTG-inducible P_{tac} promoter system. In addition, the use of the two isoenzymes of aldehyde/alcohol dehydrogenase of *C. acetobutylicum* was investigated with isoenzyme AdhE2, resulting in greater yields of *n*-butanol. Interestingly, the strain constructed by Inui *et al.* (2008) with gene *adhE2* was able to produce considerably more *n*-butanol compared to the optimized strain constructed by Atsumi *et al.* (2007), which also harboured *adhE2*, with approximately 1200 mg/l after 24 h under semi-anaerobic conditions.

The engineering of *E. coli* by Atsumi *et al.* (2007) and Inui *et al.* (2008) identified major bottlenecks and requirements for effective *n*-butanol biosynthesis in non-native hosts. Atsumi *et al.*

Table 4.3 Metabolic engineering of non-natural butanol producers with the CoA-linked pathway of clostridia for production of *n*-butanol

Reference	Organism	Strain background	Thiolase	Dehydrogenase	Crotonase	Reductase	Alcohol/aldehyde dehydrogenase	Yield (mM)
WO 2007/041269	*Escherichia coli*		*thlA (Cac)*	*hbd (Cac)*	*crt (Cac)*	*ter (Egr)*	*ald (Cbe), yqhD* overexpression	1.03
	Bacillus subtilis		*thlA (Cac)*	*hbd (Cac)*	*crt (Cac)*	*ter (Egr)*	*ald (Cbe), bdhB (Cac)*	0.19
	Saccharomyces cerevisiae		*thlA (Cac)*	*hbd (Cac)*	*crt (Cac)*	*ter (Egr)*	*ald (Cbe)*	0.02
Inui *et al.* (2008)	*Escherichia coli*		*thlA (Cac)*	*hbd (Cac)*	*crt (Cac)*	*bcd-etfAB (Cac)*	*adhE2 (Cac)*	16.2
	Escherichia coli		*thlA (Cac)*	*hbd (Cac)*	*crt (Cac)*	*bcd-etfAB (Cac)*	*adhE (Cac)*	4.2
WO 2008/010124	*Saccharomyces cerevisiae*		*thlA (Cac)*	*hbd (Cac)*	*crt (Cac)*	*bcd-etfAB (Cac)*	*adhE2 (Cac)*	0.33
WO 2008/052991	*Saccharomyces cerevisiae*		*thlA (Cac)*	*hbd (Cac)*	*crt (Cac)*	*bcd-etfAB (Cac)*	*adhE (Cac), bdhA (Cac), bdhB (Cac)*	1.92
Atsumi *et al.*, 2008; WO 2008/124523	*Escherichia coli*	*Δadh, Δldh, Δfrd, Δfnr, Δpta*	*thlA (Cac)*	*hbd (Cac)*	*crt (Cac)*	*bcd-etfAB (Cac)*	*adhE2 (Cac)*	5
	Escherichia coli		*thlA (Cac)*	*hbd (Cac)*	*crt (Cac)*	*bcd-etfAB (Cac)*	*adhE2 (Cac)*	1.9
Steen *et al.*, 2008	*Saccharomyces cerevisiae*	*Δhis3, Δleu2, Δlys2, Δura3*	*Erg10* overexpression	*hbd (Cbe)*	*crt (Cbe)*	*ccr (Sco)*	*adhE2 (Cbeij)*	0.02
	Saccharomyces cerevisiae	*Δhis3, Δleu2, Δlys2, Δura3*	*Erg10* overexpression	*hbd (Cbe)*	*crt (Cbe)*	*bcd-etfAB (Cbe)*	*adhE2 (Cbeij)*	0.02
	Saccharomyces cerevisiae	*Δhis3, Δleu2, Δlys2, Δura3*	*phaA (Reu)*	*phaB (Reu)*	*crt (Cbe)*	*ccr (Sco)*	*adhE2 (Cbeij)*	0.01
WO 2009/013160	*Saccharomyces cerevisiae*	*Δadh*	*thlA (Cac)*	*hbd (Cac)*	*crt (Cac)*	*bcd-etfAB (Cac)*	*adhE (Cac), bdhB (Cac), acdh (Lin)*	1.92

Nielsen *et al.* (2009)	*Escherichia coli*	*gapA* overexpression	*atoB overexpression*	*hbd (Cac)*	*crt (Cac)*	*bcd-etfAB (Cac)*	*adhE2 (Cac)*	43
	Escherichia coli		*atoB overexpression*	*hbd (Cac)*	*crt (Cac)*	*bcd-etfAB (Cac)*	*adhE2 (Cac)*	16
	Escherichia coli		*atoB* overexpression	*hbd (Cac)*	*crt (Cac)*	*bcd-etfAB (Cac)*	*adhE1 (Cac)*	16
	Escherichia coli		*thlA (Cac)*	*hbd (Cac)*	*crt (Cac)*	*bcd-etfAB (Cac)*	*adhE1 (Cac)*	14
	Pseudomonas putida		*thlA (Cac)*	*hbd (Cac)*	*crt (Cac)*	*bcd-etfAB (Cac)*	*adhE1 (Cac)*, *bdhB (Cac)*	9
	Bacillus subtilis	*ΔamyE*, *ΔthrC*, *ΔpydD*	*thlA (Cac)*	*hbd (Cac)*	*crt (Cac)*	*bcd-etfAB (Cac)*	*adhE2 (Cac)*	1.8
Köpke *et al.* (2010)	*Clostridium ljungdahlii*		*thlA (Cac)*	*hbd (Cac)*	*crt (Cac)*	*bcd (Cac)*, native etfAB	*adhE1 (Cac)*, *bdhB (Cac)*	2
Berezina *et al.* (2010)	*Lactobacillus brevis*		*thlA (Cac)*	*hbd (Cac)*	*crt (Cac)*	*bcd-etfAB (Cac)*	native *adhE*	4.1
Bond-Watts *et al.* (2011)	*Escherichia coli*	overexpression: *aceEF*, *lpd*	*phaA (Reu)*	*hbd (Cac)*	*crt (Cac)*	*ter (Tde)*	*adhE2 (Cac)*	63
Shen *et al.* (2011)	*Escherichia coli*	*ΔadhE*, *ΔldhA*, *ΔfrdBC*, *Δpta*, fdh overexpression	*atoB* overexpression	*hbd (Cac)*	*crt (Cac)*	*ter (Tde)*	*adhE2 (Cac)*	405
Yu *et al.* (2011)	*Clostridium tyrobutyricum*		native *thlA*	native *hbd*	native *crt*	native *bcd-etfAB*	*adhE2 (Cac)*	216
Lan and Liao (2011)	*Synechococcus elongatus*	(anaerobic)	*atoB (Eco)*	*hbd (Cac)*	*crt (Cac)*	*ter (Tde)*	*adhE2 (Cac)*	0.18
Lan and Liao (2012)	*Synechococcus elongatus*	(aerobic)	*nphT7 (SspCL190)*	*phaB (Reu)*	*phaJ (Reu)*	*ter (Tde)*	*Ald (Csa)*, *yqhD (Eco)*	0.41

*Cac, *C. acetobutylicum*; Cbe, *C. beijerinckii*; Csa, *C. saccharoperbutylacetonicum*; Eco, *E. coli;* Egr, *Euglena gracilis*; Lin, *Listeria innocua*; Reu, *Ralstonia eutropha*; Sco, *Streptomyces collinus*; Tde, *Treponema denticola*; SspCL190, *Streptomyces* sp. strain CL190.

(2007) observed highest production of *n*-butanol under semi-anaerobic conditions, followed by anaerobic, and aerobic conditions being the least productive. It has been observed that AdhE2 is inactivated by oxygen *in vitro* (Nair *et al.*, 1994), so it is important to note that *n*-butanol production occurred even under aerobic conditions at approximately 20% to that of the semi-anaerobic conditions. Overexpression of the native thiolase produced approximately 300% more *n*-butanol than over-expression of the *C. acetobutylicum* thiolase (Atsumi *et al.*, 2007). Increased activity under semi-anaerobic conditions was attributed to an increase in NADH production through slight increases in the concentration of oxygen. A lack of NADH was shown to be a key bottleneck in *n*-butanol production and the major explanation for low *n*-butanol production under aerobic conditions, due to both acetyl-CoA and NADH being funnelled into the more efficient tricarboxylic acid cycle (TCA) and electron transport chain (ETC). In contrast, Inui *et al.* (2008) were unable to detect *n*-butanol production under aerobic or anaerobic conditions, but the strain was capable of producing approximately 2.5-fold more *n*-butanol under semi-anaerobic conditions compared to the Atsumi *et al.* (2007) strain.

A number of other organisms were shortly thereafter engineered to produce *n*-butanol; *S. cerevisiae* (Steen *et al.*, 2008), *P. putida*, *B. subtilis* (Nielsen *et al.*, 2009), and *Lactobacillus brevis* (Berezina *et al.*, 2010). However, none were able to produce *n*-butanol equivalent to that of the engineered *E. coli* of Atsumi *et al.* (2007) and Inui *et al.* (2008) with the highest amount produced from engineered *L. brevis* at 300 mg/l. All of the studies identified the Bcd-complex as a major inhibiting factor in butanol production along with the availability of either NADH or acetyl-CoA.

Several of the studies attempted to address the issue of the *C. acetobutylicum* Bcd-complex (Bcd_{ca}). Atsumi *et al.* (2007) investigated replacement of Bcd_{ca} with the Bcd-complex of *Megasphaera elsdenii* (Bcd_{me}) or crotonyl-CoA reductase from *Streptomyces collinus* (Ccr). Both Ccr and Bcd_{me} have been expressed in *E. coli* and both *in vivo* and *in vitro* activities have a specificity for the crotonyl-CoA to butyryl-CoA reaction (Becker *et al.*, 1993; Wallace *et al.*, 1995). Surprisingly, Bcd_{ca} performed better, despite the fact that no *in vitro* activity could be detected, while Bcd_{me} produced substantially less *n*-butanol and Ccr performed the worst of all three with almost undetectable levels of *n*-butanol produced. Nielsen *et al.* (2009) had somewhat contradictory results to Atsumi *et al.* (2007) with the use of Ccr achieving 100 mg/l in 48 h, although yields were less than Bcd_{ca} (200 mg/l). Use of Ccr in *S. cerevisiae* yielded a slight increase in *n*-butanol production over Bcd_{ca}; however, the maximum *n*-butanol yield achieved with *S. cerevisiae* was considerably less than *E. coli* at 2.5 mg/l (Steen *et al.*, 2008).

A major breakthrough in *n*-butanol production by recombinant *E. coli* was achieved by the lab of James C. Liao (UCLA), who attempted to tackle the problems encountered with the Bcd complex and substrate/cofactor availability. A publication by Shen *et al.* (2011) developed a strain of *E. coli* that produced a final concentration of 30 g/l *n*-butanol (in 168 h with constant product removal).

The high-level production strain of *E. coli* was achieved through three major developments. The first was replacement of the Bcd complex with an alternative enzyme, *trans*-2-enoyl-CoA reductase (Ter) from the organism *Treponema denticola* (Tucci and Martin, 2007). The enzyme Ter is similar to Ccr, but utilizes NADH as an electron source as opposed to NADPH. Critically, the k_{cat}/K_m of Ter is approximately 85-fold greater than Ccr, having a 12-fold increase in k_{cat} and 6-fold decrease in K_m for crotonyl-CoA (Tucci and Martin, 2007; Wallace *et al.*, 1995). Replacement of the Bcd complex with Ter created a pathway dependent on NADH as the sole source of reducing power. In addition, Ter appears incapable of the reverse reaction of butyryl-CoA to crotonyl-CoA (*in vitro*) thus creating a driving force towards *n*-butanol synthesis. The second development was to increase the NADH to NAD^+ ratio (NADH driving force) (Shen *et al.*, 2011). The NADH driving force was created by engineering a strain of *E. coli* incapable of growing anaerobically by deleting the native *adhE*, *ldhA*, and *frdBC* genes, which encode key enzymes in ethanol, lactate, and succinate fermentation. This rendered the organism incapable of recycling NADH to NAD^+. In addition, the non-native enzyme formate dehydrogenase was overexpressed, which further increased concentrations of NADH by converting the formate generated by pyruvate:formate-lyase to carbon dioxide and NADH. The third and final development involved increasing the availability of acetyl-CoA, which had already been increased as a consequence of implementing the NADH driving force. The flux to acetyl-CoA was improved by removing the enzyme phosphotransacetylase (*pta*), and in doing so eliminated acetate production.

The final strain produced approximately 7 g/l *n*-butanol in 24 h under anaerobic conditions with 2% (w/v) glucose, and produced approximately 15 g/l in 3 days. For comparison, the original strain developed by Atsumi *et al.* (2007) produced 0.5 g/l in 24 h. A final demonstration of the strain utilized a 1-l bioreactor with continuous product removal by gas stripping. After 7 days of growth, 30 g/l of *n*-butanol was produced with a conversion efficiency of 70% g/g glucose and average productivity of 0.2 g/l/h.

A similar study by Bond-Watts *et al.* (2011) achieved 4.65 g/l after 3 days of growth. The *E. coli* strain developed had a pathway comprising PhaA, Hbd, Crt, Ter and AdhE2, expressed from plasmids under the control of promoters P_{ara} (*phaA-hbd*), P_{trc} (*crt*), and P_{tac} (*ter-adhE2*). In addition, the pyruvate dehydrogenase complex (PDHc, *aceEF-lpd*) was overexpressed from a third plasmid under the P_{lac} promoter, with a concomitant three-fold increase in PDHc activity. The use of Ccr was investigated, but maximum production achieved was only 0.12 g/l after 6 days' growth. This was primarily attributed to butyryl-CoA synthesis being a side-reaction of Ccr in the absence of sodium bicarbonate, under physiological conditions the enzyme produces mostly ethylmalonyl-CoA. Increase of *n*-butanol production with Ter was also attributed to irreversible activity of the enzyme, effectively driving *n*-butanol synthesis by preventing reversion of the pathway.

Several patent applications have been filed around this process (Table 4.3).

4.3.2 Isobutanol and *n*-butanol from amino acids via a non-fermentation pathway

Besides the classic fermentative clostridial route to *n*-butanol, a novel non-fermentative pathway for the biosynthesis of alcohols such as isobutanol or *n*-butanol has been identified by re-routing amino acid biosynthesis (Atsumi *et al.*, 2008). The novel pathway was devised to avoid the use of CoA-linked compounds and to limit the use of oxygen-sensitive enzymes A brief summary is presented below and in Table 4.4, but for details see Chapter 5.

The first generation strain produced 1-propanol (0.53 mM, 31 mg/l), *n*-butanol (16 mg/l), isobutanol (388 mg/l), 2-methyl-1-butanol (67 mg/l), 3-methyl-1-butanol (131 mg/l), and 2-phenylethanol (39 mg/l) after 40 h of growth (Atsumi *et al.*, 2008). A second-generation strain hyper-produced 2-ketoisovalerate (the precursor to isobutanol) and production was further improved with a third generation strain to 22.2 g/l after 112 h of growth (86% theoretical yield) (Atsumi *et al.*, 2008). *n*-Butanol was the second compound for which production was enhanced; however, final production was still significantly lower than isobutanol. Biosynthesis of *n*-butanol by this pathway is derived from threonine. The second generation strain produced 44 mg/l *n*-butanol after 40 h by overexpressing threonine deaminase. Production was further increased to 684 mg/l with a third generation strain by removal of *ilvD* (encoding dihydroxy-acid dehydratase) and supplementing the media with 8 g/l L-threonine (Atsumi *et al.*, 2008). Atsumi *et al.* (2008) noted that further enhancement of *n*-butanol biosynthesis via redirection of amino acid pathways would require several major modifications. Improved production of 2-ketobutyrate and 2-ketovalerate will be key to high *n*-butanol titres and development of a hyper-threonine-producing strain would ultimately improve levels of 2-ketobutyrate. Such strains have been developed for commercial production of L-threonine (Miwa and Tsuchida, 1983). Alternatively, 2-ketobutyrate can be synthesized directly from pyruvate and acetyl-CoA via the citramalate pathway of *Leptospira*

Table 4.4 Metabolic engineering for isobutanol production

Reference	Organism	Strain background	Gene expression				Yield (mM)
WO 2007/050671	*Escherichia coli*		*kivD (Lla)*	*yqhD* overexpression	*budB (Kpn), ilvC-ilvD* overexpression		9.9
	Saccharomyces cerevisiae		*kivD (Lla)*	*alsS*	*Ilv3, ilv5* overexpression		0.92
	Bacillus subtilis		*kivD (Lla)*	*yqhD* overexpression	*budB (Kpn), ilvC-ilvD* overexpression		1.0
Atsumi *et al.* (2008)	*Escherichia coli*		*kivD (Lla)*	*adh2 (Sce)*			5.24
US 2010/0209986	*Escherichia coli*	Δ*adh*, Δ*ldh*, Δ*frd*, Δ*fnr*, Δ*pta*	*kivD (Lla)*	*adh2 (Sce)*	*ilvIHCD* overexpression		30
	Escherichia coli	Δ*adh*, Δ*ldh*, Δ*frd*, Δ*fnr*, Δ*pta*, Δ*pflB*	*kivD (Lla)*	*adh2 (Sce)*	*ilvCD* overexpression	alsS (*Bsu*)	300
US 2009/0215137	*Escherichia coli*	Δ*ldh*, Δ*frn*, Δ*frd*, Δ*pta*, Δ*adhE*, Δ*pfl*, Δ*sthA*	*kivD (Lla)*	*adh2 (Sce)*	*alsS, ilvCD, pntAB*		216
Atsumi *et al.* (2009)	*Synechococcus elongatus*		*kivD (Lla)*	*adh2 (Sce)*	*ilvCD (Eco)*	*alsS (Bsu)*	6
Higashide *et al.* (2011)	*Clostridium cellulolyticum*		*kivD (Lla)*	*yqhD (Eco)*	*ilvCD (Eco)*	*alsS (Bsu)*	8.9
Blombach *et al.* (2011)	*Corynebacterium glutamicum*	Δ*aceE*, Δ*pqo*, Δ*ilvE*, Δ*ldhA*, Δ*mdh*	*kivD (Lla)*	native *adhA*	*ilvBNCD-pntAB* overexpression		175
Jia *et al.* (2012)	*Bacillus subtilis*		*kivD (Lla)*	*adh2 (Sce)*			8.1
Kondo *et al.* (2012)	*Saccharomyces cerevisiae*	Δ*pdc1*	*kivD (Lla)*	*adh2 (Sce)*	*Ilv2* overexpression		10.5
Lee *et al.* (2012)	*Saccharomyces cerevisiae*		*kivD (Lla)*	native *adh*	*Ilv1, ilv2* overexpression		2
Varman *et al.* (2013)	*Synechocystis* sp.		*kivD (Lla)*	*adhA (Lla)*			3.2

*Kpn, *Klebsiella pneumoniae*; Lla, *Lactobacillus lactis*; Sce, *Saccharomyces cerevisiae.*

interrogans (Xu *et al.*, 2004). A study by Atsumi and Liao (2008) investigated the possibility of using the citramalate pathway for *n*-butanol biosynthesis by expressing a citramalate synthase from *Methanococcus jannaschii*. Despite increasing the availability of 2-ketobutyrate, production of *n*-butanol (0.543 g/l) did not significantly increase, with the majority of the 2-ketobutyrate being converted to 1-propanol by Kdc/Adh (3.5 g/l). Both studies determined that regardless of the source of 2-ketobutyrate, directed evolution of LeuABCD for improved specificity towards the non-native substrate 2-ketobutyrate and the derivatives will be essential for increasing butanol biosynthesis from keto-acids. Likewise, Kdc would require directed evolution for enhanced specificity towards 2-ketovalerate over 2-ketobutyrate to reduce the production of the competing compound 1-propanol.

Other strains such as cellulolytic *Clostridium* strain *C. cellulolyticum* (Higashide *et al.*, 2011), industrial workhorses such as yeast (Kondo *et al.*, 2012; Lee *et al.*, 2012) and *Bacillus subtilis* (Jia *et al.*, 2012), amino acid producers such as *Corynebacterium glutamicum* (Blombach *et al.*, 2011), or cyanobacteria and *Ralstonia eutropha* have been successfully modified for isobutanol production as well (Table 4.4). Particularly, *Corynebacterium glutamicum* showed good potential (Blombach and Eikmanns, 2011). In addition, several patent applications have been filed around this route (Table 4.4).

4.3.3 *n*-Butanol via reverse beta-oxidation and other alternative pathways

The group of Ramon Gonzalez (Rice University) engineered *E. coli* to produce *n*-butanol through reversal of beta oxidation (Dellomonaco *et al.*, 2011). This was achieved by constitutive expression of the *fad* and *ato* operons, which encode the enzymes of fatty acid oxidation. For constitutive expression genes *fadR*, *atoC*, and *crp* were mutated and *arcC*, which together regulate the two operons, was deleted. The study aimed to produce a strain using only native *E. coli* enzymes and was successful in developing a strain capable of producing 14 g/l of *n*-butanol in 48 h of fermentation on 5% (w/v) glucose. In addition to the deregulation of fatty acid oxidation, the pathways for synthesis of acetate (*pta*), ethanol (*adhE*, *eutE*, *yqhD*), and fumarate/succinate (*frdA*) were removed and native enzymes 1,2-propanediol oxidoreductase (*fucO*) and aldehyde/alcohol dehydrogenase (*yqeF*) were overexpressed, which had greater specificity towards butanol over ethanol synthesis.

4.3.4 Commercialization and challenges of butanol production in *E. coli* and yeast

Two companies, Gevo (http://gevo.com/) and Butamax (http://www.butamax.com/, a joint venture between BP and DuPont) are working to commercialize isobutanol production using industrial workhorse organisms yeast and *E. coli*. Isobutanol offers the advantage over *n*-butanol that it is not as toxic for the cells. One reason for this is a lower octanol–water partition coefficient, log *P* (Table 4.1), that relates to the ability of the solvent to integrate into and disrupt a membrane.

Gevo already operates a commercial plant in Luverne, MN, with two million litre fermenters that has been retrofitted from an ethanol process (http://ir.gevo.com/phoenix.zhtml?c=238618&p=irol-newsArticle&ID=1843687&highlight=), while Butamax has announced a demonstration plant. Both companies have been in a patent litigation around aspects of the process(http://www.bloomberg.com/news/2013-05-10/butamax-and-gevo-grapple-with-biofuels-patent-case-claims.html).

4.4 Future trends – butanol production from greenhouse gases CO_2 and/or CO

It is becoming increasingly clear that a reliance on food crops by first generation technologies for the production of biofuels and commodity chemicals such as butanol is increasingly problematic. The diversion of food resources and already scarce arable land for the production of fuels and chemicals with a low greenhouse gas emissions profile will almost certainly not provide the solution the world will require over the next decades to the colliding yet growing demands for energy security and reduced greenhouse gas emissions. Several NGOs and industry leaders have voiced concerns that current biofuel policies and production methods are deepening poverty, raising food prices and may even accelerate climate change (Dürre, 2007; OEDC-FOA, 2011).

These short falls taken together with the volatility of sugar prices are prime drivers for the development of alternative fuels based on non-food resources. One technology that shows great promise is the production of biofuels and commodity chemicals directly using carbon-containing greenhouse gases such as carbon dioxide (CO_2) and/or carbon monoxide (CO). This can be achieved through photosynthetic routes, for example with cyanobacteria or through gas fermentation of CO_2/H_2 with organisms such as *Ralstonia* species or acetogenic bacteria. This latter group of non-photosynthetic organisms can convert syngas produced for example from biomass or as a by-product of established industrial processes. Reducing equivalents needed for these reactions can also come from electricity in an emerging concept called microbial electrosynthesis, that has been demonstrated with both *Ralstonia* and acetogens.

4.4.1 Photosynthetic production of butanol using cyanobacteria

Photoautotrophic organisms such as cyanobacteria allow conversion of CO_2 and light into useful products (see Chapter 8). *Synechococcus* sp. PCC 7942 was engineered to produce *n*-butanol using the CoA-dependent pathway; however, the engineered strain was only capable of producing butanol under strictly anaerobic growth conditions and even a small introduction of oxygen drastically reduced butanol production (Lan and Liao, 2011). In contrast to engineered *E. coli* and other organisms, the pathway in *Synechococcus* was integrated into the genome. A maximum of 7–14 mg/l *n*-butanol was produced (Lan and Liao, 2011). The strain was further modified and produced approximately 7 mg/l of butanol, but the butanol was biosynthesized photoautotrophically under aerobic conditions. Production was further increased to 26 mg/l by expressing an Ald from *Clostridium saccharoperbutylacetonicum* NI-4 (Ald_{Csa}) and an alcohol dehydrogenase (YqhD) from *E. coli* (Pérez et al., 2008). The most productive strain replaced enzymes Hbd and Crt with R-specific isoforms, PhaB and PhaJ from *Ralstonia eutropha* and *Aeromonas caviae*, respectively. Up to 30 mg/l *n*-butanol was produced after 20 days. *Synechococcus elongatus* PCC7942 was also engineered by Atsumi *et al.* (2009) to produce isobutyraldehyde and isobutanol via the 2-keto-acid pathway. Approximately 0.75 g/l isobutyraldehyde (in 8 days) and 0.45 g/l isobutanol (in 8 days) were produced, respectively. A minimal pathway for isobutanol production in *Synechocystis* sp. PCC 6803 by integrating genes *kivd-adhA* from *Lactococcus lactis* under the control of the P_{tac} promoter allowed to produce 240 mg/l in 21 days with product trapping (Varman *et al.*, 2013). While this approach shows potential, there are some challenges such as the requirement for large land areas and the need to improve production rates.

4.4.2 Butanol from CO_2 and electricity using *Ralstonia eutropha*

Ralstonia eutropha (*Cupriavidus necator*) is another platform that allows utilizing CO_2, in combination with hydrogen and oxygen. However, hydrogen is not available everywhere and for an aerobic process a lot of safety requirements are required. A solution could be the use of formate, which can be produced by electrosynthesis from CO_2. Production of isobutanol (up to 846 mg/l) from CO_2 and electricity has been demonstrated in an integrated approach using an engineered strain of *R. eutropha* (Li *et al.*, 2012). While this is a promising approach, no concept for scale-up has been developed yet.

4.4.3 Butanol from syngas using acetogenic microbes

Overview and advantages of (syn)gas fermentation

One of the most promising approaches to capture gases such as CO_2 or CO is the use of acetogenic bacteria. A recent study compared different routes for CO_2 conversion (also in respect of butanol production) and came to the conclusion that the Wood–Ljungdahl pathway of anaerobes is the most efficient (Fast and Papoutsakis, 2012). As for *Ralstonia*, microbial electrosynthesis is also possible with acetogens (Lovley and Nevin, 2013;

Nevin *et al.*, 2010) and growth on a cathode has been demonstrated for several acetogenic organisms (Nevin *et al.*, 2011). Besides CO_2 (with H_2 or electricity), acetogens are also able to use CO, both as a carbon and energy source.

This opens up the use of syngas (a mixture primarily of CO, CO_2, and H_2 in different ratios) and other waste gases. Syngas is a major feedstock in chemical industry and can also be generated directly from biomass (Griffin *et al.*, 2012; Daniell *et al.* 2012).

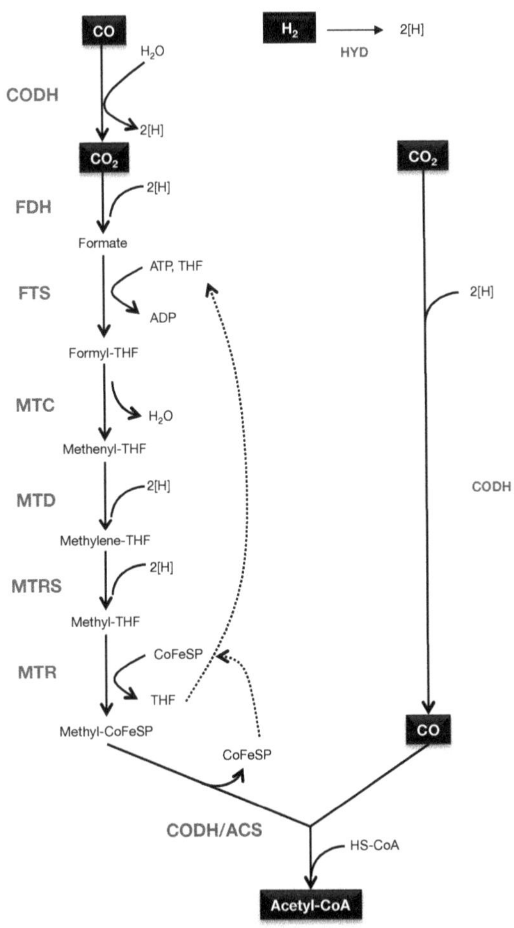

Figure 4.3 The reductive acetyl-CoA (Wood-Ljungdahl) pathway for CO/CO_2 fixation in acetogens. Single reactions shown do not represent stoichiometric fermentation balances. CODH, CO dehydrogenase; FDH, formate dehydrogenase; FTS, formyl-THF synthase; HYD, hydrogenase; MTC, methylene-THF cyclohydrase; MTD, methylene-THF dehydrogenase; MTRS, methylene-THF reductase; MTR, methyltransferase; CODH/ACS, CO dehydrogenase/acetyl-CoA synthase complex.

Acetogens use the reductive Acetyl-CoA pathway (Fig. 4.3) to ferment CO and/or CO_2 and H_2 into a range of reduced products including acetate, ethanol, butyrate, or butanol (Drake *et al.*, 2008), as well as lactate and 2,3-butanediol (Köpke *et al.*, 2011b).

The overall reactions for CO, CO_2, and H_2 to acetate and ethanol are exergonic. Equations 4.1–4.7 show the chemical reactions and the Gibbs free energy for that reaction for growth on the different substrate:

CO as sole carbon and energy source:

$$6CO + 3H_2O \rightarrow CH_3CH_2OH + 4CO_2$$

$$\Delta G^{0\prime} = -224\frac{kJ}{mol} \tag{4.1}$$

$$4CO + 2H_2O \rightarrow CH_3COOH + 2CO_2$$

$$\Delta G^{0\prime} = -175\frac{kJ}{mol} \tag{4.2}$$

CO + H_2 as carbon and energy source:

$$3CO + 3H_2 \rightarrow CH_3CH_2OH + CO_2$$

$$(CO : H_2 \text{ ratio} = 1:1) \quad \Delta G^{0\prime} = -164\frac{kJ}{mol} \tag{4.3}$$

$$2CO + 2H_2 \rightarrow CH_3COOH$$

$$\Delta G^{0\prime} = -135\frac{kJ}{mol} \tag{4.4}$$

$$2CO + 4H_2 \rightarrow CH_3CH_2OH + H_2O$$

$$(CO : H_2 \text{ ratio} = 1:2) \quad \Delta G^{0\prime} = -144\frac{kJ}{mol} \tag{4.5}$$

CO_2 as carbon source with H_2 as energy source:

$$2CO_2 + 4H_2 \rightarrow CH_3COOH + 2H_2O$$

$$\Delta G^{0\prime} = -95\frac{kJ}{mol} \tag{4.6}$$

$$2CO_2 + 6H_2 \rightarrow CH_3CH_2OH + 3H_2O$$

$$\Delta G^{0'} = -104 \frac{kJ}{mol} \quad (4.7)$$

The reductive acetyl-CoA pathway (Fig. 4.3) was first described in *Moorella thermoacetica* (initially classified as *Clostridium thermoaceticum*) by Harland G. Wood and Lars G. Ljungdahl (Drake *et al.*, 2006; Ljungdahl, 1969) in 1966, hence it is also known as the Wood–Ljungdahl pathway in their honour.

The pathway is split into two branches, in the Eastern branch (or the methyl-branch) CO_2 is converted into formate by formate dehydrogenase, and this is followed by an ATP-dependent condensation with tetrahydrofolate (THF) to form 10-formyl-THF catalysed by formyl-THF synthase. 10-Formyl-THF undergoes three reductive steps catalysed by methylene-THF cyclohydrolase, methylene-THF dehydrogenase, and methylene-THF reductase, resulting in methyl-THF (Ragsdale, 2008; Ragsdale and Pierce, 2008). The B_{12}-dependent methyltransferase then catalyses the transfer of the methyl group from methyl-THF to a corrinoid/iron-sulfur protein as the final step of the Eastern branch (Seravalli *et al.*, 1999).

The Eastern branch is ubiquitous for all one-carbon metabolizing organisms (Ragsdale and Pierce, 2008), whereas the Western branch (or the carbonyl-branch), where CO is taken up directly (or generated from CO_2 by CO dehydrogenase) is unique to anaerobes (Ragsdale, 1997). The Western and the Eastern branches merge in the multi-subunit bifunctional metalloenzyme CO dehydrogenase/Acetyl-CoA synthase (CODH/ACS), where the Eastern branch supplies the methyl group and the Western branch the carbonyl group in acetyl-CoA (Doukov *et al.*, 2002; Ragsdale, 2008). While being unique to anaerobes, the CODH/ACS enzyme is not exclusively found in acetogens, but also in other bacteria (sulfate reducers and desulfitobacteria) and archaea (methanogens and *Archaeoglobus*) (Moersdorf *et al.*, 1992; Ragsdale, 2004). However, of the micro-organisms that possess some form of the reductive acetyl-CoA pathway, only actogens have been reported to synthesize products that can be used as liquid transportation fuels. A more in-depth description of the Wood–Ljungdahl pathway can be found in several excellent reviews including those by Stephen Ragsdale (Ragsdale, 2008; Ragsdale and Pierce, 2008).

Despite being in an early stage of industrialization, this process has significant promise and shows advantages over other lignocellulosic processes. No pretreatment of lignocellulolytic biomass is required, offering significant cost savings (the enzymatic step alone can account for up to 16 c/l ethanol (Kumar and Murthy, 2011)). While gasification allows use of the complete biomass fraction including lignin (which makes up 10–25% of the biomass), this fraction needs to be removed from the cellulose and hemicellulose polymers in other processes (Sims *et al.*, 2010)). The gasification process converts the entire biomass, at high thermal conversion efficiencies (MJ of feedstock converted to MJ of syngas) (75–80%) (McKendry, 2002), into a fermentable gas stream.

Gas fermentation also has advantages over established Fischer–Tropsch (FT) conversion of syngas to liquid fuels. FT conversion was invented by Franz Fischer and Hans Tropsch in 1925 (Fischer and Tropsch, 1930) (US patent 1,746,464) and relies on passing syngas over metal catalysts (mainly the transition metals iron, cobalt, and ruthenium) (Dry, 2002) to thermochemically convert the CO and H_2 components of that gas into a mixture of liquid hydrocarbons, which are then purified to produce liquid fuels (Faaij *et al.*, 2002) and chemical products. Although the FT process is currently in use at commercial scale (Dry, 2002), it is recognized as having significant scale and economic limitations. The catalysts used in FT processes are easily poisoned by CO_2 or low concentrations of common syngas contaminants (such as sulfur species, BTEX, and tars), and also the product specificity of the FT process is highly dependent on the syngas composition. Thus, an expensive gas pretreatment step is required.

Compared to the FT process, biocatalytic conversion of syngas requires very little pretreatment and has greater flexibility with regards to gas composition, so any type of biomass can be utilized. The amount of gas pretreatment depends on the nature of the contaminants in the gas and the tolerance of the biocatalyst to the specific contaminants. Typical contaminants, which need to be removed from syngas, are fine particular matter and tars. Biocatalytic conversion of syngas is particularly advantageous in specificity over FT conversion, as it is highly specific with regard to the product.

Natural producers of butanol from syngas

Although there are many autotrophic bacteria described by now, only a few have been shown to produce *n*-butanol from CO and CO_2 and H_2. Equations 4.8 and 4.9 show the energetics and stoichiometry of autotrophic butanol fermentation:

$$12CO + 5H_2O \rightarrow CH_3CH_2CH_2CH_2OH + 8CO_2$$
$$\Delta G^{0\prime} = -635\frac{kJ}{mol} \tag{4.8}$$

$$12H_2 + 4CO_2 \rightarrow CH_3CH_2CH_2CH_2OH + 7H_2O$$
$$\Delta G^{0\prime} = -499\frac{kJ}{mol} \tag{4.9}$$

Clostridium carboxidivorans (Clostridium strain P7) was isolated from lagoon settlement in 2005 by Liou *et al.* (2005) and can grow both heterotophically on simple sugars, or autotrophically on CO or CO_2 and H_2. This organism has a temperature optimum of 37°C and a pH optimum of 5.7–5.8. Fermentation products are ethanol, acetate, butyrate, and *n*-butanol (Liou *et al.*, 2005; Rajagopalan *et al.*, 2002). Butanol production has been reported of up to 0.1% w/v (Rajagopalan *et al.*, 2002). *C. carboxidivorans* has been sequenced and the draft assembly (Paul *et al.*, 2010) shows that it contains the genes for the reductive acetyl-CoA pathway and a plasmid with the genes for butanol and butyrate production similar to that of *C. acetobutylicum* (Bruant *et al.*, 2010), but is missing the CoA transferase genes (*ctfAB*) for acetone production (Daniell *et al.*, 2012). *C. carboxidivorans* has been explored for the commercial production of both ethanol (Hurst and Lewis, 2010; Ukpong *et al.*, 2012) and butanol from syngas (Bruant *et al.*, 2010).

Butyribacterium methylotrophicum was isolated in 1980 from a sewage digester and has a temperature optimum of 37°C and a pH optimum of 6 (Zeikus *et al.*, 1980). It can grow autotrophically in both CO and on CO_2 + H_2, as well as heterotrophically on simple sugars. Its fermentation products are acetate, ethanol, butyrate, and *n*-butanol (Lynd and Zeikus, 1983; Lynd *et al.*, 1982; Shen *et al.*, 1999). So far, a butanol production of 0.08 g/l has been shown when growing autotrophically on CO (Grethlein *et al.*, 1991). *B. methylotrophicum* is, however, classified as a risk group 2 organism (DSMZ), making it significantly more difficult to use in an industrial setting.

Clostridium scatologenes and C. drakei are two very closely related organisms that both have growth optimums of 37°C. *C. scatologenes* was isolated from soil and *C. drakei* was isolated from an acidic coal mine pond. Both are found to produce butyrate, even though acetate is their main fermentation product (Küsel *et al.*, 2000; Liou *et al.*, 2005).

In addition, conversion of butyrate to *n*-butanol in presence of CO has been demonstrated with several acetogenic species such as *C. ljungdahlii* and *C. ragsdalei* (Perez et al., 2012) or *C. autoethanogenum* (WO2009/113878).

Metabolic engineering

The application of metabolic engineering technologies to acetogenic bacteria has also enabled butanol production to be demonstrated in a broader array of bacteria. Besides the natural *n*-butanol producing acetogens, metabolic engineering of acetogens for *n*-butanol production by introducing the *n*-butanol pathway has been achieved and some efforts have been made to introduce the reductive acetyl-CoA pathway into either an ABE-producing organism as *C. acetobutylicum* or an industrial workhorse as *E. coli* or yeast.

There are two described efforts to metabolically engineer acetogens to produce *n*-butanol from syngas, both of them utilizing the classic clostridial acetyl-CoA fermentation pathway (Fig. 4.1). Köpke *et al.* (2010b) introduced the pathway from *Clostridium acetobutylicum* (*thlA, hbd, crt, bcd, adhE, bdhA*) into *Clostridium ljungdahlii* and grew it on syngas as sole carbon and energy source, obtaining a maximum titre of 2 mM (~150 mg/l). Electron-transferring flavoproteins of the Bcd complex were not included on the plasmid, but *C. ljungdahlii* has several sets of native electron-transferring flavoproteins, which proved to be functional. However, at the end of the growth, the produced 1-butanol was re-assimilated and converted to butyrate. Recently, a set of butanol dehydrogenases has been identified in *C. ljungdahlii* that may be responsible for this reaction (Tan *et al.*, 2013).

The other example is the engineering of another acetogen, *Clostridium autoethanogenum* (in addition to *C. ljungdahlii*), again utilizing the acetyl-CoA fermentation pathway from *C. acetobutylicum* (WO2012/053905). Butanol has been described as main fermentation product from CO in this study. This was achieved by including the complete Bcd complex from *C. acetobutylicum* including EtfAB (rather than relying on the native electron-transferring flavoproteins). In addition, the bifunctional aldehyde/alcohol dehydrogenase was omitted, instead a route via phosphotransbutyrylase and butyrate kinase is utilized. This route has the advantage of yielding ATP via substrate level phosphorylation (SLP). The produced butyrate can then be reduced to butyraldehyde via a aldehyde:ferredoxin-oxidoreductase (AOR), which can then be reduced to *n*-butanol. In acetogens growing on CO, this AOR reaction, which relies on reduced ferredoxin, can be coupled with CO oxidation via the carbon monoxide dehydrogenase that regenerates reduced ferredoxin ($CO + Fd_{ox} \rightarrow CO_2 + Fd_{red}$) (WO2009/113878).

In addition, acetogens *C. aceticum* (Schiel-Bengelsdorf and Dürre, 2012), *C. ljundahlii*, and *C. autoethanogenum* (WO 2012/115527) have been metabolically engineered for acetone and isopropanol production.

Rather than engineering a microbe optimized for growth on the optimal feedstock to produce the desired product, another approach being considered is to engineer microbes already optimized for a given product to perform on a more desirable feedstock. To some degree this approach has previously been applied in yeast or ABE strains to broaden this organism's ability to utilize C_5 sugars from cellulose for the production of cellulosic ethanol. The laboratory of Terry Papoutsakis (University of Delaware) has adopted this approach for syngas fermentation and has engineered *Clostridium acetobutylicum* with genes of the Wood–Ljungdahl pathway (WO 2012/034023). A similar attempt has also been made to convert *E. coli* (WO 2009/094485), so far with moderate success. Major hurdles are the complexity of the involved multi-subunit enzyme complexes involving several Fe/S clusters and cofactors. For methanogens, it has been estimated that over 200 genes are involved in formation of methane from CO_2 and H_2 (Kaster *et al.*, 2011).

Commercialization and challenges

Recognition of the potential feedstock flexibility and economic viability of gas fermentation as a route for biofuel and chemical production has resulted in a growing interest in the scale-up and commercialization of this technology. The ground work for much of this activity was initially laid by the work of the team led by James L. Gaddy of the University of Arkansas during the 1980s and 1990s using strains of *C. ljungdahlii* (Gaddy and Clausen, 1992) (US patent 5,173,429). This activity resulted in the foundation of a company named Bioengineering Resources Inc. (BRI) (Klasson *et al.*, 1990). BRI was acquired in 2008 by the multinational chemical company INEOS creating INEOS Bio (www.ineosbio.com). INEOS Bio completed construction of a 8 million metric tons per annum semi-commercial plant end of 2012 (http://www.ineos.com/en/businesses/INEOS-Bio/News/Ineos-Bio-Florida-Waste-to-Biofuel--Energy-Plant-Enters-Service/?business=INEOS+Bio) and in July 2013 announced production of ethanol from waste biomass material (http://www.ineos.com/en/businesses/INEOS-Bio/News/INEOS-Bio-Produces-Cellulosic-Ethanol/?business=INEOS+Bio). In addition, the plant produces 6 megawatts of electricity from unused syngas and heat recovery.

Coskata Inc. (www.coskata.com) was incorporated in 2006 in Warrenville, IL, with the aim of commercializing a gas fermentation technology for ethanol production based on *C. ragsdalei* and *C. carboxidivorans*, licenced from the University of Oklahoma and Oklahoma State University. Coskata has also identified and developed their own ethanol-producing proprietary bacterium, '*C. coskatii*' (WO 2011/116124).

In 2005, LanzaTech (www.lanzatech.com) was founded in Auckland, New Zealand, its headquarter being in Chicago, IL. The core technology

Figure 4.4 LanzaTech's gas fermentation demonstration plant with BaoSteel in Shanghai. A colour version of this figure is available in the plate section at the back of the book.

focuses on utilizing syngas and CO-rich industrial waste gases for the production of both ethanol and 2,3-butanediol (Köpke *et al.*, 2011b) using a proprietary strain of *C. autoethanogenum* (WO 2012/015317) as their biocatalyst, with which production of *n*-butanol (WO2012/053905) and other chemicals has also been demonstrated. In 2012, LanzaTech successfully demonstrated its technology at scale in a 100,000 gallon per annum demonstration plant in Shanghai, China, through a Joint Venture with Baosteel Group using steel mill waste gas as the feedstock (http://www.lanzatech.com/sites/default/files/imce_uploads/lanzatech-baosteel_release_dec_2012.pdf; Fig. 4.4). A second demonstration unit in China is operated in a joint venture with Shougang Steel Group, and in the US LanzaTech is operating a plant in Sopperton, GA, focusing on biomass syngas conversion.

Acknowledgements

Work in PD's group was supported by grants from the BMBF GenoMikPlus (Competence Network Göttingen), BioEnergy 2021, and SysMO projects COSMIC2 (0315782A), www.sysmo.net, as well as ERA-Net IB project REACTIF.

References

Al-Hinai, M.A., Fast, A.G., and Papoutsakis, E.T. (2012). Novel system for efficient isolation of *Clostridium* double-crossover allelic exchange mutants enabling markerless chromosomal gene deletions and DNA integration. Appl. Environ. Microbiol. *78*, 8112–8121.

Alsaker, K.V., and Papoutsakis, E.T. (2005). Transcriptional program of early sporulation and stationary-phase events in *Clostridium acetobutylicum*. J. Bacteriol. *187*, 7103–7118.

Atsumi, S., and Liao, J.C. (2008). Directed evolution of *Methanococcus jannaschii* citramalate synthase for biosynthesis of 1-propanol and 1-butanol by *Escherichia coli*. Appl. Environ. Microbiol. *74*, 7802–7808.

Atsumi, S., Cann, A.F., Connor, M.R., Shen, C.R., Smith, K.M., Brynildsen, M.P., Chou, K.J.Y., Hanai, T., and Liao, J.C. (2007). Metabolic engineering of *Escherichia coli* for 1-butanol production. Metabol. Eng. *10*, 305–311.

Atsumi, S., Hanai, T., and Liao, J.C. (2008). Non-fermentative pathways for synthesis of branched-chain higher alcohols as biofuels. Nature *451*, 86–89.

Atsumi, S., Higashide, W., and Liao, J.C. (2009). Direct photosynthetic recycling of carbon dioxide to isobutyraldehyde. Nature Biotechnol. *27*, 1177–1180.

Bahl, H., Andersch, W., and Gottschalk, G. (1982). Continuous production of acetone and butanol by *Clostridium acetobutylicum* in a two-stage phosphate limited chemostat. Eur. J. Appl. Microbiol. Biotechnol. *15*, 201–205 (Erratum *17*, 73, 1983).

Ballongue, J., Amine, J., Masion, H., Petidemange, H., and Gay, R. (1985). Induction of acetoacetate decarboxylase in *Clostridium acetobutylicum*. FEMS Microbiol. Lett. *29*, 273–277.

Bao, G., Wang, R., Zhu, Y., Dong, H., Mao, S., Zhang, Y., Chen, Z., Li, Y., and Ma, Y. (2011). Complete genome sequence of *Clostridium acetobutylicum* DSM 1731, a solvent-producing strain with multireplicon genome architecture. J. Bacteriol. *193*, 5007–5008.

Becker, D.F., Fuchs, J.A., Banfield, D.K., Funk, W.D., MacGillivray, R.T., and Stankovich, M.T. (1993). Characterization of wild-type and an active-site mutant in *Escherichia coli* of short-chain acyl-CoA dehydrogenase from *Megasphaera elsdenii*. Biochemistry 32, 10736–10742.

Berezina, O.V., Zakharova, N.V., Brandt, A., Yarotsky, S.V., Schwarz, W.H., and Zverlov, V.V. (2010). Reconstructing the clostridial n-butanol metabolic pathway in *Lactobacillus brevis*. Appl. Microbiol. Biotechnol. *87*, 635–646.

Bermejo, L.L., Welker, N.E., and Papoutsakis, E.T. (1998). Expression of *Clostridium acetobutylicum* ATCC 824 genes in *Escherichia coli* for acetone production and acetate detoxification. Appl. Environ. Microbiol. *64*, 1079–1085.

Blombach, B., and Eikmanns, B.J. (2011). Current knowledge on isobutanol production with *Escherichia coli*, *Bacillus subtilis* and *Corynebacterium glutamicum*. Bioeng. Bugs *2*, 346–350.

Blombach, B., Riester, T., Wieschalka, S., Ziert, C., Youn, J.-W., Wendisch, V.F., and Eikmanns, B.J. (2011). *Corynebacterium glutamicum* tailored for efficient isobutanol production. Appl. Environ. Microbiol. *77*, 3300–3310.

Bond-Watts, B.B., Bellerose, R.J., and Chang, M.C.Y. (2011). Enzyme machanism as a kinetic control element for designing synthetic biofuel pathways. Nature Chem. Biol. *7*, 222–227.

Boynton, Z.L., Bennett, G.N., and Rudolph, F.B. (1994). Intracellular concentrations of Coenzyme A and its derivatives from *Clostridium acetobutylicum* ATCC 824 and their roles in enzyme regulation. Appl. Environ. Microbiol. *60*, 39–44.

Bruant, G., Lévesque, M.-J., Peter, C., Guiot, S.R., and Masson, L. (2010). Genomic analysis of carbon monoxide utilization and butanol production by *Clostridium carboxidivorans* strain P7. PloS One *5*, e13033.

Chiao, J., and Sun, Z. (2007). History of the acetone-butanol-ethanol fermentation industry in China: development of continuous production technology. J. Mol. Microbiol. Biotechnol. *13*, 12–14.

Dai, Z., Dong, H., Zhu, Y., Zhang, Y., Li, Y., and Ma, Y. (2012). Introducing a single secondary alcohol dehydrogenase into butanol-tolerant *Clostridium acetobutylicum* Rh8 switches ABE fermentation to high level IBE fermentation. Biotechnol. Biofuels *5*, 44.

Daniell, J., Köpke, M., and Simpson, S. (2012). Commercial biomass syngas fermentation. Energies *5*, 5372–5417.

Dellomonaco, C., Clomburg, J.M., Miller, E.N., and Gonzalez, R. (2011). Engineered reversal of the β-oxidation cycle for the synthesis of fuels and chemicals. Nature *476*, 355–359.

Doukov, T.I., Iverson, T.M., Seravalli, J., Ragsdale, S.W., and Drennan, C.L. (2002). A Ni–Fe-Cu center in a bifunctional carbon monoxide dehydrogenase/acetyl-CoA synthase. Science *298*, 567–572.

Drake, H.L., Küsel, K., Matthies, C., Wood, H.G., and Ljungdahl, L.G. (2006). Acetogenic prokaryotes. In The Prokaryotes, M. Dworkin, S. Falkow, E. Rosenberg, K.-H. Schleifer, and E. Stackebrandt, eds. (New York, NY: Springer), pp. 354–420.

Drake, H.L., Gössner, A.S., and Daniel, S.L. (2008). Old acetogens, new light. Ann. N. Y. Acad. Sci. *1125*, 100–128.

Dry, M.E. (2002). The Fischer–Tropsch process: 1950–2000. Catalysis Today *71*, 227–241.

Dunlop, M.J., Dossani, Z.Y., Szmidt, H.L., Chu, H.C., Lee, T.S., Keasling, J.D., Hadi, M.Z., and Mukhopadhyay, A. (2011). Engineering microbial biofuel tolerance and export using efflux pumps. Mol. Systems Biol. *7*, 487.

Dürre, P. (2005). Formation of solvents in clostridia. In Handbook on Clostridia, P. Dürre, ed. (Boca Raton, FL: CRC Press), pp. 673–695.

Dürre, P. (2007). Biobutanol: an attractive biofuel. Biotechnol. J. *2*, 1525–1534.

Dürre, P. (2011). Fermentative production of butanol – the academic perspective. Curr. Op. Biotechnol. *22*, 331–336.

Dürre, P., and Bahl, H. (1996). Microbial production of acetone/butanol/isopropanol. In Biotechnology 6, M. Roehr, ed. (Weinheim, Germany: VCH Verlagsgesellschaft), pp. 229–268.

Ezeji, T.C., Qureshi, N., and Blaschek, H.P. (2005). Industrially fermentations. In: P. Dürre, ed. Handbook on Clostridia. Boca Raton, FL: CRC Press, pp. 799–814.

Faaij, P.C., Hamelinck, C.N., and van Hardeveld, M.R.M. (2002). Exploration of the possibilities for production of Fischer Tropsch liquids and power via biomass gasification. Biomass Bioener. *23*, 129–152.

Fast, A.G., and Papoutsakis, E.T. (2012). Stoichiometric and energetic analyses of non-photosynthetic CO_2-fixation pathways to support synthetic biology strategies for production of fuels and chemicals. Curr. Op. Chem. Eng. *1*, 380–395.

George, H., Johnson, J.L., Moore, W.E., Holdeman, L.V., and Chen, J.S. (1983). Acetone, isopropanol, and butanol production by *Clostridium beijerinckii* (syn. *Clostridium butylicum*) and *Clostridium aurantibutyricum*. Appl. Environ. Microbiol. *45*, 1160–1163.

Gottwald, M., and Gottschalk, G. (1985). The internal pH of *Clostridium acetobutylicum* and its effect on the shift from acid to solvent formation. Arch. Microbiol. *143*, 42–46.

Green, E.M. (2011). Fermentative production of butanol – the industrial perspective. Curr. Op. Biotechnol. *22*, 337–343.

Grethlein, A.J., Worden, R.M., Jain, M.K., and Datta, R. (1991). Evidence for production of n-butanol from carbon monoxide by *Butyribacterium methylotrophicum*. J. Ferment. Bioeng. *72*, 58–60.

Griffin, D.W., Schultz, M.A., Irving, E., and Road, P. (2012). Fuel and chemical products from biomass syngas: a comparison of gas fermentation to thermochemical conversion routes. Environ. Progress Sustain. Ener. *31*, 1–6.

Grimmler, C., Janssen, H., Krausse, D., Fischer, R.-J., Bahl, H., Dürre, P., Liebl, W., and Ehrenreich, A. (2011). Genome-wide gene expression analysis of the switch between acidogenesis and solventogenesis in continuous cultures of *Clostridium acetobutylicum*. J. Mol. Microbiol. Biotechnol. *20*, 1–15.

Grupe, H., and Gottschalk, G. (1992). Physiological events in *Clostridium acetobutylicum* during the shift from acidogenesis to solventogenesis in continuous culture and presentation of a model for shift induction. Appl. Environ. Microbiol. *58*, 3896–3902.

Gu, Y., Li, J., Zhang, L., Chen, J., Niu, L., Yang, Y., Yang, S., and Jiang, W. (2009). Improvement of xylose utilization in *Clostridium acetobutylicum* via expression of the *talA* gene encoding transaldolase from *Escherichia coli*. J. Biotechnol. *143*, 284–287.

Gu, Y., Jiang, Y., Wu, H., Liu, X., Li, Z., Li, J., Xiao, H., Shen, Z., Dong, H., Yang, Y., *et al.* (2011). Economical challenges to microbial producers of butanol: feedstock, butanol ratio and titer. Biotechnol. J. *6*, 1348–1357.

Han, B., Ujor, V., Lai, L.B., Gopalan, V., and Ezeji, T.C. (2013). Use of proteomic analysis to elucidate the role of calcium on acetone-butanol-ethanol (ABE) fermentation in *Clostridium beijerinckii* NCIMB 8052. Appl. Environ. Microbiol. *79*, 282–293.

Harris, L.M., Welker, N.E., and Papoutsakis, E.T. (2002). Northern, morphological, and fermentation analysis of *spo0A* inactivation and overexpression in *Clostridium acetobutylicum* ATCC 824. J. Bacteriol. 184, 3586–3697.

Haus, S., Jabbari, S., Millat, T., Janssen, H., Fischer, R.-J., Bahl, H., King, J.R., and Wolkenhauer, O. (2011). A systems biology approach to investigate the effect of pH-induced gene regulation on solvent production by *Clostridium acetobutylicum* in continuous culture. BMC Systems Biol. *5*, 10.

Heap, J.T., Kuehne, S. a, Ehsaan, M., Cartman, S.T., Cooksley, C.M., Scott, J.C., and Minton, N.P. (2010). The ClosTron: mutagenesis in *Clostridium* refined and streamlined. J. Microbiol. Methods *80*, 49–55.

Heap, J.T., Ehsaan, M., Cooksley, C.M., Ng, Y.-K., Cartman, S.T., Winzer, K., and Minton, N.P. (2012). Integration of DNA into bacterial chromosomes from plasmids without a counter-selection marker. Nucl. Acids Res. 40, 1–10.

Higashide, W., Li, Y., Yang, Y., and Liao, J.C. (2011). Metabolic engineering of *Clostridium cellulolyticum* for production of isobutanol from cellulose. Appl. Environ. Microbiol. 77, 2727–2733.

Hillmann, F., Fischer, R.-J., Saint-Prix, F., Girbal, L., and Bahl, H. (2008). PerR acts as a switch for oxygen tolerance in the strict anaerobe *Clostridium acetobutylicum*. Mol. Microbiol. *68*, 848–860.

Hu, S., Zheng, H., Gu, Y., Zhao, J., Zhang, W., Yang, Y., Wang, S., Zhao, G., Yang, S., and Jiang, W. (2011). Comparative genomic and transcriptomic analysis revealed genetic characteristics related to solvent formation and xylose utilization in *Clostridium acetobutylicum* EA 2018. BMC Genomics *12*, 93.

Hurst, K.M., and Lewis, R.S. (2010). Carbon monoxide partial pressure effects on the metabolic process of syngas fermentation. Biochem. Eng. J. *48*, 159–165.

Hüsemann, M.H., and Papoutsakis, E.T. (1990). Effects of propionate and acetate additions on solvent production in batch cultures of *Clostridium acetobutylicum*. Appl. Environ. Microbiol. *56*, 1497–1500.

Inui, M., Suda, M., Kimura, S., Yasuda, K., Suzuki, H., Toda, H., Yamamoto, S., Okino, S., Suzuki, N., and Yukawa, H. (2008). Expression of *Clostridium acetobutylicum* butanol synthetic genes in *Escherichia coli*. Appl. Environ. Microbiol. 77, 1305–1316.

Ismaiel, A.A., Zhu, C.X., Colby, G.D., and Chen, J.S. (1993). Purification and characterization of a primary-secondary alcohol dehydrogenase from two strains of *Clostridium beijerinckii*. J. Bacteriol. *175*, 5097–5105.

Jang, Y.-S., Lee, J.Y., Lee, J., Park, J.H., Im, J.A., Eom, M.-H., Lee, J., Lee, S.-H., Song, H., Cho, J.-H., *et al.* (2012). Enhanced butanol production obtained by reinforcing the direct butanol-forming route in *Clostridium acetobutylicum*. mBio 3, e00314–12.

Janssen, H., Döring, C., Ehrenreich, A., Voigt, B., Hecker, M., Bahl, H., and Fischer, R.-J. (2010). A proteomic and transcriptional view of acidogenic and solventogenic steady-state cells of *Clostridium acetobutylicum* in a chemostat culture. Appl. Microbiol. Biotechnol. *87*, 2209–2226.

Jia, X., Li, S., Xie, S., and Wen, J. (2012). Engineering a metabolic pathway for isobutanol biosynthesis in *Bacillus subtilis*. Appl. Biochem. Biotechnol. *168*, 1–9.

Jojima, T., Inui, M., and Yukawa, H. (2008). Production of isopropanol by metabolically engineered *Escherichia coli*. Appl. Microbiol. Biotechnol. 77, 1219–1224.

Jones, D.T. (2001). Applied acetone-butanol fermentation. In Clostridia. Biotechnology and Medical Applications, P. Dürre, and H. Bahl, eds. (Weinheim: Wiley-VCH Verlag), pp. 125–168.

Jones, D.T., and Keis, S. (1995). Origins and relationships of industrial solvent-producing clostridial strains. FEMS Microbiol. Rev. *17*, 223–232.

Jones, D.T., and Woods, D.R. (1986). Acetone-butanol fermentation revisited. Microbiol. Rev. *50*, 484–524.

Jones, S.W., Paredes, C.J., Tracy, B., Cheng, N., Sillers, R., Senger, R.S., and Papoutsakis, E.T. (2008). The transcriptional program underlying the physiology of clostridial sporulation. Genome Biol. *9*, R114.

Kanchanatawee, S., and Maddox, I.S. (1990). Effect of phosphate and ammonium ion concentrations on solvent production in defined medium by *Clostridium acetobutylicum*. J. Ind. Microbiol. *5*, 277–282.

Kaster, A.-K., Goenrich, M., Seedorf, H., Liesegang, H., Wollherr, A., Gottschalk, G., and Thauer, R.K. (2011). More than 200 genes required for methane formation from H_2 and CO_2 and energy conservation are present in *Methanothermobacter marburgensis* and *Methanothermobacter thermautotrophicus*. Archaea (Vancouver, B.C.) *2011*, 973848.

Keis, S., Bennett, C.F., Ward, V.K., and Jones, D.T. (1995). Taxonomy and phylogeny of industrial solvent-producing clostridia. Int. J. Syst. Bacteriol. *45*, 693–705.

Keis, S., Shaheen, R., and Jones, D.T. (2001). Emended descriptions of *Clostridium acetobutylicum* and *Clostridium beijerinckii*, and descriptions of *Clostridium saccharoperbutylacetonicum* sp. nov. and *Clostridium saccharobutylicum* sp. nov. Int. J. Syst. Evol. Microbiol. *51*, 2095–2103.

Klasson, K.T., Elmore, B.B., Vega, J.L., Ackerson, M.D., Clausen, E.C., and Gaddy, J.L. (1990). Biological production of liquid and gaseous fuels from synthesis gas. Appl. Biochem. Biotechnol. *24–25*, 857–873.

Kondo, T., Tezuka, H., Ishii, J., Matsuda, F., Ogino, C., and Kondo, A. (2012). Genetic engineering to enhance the Ehrlich pathway and alter carbon flux for increased isobutanol production from glucose by *Saccharomyces cerevisiae*. J. Biotechnol. *159*, 32–37.

Köpke, M., Held, C., Hujer, S., Liesegang, H., Wiezer, A., Wollherr, A., Ehrenreich, A., Liebl, W., Gottschalk, G., and Dürre, P. (2010). *Clostridium ljungdahlii* represents a microbial production platform based on syngas. Proc. Natl. Acad. Sci. U.S.A. *107*, 13087–13092.

Köpke, M., Noack, S., and Dürre, P. (2011a). The past, present, and future of biofuels – biobutanol as promising alternative. In: M.A. dos Santos Bernades, ed. Biofuel Production – Recent Developments and Prospects, Rijeka, Croatia: InTech, pp. 451–486.

Köpke, M., Mihalcea, C., Liew, F., Tizard, J.H., Ali, M.S., Conolly, J.J., Al-Sinawi, B., and Simpson, S.D. (2011b). 2,3-Butanediol production by acetogenic bacteria, an alternative route to chemical synthesis, using industrial waste gas. Appl. Env. Microbiol. *77*, 5467–5475.

Kovács, K., Willson, B.J., Schwarz, K., Heap, J.T., Jackson, A., Bolam, D.N., Winzer, K., and Minton, N.P. (2013). Secretion and assembly of functional mini-cellulosomes from synthetic chromosomal operons in *Clostridium acetobutylicum* ATCC 824. Biotechnol. Biofuels *20*, 117. doi:10.1186/1754-6834-6-117

Kumar, D., and Murthy, G.S. (2011). Impact of pretreatment and downstream processing technologies on economics and energy in cellulosic ethanol production. Biotechnol. Biofuels *4*, 27.

Küsel, K., Dorsch, T., Acker, G., Stackebrandt, E., and Drake, H.L. (2000). *Clostridium scatologenes* strain SL1 isolated as an acetogenic bacterium from acidic sediments. Int. J. Syst. Evol. Microbiol. *50*, 537–546.

Lan, E.I., and Liao, J.C. (2011). Metabolic engineering of cyanobacteria for 1-butanol production from carbon dioxide. Metabolic Eng. *13*, 353–363.

Lan, E.I., and Liao, J.C. (2012). ATP drives direct photosynthetic production of 1-butanol in cyanobacteria. Proc. Nat. Acad. Sci. USA *109*, 6018–6023.

Lee, J., Jang, Y.-S., Choi, S.J., Im, J.A., Song, H., Cho, J.H., Seung, D.Y., Papoutsakis, E.T., Bennett, G.N., and Lee, S.Y. (2011). Metabolic engineering of *Clostridium acetobutylicum* ATCC 824 for isopropanol-butanol-ethanol fermentation. Appl. Environ. Microbiol. *78*, 1416–1423.

Lee, S.Y., Park, J.H., Jang, S.H., Nielsen, L.K., Kim, J., and Jung, K.S. (2008). Fermentative butanol production by clostridia. Biotechnol. Bioeng. *101*, 209–228.

Lee, W.-H., Seo, S.-O., Bae, Y.-H., Nan, H., Jin, Y.-S., and Seo, J.-H. (2012). Isobutanol production in engineered *Saccharomyces cerevisiae* by overexpression of 2-ketoisovalerate decarboxylase and valine biosynthetic enzymes. Bioprocess Biosystems Eng. *35*, 1467–1475.

Li, H., Opgenorth, P.H., Wernick, D.G., Rogers, S., Wu, T.-Y., Higashide, W., Malati, P., Huo, Y.-X., Cho, K.M., and Liao, J.C. (2012). Integrated electromicrobial conversion of CO_2 to higher alcohols. Science *335*, 1596.

Li, R.-D., Li, Y.-Y., Lu, L.-Y., Ren, C., Li, Y.-X., and Liu, L. (2011). An improved kinetic model for the acetone-butanol-ethanol pathway of *Clostridium acetobutylicum* and model-based perturbation analysis. BMC Systems Biol. *5*, S12.

Liou, J.S.-C., Balkwill, D.L., Drake, G.R., and Tanner, R.S. (2005). *Clostridium carboxidivorans* sp. nov., a solvent-producing clostridium isolated from an agricultural settling lagoon, and reclassification of the acetogen *Clostridium scatologenes* strain SL1 as *Clostridium drakei* sp. nov. Int. J. Syst. Evol. Microbiol. *55*, 2085–2091.

Ljungdahl, L.G. (1969). Total synthesis of acetate from CO_2 by heterotrophic bacteria. Annu. Rev. Microbiol. *23*, 515–538.

Lovley, D.R., and Nevin, K.P. (2013). Electrobiocommodities: powering microbial production of fuels and commodity chemicals from carbon dioxide with electricity. Curr. Op. Biotechnol. *24*, 385–390.

Lütke-Eversloh, T., and Bahl, H. (2011). Metabolic engineering of *Clostridium acetobutylicum*: recent advances to improve butanol production. Curr. Op. Biotechnol. *22*, 634–647.

Lutz, R., and Bujard, H. (1997). Independent and tight regulation of transcriptional units in *Escherichia coli* via the LacR/O, the TetR/O and AraC/I regulatory elements. Nucl. Acids Res. *25*, 1203–1210.

Lynd, L.H., and Zeikus, J.G. (1983). Metabolism of H_2-CO_2, methanol, and glucose by *Butyribacterium methylotrophicum*. J. Bacteriol. *153*, 1415–1423.

Lynd, L., Kerby, R., and Zeikus, J.G. (1982). Carbon monoxide metabolism of the methylotrophic acidogen *Butyribacterium methylotrophicum*. J. Bacteriol. *149*, 255–263.

McAnulty, M.J., Yen, J.Y., Freedman, B.G., and Senger, R.S. (2012). Genome-scale modeling using flux ratio constraints to enable metabolic engineering of clostridial metabolism in silico. BMC Systems Biol. *6*, 42.

McKendry, P. (2002). Energy production from biomass (Part 3): Gasification technologies. Biores. Technol. *83*, 55–63.

Mao, S., Luo, Y., Zhang, T., Li, J., Bao, G., Zhu, Y., Chen, Z., Zhang, Y., Li, Y., and Ma, Y. (2010). Proteome reference map and comparative proteomic analysis between a wild type *Clostridium acetobutylicum* DSM 1731 and its mutant with enhanced butanol tolerance and butanol yield. J. Proteome Res. *9*, 3046–3061.

May, A., Fischer, R.-J., Maria Thum, S., Schaffer, S., Verseck, S., Dürre, P., and Bahl, H. (2013). A modified pathway for the production of acetone in *Escherichia coli*. Metabolic Eng. *15*, 218–225.

Millat, T., Janssen, H., Bahl, H., Fischer, R.-J., and Wolkenhauer, O. (2013). Integrative modelling of pH-dependent enzyme activity and transcriptomic regulation of the acetone-butanol-ethanol fermentation of *Clostridium acetobutylicum* in continuous culture. Microbial Biotechnol. *6*, 526–539.

Milne, C.B., Eddy, J.A., Raju, R., Ardekani, S., Kim, P.-J., Senger, R.S., Jin, Y.-S., Blaschek, H.P., and Price, N.D. (2011). Metabolic network reconstruction and genome-scale model of butanol-producing strain *Clostridium beijerinckii* NCIMB 8052. BMC Systems Biol. *5*, 130.

Miwa, K., and Tsuchida, T. (1983). Construction of L-threonine overproducing strains of *Escherichia coli* K-12 using recombinant DNA techniques. Agr. Biol. Chem. *47*, 2329–2334.

Moersdorf, G., Frunzke, K., Gadkari, D., and Meyer, O. (1992). Microbial growth on carbon monoxide. Biodegradation 3, 61–82.

Nair, R.V., Bennett, G.N., and Papoutsakis, E.T. (1994). Molecular characterization of an aldehyde/alcohol dehydrogenase gene from *Clostridium acetobutylicum* ATCC 824. J. Bacteriol. *176*, 871–885.

Nair, R.V., Green, E.M., Watson, D.E., Bennett, G.N., and Papoutsakis, E.T. (1999). Regulation of the *sol* locus genes for butanol and acetone formation in *Clostridium acetobutylicum* ATCC 824 by a putative transcriptional repressor. J. Bacteriol. *181*, 319–330.

Nevin, K.P., Woodard, T.L., Franks, A.E., Summers, Z.M., and Lovley, D.R. (2010). Microbial electrosynthesis: feeding microbes electricity to convert carbon dioxide and water to multicarbon extracellular organic compounds. mBio *1*, e00103–10.

Nevin, K.P., Hensley, S.A., Franks, A.E., Summers, Z.M., Ou, J., Woodard, T.L., Snoeyenbos-West, O.L., and Lovley, D.R. (2011). Electrosynthesis of organic compounds from carbon dioxide is catalyzed by a diversity of acetogenic microorganisms. Appl. Environ. Microbiol. *77*, 2882–2886.

Ni, Y., and Sun, Z. (2009). Recent progress on industrial fermentative production of acetone-butanol-ethanol by *Clostridium acetobutylicum* in China. Appl. Microbiol. Biotechnol. *83*, 415–423.

Nielsen, D.R., Leonard, E., Yoon, S.-H., Tseng, H.-C., Yuan, C., and Prather, K.L.J. (2009). Engineering alternative butanol production platforms in heterologous bacteria. Metabolic Eng. *11*, 262–273.

Nölling, J., Breton, G., Omelchenko, M.V., Makarova, K.S., Zeng, Q., Gibson, R., Lee, H.M., Dubois, J., Qiu, D., Hitti, J., *et al.* (2001). Genome sequence and comparative analysis of the solvent-producing bacterium *Clostridium acetobutylicum*. J. Bacteriol. *183*, 4823–4838.

OEDC-FOA (2011). OECD-FAO Agricultural Outlook 2011–2022.

Pasteur, L. (1862). Quelques résultats nouveaux relatifs aux fermentations acétique et butyrique. Bull. Soc. Chim. Paris *May*, 52–53.

Paul, D., Austin, F.W., Arick, T., Bridges, S.M., Burgess, S.C., Dandass, Y.S., and Lawrence, M.L. (2010). Genome sequence of the solvent-producing bacterium *Clostridium carboxidivorans* strain P7T. J. Bacteriol. *192*, 5554–5555.

Pérez, J.M., Arenas, F.A., Pradenas, G.A., Sandoval, J.M., and Vásquez, C.C. (2008). *Escherichia coli* YqhD exhibits aldehyde reductase activity and protects from the harmful effect of lipid peroxidation-derived aldehydes. J. Biol. Chem. *283*, 7346–7353.

Pérez, J.M., Richter, H., Loftus, S.E., and Angenent, L.T. (2012). Biocatalytic reduction of short-chain carboxylic acids into their corresponding alcohols with syngas fermentation. Biotechnol. Bioeng. *110*, 1–30.

Ragsdale, S.W. (1997). The eastern and western branches of the Wood/Ljungdahl pathway: how the east and west were won. BioFactors *6*, 3–11.

Ragsdale, S.W. (2004). Life with carbon monoxide. Crit. Rev. Biochem. Mol. Biol. *39*, 165–195.

Ragsdale, S.W. (2008). Enzymology of the Wood-Ljungdahl pathway of acetogenesis. Ann. N. Y. Acad. Sci. *1125*, 129–136.

Ragsdale, S.W., and Pierce, E. (2008). Acetogenesis and the Wood-Ljungdahl pathway of CO_2 fixation. Biochim. Biophys. Acta *1784*, 1873–1898.

Rajagopalan, S., Datar, R.P., and Lewis, R.S. (2002). Formation of ethanol from carbon monoxide via a new microbial catalyst. Biomass Bioener. *23*, 487–493.

Ravagnani, A., Jennert, K.C., Steiner, E., Grünberg, R., Jefferies, J.R., Wilkinson, S.R., Young, D.I., Tidswell, E.C., Brown, D.P., Youngman, P., Morris, J.G., and Young, M. (2000). Spo0A directly controls the switch from acid to solvent production in solvent-forming clostridia. Mol. Microbiol. 37, 1172–1185.

Ren, C., Gu, Y., Wu, Y., Zhang, W., Yang, C., Yang, S., and Jiang, W. (2012). Pleiotropic functions of catabolite control protein CcpA in butanol-producing *Clostridium acetobutylicum*. BMC Genomics *13*, 349.

Sabathé, F., Bélaïch, A., and Soucaille, P. (2002). Characterization of the cellulolytic complex (cellulosome) of *Clostridium acetobutylicum*. FEMS Microbiol. Lett. *217*, 15–22.

Schiel-Bengelsdorf, B., and Dürre, P. (2012). Pathway engineering and synthetic biology using acetogens. FEBS Lett. *586*, 2191–2198.

Senger, R.S., and Papoutsakis, E.T. (2008). Genome-scale model for *Clostridium acetobutylicum*: Part I. Metabolic network resolution and analysis. Biotechnol. Bioeng. *101*, 1036–1052.

Seravalli, J., Zhao, S., and Ragsdale, S.W. (1999). Mechanism of transfer of the methyl group from (6S)-methyltetrahydrofolate to the corrinoid/iron-sulfur protein catalyzed by the methyltransferase from *Clostridium thermoaceticum*: a key step in the Wood-Ljungdahl pathway of acetyl-CoA synthesis. Biochemistry *38*, 5728–5735.

Shaheen, R., Shirley, M., and Jones, D.T. (2000). Comparative fermentation studies of industrial strains belonging to four species of solvent-producing clostridia. J. Mol. Microbiol. Biotechnol. 2, 115–124.

Shen, C.R., Lan, E.I., Dekishima, Y., Baez, A., Cho, K.M., and Liao, J.C. (2011). Driving forces enable high-titer anaerobic 1-butanol synthesis in *Escherichia coli*. Appl. Environ. Microbiol. 77, 2905–2915.

Shen, G.-J., Shieh, J.-S., Grethlein, a. J., Jain, M.K., and Zeikus, J.G. (1999). Biochemical basis for carbon monoxide tolerance and butanol production by *Butyribacterium methylotrophicum*. Appl. Microbiol. Biotechnol. *51*, 827–832.

Shi, Z., and Blaschek, H.P. (2008). Transcriptional analysis of *Clostridium beijerinckii* NCIMB 8052 and the hyper-butanol-producing mutant BA101 during the shift from acidogenesis to solventogenesis. Appl. Environ. Microbiol. *74*, 7709–7714.

Siemerink, M.A., Kuit, W., López Contreras, A.M., Eggink, G., van der Oost, J., and Kengen, S.W.M. (2011). D-2,3-butanediol production due to heterologous expression of an acetoin reductase in *Clostridium acetobutylicum*. Appl. Environ. Microbiol. 77, 2582–2588.

Sims, R.E.H., Mabee, W., Saddler, J.N., and Taylor, M. (2010). An overview of second generation biofuel technologies. Biores. Technol. *101*, 1570–1580.

Sivagnanam, K., Raghavan, V.G.S., Shah, M., Hettich, R.L., Verberkmoes, N.C., and Lefsrud, M.G. (2012). Shotgun proteomic monitoring of *Clostridium acetobutylicum* during stationary phase of butanol fermentation using xylose and comparison with the exponential phase. J. Ind. Microbiol. Biotechnol. *39*, 949–955.

Steen, E.J., Chan, R., Prasad, N., Myers, S., Petzold, C.J., Redding, A., Ouellet, M., and Keasling, J.D. (2008). Metabolic engineering of *Saccharomyces cerevisiae* for the production of n-butanol. Microbial Cell Factories *7*, e36.

Steiner, E., Dago, A.E., Young, D.I., Heap, J.T., Minton, N.P., Hoch, J.A., and Young, M. (2011). Multiple orphan histidine kinases interact directly with Spo0A to control the initiation of endospore formation in *Clostridium acetobutylicum*. Mol. Microbiol. *80*, 641–654.

Tan, Y., Liu, J., Liu, Z., and Li, F. (2013). Characterization of two novel butanol dehydrogenases involved in butanol degradation in syngas-utilizing bacterium *Clostridium ljungdahlii* DSM 13528. J. Basic Microbiol., doi: 10.1002/jobm.201300046.

Terracciano, J.S., and Kashket, E.R. (1986). Intracellular conditions required for initiation of solvent production by *Clostridium acetobutylicum*. Appl. Environ. Microbiol. *52*, 86–91.

Thormann, K., and Dürre, P. (2001). Orf5/SolR: a transcriptional repressor of the *sol* operon of *Clostridium acetobutylicum*? J. Ind. Microbiol. Biotechnol. *27*, 307–313.

Thormann, K., Feustel, L., Lorenz, K., Nakotte, S., and Dürre, P. (2002). Control of butanol formation in *Clostridium acetobutylicum* by transcriptional activation. J. Bacteriol. *184*, 1966–1973.

Tomas, C.A., Welker, N.E., and Papoutsakis, E.T. (2003). Overexpression of *groESL* in *Clostridium acetobutylicum* results in increased solvent production and tolerance, prolonged metabolism, and changes in the cell's transcriptional program. Appl. Environ. Microbiol. *69*, 4951–4965.

Tracy, B.P. (2012). Improving butanol fermentation to enter the advanced biofuel market. mBio 3, e00518–12.

Tucci, S., and Martin, W. (2007). A novel prokaryotic trans-2-enoyl-CoA reductase from the spirochete *Treponema denticola*. FEBS Lett. *581*, 1561–1566.

Ukpong, M.N., Atiyeh, H.K., De Lorme, M.J.M., Liu, K., Zhu, X., Tanner, R.S., Wilkins, M.R., and Stevenson, B.S. (2012). Physiological response of *Clostridium carboxidivorans* during conversion of synthesis gas to solvents in a gas-fed bioreactor. Biotechnol. Bioeng. *109*, 2720–2728.

Varman, A.M., Xiao, Y., Pakrasi, H.B., and Tang, Y.J. (2013). Metabolic engineering of *Synechocystis* sp. strain PCC 6803 for isobutanol production. Appl. Environ. Microbiol. *79*, 908–914.

Wallace, K.K., Bao, Z.-Y., Dai, H., Digate, R., Schuler, G., Speedie, M.K., and Reynolds, K.A. (1995). Purification of crotonyl-CoA reductase from *Streptomyces collinus* and cloning, sequencing and expression of the corresponding gene in *Escherichia coli*. Eur. J. Biochem. *233*, 954–962.

Wallenius, J., Viikilä, M., Survase, S., Ojamo, H., and Eerikäinen, T. (2013). Constraint-based genome-scale metabolic modeling of *Clostridium acetobutylicum* behavior in an immobilized column. Biores. Technol. *142*, 603–610.

Waltz, E. (2012). Cathay industrial biotech. Nature Biotechnol. *30*, 480.

Wang, Y., Li, X., Mao, Y., and Blaschek, H.P. (2012). Genome-wide dynamic transcriptional profiling in *Clostridium beijerinckii* NCIMB 8052 using single-nucleotide resolution RNA-Seq. BMC Genomics *13*, 102.

Weyer, E.R., and Rettger, L.F. (1927). A comparative study of six different strains of the organism commonly concerned in large-scale production of butyl alcohol and acetone by the biological process. J. Bacteriol. *14*, 399–424.

Wietzke, M., and Bahl, H. (2012). The redox-sensing protein Rex, a transcriptional regulator of solventogenesis in *Clostridium acetobutylicum*. Appl. Microbiol. Biotechnol. *96*, 749–761.

Xiao, H., Gu, Y., Ning, Y., Yang, Y., Mitchell, W.J., Jiang, W., and Yang, S. (2011). Confirmation and elimination of xylose metabolism bottlenecks in glucose phosphoenolpyruvate-dependent phosphotransferase system-deficient *Clostridium acetobutylicum* for simultaneous utilization of glucose, xylose, and arabinose. Appl. Environ. Microbiol. *77*, 7886–7895.

Xu, H., Zhang, Y., Guo, X., and Ren, S. (2004). Isoleucine biosynthesis in *Leptospira interrogans* serotype lai strain 56601 proceeds via a threonine-independent pathway. J. Bacteriol. *186*, 5400–5409.

Yu, M., Zhang, Y., Tang, I.-C., and Yang, S.-T. (2011). Metabolic engineering of *Clostridium tyrobutyricum* for n-butanol production. Metabolic Eng. *13*, 373–382.

Zeikus, J.G., Lynd, L.H., Thompson, T.E., Krzycki, J.A., Weimer, P.J., and Hegge, P.W. (1980). Isolation and characterization of a new, methylotrophic, acidogenic anaerobe, the marburg strain. Curr. Microbiol. 3, 381–386.

Zhao, Y., Tomas, C.A., Rudolph, F.B., Papoutsakis, E.T., and Bennett, G.N. (2005). Intracellular butyryl phosphate and acetyl phosphate concentrations in *Clostridium acetobutylicum* and their implications for solvent formation. Appl. Environ. Microbiol. *71*, 530–537.

Zhu, L., Dong, H., Zhang, Y., and Li, Y. (2011). Engineering the robustness of *Clostridium acetobutylicum* by introducing glutathione biosynthetic capability. Metabolic Eng. *13*, 426–434.

Zverlov, V.V., Berezina, O., Velikodvorskaya, G.A., and Schwarz, W.H. (2006). Bacterial acetone and butanol production by industrial fermentation in the Soviet Union: use of hydrolyzed agricultural waste for biorefinery. Appl. Microbiol. Biotechnol. *71*, 587–597.

Higher Chain Alcohols from Non-fermentative Pathways

5

Jordan T. McEwen, Yohei Tashiro and Shota Atsumi

Abstract

Utilizing microorganisms for production of biofuels is one approach being explored to accommodate energy need of an increasingly modern population worldwide and to remedy heightened CO_2 emissions. Although ethanol production is well established, higher (C_3–C_{10}) alcohols possess chemical properties that are more similar to gasoline. However, natural microorganisms do not naturally produce these alcohols efficiently. Diverse approaches have been applied to microorganisms for the production of higher chained alcohols. These include: design and construction of biosynthetic pathways, optimization of genetic and regulatory processes, and engineering of enzymes for non-native substrate activity. This chapter will focus on the current approaches and techniques for engineering microorganisms for production of C_3–C_{10} alcohols.

5.1 Introduction

Energy demand to support a continuing modernization of society has steadily increased while energy sources have primarily continued to rely heavily on non-renewable fossil fuels. The US Energy Information Administration reports worldwide energy consumption of oil as steadily increasing to 87 million barrels per day as of 2013; roughly 81% of the world's energy demand is met from fossil fuels (Doman *et al.*, 2013). Furthermore, worldwide energy consumption has been projected to increase to as high as 115 million barrels per day by 2040. Many nations throughout the world have growing concern about the far-reaching geopolitical and environmental effects of a society built upon non-renewable energy resources (DOE, 2013). Therefore, a variety of renewable energy sources are becoming increasingly important in many of the world's largest energy markets. One of these renewable energy sources particular suited to replace petroleum based transport fuels is biofuel, a variety of organic molecules with traditional fuel-like properties that are produced from renewable, biological sources.

Current biofuel production has relied heavily upon fermentative ethanol with chemical properties that greatly differ from conventional gasoline. This hydroscopic hydrocarbon provides about 30% less energy per volume than gasoline and can lead to corrosion in current transportation infrastructure (Li *et al.*, 2005). To circumvent these problems, investigation into the efficient, biological production of higher alcohols containing three or more carbons has been increasing. These target alcohols have energy densities and physical characteristics more similar to current fossil fuel molecules compared to ethanol.

A variety of species in the genus *Clostridium* are known to naturally produce isopropanol and 1-butanol in relatively limited quantities (Jones and Woods, 1986). Owing to the difficulty in culturing and genetically manipulating these particular bacterial strains, improvements to *Clostridium* based production of these alcohols have proven to be difficult to attain. To circumvent the difficulties of using these organisms for production of higher chain alcohols, the metabolic pathways from these producers, as well as other novel pathways, have been constructed in user-friendly microbes for increased production. This chapter outlines the engineering approaches for developing and increasing the production of higher chain alcohols.

5.2 Steps to production

Developing a biological fuel production platform, regardless of the fuel molecules of interest, involves common choices that include the selection of the host organism, metabolic pathways and production conditions. These steps are most often followed by system optimization to improve the titre, yield, and productivity of the metabolic process (Stephanopoulos, 2007). In a bottom up approach, the fuel-like metabolite is chosen and a common microbial system utilizing traditional carbon sources is engineered for the production of the target (Dellomonaco *et al.*, 2011; Peralta-Yahya *et al.*, 2012; Steen *et al.*, 2010). Conversely, in a top down approach, an unconventional host organism with interesting characteristics is chosen for installation of a well-developed metabolic pathways and process systems (Atsumi *et al.*, 2009a; Li *et al.*, 2012; Oliver *et al.*, 2013).

In addition to demonstrating scientific proof-of-concept, the titre, yield, and productivity of the process are considered the performance criteria for feasible biofuel production (Stephanopoulos, 2007). Product titre is the concentration of product accumulated in the culture. It is often the initial performance index of interest in the early stages of research. As the titre of a particular metabolite increases so does the ease of isolation, however, the possibility of toxicity due to the metabolite also increases. Yield is defined as the amount of product produced per unit of substrate consumed. Current competition to biofuel in the form of traditional petro fuels, mandates the actual production yield be close to the maximum theoretical yield. Productivity refers to the rate of production per unit volume of the culture. Using these performance factors, metabolic and process engineers are able to measure, compare and report the effects of perturbations to the microbial system.

Production platform organisms are either native producers or non-native hosts. Native strains are useful in understanding basic metabolic pathways necessary for biochemical production. These producers may provide helpful insight into the production of molecules where metabolic pathways are largely uncharacterized, and regulatory effects pertaining to the production of a specific metabolite may be explored. Native producers may lack well-established genetic manipulation techniques. Non-native hosts, such as *Escherichia coli* and *Saccharomyces cerevisiae*, have been well characterized, grow quickly, and have extensive genetic engineering techniques available. Workhorse organisms lend themselves to more sophisticated design techniques and complex genetic constructs. The availability of genome sequencing, computational biochemical modelling, and other bioinformatic tools has greatly accelerated the discovery of useful genetic variants and allowed for rational refinement of heterologous metabolic pathways (Palsson, 2009; Stephanopoulos, 2000; Tomar and De, 2013) Applying these tools to metabolically engineered pathways allows for the combinatorial replacement of native pathway elements with superior genetic elements, derived from a diverse range of organisms. While both types of hosts have specific advantages and challenges, all production hosts require a redox balanced, high flux, non-toxic metabolic pathway for efficient chemical production.

5.3 Fermentative alcohol production

As reviewed in Chapter 4, both isopropanol and 1-butanol have long been known to be naturally produced in various strains of *Clostridium* via the acetone-butanol-ethanol (ABE) fermentation pathway (Jones and Woods, 1986). This dates back to over century where both compounds were recognized as byproducts from the fermentative process for acetone production (Liu and Khosla, 2010). The strain *Clostridium beijerinckii* BA101 was engineered for organic solvent production and was found to produce 33 g/l of total solvent, including isopropanol, 1-butanol and acetone among other compounds (Annous and Blaschek, 1991; Formanek *et al.*, 1997; Papoutsakis, 2008). Transcriptomic studies of another natural producer, *Clostridium acetobutylicum* gave further insight into strain improvement, such as widening the solvent production window by manipulating the sporulation programme (Jones and Woods,

1986; Jones *et al.*, 2008; Papoutsakis, 2008). In addition to upstream optimization of the *Clostridium* host, downstream fermentation techniques showed improvement, such as utilization of a fibrous bed bioreactor that immobilizes cells during continuous production (Huang *et al.*, 2004).

These *Clostridium* species naturally produce desirable higher alcohols as well as additional organics such as butyrate, acetone, and ethanol (Jones and Woods, 1986). This coproduction diminishes the effectiveness of the production. Engineering *Clostridium* strains for optimal production has additional challenges including: relatively slow growth, spore-forming life cycle, and infantile genetic systems (Yu *et al.*, 2011). To overcome these difficulties with the native producers, the *Clostridium* strain fermentation system was transferred to a user-friendly host, *E. coli* (Atsumi *et al.*, 2008a). *E. coli* is a well-characterized microorganism with a set of readily available tools for genetic manipulation proven to aptly produce a variety of biomolecules through heterologous metabolic pathways (Farmer and Liao, 2000; Kim *et al.*, 2007; Martin *et al.*, 2003). Through the use of specialized genetic parts adapted from a variety of microorganisms and other genomic modification techniques, several strains of *E. coli* have been engineered to efficiently produce both 1-butanol and isopropanol from these pathways (Fig. 5.1).

5.3.1 Isopropanol production

Isopropanol is an important solvent in industrial, consumer and medical products, highly used for antimicrobial characteristics. This chemical may be used in a biofuel industry as either a fuel additive or a solvent for esterification of oils into biodiesel. The natural producer utilizes a biosynthetic path starting at a key intermediate metabolite, acetyl-CoA, consisting of four enzymes: acetyl-CoA acetyltransferase, acetoacetyl-CoA transferase, acetoacetate decarboxylase (ADC) and secondary alcohol dehydrogenase (sADH) (Fig. 5.1). From pyruvate, *E. coli* is capable of naturally producing acetate and acetoacetate, lacking only the enzymes for the final two steps of the isopropanol pathway. Acetone production has been achieved in *E. coli* by addition of four heterologous enzymes from *Clostridium* (Bermejo *et al.*, 1998). This system was augmented, improved and optimized utilizing a combination of *E. coli* and *Clostridium* genes.

Several different gene combinations were tested for isopropanol production in addition to testing *C. beijerinckii* sADH and *Thermoanaerobacter brockii* HTD4 sADH. The greatest amount of production was observed from a strain bearing the *thl* gene from *C. acetobutylicum*, *atoAD* genes from *E. coli* (encoding both subunits of the acetyl-CoA:acetoacetyl- CoA transferase), ADC from *C. acetobutylicum*, and ADH from *C. beijerinckii*. This recombinant strain produced 5 g/l isopropanol with a volumetric productivity of 0.4 g/l/h; both parameters exceeding that of the native Clostridium strains (Hanai *et al.*, 2007). The engineered strain also achieved 44% of the theoretical maximum carbon yield. As isopropanol is a toxic solvent for microbes at high concentrations, an *in situ* removal process was used to drive the production forward without limiting the fitness of the host. Using a gas stripping apparatus among other modifications, maximum titres were observed as a high 143 g/l after 240 hours, with a carbon yield of 67% of the theoretical maximum (Inokuma *et al.*, 2010).

5.3.2 Fermentative 1-butanol production

1-Butanol exemplifies the advantages of an advanced biofuel allowing storage and having transportation using existing infrastructure and an energy content (27 MJ/l) similar to gasoline (32 MJ/l). It can completely replace gasoline or mix with gasoline at any ratio. The native pathway *C. acetobutylicum* was introduced into *E. coli*. This biosynthetic pathway was optimized by installation of a combination of native and heterologous genes: *thl*, *hbd*, *crt*, *bcd*, *etfAB*, and *adhE2* (Fig. 5.1) (Atsumi *et al.*, 2008a). With the *E. coli atoB* gene introduced in place of the *C. acetobutylicum thl* gene, 1-butanol productivity increased more than threefold. Additionally, competing host pathways for carbon and reducing equivalents were deleted. The best strain harboured deletions in the *ldhA*, *adhE*, *frdBC*, *pta*, and *fnr* genes, and therefore produced markedly lower quantities of acetate, lactate, ethanol, and succinate. The highest butanol productivity observed in *E. coli* was 1.2 g/l over a 60-hour period, with a carbon yield that

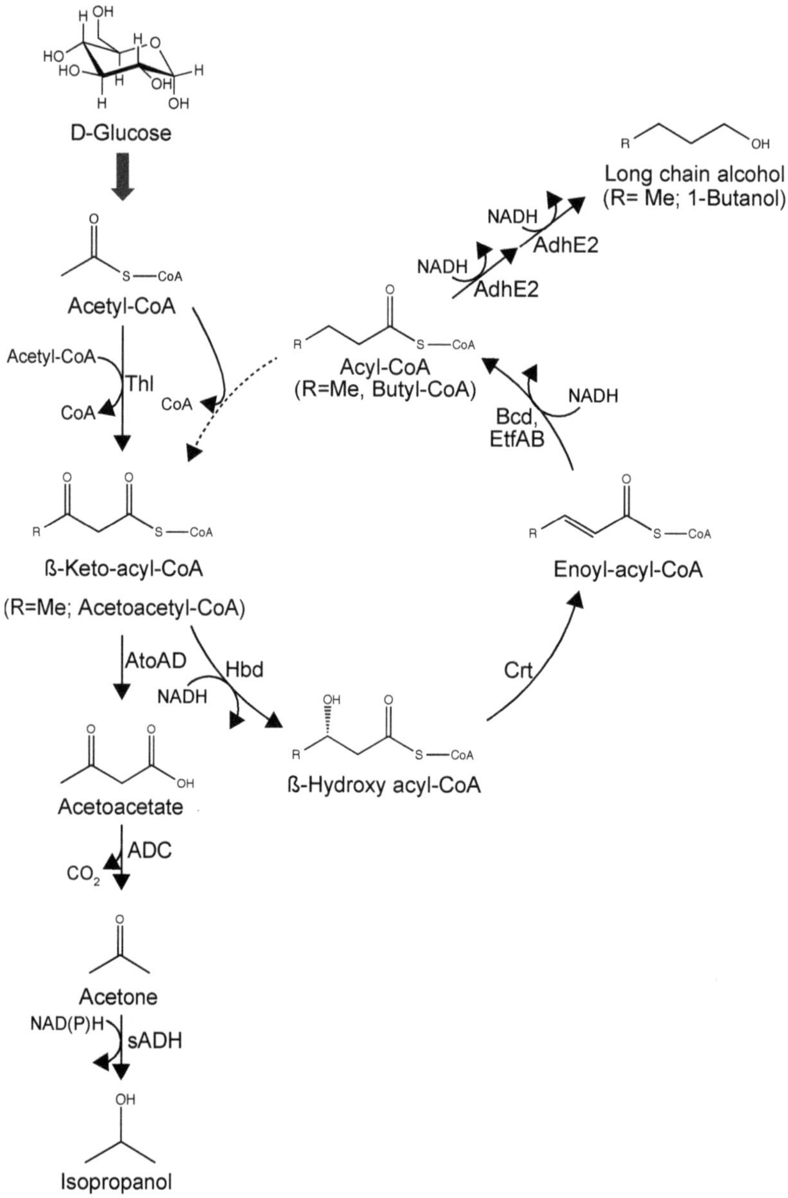

Figure 5.1 Fermentative alcohol pathway. The initial step is condensation of two acetyl-CoA to generate ß-keto-acyl-CoA. ß-keto-acyl-CoA is converted to isopropanol via three enzymatic steps catalysed by acetyl-CoA:acetoacetyl-CoA transferase (cf. AtoAD), ADC, and sADH. For 1-butanol, the engineered pathway consists of six enzymatic steps from acetyl-CoA. Thl, acetoacetyl-CoA thiolase; Hbd, 3-hydroxybutyryl-CoA dehydrogenase; Crt, crotonase; Bcd, butyryl-CoA dehydrogenase; Etf, electron transfer flavoprotein; AdhE2 aldehyde/alcohol dehydrogenase. The synthesized butyl-CoA is used as an initial substrate instead of acetyl-CoA and go through same reactions in reverse ß-oxidation. Each turn of the cycle generate an acyl-CoA that is two carbons longer than the initial acyl-CoA thioester.

was 15% of the theoretical maximum. Additional optimizations of these fermentative pathways were made by combining irreversible reactions with a substrate driving force made by knocking out competing reactions that utilize NADH and acetyl-CoA (Bond-Watts *et al.*, 2011). Using an optimized strain with *in situ* product removal allowed for a final titre of 30 g/l (Shen *et al.*, 2011).

While *E. coli* has been generally considered a user-friendly host for genetic manipulation, *S. cerevisiae* is also an attractive host. *S. cerevisiae* is used extensively in ethanol production, and has proven to be resistant to higher concentrations of the solvent (Thomas *et al.*, 1978). For these properties, the fermentative pathway for 1- butanol production has been constructed in *S. cerevisiae* (Steen *et al.*, 2008). However, initial 1- butanol production was very low (2.5 mg/l), and the optimization of the pathway in yeast has not since been reported.

The CoA-dependent pathway for 1-butanol has also been expressed in *Pseudomonas putida* and *Bacillus subtilis* (Nielsen *et al.*, 2009). Similarly to *S. cerevisiae*, these strains have higher solvent tolerance than *E. coli*. Expression of the genes *thiL*, *crt*, *bcd*, *etfAB*, *hbd*, and *adhE1* in *P. putida* achieved 1-butanol production of 44 mg/l from glucose and 122 mg/l from the more reduced substrate, glycerol (Nielsen *et al.*, 2009). The same genes were expressed in *B. subtilis*, and generated 23 mg/l 1-butanol from glucose and 24 mg/l from glycerol in anaerobic conditions. No 1-butanol was detected from *B. subtilis* in aerobic conditions. 1-Butanol has also been generated from a recombinant strain of *Lactobacillus brevis*, expressing *Clostridia thl*, *crt*, *bcd*, *etfAB*, and *hbd*, while utilizing its native thiolase and aldehyde/alcohol dehydrogenase (Berezina *et al.*, 2010). The engineered *L. brevis* produced 300 mg/l 1-butanol from glucose under semi-anaerobic conditions.

5.3.3 Reverse β-oxidation

A synthetic fermentative pathway was engineered by reversing the ubiquitous β-oxidation cycle for the production of 1-butanol among other alcohols (Fig. 5.1) (Dellomonaco *et al.*, 2011; Gulevich *et al.*, 2012). In order for this pathway to run in reverse effectively, numerous gene knockouts, gene replacements, as well as careful selection of production conditions must be considered. Important gene knockouts included: *arcA* (transcriptional dual regulator), *pta* (phosphate acetyltransferase) *frdA* (furmarate reductase flavoprotein), *adhE* (alcohol dehydrogenase), *eutE* (aldehyde dehydrogenase), and *yqhD* (aldehyde reductase). This knockout strain was combined with the constitutive expression of a feed back inhibition resistant mutant, Crp*, and the complete fad-ato regulon as well as overexpression of *fucO* and *ypeF* for the production of 1-butanol.

Owing to the cyclic nature of this pathway, the carbon backbone of the metabolic substrate may be elongated by two for every cycle. This number of revolution through the cycle for each substrate may be modulated by expression of more specific terminal thioesterases. By combining these thioesterases with expression of various native alcohol dehydrogenases, straight C_4–C_{10} even and odd chain alcohols were produced with a maximum titre 14 g/l 1-butanol (Dellomonaco *et al.*, 2011). Longer chain alcohols were produced between 100–200 mg/l, only slightly less than other more recent production schemes for longer chain alcohols from fatty acids (Akhtar *et al.*, 2013).

5.4 2-Keto acid-based alcohols

The well-understood ethanol pathway provides a different strategy to produce alcohols. Ethanol production by traditional fermentation, mainly from *S. cerevisiae* and *Z. mobilis*, is achieved using a non-oxidative decarboxylation pathway where pyruvate, which is the smallest 2-keto acid, is converted into acetaldehyde by decarboxylation and subsequently converted to ethanol via reduction by an aldehyde reductase (Fig. 5.2A) (Ingram *et al.*, 1987). This pathway removes the need for carbon elongation through a CoA intermediate. Various 2-keto acids are produced from pyruvate for amino acid biosynthesis. The Ehrlich pathway is a catabolic pathway where amino acids metabolized through transamination, and the resulting 2-keto acid chains may be decarboxylated and oxidized to form alcohols (Elsden and Sentheshanmuganathan, 1958). Utilizing the components of the Ehrlich pathway through metabolic engineering, a variety of non-fermentative longer chain alcohols may be efficiently produced (Fig. 5.2B) (Atsumi *et al.*, 2008b). First, 2-keto acid intermediates are generated from pyruvate, using amino acid biosynthesis pathways. Then decarboxylation drives the reaction forward, generating an aldehyde. Lastly, the aldehyde is reduced using an aldehyde reductase to generate the final product, the longer chain alcohol. Since

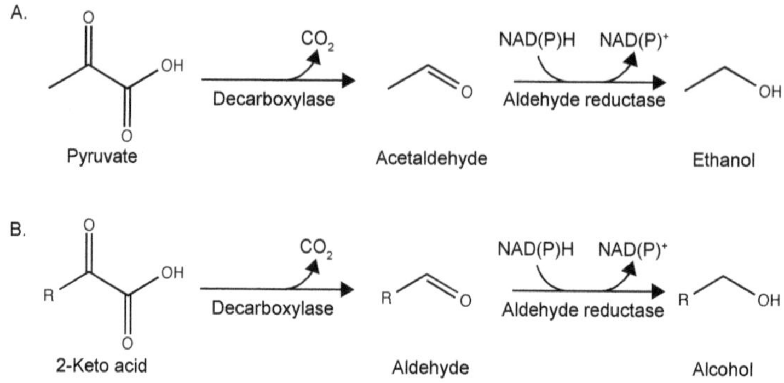

Figure 5.2 2-Keto acid based alcohol pathways. (A) Ethanol production from pyruvate. (B) Higher alcohol production from 2-keto acids.

amino acid biosynthesis pathways are ubiquitous among organisms, this strategy may avoid many of the problems that were observed from adaptation of more complex pathways from heterologous organisms.

5.4.1 Isobutanol

Utilizing the precursor from the natural valine biosynthetic pathway, 2-ketoisovalerate can be converted to isobutanol. The generation of 2-ketoisovalerate from pyruvate was done by three enzymes: acetolactate synthase (ALS) for the condensation of two pyruvates, 2-ketol-acid reductoisomerase (IlvC) for the reduction of the β hydroxyl group, and dihydroxy acid dehydratase (IlvD) for dehydration to generate 2-ketoisovalerate (Atsumi *et al.*, 2008b). Applying 2-ketoacid decarboxylase (KDC) and aldehyde reductase/alcohol dehydrogenase (ADH) to this pathway converts 2-ketoisovalerate to isobutanol (Fig. 5.3).

In order to drive flux towards the corresponding alcohol it is important to choose a suitable 2-keto acid decarboxylase. Atsumi and coworkers screened five KDCs from different organisms: Pdc6, Aro10, Thi3 from *S. cerevisiae*; Kivd from *Lactococcus lactis*; and Pdc from *C. acetobutylicum* (Atsumi *et al.*, 2008b). The results show that Kivd from *L. lactis* is the most active, versatile and suitable of these decarboxylases for the pathway. By overexpressing the three enzymes that convert pyruvate to 2-ketoisovalerate in conjunction with KDC and ADH, the engineered strain produced 1.7 g/l isobutanol.

Metabolic engineering strategies of overexpressing higher activity enzymes in the pathway and eliminating competing pathways were utilized for enhancing the production of isobutanol. Increasing carbon flux to 2-ketoisovalerate from pyruvate is an important consideration for efficient production. Acetolactate synthase encoded by *alsS* from *B. subtilis* was used for pyruvate condensation instead of IlvIH of *E. coli*. This particular synthase has been shown to exhibit high activity and affinity for pyruvate, whereas the *E. coli* IlvIH has higher preference for 2-ketobutyrate (Vyazmensky *et al.*, 1996). This addition increased isobutanol production (Atsumi *et al.*, 2008b). Further optimization was achieved by deleting the following non-essential genes: *adhE, ldhA, frdAB, fnr, pta,* and *pflB*, thereby increasing metabolic flux towards isobutanol more effectively. Each of these strategies cumulatively increased isobutanol titre to more than 20 g/l from glucose, with 85% of the theoretical maximum yield in *E. coli*.

Decreasing the number of overexpressed enzymes may be useful in maintaining an advantageous balance between overall production and cellular fitness. AlsS from *B. subtilis* was shown to perform the decarboxylation of 2-ketoisovalerate and produce isobutyraldehyde in addition to facilitating the nucleophilic attack of pyruvate (Atsumi *et al.*, 2009b). Comparison of the *Klebsiella pneumoniae* AlsS X-ray crystal structure shows that the side chain of Gln483 (Gln487 in *B. subtilis* AlsS) causes steric hindrance with a second larger substrate (Pang *et al.*, 2002). Thus AlsS poorly allows the use of 2-ketobutyrate as the

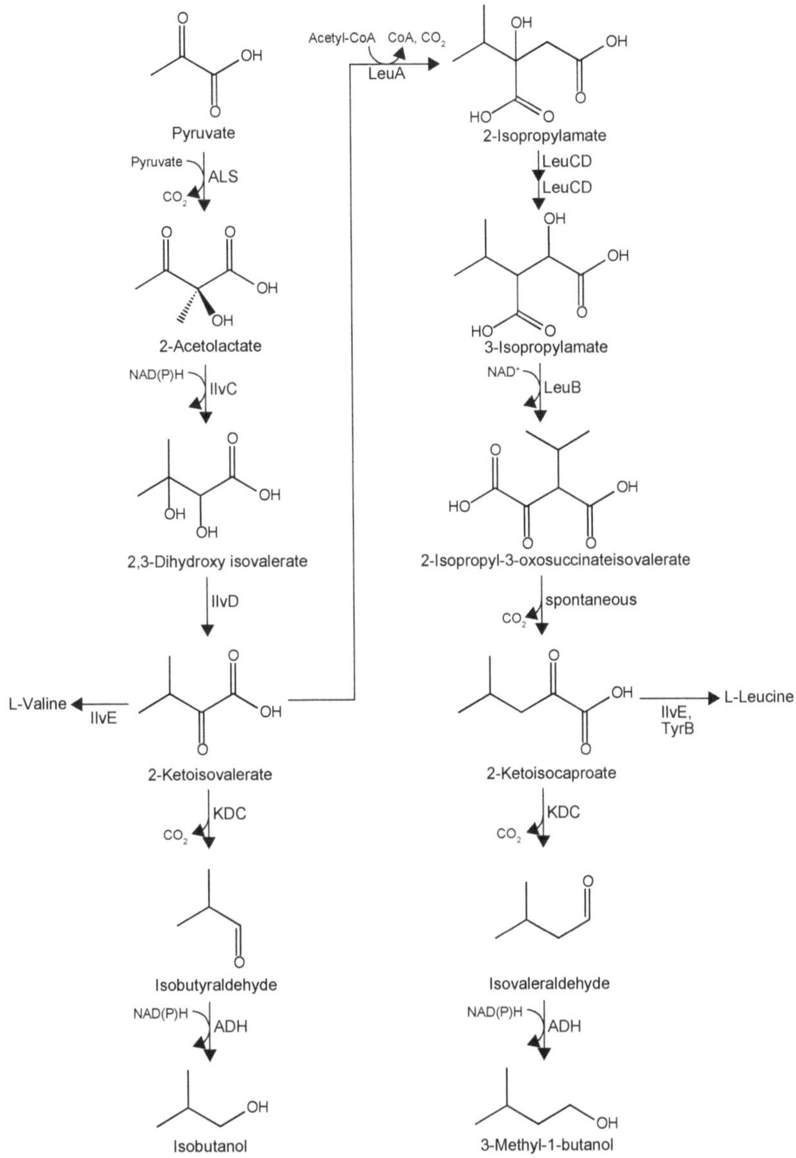

Figure 5.3 Isobutanol and 3-methyl-1-butanol production via L-valine/L-leucine biosynthesis pathway. 2-Ketoisovalerate and 2-ketoisocaproate are precursors of L-valine and L-leucine, respectively. These 2-keto acids are converted to isobutanol and 3-methyl-1-butanol by KDC and ADH.

attacking substrate. Characterization efforts of the decarboxylase activity of AlsS led to the systematic analysis point mutations at position 487. The affinity of all variants toward 2-ketoisovalerate was an order of magnitude less than the affinity for pyruvate. In the interest of using AlsS for both the synthesis of 2-acetolactate as well the decarboxylation of 2-ketoisovalerate, each of these variants were tested for isobutanol production in strains overexpressing *ilvC* and *ilvD*. No variants were found that had any increase over wild type AlsS in isobutanol production.

In order to identify unexpected mutations that may improve isobutanol production, an *E. coli* strain was mutagenized with N′-nitro-N-nitrosoguanidine (NTG) (Smith and Liao, 2011). Isolates were then subject to growth conditions with the toxic L-valine analogue, norvaline. (Ikeda, 2002). This strategy selects for a strain better able to survive high norvaline concentrations ideally by up-regulating biosynthesis of the natural amino acid L-valine and precursors. Increased activity allows for the ability to overcome the toxic effects of norvaline. This selection results is an increase in carbon flux to the desired L-valine precursor, 2-ketoisovalerate, subsequently increasing isobutanol production. Several isolates that showed tolerance to norvaline exposure were transformed with vectors containing *kivd* and *adh* genes. Isobutanol production was measured after 24 hours and compared to the production of the parent strain. The most improved mutant showed increased isobutanol production from 5.3 g/l to 8.0 g/l (Smith and Liao, 2011). The genome from the winning strain was analysed and negative mutations were repaired. Additional rounds of mutations allowed an increase of isobutanol production equal to the rationally designed production strain.

Anaerobic conditions are generally preferred for large-scale production due to lower operating costs and higher yields. The previously elucidated pathway, although highly efficient, is most suitable for aerobic or micro-aerobic isobutanol production. An imbalance in cofactor utilization can occur in anaerobic conditions. One of the pathway enzymes, IlvC requires nicotinamide dinucleotide phosphate (NADPH); however, glycolysis produces only nicotinamide dinucleotide (NADH). Two strategies explored to overcome this limitation are: (1) generation of IlvC variants with dependence on NADH for catalysis and (2) expression of a transhydrogenase (PntAB from *E. coli*) to convert NADH to NADPH (Shi *et al.*, 2013). The engineered strain with fully NADH-dependent pathway increased the yield to 100% of the theoretical maximum yield (Bastian *et al.*, 2011).

5.4.2 3-Methyl-1-butanol

3-Methyl-1-butanol is a potential fuel additive or substitute with a vapour pressure more than 20-fold lower than that of ethanol and an energy density is more than 80% (28.2 MJ/litre) of that of gasoline (34.8 MJ/litre). The 3-methyl-1-butanol pathway utilizes 2-ketoisocaproate in the L-leucine biosynthesis pathway. Since both isobutanol and 3-methyl-1-butanol production pathways utilize 2-ketoisovalerate as a key intermediate, many of the same strategies aim to increase production of both alcohols. Isobutanol production is inevitably connected with 3-methyl-1-butanol production with the competition for 2-ketoisovalerate between KDC for decarboxylation and LeuA for carbon chain elongation (Fig. 5.3). This leads to a co-production of the two alcohols, making specific production of 3-methyl-1-butanol more difficult. However the overexpression of a feedback-insensitive mutant of LeuA, manipulating ribosomal binding sites in key genes, and removing two competing enzymes (IlvE, TyrB) led to a significant increase of production and a titre of 1.3 g/l (Connor and Liao, 2008).

Further optimizations for the enhanced production of 3-methyl-1-butanol utilized the amino acid analogue selection (Ikeda, 2002). By addition of ever-increasing amounts of the toxic leucine analogue, 4-aza-D,L-leucine to chemically mutated strains, the L-leucine production capability can be incrementally enhanced. This strategy improved 3-methyl-1-butanol production to 4.4 g/l (Connor *et al.*, 2010). At these titres, toxicity to the cells becomes a limiting factor for the production. By applying a two-phase production scheme with a layer of oleyl alcohol on top of the culture medium, 3-methyl-1-butanol was continuously extracted from the production media. The two-phase production scheme increased a final titre to 9.5 g/l (Connor *et al.*, 2010). The top oleyl alcohol phase collected more than 90% of the 3-methyl-1-butanol, thereby relieving the cells from the toxic effects of the chemical.

5.4.3 1-Propanol and 1-butanol biosynthesis in *Escherichia coli*

Production of 1-propanol and 1-butanol was demonstrated in *E. coli* by taking advantage of the Ehrlich pathways (Atsumi *et al.*, 2008b; Shen and Liao, 2008). Each alcohol produced from L-isoleucine biosynthesis relies on the intermediate, 2-ketobutyrate (Fig. 5.4). Kivd and Adh2 may act directly on 2-ketobutyrate for the generation of 1-propanol. A side reaction of LeuABCD also consumes 2-ketobutyrate and produces 2-ketovalerate for the biosynthesis of

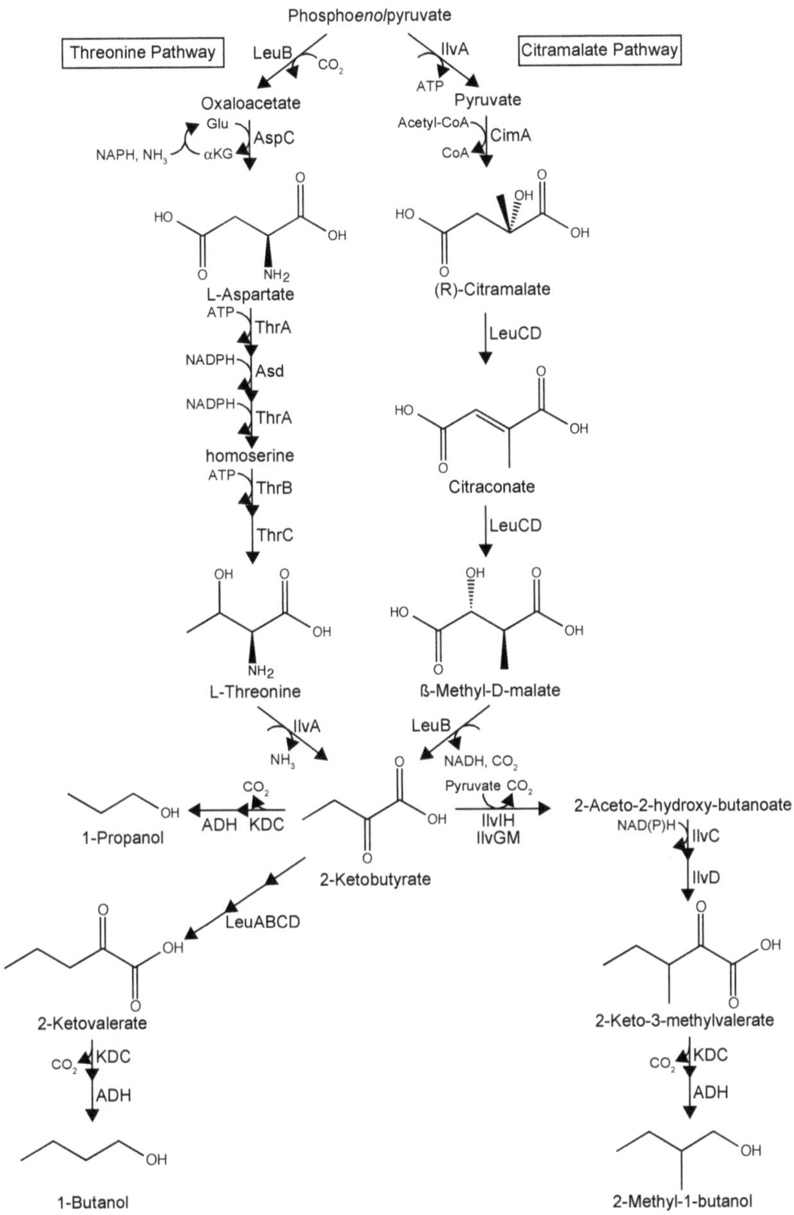

Figure 5.4 1-Propanol, 1-butanol and 2-methyl-1-butanol production via 2-ketobutyrate. 2-Ketobutyrate can be produced via either the threonine pathway or the citramalate pathway and is a precursor for 1-propanol, 1-butanol, and 2-methyl-1-butanol.

the unnatural amino acid norvaline (Guymon *et al.*, 1961; Ingraham *et al.*, 1961; Kisumi *et al.*, 1976). 2-Ketovalerate is the 2-keto acid that may be converted to 1-butanol.

A variety of strategies were utilized in order to optimize the production of 1-propanol and 1-butanol. First adding recombinant DNA for overexpression of key genes in the cell allows for the increase in the rate of key reactions for the control of carbon flow through metabolism (Shen and Liao, 2008). *ilvA* and *leuABCD* were overexpressed to direct metabolic flux towards 2-ketobutyrate and 2-ketovalerate respectively. In addition, the genes encoding Kivd and Adh2 were also over expressed. The engineered strain produced 60 and 30 mg/l 1-butanol and 1-propanol respectively. Although overexpression showed improved titres from non-detectable amounts of production, much room for improvement remained.

L-threonine, a direct precursor of 2-ketobutyrate, was added to the culture medium at induction, resulting in a 20-fold increase in titre for both alcohols (Shen and Liao, 2008). This improved production suggests that L-threonine levels may limit carbon flux towards higher chain alcohols. ThrA is major regulatory element using transcriptional attenuation and allosteric feedback inhibition (Debabov, 2003; Szczesiul and Wampler, 1976). By expressing *thrA^fbrBC^* encoding a feedback-resistant mutant of ThrA the production of both 1-propanol and 1-butanol increased three to four-fold higher relative to the parent strain without $ThrA^{fbrBC}$ (Shen and Liao, 2008). Also, wild type *thrABC* overexpression showed a 10–20% decrease in the two alcohols compared with the *thr^fbrBC^*-containing strain. These results showed that the intracellular L-threonine accumulation was minor but still affected the wild-type activity. Nevertheless, this strategy effectively directed the metabolic flux more towards the L-threonine pathway thus improving the overall 1-propanol and 1-butanol productivity.

In order to further increase the carbon flux for the desired product, eliminating competing pathways through a series of gene knockouts is an important set of modifications for any production strain. Based on pathway analysis, the genes for homoserine O-succinyltransferase, *metA* and threonine dehydrogenase, *tdh* were first deleted to minimize the loss of desired precursors to methionine biosynthesis and to block the catabolism of threonine into 2-amino-3-ketobutyrate. With *metA* and *tdh* disrupted, the combined production of 1-propanol and 1-butanol increased to 1.2 g/l, with the major contribution primarily coming from 1-propanol (Shen and Liao, 2008). Increases to 1-butanol production owing to these genes deletions were modest.

To reduce the loss of two key precursors, acetyl-CoA and 2-ketobutyrate, three gene deletions were preformed (Shen and Liao, 2008). The *E. coli adhE* gene was also deleted to avoid the production of ethanol from acetyl-CoA. While ethanol production was significantly reduced, this change in carbon flux did not show any benefit for the production of higher chain alcohols. By disrupting the genes *ilvB* and *ilvI*, carbon flux towards the other branched amino acid was hindered, increasing the availability of 2-ketobutyrate. These knockouts essentially eliminated the off-target production of other ketoacid-based alcohols and doubled the production of 1-butanol, though left the strain reliant on the supplementation of branched amino acids to all cell growth. These production improvements showed the value of eliminating competing pathways as a common strategy for heightening production of alcohols.

One possible downside of utilizing ubiquitous amino acid biosynthesis pathways for the production of valuable chemicals is that these natural pathways are universally regulated to avoid overproduction. Transcriptional attenuation is a native bacterial regulation mechanism whereby a leader sequence in an mRNA causes ribosomal stalling and hinders proper translation (Henkin and Yanofsky, 2002). This attenuation may be avoided by addition of non-native leader sequences to all essential enzymes. However, feedback inhibition cannot be controlled in this manner since essential amino acid products directly interact with the upstream enzymes causing inhibition, in this case, specifically upon the protein products of the *ilvA* and *leuA* genes. To nullify possible inhibitory effects, previously studied, feedback insensitive enzymes replaced IlvA and LeuA. TdcB is a feedback resistant, native enzyme from *E. coli*, capable of mimicking IlvA by facilitating the deamination of threonine to 2-ketobutyrate (Umbarger, 1956). $LeuA^{fbr}$, a leuA feedback insensitive mutant (Gusyatiner *et al.*, 2002) was introduced into the alcohol production strain. However these alterations failed to increase the production of 1-propanol and 1-butanol, suggesting that the intracellular concentration of amino acids did not cause major inhibitory effects in catalytic function.

The previously reported pathway for 1-propanol and 1-butanol production utilized the L-L-threonine biosynthesis pathway. A new pathway was applied for production of these longer chained alcohol that bypassed L-threonine biosynthesis and directly converted pyruvate to 2-ketobutyrate (Fig. 5.4). Citramalate synthase (CimA) from *Methanococcus jannaschii* may catalyse the addition of an acetate group from acetyl-CoA to pyruvate to synthesize citramalate, a direct precursor of 2-ketobutryate through the reactions catalysed by LeuBCD (Barak *et al.*, 1987; Gollop *et al.*, 1990; Howell *et al.*, 1999). This leads to a shorter, L-threonine-independent keto-acid mediated pathway for the production of 1-propanol and 1-butanol from glucose. Because 2-ketobutyrate is an essential precursor in the biosynthesis of L-isoleucine, an *E. coli* strain deficient in *ilvA* and *tdcB* cannot synthesize 2-ketobutyrate and becomes auxotrophic for L-isoleucine. By leveraging a growth selection based on a requirement for 2-ketobutyrate in such a strain, the directed evolution of CimA enhanced the activity of this enzyme in *E. coli*, allowing the production of 1-propanol and 1-butanol to titres 3.5 g/l and 0.5 g/l, respectively, thus increasing the total alcohol production to 4 g/l (Atsumi and Liao, 2008). Synergistically combining the L-threonine biosynthesis and CimA pathways for the redox balanced production of 1-propanol and 1-butanol leads to a higher yield and productivity than either pathway alone (Shen and Liao, 2013).

5.4.4 2-Methyl-1-butanol

2-Methyl-1-butanol is a five-carbon alcohol, and may play an important role in future advanced biofuel mixtures. Using 2-ketobutyrate as a precursor, additional carbon chain elongation via IlvIHCD allows the synthesis of 2-keto-3-methylvalerate (Fig. 5.4). IlvIH, acetohydroxy acid synthase, catalyses the condensation of 2KB and pyruvate to 2-aceto-2-hydroxybutyrate. Two more enzymes, IlvC and IlvD, complete the elongation mechanism and lead to the formation of 2-keto-3-methylvalerate, the precursor for 2-methyl-1-butanol (Myers, 1961; Ramakrishnan and Adelberg, 1964).

Owing to an overlap in biosynthetic pathways, many strategies for maximizing 2-methyl-1-butanol production were similar to the optimization of 1-propanol and 1-butanol. However, two unique challenges were presented with 2-methyl-1-butanol optimization. The first is that products of glycolysis are split between two upstream pathways for essential precursors (pyruvate and 2-ketobutyrate). The difference in carbon flux between the two branches may lead to an imbalanced supply of precursors that facilitates side production. The second challenge is that 2-ketobutyrate and pyruvate can be converted to other metabolites leading to 1-propanol and isobutanol, respectively with essential enzymes for 2-methyl-1-butanol production. Therefore, a simple gene deletion cannot solve the problem. To alleviate these problems each enzyme in the production path was characterized and chosen for specificity toward the desire reactants and products. Due to the promiscuous nature of acetohydroxy acid synthase (AHAS) multiple AHAS enzymes were considered to identify an isozyme with the greatest activity towards isoleucine biosynthesis. AHAS II encoded by *ilvGM* from *E. coli* proved to be more specific for 2-ketobutyrate than the other AHAS isozymes (Atsumi *et al.*, 2008b; Barak *et al.*, 1987; Gollop *et al.*, 1990). Thus, this isozyme was chosen for the 2-methyl-1-butanol production strain (Cann and Liao, 2008). Three threonine deaminases were tested, and while all overexpressed genes products increased 2-methyl-1-butanol, the best isozyme tested for 2-methyl-1-butanol production was the *C. glutamicum* threonine deaminase. This deaminase converted 88% of the supplied L-threonine to 2-methyl-1-butanol and 1-propanol. Knocking out genes *metA*, *tdh*, *leuA* and *ilvE* allowed for the removal of competitive pathways for the precursors of the 2-methyl-1-butanol pathway (Cann and Liao, 2008).

The final strain was reported to produce 1.25 g/l 2-methyl-1-butanol in 24 h with 44% of the maximum theoretical yield (Cann and Liao, 2008). While more optimization may allow for greater production, 1-propanol and isobutanol are necessarily formed using the 2-methyl-1-butanol pathway, thereby limiting the potential for specific production. Future production may benefit from metabolic engineering strategies that focus on reducing the promiscuity of enzymes thus reducing the production of side products.

5.4.5 Further usage of the isoleucine biosynthesis pathway C_6–C_8

While a variety of interesting alcohols have been produced with this approach, alcohols with carbon chain lengths longer than five are attractive biofuel targets owing to higher energy density and lower water solubility. These characteristics even more closely replicate many current fuel molecules and may facilitate postproduction purification from culture medium through efficient separation processes.

By leveraging the ability for carbon elongation through the catalysed reactions of LeuABCD to a variety of 2-ketoacids, production of C_6 to C_8 alcohols has been shown (Marcheschi *et al.*, 2012; Zhang *et al.*, 2010). This strategy revolves around the engineering of LeuA to recognize a larger variety of substrates, thus enabling the carbon additions leading to longer chained alcohols (Fig. 5.5). Initial enzyme engineering efforts focused on the recognition of the isoleucine precursor 2-keto-3-methylvalerate allowing for production of 3-methyl-1-pentanol (Zhang *et al.*, 2010). Since the final reaction of LeuABCD produces a ketoacid, the process has been shown to be iterative, allowing for the product to be fed back into the LeuABCD reactions for a stepwise carbon elongation mechanism (Marcheschi *et al.*, 2012). Utilizing additional enzyme engineering, mutant variants of LeuA have been discovered that allow the production of variety of C_6–C_8 alcohols (Marcheschi *et al.*, 2012). While the titres for these longer chain alcohols are substantially lower than those directly corresponding to natural amino acids, the ability to engineer novel products is a valuable starting point for increasingly diverse production schemes.

5.4.6 2-Ketoacid-based alcohols in other hosts

E. coli has been shown to be an excellent host for proof of concept of high production of a variety of higher chain alcohols (Atsumi *et al.*, 2008b; Cann and Liao, 2008; Connor *et al.*, 2010; Shen *et al.*, 2011). However, other organisms may be used in industrial production of chemicals. All organisms have a specific niche that may or may not be useful for industrial purposes, so a range of organisms other than *E. coli* have been engineered for production of these alcohols with wide ranging results. This exploration is possible because of the

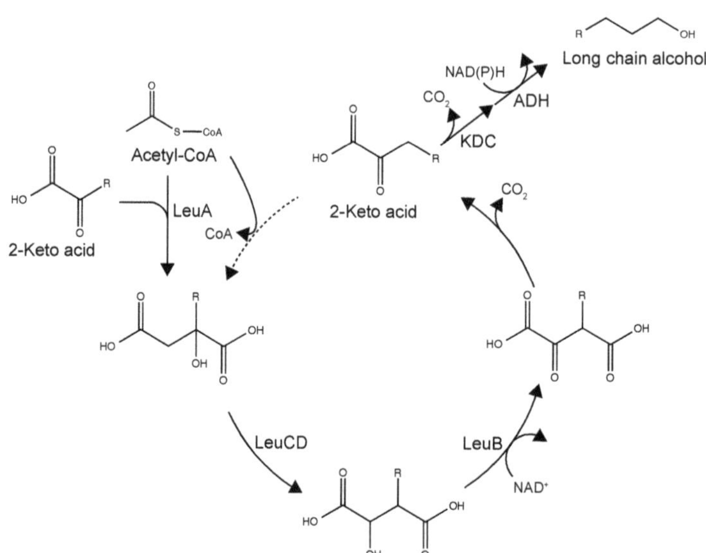

Figure 5.5 LeuABCD carbon elongation cycle. 2-Keto acid is combined with acyl-CoA by LeuA, then the produced molecule is converted to a 2-keto acid one carbon longer than the initial 2-keto acid via three enzymatic steps (isomerization, oxidation and decarboxylation). Long chain alcohols are produced from the elongated 2-keto acids by KDC and ADH.

continued advancement in genetic manipulation tools of diverse microorganisms and the ubiquity of amino acid biosynthetic pathways.

S. cerevisiae has a relatively high tolerance for alcohols, well-characterized genetic tractability, naturally produces isobutanol anaerobically from L-valine in small quantities (Hazelwood *et al.*, 2008). This organism has been explored as a host for augmented isobutanol production by increased flux through the overexpression of native L-valine biosynthetic pathway (Chen *et al.*, 2011). Additional overexpression of the genes for KDC from *L. lactis* and the endogenous ADH6 as well as removal of a pyruvate decarboxylase resulted in a maximum titre of 143 mg/l and a yield of 6.6 mg isobutanol/g glucose (Kondo *et al.*, 2012). One difficulty in the production of isobutanol in *S. cerevisiae* is the separation of the metabolic pathway into organelles. By engineering the localization of the pathway enzymes, the engineered strains increase the relative concentration of pathway specific enzymes and metabolites. Isolating and overexpressing the Ehrlich pathway in either the mitochondria or the cytosol increases isobutanol titre to about 630 mg/l (Avalos *et al.*, 2013; Brat *et al.*, 2012). This allows for more efficient production as well as reduced chances of substrate intermediates to be lost to competing pathways. While these strategies have greatly increased the viability of isobutanol in *S. cerevisiae*, this production host lags behind many prokaryotic strains in this area.

Corynebacterium glutamicum is a well-characterized model host for the production of a variety of biologically derived chemicals. Industrial pursuits have optimized this strain for the efficient production amino acids, possibly making *C. glutamicum* an ideal strain for the production of higher chained alcohols from amino acid precursors. One industrially useful characteristic of *C. glutamicum* is the tendency to shift metabolic priorities from growth to metabolite formation when put under anaerobic conditions (Yamamoto *et al.*, 2011, 2013). This particular strain has been shown to be less susceptible to the toxic effects of isobutanol accumulation in culture and produced up to 4.9 g/l after 120 h in aerobic conditions (Smith *et al.*, 2010). Through further identification of genetic parts better suited for anaerobic conditions, additional genetic disruptions and continuous *in situ* production removal through a two-phase design, efficient isobutanol in *C. glutamicum* was achieved at a titre of 73 g/l with 78.1% of the theoretical maximum yield (Yamamoto *et al.*, 2013).

While glucose has been the most popular source of carbon in longer chain alcohol production studies, more economically feasible carbon sources are of interest to compete with petroleum-derived fuels. Utilizing CO_2 is an attractive approach since the devised system directly recycles the waste product of combustion thus may help mitigate current environmental concerns. As reviewed in Chapter 8, higher chain alcohols can be produced by genetically modifying cyanobacterial hosts. *Synechococcus elongatus* PCC7942 is a model cyanobacterium, capable of using light energy to fix CO_2 into reduce compounds. Isobutanol has been produced from CO_2 by expression of five genes of the isobutanol pathway (*alsS, ilvC, ilvD, kivd,* and *yqhD*) in *S. elongatus* with a maximum titre 450 mg/l (Atsumi *et al.*, 2009a). Due to toxic effects of isobutanol on the cyanobacterial production culture, the precursor isobutyraldehyde was also produced, optimized, and extracted at a maximum titre of 1.1 g/l (Atsumi *et al.*, 2009a). Isobutyraldehyde is a valuable feedstock for a range of useful chemicals, can easily be converted to isobutanol through chemical means, and can be more easily extracted from culture owing to a lower boiling point and higher vapour pressure than isobutanol. Another cyanobacterium, *Synechocystis sp.* PCC6803, has also been engineered for isobutanol production from a mixed feedstock of glucose and CO_2; however, this strain produced more from CO_2 alone (Varman *et al.*, 2013).

5.5 Conclusion

Here we have elaborated on the relatively simple method of utilizing key enzymes from the native host pathway for the increased production of various higher chain alcohols in a variety of organisms. Future exploration of these pathways may yield many necessary advances for the economic feasibility of large-scale production of longer chain alcohols from biological organisms.

Further advances will more accurately mimic natural metabolic pathways including: enzyme engineering for more specific, highly active proteins; tightly controlled regulation schemes for fine-tune adjustments of metabolite concentrations; and continued expansion into even longer and more complex fuel-like molecule production. Such future progress will help to alleviate current environmental concerns and promote stability in the energy market. Additionally, these advancements may increase the ability of society to move towards a more renewable energy market globally, thus allowing a more secure economic and environmental outlook for future generations.

References

Akhtar, M.K., Turner, N.J., and Jones, P.R. (2013). Carboxylic acid reductase is a versatile enzyme for the conversion of fatty acids into fuels and chemical commodities. Proc. Natl Acad. Sci. U.S.A. *110*, 87–92.

Annous, B.A., and Blaschek, H.P. (1991). Isolation and characterization of *Clostridium acetobutylicum* mutants with enhanced amylolytic activity. Appl Environ Microbiol 57, 2544–2548.

Atsumi, S., and Liao, J.C. (2008). Directed evolution of *Methanococcus jannaschii* citramalate synthase for biosynthesis of 1-propanol and 1-butanol by *Escherichia coli*. Appl Environ Microbiol 74, 7802–7808.

Atsumi, S., Cann, A.F., Connor, M.R., Shen, C.R., Smith, K.M., Brynildsen, M.P., Chou, K.J.Y., Hanai, T., and Liao, J.C. (2008a). Metabolic engineering of *Escherichia coli* for 1-butanol production. Metab Eng *10*, 305–311.

Atsumi, S., Hanai, T., and Liao, J.C. (2008b). Non-fermentative pathways for synthesis of branched-chain higher alcohols as biofuels. Nature *451*, 86–89.

Atsumi, S., Higashide, W., and Liao, J.C. (2009a). Direct photosynthetic recycling of carbon dioxide to isobutyraldehyde. Nat Biotechnol 27, 1177–1180.

Atsumi, S., Li, Z., and Liao, J.C. (2009b). Acetolactate synthase from *Bacillus subtilis* serves as a 2-ketoisovalerate decarboxylase for isobutanol biosynthesis in *Escherichia coli*. Appl. Environ. Microbiol. 75, 6306–6311.

Avalos, J.L., Fink, G.R., and Stephanopoulos, G. (2013). Compartmentalization of metabolic pathways in yeast mitochondria improves the production of branched-chain alcohols. Nat. Biotechnol. *31*, 335–341.

Barak, Z., Chipman, D.M., and Gollop, N. (1987). Physiological implications of the specificity of acetohydroxy acid synthase isozymes of enteric bacteria. J. Bacteriol. *169*, 3750–3756.

Bastian, S., Liu, X., Meyerowitz, J.T., Snow, C.D., Chen, M.M.Y., and Arnold, F.H. (2011). Engineered ketol-acid reductoisomerase and alcohol dehydrogenase enable anaerobic 2-methylpropan-1-ol production at theoretical yield in *Escherichia coli*. Metab. Eng. *13*, 345–352.

Berezina, O.V., Zakharova, N.V., Brandt, A., Yarotsky, S.V., Schwarz, W.H., and Zverlov, V.V. (2010). Reconstructing the clostridial n-butanol metabolic pathway in *Lactobacillus brevis*. Appl. Microbiol. Biotechnol. *87*, 635–646.

Bermejo, L.L., Welker, N.E., and Papoutsakis, E.T. (1998). Expression of *Clostridium acetobutylicum* ATCC 824 genes in *Escherichia coli* for acetone production and acetate detoxification. Appl. Environ. Microbio.l *64*, 1079–1085.

Bond-Watts, B.B., Bellerose, R.J., and Chang, M.C.Y. (2011). Enzyme mechanism as a kinetic control element for designing synthetic biofuel pathways. Nat. Chem. Biol. 7, 222–227.

Brat, D., Weber, C., Lorenzen, W., Bode, H.B., and Boles, E. (2012). Cytosolic re-localization and optimization of valine synthesis and catabolism enables increased isobutanol production with the yeast *Saccharomyces cerevisiae*. Biotechnol Biofuels 5, 65.

Cann, A.F., and Liao, J.C. (2008). Production of 2-methyl-1-butanol in engineered *Escherichia coli*. Appl Microbiol Biotechnol *81*, 89–98.

Chen, X., Nielsen, K.F., Borodina, I., Kielland-Brandt, M.C., and Karhumaa, K. (2011). Increased isobutanol production in *Saccharomyces cerevisiae* by overexpression of genes in valine metabolism. Biotechnol Biofuels *4*, 2089–2090.

Connor, M.R., and Liao, J.C. (2008). Engineering of an *Escherichia coli* strain for the production of 3-methyl-1-butanol. Appl. Environ. Microbiol. *74*, 5769–5775.

Connor, M.R., Cann, A.F., and Liao, J.C. (2010). 3-Methyl-1-butanol production in *Escherichia coli*: random mutagenesis and two-phase fermentation. Appl. Microbiol. Biotechnol. *86*, 1155–1164.

Debabov, V.G. (2003). The threonine story. In: R. Faurie, and J. Thommel, eds. Microbial Production of L-Amino Acids. Springer, pp. 113–136.

Dellomonaco, C., Clomburg, J.M., Miller, E.N., and Gonzalez, R. (2011). Engineered reversal of the β-oxidation cycle for the synthesis of fuels and chemicals. Nature *476*, 355–359.

DOE (2013). Report on the First Quadrennial Technology Review.

Doman, L.E., Arora, V., Metelitsa, A., Leahy, M., Barden, J.L., Ford, M., Mellish, M.L., Jones, A., Murphy, B.T., Lillis, K., *et al.* (2013). International Energy Outlook 2013.

Elsden, S.R., and Sentheshanmuganathan, S. (1958). The mechanism of the formation of tyrosol by *Saccharomyces cerevisiae*. Biochem. J. *69*, 210–218.

Farmer, W.R., and Liao, J.C. (2000). Improving lycopene production in *Escherichia coli* by engineering metabolic control. Nat. Biotechnol. *18*, 533–537.

Formanek, J., Mackie, R., and Blaschek, H.P. (1997). Enhanced butanol production by *Clostridium beijerinckii* BA101 grown in semidefined P2 medium containing 6 percent maltodextrin or glucose. Appl. Environ. Microbiol. 63, 2306–2310.

Gollop, N., Damri, B., Chipman, D.M., and Barak, Z. (1990). Physiological implications of the substrate specificities of acetohydroxy acid synthases from varied organisms. J Bacteriol *172*, 3444–3449.

Gulevich, A.Y., Skorokhodova, A.Y., Sukhozhenko, A.V., Shakulov, R.S., and Debabov, V.G. (2012). Metabolic engineering of *Escherichia coli* for 1-butanol biosynthesis through the inverted aerobic fatty acid β-oxidation pathway. Biotechnol. Lett. *34*, 463–469.

Gusyatiner, M.M., Lunts, M.G., Kozlov, Y.I., Ivanovskaya, L.V., and Voroshilova, E.B. (2002). DNA coding for mutant isopropylmalate synthase L-leucine-producing microorganism and method for producing L-leucine. US Patent No. 6,403,342.

Guymon, J.F., Ingraham, J.L., and Crowell, E.A. (1961). The formation of n-propyl alcohol by *Saccharomyces cerevisiae*. Arch. Biochem. Biophys. *95*, 163–168.

Hanai, T., Atsumi, S., and Liao, J.C. (2007). Engineered Synthetic Pathway for Isopropanol Production in *Escherichia coli*. Appl. Environ. Microbiol. *73*, 7814–7818.

Hazelwood, L.A., Daran, J.-M., van Maris, A.J.A., Pronk, J.T., and Dickinson, J.R. (2008). The Ehrlich pathway for fusel alcohol production: a century of research on *Saccharomyces cerevisiae* metabolism. Appl. Environ. Microbiol. *74*, 2259–2266.

Henkin, T.M., and Yanofsky, C. (2002). Regulation by transcription attenuation in bacteria: how RNA provides instructions for transcription termination/antitermination decisions. Bioessays *24*, 700–707.

Howell, D.M., Xu, H., and White, R.H. (1999). (R)-citramalate synthase in methanogenic archaea. J. Bacteriol. *181*, 331–333.

Huang, W.-C., Ramey, D.E., and Yang, S.-T. (2004). Continuous production of butanol by *Clostridium acetobutylicum* immobilized in a fibrous bed bioreactor. Appl. Biochem. Biotechnol. *113–116*, 887–898.

Ikeda, M. (2002). Amino Acid Production Processes. Berlin.: Springer-Verlag, pp. 1–35.

Ingraham, J.L., Guymon, J.F., and Crowell, E.A. (1961). The pathway of formation of n-butyl and n-amyl alcohols by a mutant strain of *Saccharomyces cerevisiae*. Arch. Biochem. Biophy.s 95, 169–175.

Ingram, L.O., Conway, T., Clark, D.P., Sewell, G.W., and Preston, J.F. (1987). Genetic engineering of ethanol production in *Escherichia coli*. Appl Environ Microbiol 53, 2420–2425.

Inokuma, K., Liao, J.C., Okamoto, M., and Hanai, T. (2010). Improvement of isopropanol production by metabolically engineered *Escherichia coli* using gas stripping. J. Biosci. Bioeng. *110*, 696–701.

Jones, D.T., and Woods, D.R. (1986). Acetone-butanol fermentation revisited. Microbiol. Rev. *50*, 484–524.

Jones, S.W., Paredes, C.J., Tracy, B., Cheng, N., Sillers, R., Senger, R.S., and Papoutsakis, E.T. (2008). The transcriptional program underlying the physiology of clostridial sporulation. Genome Biol. *9*, R114.

Kim, Y., Ingram, L.O., and Shanmugam, K.T. (2007). Construction of an *Escherichia coli* K-12 mutant for homoethanologenic fermentation of glucose or xylose without foreign genes. Appl. Environ. Microbiol. *73*, 1766–1771.

Kisumi, M., Sugiura, M., and Chibata, I. (1976). Biosynthesis of norvaline, norleucine, and homoisoleucine in *Serratia marcescens*. J. Biochem. *80*, 333–339.

Kondo, T., Tezuka, H., Ishii, J., Matsuda, F., Ogino, C., and Kondo, A. (2012). Genetic engineering to enhance the Ehrlich pathway and alter carbon flux for increased isobutanol production from glucose by *Saccharomyces cerevisiae*. J Biotechnol. *159*, 32–37.

Li, D.-G., Zhen, H., Xingcai, L., Wu-gao, Z., and Jian-guang, Y. (2005). Physico-chemical properties of ethanol–diesel blend fuel and its effect on performance and emissions of diesel engines. Renewable Energy *30*, 967–976.

Li, H., Opgenorth, P.H., Wernick, D.G., Rogers, S., Wu, T.Y., Higashide, W., Malati, P., Huo, Y.X., Cho, K.M., and Liao, J.C. (2012). Integrated electromicrobial conversion of CO_2 to higher alcohols. Science *335*, 1596–1596.

Marcheschi, R.J., Li, H., Zhang, K., Noey, E.L., Kim, S., Chaubey, A., Houk, K.N., and Liao, J.C. (2012). A synthetic recursive "+1" pathway for carbon chain elongation. ACS Chem. Biol. *7*, 689–697.

Martin, V.J.J., Pitera, D.J., Withers, S.T., Newman, J.D., and Keasling, J.D. (2003). Engineering a mevalonate pathway in *Escherichia coli* for production of terpenoids. Nat. Biotechnol. *21*, 796–802.

Myers, J.W. (1961). Dihydroxy acid dehydrase: an enzyme involved in the biosynthesis of isoleucine and valine. J. Biol. Chem. *236*, 1414–1418.

Nielsen, D.R., Leonard, E., Yoon, S.-H., Tseng, H.-C., Yuan, C., and Prather, K.L.J. (2009). Engineering alternative butanol production platforms in heterologous bacteria. Metab. En.g *11*, 262–273.

Oliver, J.W.K., Machado, I.M.P., Yoneda, H., and Atsumi, S. (2013). Cyanobacterial conversion of carbon dioxide to 2,3-butanediol. Proc. Natl Acad. Sci. U.S.A. *110*, 1249–1254.

Palsson, B. (2009). Metabolic systems biology. FEBS Lett. *583*, 3900–3904.

Pang, S.S., Guddat, L.W., and Duggleby, R.G. (2002). Crystallization of the FAD-independent acetolactate synthase of *Klebsiella pneumoniae*. Acta Crystallogr. D Biol. Crystallogr. *58*, 1237–1239.

Papoutsakis, E.T. (2008). Engineering solventogenic clostridia. Curr. Opin. Biotech. *19*, 420–429.

Peralta-Yahya, P.P., Zhang, F., del Cardayre, S.B., and Keasling, J.D. (2012). Microbial engineering for the production of advanced biofuels. Nature *488*, 320–328.

Ramakrishnan, T., and Adelberg, E.A. (1964). Regulatory mechanisms in the biosynthesis of isoleucine and valine I. Genetic derepression of enzyme formation. J Bacteriol 87, 566–573.

Shen, C.R., and Liao, J.C. (2008). Metabolic engineering of *Escherichia coli* for 1-butanol and 1-propanol production via the keto-acid pathways. Metab Eng *10*, 312–320.

Shen, C.R., and Liao, J.C. (2013). Synergy as design principle for metabolic engineering of 1-propanol production in *Escherichia coli*. Metab. Eng. *17*, 12–22.

Shen, C.R., Lan, E.I., Dekishima, Y., Baez, A., Cho, K.M., and Liao, J.C. (2011). Driving forces enable high-titer anaerobic 1-butanol synthesis in *Escherichia coli*. Appl. Environ. Microbiol. 77, 2905–2915.

Shi, A., Zhu, X., Lu, J., Zhang, X., and Ma, Y. (2013). Activating transhydrogenase and NAD kinase in combination for improving isobutanol production. Metab. Eng. *16*, 1–10.

Smith, K.M., and Liao, J.C. (2011). An evolutionary strategy for isobutanol production strain development in *Escherichia coli*. Metab Eng *13*, 674–681.

Smith, K.M., Cho, K.M., and Liao, J.C. (2010). Engineering *Corynebacterium glutamicum* for isobutanol production. Appl Microbiol Biotechnol *87*, 1045–1055.

Steen, E.J., Chan, R., Prasad, N., Myers, S., Petzold, C.J., Redding, A., Ouellet, M., and Keasling, J.D. (2008). Metabolic engineering of *Saccharomyces cerevisiae* for the production of n-butanol. Microb. Cell Fact *7*, 36.

Steen, E.J., Kang, Y., Bokinsky, G., Hu, Z., Schirmer, A., McClure, A., del Cardayre, S.B., and Keasling, J.D. (2010). Microbial production of fatty-acid-derived fuels and chemicals from plant biomass. Nature *463*, 559–562.

Stephanopoulos, G. (2000). Bioinformatics and metabolic engineering. Metab. Eng. *2*, 157–158.

Stephanopoulos, G. (2007). Challenges in engineering microbes for biofuels production. Science *315*, 801–804.

Szczesiul, M., and Wampler, D.E. (1976). Regulation of a metabolic system *in vitro*: synthesis of threonine from aspartic acid. Biochemistry *15*, 2236–2244.

Thomas, D.S., Hossack, J.A., and Rose, A.H. (1978). Plasma-Membrane lipid composition and ethanol tolerance in *Saccharomyces cerevisiae.* Archiv. Mikrobiol. *117*, 239–245.

Tomar, N., and De, R.K. (2013). Comparing methods for metabolic network analysis and an application to metabolic engineering. Gene *521*, 1–14.

Umbarger, H.E. (1956). Evidence for a negative-feedback mechanism in the biosynthesis of isoleucine. Science *123*, 848–848.

Varman, A.M., Xiao, Y., Pakrasi, H.B., and Tang, Y.J. (2013). Metabolic engineering of *Synechocystis* sp. strain PCC 6803 for isobutanol production. Appl. Environ. Microbiol 79, 908–914.

Vyazmensky, M., Sella, C., Barak, Z., and Chipman, D.M. (1996). Isolation and characterization of subunits of acetohydroxy acid synthase isozyme III and reconstitution of the holoenzyme. Biochemistry 35, 10339–10346.

Yamamoto, S., Sakai, M., Inui, M., and Yukawa, H. (2011). Diversity of metabolic shift in response to oxygen deprivation in *Corynebacterium glutamicum* and its close relatives. Appl. Microbiol. Biotechnol. *90*, 1051–1061.

Yamamoto, S., Suda, M., Niimi, S., Inui, M., and Yukawa, H. (2013). Strain optimization for efficient isobutanol production using *Corynebacterium glutamicum* under oxygen deprivation. Biotechnol. Bioeng. doi: 10.1002/bit.24961

Yu, M., Zhang, Y., Tang, I.-C., and Yang, S.-T. (2011). Metabolic Engineering. Metab. Eng. *13*, 373–382.

Zhang, K., Li, H., Cho, K.M., and Liao, J.C. (2010). Expanding metabolism for total biosynthesis of the nonnatural amino acid L-homoalanine. Proc. Natl Acad. Sci. U.S.A. *107*, 6234–6239.

Isoprene-derived Biofuels from Engineered Microbes

6

Han Min Woo and Taek Soon Lee

Abstract

Recently, biofuel researches have shown an increased emphasis on the advanced biofuels which are more direct replacements to petroleum-derived transportation fuels, and more compatible to the existing fuel infrastructure than the most popular biofuel, ethanol. Advanced biofuels are mostly produced from the existing hydrocarbon biosynthetic pathways such as fatty acid biosynthesis and isoprenoid biosynthesis. These hydrocarbon biosynthetic pathways generate a range of potential biofuels with characteristics suitable for gasoline, diesels, or jet-fuels, and among them, isoprenoid pathways are unique for their richness in the type of compounds they can generate. Here, we provide the insights of a class of isoprenoid compounds that can be produced in microbes by adequate microbial engineering. Subsequently, microbial cells have been constructed to produce isoprenoids using tools of pathway and host engineering, and optimized through metabolic engineering. To broaden the spectrum of microbial synthesis of target isoprenoids, synthetic biology and systems biology also have been applied to engineered microbes. Several engineered *E. coli* and yeast strains have been constructed and being optimized for industrial applications. Downstream-modification of isoprenoids produced from engineered microbes is necessary to satisfy the current fuel properties, and catalytic processes of target compounds must be combined with the development of isoprenoids-producing strains.

6.1 Classes of isoprenoid compounds

Isoprenoids are one of the most diverse groups of natural products in nature. Comprising over 50,000 compounds, isoprenoids include primary metabolites (sterols, carotenoids, quinones, and hormones) and particular secondary metabolites in plants and microbes which have frequently been used for medical purposes. Natural products chemists have long marvelled at the structural diversity of terpenes and its biosynthetic pathways, and have been developing commercial drugs based on these natural products. Usually, isoprenoids are classified into groups according to the number of carbons: hemiterpenes (C_5), monoterpenes (C_{10}), sesquiterpenes (C_{15}), diterpenes (C_{20}) and triterpenes (C_{30}). Isopentenyl diphosphate (IPP) and its isomer dimethylallyl diphosphate (DMAPP) are two universal C_5 building blocks to synthesize all isoprenoids. Some examples of isoprenoids are shown in Fig. 6.1. These starting precursors can be synthesized by two routes: either 1-deoxy-D-xylulose-5-phosphate (DXP) or the mevalonate pathway. The detail of these two pathways will be discussed in section 6.2. Isoprenoids generally contain between 5 and 30 carbons, but they also include highly modified compounds from phytoene (C_{40}). Many isoprenoids have been found to have significant pharmaceutical importance. Often isoprenoids attached to regulatory proteins have been shown to regulate cellular development. The wide scope of terpene synthase activity was first studied with crude plant extracts and purified preparation of native enzymes. More recently, the isolation of terpene synthase genes

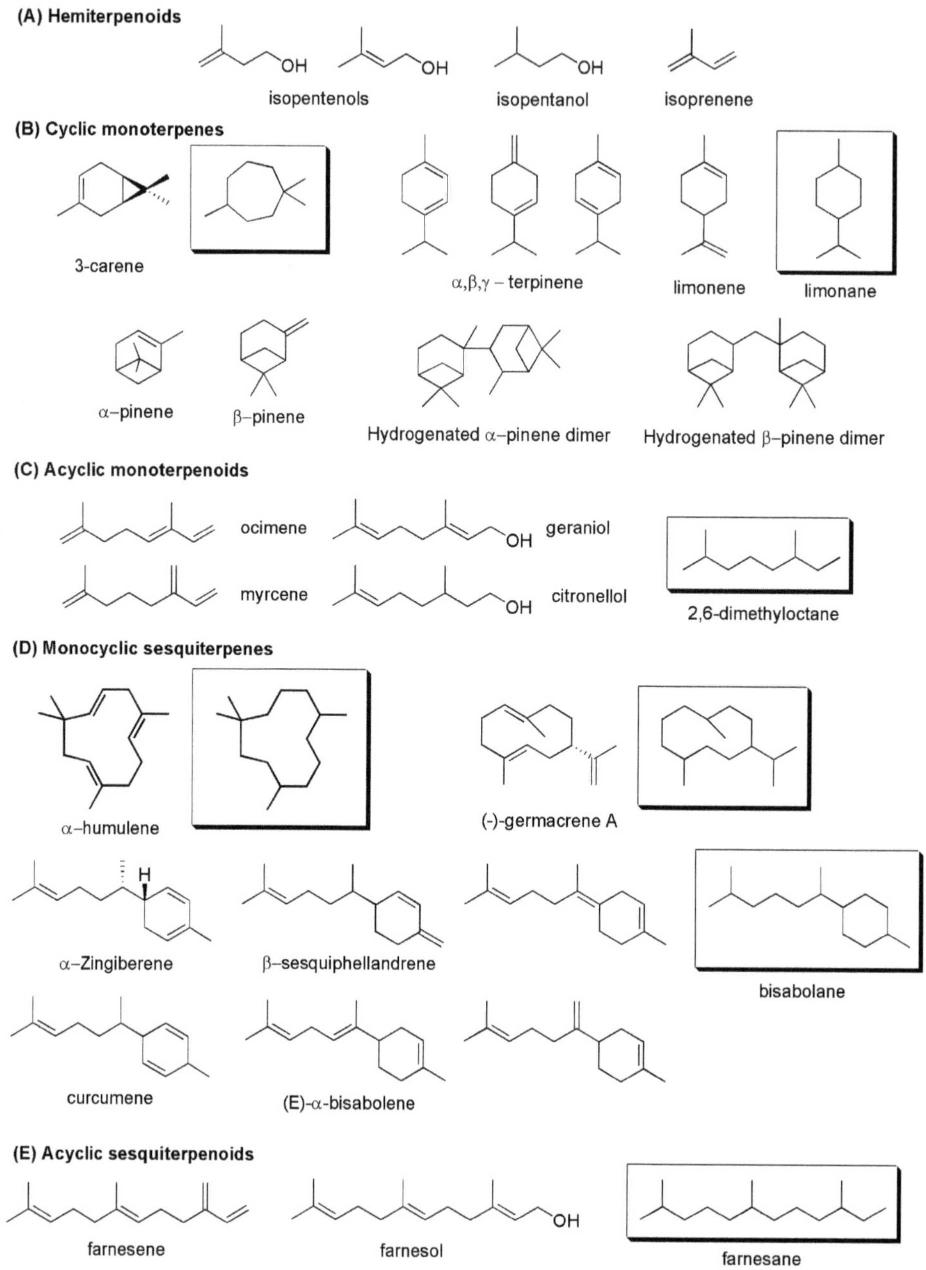

Figure 6.1 Example of various isoprenoid compounds. (A) Hemiterpenoids (C_5). (B) Cyclic monoterpenes (C_{10}) and hydrogenated pinene dimers. (C) Acyclic monoterpenoids (C_{10}). (D) Monocyclic sesquiterpenes (C_{15}). (E) Acyclic sesquiterpenoids (C_{15}).

and their heterologous expression has provided more knowledge for these enzymes and their catalytic functions. In this section, the classes and structures of isoprenoids will be discussed and the biosynthetic pathway and its engineering will be discussed in the section.

6.1.1 Hemiterpenoids

Hemiterpenoids have a single isoprene unit and are made from IPP or DMAPP. Isoprenols, or amyl alcohols, have long been known as flavour compounds produced during fermentation in food industry. 3-Methyl-2-butenol, for example, is readily converted to citral, an intermediate in the synthesis of pharmaceuticals, vitamins, carotenoids, flavours and fragrances. Recently isoprenols have been tested for fuel properties as a gasoline replacement in various engines and showed better combustion efficiencies than ethanol which is widely used as a gasoline substitute (Yang *et al.*, 2010).

Isoprene (2-methyl-1,3-butadiene) is the simplest hemiterpene compound and produced mostly from plants. It is a volatile compound at room temperature with a boiling temperature of 34°C, and recently microbial production of isoprene has been also reported. (Yang *et al.*, 2012; Zhao *et al.*, 2011) Isoprene is a monomer of synthetic rubber, and it also has been suggested as a precursor for diesel and jet fuel through chemical transformation (McAuliffe *et al.*, 2013).

There is an alternative pathway to produce hemiterpenoids. For example, isopentanol (3-methyl-butanol) has been produced in yeast with the Ehrlich pathway from amino acid leucine. The amino acid first is transaminated by aminotransferase to form a 2-ketoacid, and decarboxylated by 2-ketoacid decarboxylase to form aldehyde which is reduced to alcohol by alcohol dehydrogenase. Recently, metabolically engineered *Escherichia coli* with the Ehrlich pathway from *Saccharomyces cerevisiae* and the leucine biosynthetic pathway from *Bacillus subtilis* was used for the production of isopentanol (Cann and Liao, 2010; Connor *et al.*, 2010).

6.1.2 Monoterpenes

Monoterpenes and monoterpenoids have widespread economic value with uses ranging from flavours (e.g. limonene) and fragrances (e.g. geraniol, citranellol) to anticancer activity (e.g. perillyl alcohol). They also have diverse functions in chemical defence system (Zulak and Bohlmann, 2010). Monoterpenes are generally synthesized in different glandular structures of plants. Most monoterpenes were identified from plant extracts and for the industrial applications plant cell cultures are valuable tools for effecting biotransformations of monoterpenes. For instance, in vitro cultures of *Achillea millefolium* spp. and *Eucalyptus* spp. were used for the production of essential oils (Figueiredo *et al.*, 1996). Five-year-old transformed culture of *Mentha citrate* afforded oil closely related to the parental plant. However, recent development of microbial biotechnology has showed the potential of monoterpene as feedstocks for the future bio-catalytic production of oils, flavours and pharmaceuticals using many new identified monoterpene structures and synthases in recent years (Peralta-Yahya *et al.*, 2012; Ro *et al.*, 2006).

Diversity of terpene originated not only from the number of terpene synthase but also the ability of catalytic mechanisms to produce multiple chemicals from a single precursor. Over many years, a common carbocationic reaction mechanism has been established for all monoterpene synthases initiated by the divalent metal ion-dependent ionization of the substrate. The cationic intermediate follows a series of cyclization, hydride shift or other rearrangement until proton loss or the addition of a nucleophile for the termination. Characterized monoterpene synthases mainly convert geranyl diphosphate (GPP) as the substrate to either cyclic or acyclic compounds. The formation of cyclic monoterpene requires the preliminary isomerization of the geranyl cation to a linalyl intermediate capable of cyclization (Fig. 6.2) (Degenhardt *et al.*, 2009). The resulting α-terpinyl cation intermediate is ready to the secondary cyclization to diversify the formation of monoterpene. For the acyclic compounds, the elimination of the diphosphate group from GPP leads to either the geranyl cation or the linalyl cation, and they directly forms monoterpene compounds such as myrcene and (E)-β-ocimene or hydrolysis product such as geraniol or linalool. For cyclic compounds, monoterpene synthases or terpene cylases are needed. For example, limonene synthase catalyses the isomerization-cyclization of GPP to limonene, which has commercial value as flavour and a possible anti-carcinogen. It can be further converted to transportation fuels that will be discussed in the following section. Here, we introduce several cyclic monoterpenes and the mechanism of the cyclization in Fig. 6.2.

Lemon tree (*Citrus limon*) provides many important fruit and essential oils that are widely used as flavouring agent in bakery, as fragrance in perfumery and for pharmaceutical application as well. The essential oil is produced

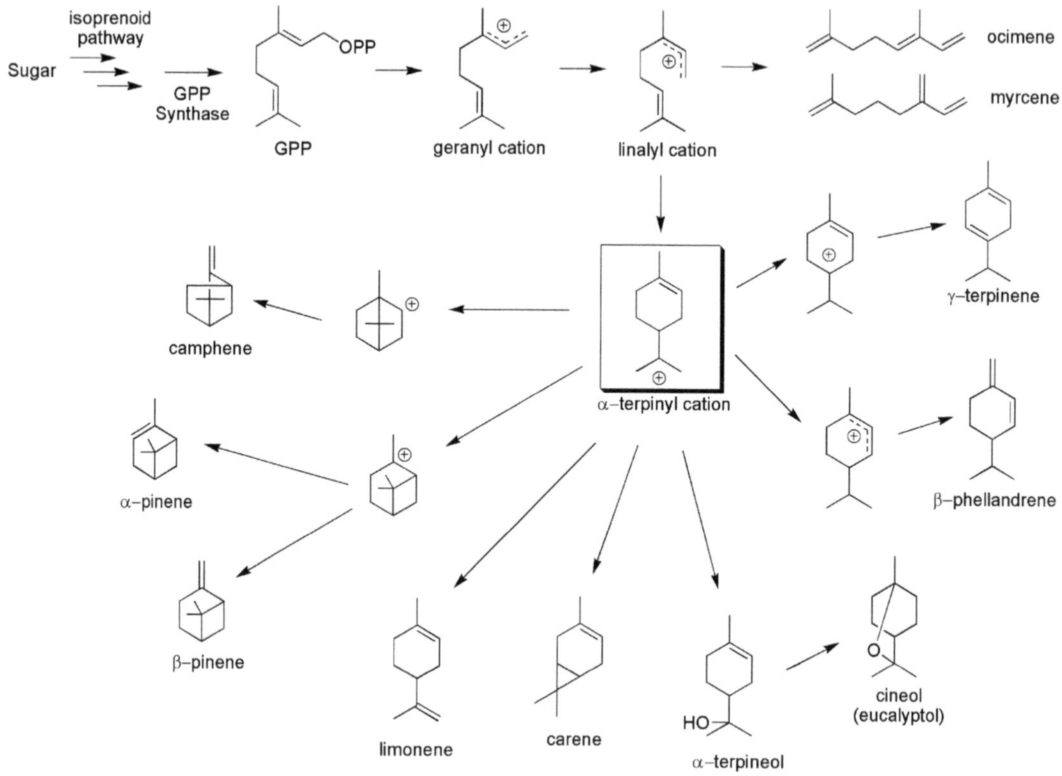

Figure 6.2 The reaction mechanism of monoterpene synthase with the ionization of the geranyl diphosphate (GPP) substrate.

from the peel of flavedo of the fruit. From the extraction of compressed lemon peel oil, 61% of the total monoterepene content consists of 4R- (+)-limonene (96% enantiomeric excess (e.e.)) and 17% of (−)- (1S,5S)-β-pinene (88% e.e.) (Lücker *et al.*, 2002). Recently, cDNAs encoding (−)-limonene synthase were isolated from *Mentha* species, *Abies grandis* and *Perilla frutescens*, and the enzyme was sequenced and expressed in *E. coli* for the further characterization (Colby *et al.*, 1993). (−)-Limonene can be further converted to (−)-menthol, which is the most famous monoterpene for its pleasant flavour and aroma and its anaesthetic effect, via a series of tailoring enzyme reactions (Croteau *et al.*, 2005). Menthol is mainly produced from the steam-distilled essential oil of corn-mint, and used for many confectionary goods, health care products, pharmaceutical products, and other daily life products. Recent studies of molecular genetics have been identified some of the terpene biosynthesis for menthol (Croteau *et al.*, 2005; Davis *et al.*, 2005; Ringer *et al.*, 2005).

Another interesting monoterpene is pinene. Pinene is a bicyclic monoterpene and one of the most volatile organic compounds in plant. It is also an important constituent of pine resins of many conifers. Volatile monoterpene compounds play ecological roles in host selection and fungal pathogens. For example, *Artemisia annua* is an annual herb widely distributed in Asia, Europe and North America and has rich in terpenoids including monoterpenoids such as camphor, 1,8-cineole, and pinene, and the sesquiterpenoid lactone, artemisinin. Some plants have no special storage organelles for these monoterpenes and have developed the controllable transcriptional and translational mechanisms for the light switching environment (Loreto *et al.*, 1996). Most pine (*Pinus*) species produce turpentine containing a mixture of α-pinene and β-pinene. Each pinene synthase has been characterized in loblolly pine and functionalized in *Escherichia coli* (Hyatt and Croteau, 2005; Phillips *et al.*, 1999, 2003).

6.1.3 Sesquiterpenes

Sesquiterpenes are derived from farnesyl diphosphate (FPP) which contains three isoprene units. FPP is either cyclized or dephosphorylated to form different sesquiterpenes and sesquiterpenoids as shown in Fig. 6.3 (Degenhardt *et al.*, 2009). While monoterpene synthases are localized in the plastids, sesquiterpene synthase are observed in the cytosol. The formation of sesquiterpenes from FPP also undergoes a carbocation-based reaction mechanism to produce the intermediate, *trans-* or *cis*-farnesyl cations. After the initial cyclization, deprotonation to a neutral intermediate, hydride shifts, methyl shifts and Wagner–Meerwein rearrangements occur (Fig. 6.3). FPP with the presence of three double bonds increases structural diversity of sesquiterpenes. Depending the structure of sesquiterpenes, it can be divided into different subgroup such as acyclic-, monocyclic-, bicyclic-, tricyclic-sesquiterpenes, and so on. Some examples are listed in detail here.

Farnesene is one of the simplest acyclic sesquiterpenes and was identified from apple peels in the plant defence mechanisms. To understand the formation of farnesene is important because the subsequent tail-to-tail dimerization of FPP leads to the key triterpenes formation. Most knowledge on biosynthesis of farnesene is derived from the biosynthesis of cholesterol with clear implication that the farnesyl units are used in the biosynthesis. α- and β-farnesene were identified in the plant volatiles as a defence against insect herbivores (Pare and Tumlinson, 1999). Recently, farnesene has been produced from genetically engineered yeast for the use as diesel and jet fuel precursor (Renninger, 2010).

Terpenoids from ginger have been reported to have important human health roles, by arresting the cell cycle, inducing cellular apoptosis in cancer cell and aiding patients with chylothorax.

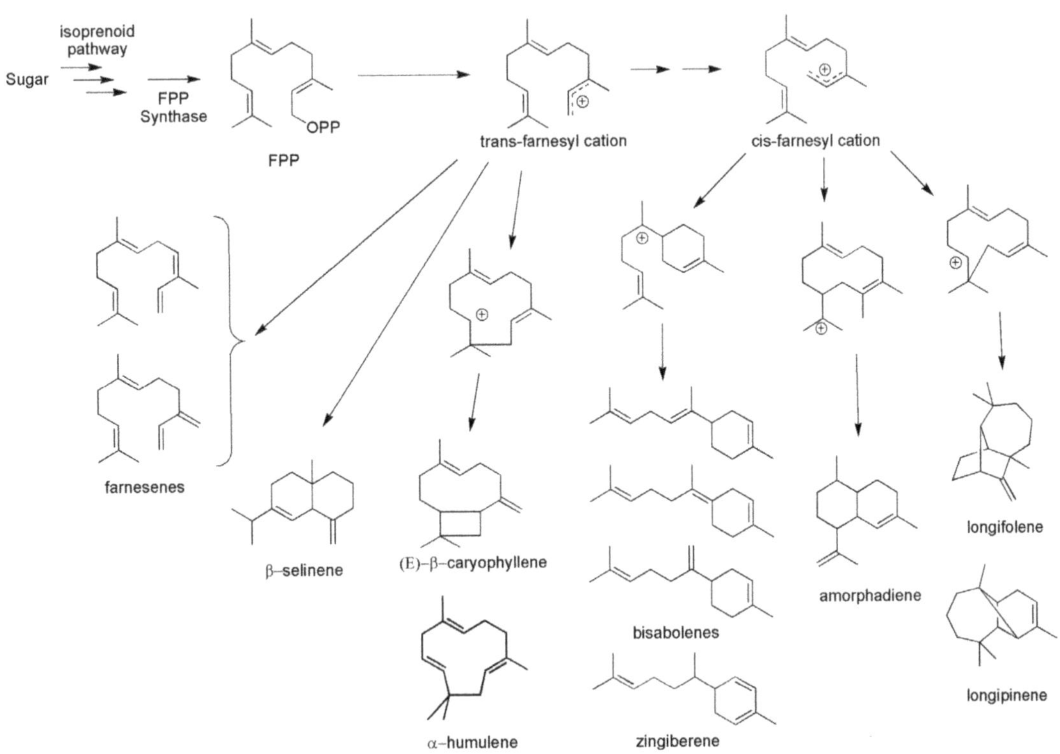

Figure 6.3 The reaction mechanisms of sesquiterpene synthase with the ionization of the farnesyl diphosphate (FPP) substrate.

Major constituent of the ginger oil is zingiberene, a monocyclic sesquiterpene, formed by (−)-caryolan-1-ol synthase and α-zingiberen e/β-sesquiphellandrene synthase in ginger (Koo and Gang, 2012). Similar monocyclic sesquiterpene products produced from various plants are bisabolenes that upon hydrogenation yields bisabolane, a recently reported diesel substitute (Peralta-Yahya *et al.*, 2011).

Artemisinin is a sesquiterpenoid lactone synthesized from amorpha-4,11-diene, a bicyclic sesquiterpene and has been extracted from sweet wormwood (*Artemisia annua*), a shrub known to Chinese herbalist. The World Health Organization has endorsed the use of artemisinin-based combination therapies to treat drug-resistant strains of the malaria-causing agent. However, the artemisinin yields from the natural producer are fairly low as 0.01% to 2% of dry cell weight biomass. Since the total biosynthesis of artemisinin is not commercialized yet, recent biotechnology has been progressed to semi-synthesize the artemisinin using metabolically engineered microbial host, which will be discussed later in detail (Paddon *et al.*, 2013).

6.1.4 Diterpenes and triterpenes

Geranylgeranly diphosphate (GGPP) is the precursor of the diterpene (C_{20}) and is usually produced by a cyclase, GGPP synthase. They are originated from fungal or plant resin, gummy exudates and are extracted in high boiling fractions after distillation of essential oils. Diterpenes include physiologically active compounds such as vitamin A (retinol), phytohormones for plant cell growth and germination, fungal hormone for the sexual reproduction, disease resistance agents (phytoalexin), casbene, and taxadiene, a precursor of taxol. Casbene is a macrocyclic diterpene hydrocarbon that is produced in young castor bean seeding after the explosion to fungi (Dudley *et al.*, 1986). Taxadiene is a precursor of taxol, which is an anticancer drug isolated from Pacific yew tree. Because of the low amount of taxol in the extracts of the dry biomass of Pacific yew (0.01–0.04%), yew cell suspension cultures have been developed the main source of taxol for pharmaceutical company, Bristol-Myers Squibb and Phyton Biotech. Inc.. Taxol is one of the structurally more complex representatives the approximately 400 defined taxoids of *Taxus* species, all of which are based on the unique taxane skeleton. Recently the precursor of taxol (taxadiene) was produced in engineered *E. coli* with a relatively high titre by expressing a taxadiene synthase from *Taxus brevifolia* (Ajikumar *et al.*, 2010).

Triterpene (C_{30}) consists of six isoprene units and is formed from the condensation of two C_{15} precursors. Squalene, a precursor of the whole family of steroids, is a natural organic compound obtained from shark liver oil and plant oil and it is widely used for a diet ingredient that may be a chemopreventive substance. Squalene is synthesized from an initial condensation of two molecules of farnesyl diphosphate (FPP) to form presqualene disphosphate (PSPP), and a reductive rearrangement of PSPP takes place to finally complete squalene biosynthesis. Famous carotenoids such as lycopene and β-carotenes, which are organic pigments found in the chloroplast, are categorized as the tetraterpenes (C_{40}) consisting of eight isoprene units.

These large isoprenoids can serve as feedstocks for oil refineries. Their high molecular weight (and higher number of carbon) precludes them from being used directly as fuel such as diesel or gasoline. These isoprenoids can be catalytically cracked using existing refinery operations to make a variety of shorter chain length or aromatic fuels (Tracy *et al.*, 2010, 2011).

6.2 Metabolic pathway and host engineering to optimize isoprenoid-precursors biosynthetic pathways

Isoprenoids require an intermediate isopentenyl diphosphate (IPP) and its isomer dimethylallyl diphosphate (DMAPP) as the universal building blocks. There are two biosynthetic pathways to produce IPP and DMAPP via either mevalonic acid (MVA) or 2-methyl-d-erythrito-4-phosphate (MEP) as an intermediate (Fig. 6.4). The mevalonate pathway was discovered in fungi, plant cytoplasm, other eukaryotes and mammalian cells where IPP and DMAPP are formed from the condensation of three molecules of acetyl-CoA to mevalonate. For decades, the mevalonate pathway,

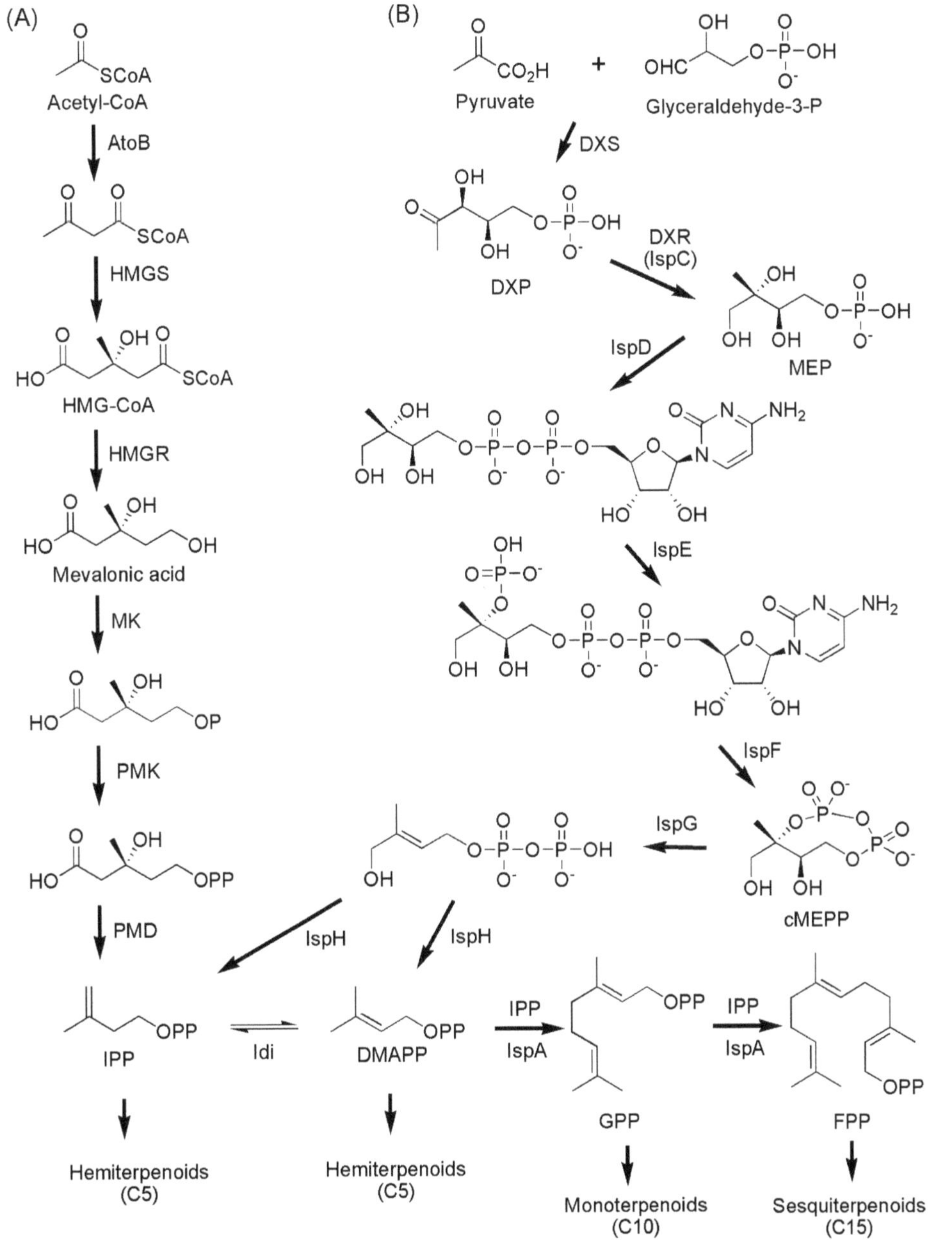

Figure 6.4 Two major pathways for isoprenoids biosynthesis: (A) Mevalonate pathway and (B) MEP pathway (or DXP pathway). Both pathways produce IPP and DMAPP as C5 building blocks for the isoprenoids biosynthesis. In mevalonate pathway, three molecules of acetyl-CoA condensed to form HMG-CoA and additional four steps are involved to produce IPP which is isomerized to DMAPP by the isomerase (IDI). The MEP pathway initiated with the condensation of pyruvate and glyceraldehyde 3-phopshate by DXS enzyme and additional six steps lead DXP to IPP and DMAPP. The details of both pathways are in the text.

representing the series of enzymatic reactions discussed later in this section, has been established to understand the production of isoprenoids. On the other hand, often found in plastid or prokaryotes, the MEP pathway ('Rohmer' pathway, DXP pathway or non-mevalonate pathway) (Rohmer *et al.*, 1996) does not use mevalonate as an intermediate but IPP is formed from

2-methyl-D-erythrito-4-phosphate (MEP) via the condensation of pyruvate and glyceraldehyde 3-phosphate (G3P). In plants, these two isoprenoid biosynthesis pathways exist in parallel for primary and secondary metabolisms, which could be useful for plants to adapt their environments for more efficient carbon utilization (Wagner and Krab, 1995). However, in this section, we focus the microbial metabolism for the isoprenoid biosynthesis.

Mevalonate pathway initiates with the Claisen condensation and cleavage reaction catalysed by thiolase to extend carbon backbone by acetyl-CoA units to produce acetoacetyl-CoA. This thiolytic cleavage is most familiar as a key step of β-oxidation of fatty acids. Subsequently, another acetyl-CoA is condensed with acetoacetyl-CoA to produce 3-hydroxy-3-methylglutaryl-CoA (HMG-CoA), which is physiologically almost irreversible by showing slow catalysis of the cleavage of HMG-CoA. Then, mevalonic acid, a stable precursor of isoprenoid/sterol biosynthesis, is produced by the HMG-CoA reductase from HMG-CoA using mostly NADPH as a cofactor. This catalytic reaction occurs through the successive reduction steps to produce first mevaldyl-CoA and release CoASH to form mevaldehyde; the second reduction step undergoes to form mevalonate. Two kinases, mevalonate kinase (MK) and phosphomevalonate kinase (PMK), catalyse the phosphorylation of mevalonate to produce mevalonate 5-phosphate and mevalonate 5-diphosphate, respectively. Mevalonate kinase is subject to the feedback inhibition by geranyl diphosphate (GPP) and farnesyl diphosphate (FPP) in the isoprenoid biosynthetic pathway (Voynova *et al.*, 2004). For the final step of the mevalonate pathway, mevalonate diphosphate decarboxylase (PMD) catalyses the ATP-driven decarboxylative elimination reaction of mevalonate diphosphate to form IPP, which is an essential step of mevalonate pathway (Fig. 6.4A).

The first step in the MEP pathway is the formation of 1-deoxy-D-xylulose 5-phosphate (DXP) by the condensation of pyruvate and D-glyceraldehyde 3-phosphate, catalysed by DXP synthase. This enzymatic step is crucial for entering the MEP pathway and known as the rate-limiting step of the whole pathway. DXP synthase encoded by the *dxs* gene has a typical thiamine-binding motif with a divalent cation (Mg^{2+} or Mn^{2+}). Several studies identified the formation of 1-deoxy-D-xylulose from pyruvate and G3P (Kuzuyama *et al.*, 2002; Kuzuyama *et al.*, 1998; Yokota and Sasajima, 1986). The second step of MEP pathway is catalysed by DXP reductoisomerase (encoded by the *dxr* gene) to convert 1-deoxy-D-xylulose 5-phosphate to MEP. However, this enzyme may be involved in the biosynthesis of thiamine and pyridoxol (Arfin and Umbarger, 1969). MEP is converted to 4-(cytidine-5′-diphospho)–2-methyl-D-erythritol (CDP-ME), 2-phospho-CDP-ME (CDP-ME2P), 2-methyl-D-erythritol 2,4-cyclodiphosphate (cMEPP) via the series of enzymatic reactions of the transferase encoded by the *ygbP* gene, the kinase encoded by the *ychB* gene, and cMEPP synthase encoded by the *ygbB* gene. Two additional steps to convert cMEPP to IPP are required; however, the functional enzymes encoded by the putative *gcpE* and *lytB* gene were not characterized yet (Fig. 6.4B).

Early stage work in the metabolic engineering of isoprenoids production has been focused on possibility of carotenogenic gene expressions in non-carotenogenic hosts. Identification and quantification of carotenoids (C40) products is readily achievable by colour screening. This colour-based screening helps to decipher isoprenoids pathways in early studies. Chromatographic and spectroscopic analysis of the pigments synthesized in *E. coli* carrying the heterologous isoprenoid pathway genes showed the several isoprenoids production (Albrecht *et al.*, 1998; Cunningham *et al.*, 1993). For example, genes of the *Erwinia uredovora* carotenoid biosynthetic pathways were expressed in the heterologous host *E. coli* to characterize their functions and six open reading frames (ORFs) were found and designated the *crtE, crtX, crtY, crtI, crtB* and *crtZ* genes which are corresponding to those in *Rhodobacter capsulatus* (Misawa *et al.*, 1990). *E. coli* has endogenous FPP synthase gene (*ispA*) and is able to produce the FPP, the common precursor of these bacterial isoprenoids. Subsequently, several studies of re-directing the carbon flux for the biosynthesis of the isoprenoids compounds to the pathway for carotenoid production were performed by introducing the carotenogenic genes starting with the *crtE* gene (encoding for GGPP synthase; converting FPP to GGPP) and there were successful heterotrophic production of lycopene, β-carotene and zeaxanthin in *E. coli* or other bacteria (Linden *et al.*, 1991; Misawa *et al.*, 1991; Miura *et al.*, 1998; Yamano *et al.*, 1994). Also, the uncommon carotenoids, hydroxyl carotenoids were also produced in the non-carotenogenic bacterium and its growth conditions were optimized for the isolation and purification (Albrecht *et al.*, 1998).

The rate-limiting step of the non-mevalonate (or DXP or MEP) pathway was revealed by co-overexpression of the *E. coli dxs* with *Erwinia uredovora* genes encoding lycopene, or zeaxynthin (Matthews and Wurtzel, 2000). The co-expressed recombinant strain showed 10-fold increased carotenoid contents over the level of controls not expressing the *dxs* gene and 100-fold increase in DXS activity in *E. coli* cell extract. In addition, IPP isomerase were also expressed with GGPP synthase (Wang *et al.*, 1999) and the *ispA* gene encoding farnesyl diphosphate (FPP) synthase was combined to remove the bottleneck of isoprenoid biosynthesis. To expand the profile of carotenoid productions via rational pathway design, molecular breeding was applied to produce the library of novel carotenoids in *E. coli* (Schmidt-Dannert *et al.*, 2000).

The strain development for lycopene and carotenoid production has been adapted for the proof-of-concept platform in novel metabolic engineering tool development. For example, an innovative work for production of lycopene was achieved by reconfiguring genetic control and central metabolism to balance the precursor, glyceraldehyde-3-phosphate and pyruvate for the MEP pathway (Farmer and Liao, 2000). In this synthetic network, the artificial regulon was adapted to control the expression of the key enzymes upon the regulation of response regulator NRI by sensing the level of acetyl-phosphate as the indicator of glycolytic flux. As the result, dynamic control of gene expression was possible for lycopene biosynthesis with the metabolic state and produced lycopene with productivity of 0.16 mg/ml/h. Another interesting metabolic engineering for lycopene production has been applied using global transcription machinery engineering (gTME) which enables broad perturbations of the transcriptome (Alper and Stephanopoulos, 2007). Unknown regulatory targets through gTME were combined with single and multiple-gene target obtained from the computation of genome-scale metabolic flux analysis and applied to enhance lycopene production in *E. coli*. Choi and the colleagues also has developed the systematic approach based on genome-scale metabolic model (FSEOF; flux scanning based on enforced object flux) to identify the gene targets for either overexpression or knock-outs (single and multiple) and increased the production yields by 4-fold (12.32 mg/g dry cell weight) (Choi *et al.*, 2010). In addition, lycopene production was adopted to accelerate the genetic evolution in the lab scale using multiplex automated genome engineering (MAGE) (Wang *et al.*, 2009). The MAGE technology facilitates rapid and continuous generation of a diverse set of genetic changes to optimize the MEP pathway for the overproduction of lycopene as a model pathway, resulting in the modification of 24 genetic components at once from a degenerated pool of synthetic DNA in 3 days. Modular control of multiple pathway was tested in lycopene production with several synthetic sensory controllers and engineered polymerase (Temme *et al.*, 2012a).

Lycopene production in an engineered host was mainly employed to decipher the heterologous gene expressions and develop the novel metabolic engineering tools. On the other hand, amorpha-4,11-diene (sesquiterpene; C15), a precursor of anti-malarial drug artemisinin was selected for mass production of target chemicals in an engineered host to replace the current production process. The engineering of the MEP pathway has been successfully performed by overexpressing the rate-limiting enzymes in the pathway. However, the endogenous regulation of the MEP pathway in microbial hosts has limited the optimization of the metabolic fluxes toward the product. The introduction of the heterologous mevalonate pathway to *E. coli* host has massively increased the pool of universal precursor IPP of isoprenoid and the final sesquiterpene product without feedback regulation by endogenous controlling machinery (Martin *et al.*, 2003).

Even in the host with endogenous mevalonate

pathway, the cellular toxicity of intermediates was alleviated by overexpressing the corresponding enzymes to metabolize the toxic intermediate and produce the final product (Ro *et al.*, 2006). Furthermore a two-phase portioning bioreactor of metabolically engineered strain has developed to prevent the rapid loss of amorphadiene from the culture by adding an organic layer of dodecane to the medium (Newman *et al.*, 2006).

Controlling the gene expressions in a pathway and balancing intermediate metabolites are most important the metabolic engineering strategy to maximum production of the desired chemicals. Robust and responsive genetic controlling system for chosen pathway and host is required and several strategies for the control of gene expression have been developed. As heterologous pathways including isoprenoid-derived chemicals are required for novel target compounds, the pathways become larger and more sophisticated to be controlled and optimized. Tunable intergenic regions (TIGRs) were generated and applied to the heterologous mevalonate pathway in *E. coli* to optimize multiple gene expressions for the amorphadiene production. With TIGRs library, a dynamic gene control was possible and it facilitated the strain development (Pfleger *et al.*, 2006). Modular engineering of mevalonate pathway has been performed to identify the bottleneck reactions, based on the reports that the 3-hydroxy-3-methylglutaryl-CoA reductase showed the low activity (Pitera *et al.*, 2007) and that the plasmid copy number and promoter strengths influenced the production titre of amorphadiene (Anthony *et al.*, 2009). Targeted proteomics also has been applied to the optimization of the mevalonate pathway by changing the gene expression and measuring the level of targeted enzymes. High expressions of mevalonate kinase (MK) and mevalonate diphosphate kinase (PMK) have led the high titre of amorphadiene (Redding-Johanson *et al.*, 2011).

A lot of isoprenoid compounds, especially monoterpenes, are considered toxic to microorganisms, and this makes an undesired drawback when engineering metabolic pathways for biofuel production in microbes. Efflux pumps that export the toxic products to extracellular space can provide a direct rescue of the producing host from toxic effect of the product they are making. Recently, several efflux pumps that selectively transport toxic monoterpene products have been reported (Dunlop *et al.*, 2011). Interestingly, the result showed that the engineered *E. coli* host with overexpression of a specific efflux pump produces the target monoterpene biofuel with a higher titre, suggesting the relief from the toxicity of the product affects the overall production efficiency.

6.3 Conversions of isoprenoid precursors to fuel compounds

The chemical structure of isoprenoids implies some beneficial aspects of isoprenoids as fuel compounds. For example, the methyl branching is a common structural feature of isoprenoids and it lowers the freezing temperature of these compounds significantly, which makes these fuels great cold weather additives or alternatives. Also the cyclic structure is often observed in isoprenoids and it is generally considered as a good feature as a jet fuel. Compared to the linear hydrocarbon with same carbon number, cyclic hydrocarbons usually show higher density, which results in a higher volumetric energy density of these cyclic compounds. The octane number and the cetane number are important fuel properties for fuels of spark-ignition (gasoline) and compression-ignition (diesel) engines, respectively. The octane and cetane numbers of hydrocarbons are dependent on several structural features, including the number of carbons, degree of unsaturation, branching, cyclization, and aromaticity. Compared to the linear hydrocarbons, branching methyl groups of isoprenoids usually improve the octane number of the fuel, and olefins and aromatics generally show beneficial effects on the octane number when the fuel is used as a gasoline alternative. However, the presence of branches and unsaturated bonds tend to decrease the cetane number when isoprenoids are used as a diesel alternative.

There are several important things to consider when proposing an isoprenoid compound as a biofuel and designing the pathway for this molecule. The first thing to consider is what types of fuel we want to produce. This can be usually determined by the carbon chain length of the isoprenoid compound to biosynthesize. In the isoprenoids pathway, regardless of what pathway is used between the MEP pathway and the mevalonate pathway, the number of carbons in the fuel molecule is determined by adequate choice of the pathway genes, especially what isoprenyl diphosphate synthase and the terpene synthase are used in the pathway. For example, C_5 is more adequate as a gasoline fuel, and they can be prepared from IPP or DMAPP directly without condensation by the GPP synthase or the FPP synthase. So the pathway should be designed to minimize or knock-out the expression of these enzymes and to maximize the downstream enzyme such as pyrophosphatase for alcohol production. For C_{10} isoprenoids, which can be used as gasoline or jet fuel, the condensation of IPP and DMAPP is limited to GPP formation catalysed by the GPP synthase without further condensation by the FPP synthase. So the expression of the FPP synthase should be minimized or knocked-out while the target monoterpene synthase should be overexpressed to maximize the monoterpene production. For C_{15} isoprenoids, which are in a diesel or jet fuel range, the FPP synthase should be overexpressed as well as the adequate sesquiterpene synthase that produces the C_{15} compound of interest.

The vast range of hydrocarbon architecture accessible from the same pathway is a particular advantage, as incorporation of different terpene synthases into the same isoprenoid-producing host strain can yield completely different products. Bisabolene and farnesene, for example, are produced from the same precursor FPP, which is also used for antimalarial drug precursor amorphadiene biosynthesis (Martin *et al.*, 2003), using different terpene synthases. Farnesene synthases and bisabolene synthases are known from various plants. A bioprospecting of potential terpene synthase candidates and the precursor pathway engineering, which has been performed for antimalarial drug production, are combined and resulted in the production of alternative diesel transportation fuel which has similar cetane number of current diesel (40–60) (Peralta-Yahya *et al.*, 2011).

One of the structural characteristics of isoprenoids is the unsaturated carbon–carbon double bonds shown in most isoprenoids structure. However, these carbon–carbon double bonds are a source of instability or reactivity which makes the long-term storage of these compounds unfavourable and also can lead to engine deposits when used as fuels. Furthermore, double bonds are another source of lowering the cetane number in diesel (Murphy *et al.*, 2004) while slightly improving the octane number in gasoline. Therefore, it is important to reduce or hydrogenate the double bonds when isoprenoids are used as fuel precursors. Chemical hydrogenation of isoprenoids using gaseous hydrogen and heterogeneous catalysts has been reported for biofuels production (Harvey *et al.*, 2010; Peralta-Yahya *et al.*, 2011; Tracy *et al.*, 2009). Biological double bond reduction, however, is not a trivial task. Most known biological hydrogenation of double bonds requires a carbonyl group adjacent to the double bond to reduce for the stabilization of the carbanion formed by hydride addition to the double bond. The reduction of isolated (i.e. not conjugated or not stabilized by the adjacent electron-withdrawing group) double bonds is a quite rare reaction in nature, and the geranylgeranyl reductases (GGRs) are almost the only known enzymes that reduce the isolated double bonds (Keller *et al.*, 1998; Murakami *et al.*, 2007; Nishimura and Eguchi, 2006). (Fig. 6.5A) These enzymes are mostly found in archaea and responsible for archaeal membrane biosynthesis. Recently the X-ray crystal structure of these enzymes were reported and provided the insight on mechanism and substrate specificity (Sasaki *et al.*, 2011; Xu *et al.*, 2010), but it is still very challenging topic that needs more work to be done.

Various chemical modifications of isoprenoids for their conversion to adequate fuels have been reported (Fig. 6.5B–D). For example, monoterpenes pinenes are hydrogenated and dimerized chemically using heterogeneous catalysts and these fuels showed a density and net heat of combustion comparable to a jet fuel JP-10 (Harvey *et al.*, 2010). Also hydrogenated form of limonene was reported as blending agents in diesel fuels without significant detrimental effects on the properties of the fuels (Tracy *et al.*, 2009). A ruthenium-catalysed chemical conversion of a linear monoterpene alcohol linalool to dimethyldicyclopentadiene, a precursor of high-density

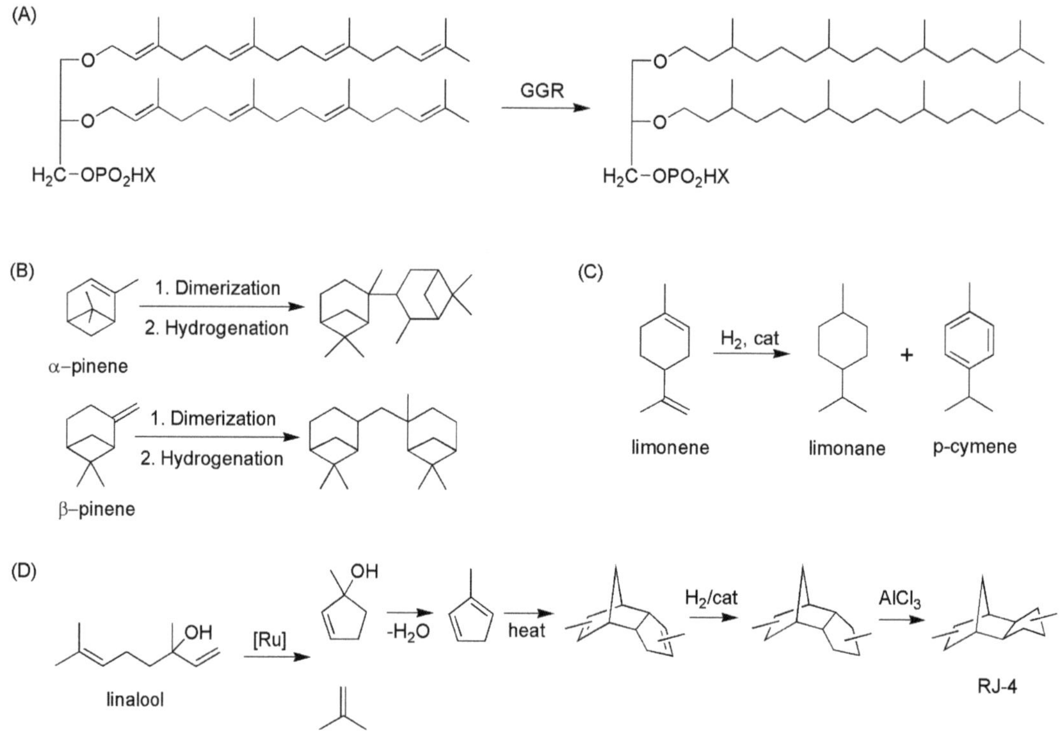

Figure 6.5 Biosynthetic and synthetic conversion of isoprenoids precursors to biofuels. (A) Double bond reduction by geranylgeranyl reductase (GGR). (B) Hydrogenated pinene dimer production. (C) Limonene hydrogenation. (D) Jet fuel production from monoterpenoid linalool.

jet fuel RJ-4, was reported too (Meylemans *et al.*, 2011). These represent a potential usefulness of monoterpenoids as renewable jet and diesel fuel additives.

6.4 Future trends in isoprene-derived biofuels

Isoprenoids biosynthesis is one of the major pathways to generate hydrocarbons in nature. Isoprenoid-derived biofuels produced from microbial system is an alternative and also promising choice of biofuels due to their superb fuel properties adequate for all three major types of fuels (gasoline, diesel, and jet fuel). So far, however, microbial strains that can convert sugars to isoprenoids at yields similar to the most popular biofuel, ethanol, have not been reported. The theoretical maximum yield of fermentative ethanol production is over 50% (i.e. over 0.5 g of ethanol is produced from 1 g of glucose), but the theoretical yields of isoprenoids biosynthesis are only about 29% when the MEP pathway is used and even lower when the mevalonate pathway is used to 25% (Rude and Schirmer, 2009). This suggests that it is extremely important to improve the yield, titre, and productivity of isoprene-derived fuels production to make this class of biofuels be more economically attractive.

In many cases, isoprenoids themselves are fuel precursors and they need to be hydrogenated chemically or biologically to produce the actual fuel compounds. Chemical hydrogenation is relatively easy reaction and widely used in oil refinery, but it can add significant cost to its manufacturing. Biological hydrogenation is quite challenging toward isoprenoid substrates, but it is an important research topic which can improve the overall yield of isoprene-based fuels. Along with biological hydrogenation, the production of isoprenoids under anaerobic condition is another highly desired future direction to make isoprene-derived biofuel more attractive, since the requirement for oxygen for isoprenoid production is an obvious disadvantage when considering the process cost for large scale aerobic fermentation (Fischer *et al.*, 2008).

The cost of the carbon source is a very important determinant in bio-based chemical or fuel production that affects the economic feasibility of the process. One of the cheapest and most attractive carbon sources currently used in the bio-based chemical and fuel industry is sucrose. There are some industrial host strains that utilize sucrose as a carbon source, but about 50% of native *E. coli* hosts cannot utilize it. The engineering for more efficient sucrose-utilization in *E. coli* is an important research topic for industrial application. There have been several approaches to achieve sucrose utilization in *E. coli*. One of the simplest approaches includes heterologous expression of β-fructofuranosidase, which hydrolyses sucrose to glucose and fructose in the extracellular space (Lee *et al.*, 2010). Other approaches include the overexpression of sucrose utilization genes from native sucrose-utilizing strains, which include sucrose kinases, transtransporters and hydrolase (scrK, Y, A, B, R) as previously reported (Schmid *et al.*, 1988), and *csc* genes from *E. coli* W strains (Bruschi *et al.*, 2011).

The microbial conversion of lignocellulosic biomass into biofuels could provide truly renewable isoprene-based fuels. Unfortunately, there are several challenges to overcome before lignocellulose can be considered as an economically competitive feedstock for biofuel production. The enzymes that convert lignocellulosic biomass into fermentable sugars are not economically competitive yet when compared to the process using sucrose or starch-based glucose. Improving the catalytic activity of these enzymes on pretreated biomass and also increasing the production scale of these enzymes are important tasks to achieve economically favourable lignocellulosic biofuel production. One of the recent approaches to improve the economics of the isoprene-based biofuels production is to engineer plants that can produce isoprene-based biofuel directly from plants. Direct production of biofuels in plants is an economically attractive route, and isoprene-based fuels are one of the best targets in this approach (Borak *et al.*, 2013). Isoprenoid pathways are native in most plants, especially in pine and peppermint trees, and by genetically engineering the plant to introduce or augment isoprene-based fuel pathway, we can generate and recover fuels directly from plant biomass during the deconstruction process of lignocellulosic biomass to generate fermentable sugars (Lange *et al.*, 2011).

The availability of various engineering tools to improve titre and yield in the producing host is a huge advantage of biofuel production using microbial system. Development of microbial systems has implemented 'static control' and 'dynamic control' for the effective gene expression. 'Static controls' of metabolic pathway are the genetically encoded components to statically apply the pathways and use the natural or slightly modified promoters or constitutive promoters with various expression strengths. A technique for measuring the relative promoter strength has been widely developed with fluorescent proteins or luciferase or β-galactosidase as reporters, and recently, a standard reference plasmid library has been developed with a range of protein expression strengths achieved with the combination of promoter strength and plasmid copy number (Lee *et al.*, 2011). The ratio of two expression level was also used to define the strength of any promoters (Ajikumar *et al.*, 2010). Those defined promoters allowed to control the target pathways effective and straightforward in terms of controlling transcription as nature show the polycistronic gene expression. Besides of static controlling transcription levels, the control of translation has also shown the effect in the level of gene expression. Variety of library for static controllers of translational initiation has been constructed (Liu *et al.*, 2011; Qi *et al.*, 2012).

A global controlling method is necessary to make sure efficient use of limited cellular resources. Natural microbial systems also solve these issues by optimizing genetic components under 'dynamic controls' that allows allosteric regulation of metabolic pathway. For example, the lycopene production in *E. coli* harbouring heterologous pathway genes was enhanced using dynamic controllers that sense excess glucose and regulating the key enzyme levels with a transcriptional factor (Temme *et al.*, 2012a,b). Also, a synthetic quorum sensing (QS) system based on the Lux system linking genetic toggle switches was applied gene expression system with bimodal distribution under low intermediates or full expressions under high densities of intermediates (Kobayashi *et al.*, 2004). Quorum-sensing (QS) mediated *E. coli* enables to induce neighbouring cells as well as itself to initiate the carbon flux to the designated metabolic pathway without using expensive exogenous inducers such as isopropyl-D-1-thiogalactopyranoside (IPTG), and showed the similar capability of genetic controls compared to the IPTG-induced system. Several genetic modules were constructed to link QS-system and biofuel pathway. Synthetic biology has a capability of controlling metabolic reactions in metabolic pathways and recent Omics technologies track the level of RNAs, proteins, and metabolites in the cell to identify the bottleneck of metabolism. In addition, the detail synthetic and computational models assist the design of enzymes and metabolic pathways.

Furthermore, the power of next generation genome sequencing have revealed new metabolic reactions and enzymes from many different microorganisms and generated extensive databases of gene expression, metabolic reactions, and enzyme structures which would allow the design of new reactions and new metabolic enzymes based on these data (Ramos *et al.*, 2009). Cutting-edge technology of genome sequencing and smart microbial cell based on synthetic biology will identify the new metabolic engineering targets and eventually it will contribute the development of an efficient microbial cell factory for the production of isoprene-derived biofuels.

References

Ajikumar, P.K., Xiao, W.-H., Tyo, K.E.J., Wang, Y., Simeon, F., Leonard, E., Mucha, O., Phon, T.H., Pfeifer, B., and Stephanopoulos, G. (2010). Isoprenoid pathway optimization for Taxol precursor overproduction in *Escherichia coli*. Science (New York, NY) 330, 70–74.

Albrecht, M., Takaichi, S., Misawa, N., Schnurr, G., Böger, P., and Sandmann, G. (1998). Synthesis of atypical cyclic and acyclic hydroxy carotenoids in *Escherichia coli* transformants. J. Biotechnol, *58*, 177–185.

Alper, and Stephanopoulos, G.N. (2007). Global transcription machinery engineering: A new approach for improving cellular phenotype. Metab. Eng. 9, 258–267.

Anthony, J.R., Anthony, L.C., Nowroozi, F., Kwon, G., Newman, J.D., and Keasling, J.D. (2009). Optimization of the mevalonate-based isoprenoid biosynthetic pathway in *Escherichia coli* for production of the anti-malarial drug precursor amorpha-4,11-diene. Meta. Eng. *11*, 13–19.

Arfin, S.M., and Umbarger, H.E. (1969). Purification and properties of the acetohydroxy acid isomeroreductase of Salmonella typhimurium. The Journal of biological chemistry 244, 1118–1127.

Borak, B., Ort, D.R., and Burbaum, J.J. (2013). Energy and carbon accounting to compare bioenergy crops. Curr. Opin. Biotechnol. *24*, 369–375.

Bruschi, M., Boyes, S.J., Sugiarto, H., Nielsen, L.K., and Vickers, C.E. (2011). A transferable sucrose utilization approach for non-sucrose-utilizing *Escherichia coli* strains. Biotechnol. Adv. *30*, 1001–1010.

Cann, A.F., and Liao, J.C. (2010). Pentanol isomer synthesis in engineered microorganisms. Appl Microbiol Biotechnol 85, 893–899.

Choi, H.S., Lee, S.Y., Kim, T.Y., and Woo, H.M. (2010). In silico identification of gene amplification targets for improvement of lycopene production. Appl. Environ. Microbiol. *76*, 3097–3105.

Colby, S.M., Alonso, W.R., Katahira, E.J., McGarvey, D.J., and Croteau, R. (1993). 4S-limonene synthase from the oil glands of spearmint (Mentha spicata). cDNA isolation, characterization, and bacterial expression of the catalytically active monoterpene cyclase. J. Biol. Chem. *268*, 23016–23024.

Connor, M.R., Cann, A.F., and Liao, J.C. (2010). 3-Methyl-1-butanol production in *Escherichia coli*: random mutagenesis and two-phase fermentation. Appl. Microbiol. Biotechnol. *86*, 1155–1164.

Croteau, R.B., Davis, E.M., Ringer, K.L., and Wildung, M.R. (2005). (-)-Menthol biosynthesis and molecular genetics. Die Naturwissenschaften 92, 562–577.

Cunningham, F.X., Chamovitz, D., Misawa, N., Gantt, E., and Hirschberg, J. (1993). Cloning and functional expression in *Escherichia coli* of a cyanobacterial gene for lycopene cyclase, the enzyme that catalyzes the biosynthesis of beta-carotene. FEBS Lett. *328*, 130–138.

Davis, E.M., Ringer, K.L., McConkey, M.E., and Croteau, R. (2005). Monoterpene metabolism. Cloning, expression, and characterization of menthone reductases from peppermint. Plant Physiol. *137*, 873–881.

Degenhardt, J., Kollner, T.G., and Gershenzon, J. (2009). Monoterpene and sesquiterpene synthases and the origin of terpene skeletal diversity in plants. Phytochemistry *70*, 1621–1637.

Dudley, M.W., Green, T.R., and West, C.A. (1986). Biosynthesis of the macrocyclic diterpene casbene in castor bean (ricinus-communis l) seedlings – the purification and properties of farnesyl transferase from elicited seedlings. Plant Physiol. *81*, 343–348.

Dunlop, M.J., Dossani, Z.Y., Szmidt, H.L., Chu, H.C., Lee, T.S., Keasling, J.D., Hadi, M.Z., and Mukhopadhyay, A. (2011). Engineering microbial biofuel tolerance and export using efflux pumps. Mol. Syst. Biol. 7, 487.

Farmer, W.R., and Liao, J.C. (2000). Improving lycopene production in *Escherichia coli* by engineering metabolic control. Nature biotechnology 18, 533–537.

Figueiredo, A.C., Almendra, M.J., Barroso, J.G., and Scheffer, J.J.C. (1996). Biotransformation of monoterpenes and sesquiterpenes by cell suspension cultures of *Achillea millefolium* L ssp. *millefolium*. Biotechnol. Lett. *18*, 863–868.

Fischer, C.R., Klein-Marcuschamer, D., and Stephanopoulos, G. (2008). Selection and optimization of microbial hosts for biofuels production. Metab. Eng. *10*, 295–304.

Harvey, B.G., Wright, M.E., and Quintana, R.L. (2010). High-density renewable fuels based on the selective dimerization of pinenes. Energ. Fuel *24*, 267–273.

Hyatt, D.C., and Croteau, R. (2005). Mutational analysis of a monoterpene synthase reaction: altered catalysis through directed mutagenesis of (-)-pinene synthase from Abies grandis. Archives of biochemistry and biophysics 439, 222–233.

Keller, Y., Bouvier, F., d'Harlingue, A., and Camara, B. (1998). Metabolic compartmentation of plastid prenyllipid biosynthesis—evidence for the involvement of a multifunctional geranylgeranyl reductase. Eur J Biochem 251, 413–417.

Kobayashi, H., Kaern, M., Araki, M., Chung, K., Gardner, T.S., Cantor, C.R., and Collins, J.J. (2004). Programmable cells: interfacing natural and engineered gene networks. Proc. Natl. Acad. Sci. U.S.A. *101*, 8414–8419.

Koo, H.J., and Gang, D.R. (2012). Suites of terpene synthases explain differential terpenoid production in ginger and turmeric tissues. PLoS ONE 7, e51481.

Kuzuyama, T., Takahashi, S., Watanabe, H., and Seto, H. (1998). Direct formation of 2-C-methyl-D-erythritol 4-phosphate from 1-deoxy-D-xylulose 5-phosphate by 1-deoxy-D-xylulose 5-phosphate reductoisomerase, a new enzyme in the non-mevalonate pathway to isopentenyl diphosphate. Tetrahedron Lett. i, 4509–4512.

Kuzuyama, T., Takahashi, S., Dairi, T., and Seto, H. (2002). Detection of the mevalonate pathway in streptomyces species using the 3-hydroxy-3-methylglutaryl coenzyme A reductase gene. J. Antibiot. *55*, 919–923.

Lange, B.M., Mahmoud, S.S., Wildung, M.R., Turner, G.W., Davis, E.M., Lange, I., Baker, R.C., Boydston, R.A., and Croteau, R.B. (2011). Improving peppermint essential oil yield and composition by metabolic engineering. Proc. Natl. Acad. Sci. U.S.A. *108*, 16944–16949.

Lee, J.W., Choi, S., Park, J.H., Vickers, C.E., Nielsen, L.K., and Lee, S.Y. (2010). Development of sucrose-utilizing *Escherichia coli* K-12 strain by cloning beta-fructofuranosidases and its application for L-threonine production. Appl.Microbiol. Biotechnol. *88*, 905–913.

Lee, T.S., Krupa, R.A., Zhang, F., Hajimorad, M., Holtz, W.J., Prasad, N., Lee, S.K., and Keasling, J.D. (2011). BglBrick vectors and datasheets: A synthetic biology platform for gene expression. Journal Biol. Eng. *5*, 12.

Linden, H., Misawa, N., Chamovitz, D., Pecker, I., Hirschberg, J., and Sandmann, G. (1991). Functional complementation in *Escherichia coli* of different phytoene desaturase genes and analysis of accumulated carotenes. Z. Naturforsch. C *46*, 1045–1051.

Liu, C.C., Qi, L., Yanofsky, C., and Arkin, A.P. (2011). Regulation of transcription by unnatural amino acids. Nature biotechnology 29, 164–168.

Loreto, F., Ciccioli, P., Cecinato, A., Brancaleoni, E., Frattoni, M., and Tricoli, D. (1996). Influence of Environmental Factors and Air Composition on the Emission of [alpha]-Pinene from Quercus ilex Leaves. Plant physiology 110, 267–275.

Lücker, J., El Tamer, M.K., Schwab, W., Verstappen, F.W.A., van der Plas, L.H.W., Bouwmeester, H.J., and Verhoeven, H.A. (2002). Monoterpene biosynthesis in lemon (Citrus limon). cDNA isolation and functional analysis of four monoterpene synthases. Eur. J. Biochem./FEBS *269*, 3160–3171.

McAuliffe, J.C., Paramonov, S.E., and Sanford, K.J. (2013). Fuel compositions comprising isoprene derivatives. Patent Application number: US 12/818, 090, Publication number: US8450549 B2.

Martin, V.J.J., Pitera, D.J., Withers, S.T., Newman, J.D., and Keasling, J.D. (2003). Engineering a mevalonate pathway in *Escherichia coli* for production of terpenoids. Nat. Biotechnol. *21*, 796–802.

Matthews, P.D., and Wurtzel, E.T. (2000). Metabolic engineering of carotenoid accumulation in *Escherichia coli* by modulation of the isoprenoid precursor pool with expression of deoxyxylulose phosphate synthase. Appl. Microbiol Biotechnol. *53*, 396–400.

Meylemans, H.A., Quintana, R.L., Goldsmith, B.R., and Harvey, B.G. (2011). Solvent-Free Conversion of Linalool to Methylcyclopentadiene Dimers: A Route To Renewable High-Density Fuels. ChemSusChem *4*, 465–469.

Misawa, N., Nakagawa, M., Kobayashi, K., Yamano, S., Izawa, Y., Nakamura, K., and Harashima, K. (1990). Elucidation of the Erwinia uredovora carotenoid biosynthetic pathway by functional analysis of gene products expressed in *Escherichia coli*. J. Bacteriol. *172*, 6704–6712.

Misawa, N., Yamano, S., and Ikenaga, H. (1991). Production of beta-carotene in Zymomonas mobilis and Agrobacterium tumefaciens by introduction of the biosynthesis genes from Erwinia uredovora. Appl. Environ. Microbiol. *57*, 1847–1849.

Miura, Y., Kondo, K., Saito, T., Shimada, H., Fraser, P.D., and Misawa, N. (1998). Production of the carotenoids lycopene, beta-carotene, and astaxanthin in the food yeast Candida utilis. Appl. Environ. Microbiol. 64, 1226–1229.

Murakami, M., Shibuya, K., Nakayama, T., Nishino, T., Yoshimura, T., and Hemmi, H. (2007). Geranylgeranyl reductase involved in the biosynthesis of archaeal membrane lipids in the hyperthermophilic archaeon Archaeoglobus fulgidus. FEBS J. *274*, 805–814.

Murphy, M.J., Taylor, J.D., and McCormick, R.L. (2004). Compendium of experimental cetane number data. Golden, Colorado, National Renewable Energy Laboratory.

Newman, J.D., Marshall, J., Chang, M., Nowroozi, F., Paradise, E., Pitera, D., Newman, K.L., and Keasling, J.D. (2006). High-level production of amorpha-4,11-diene in a two-phase partitioning bioreactor of metabolically engineered *Escherichia coli*. Biotechnol. Bioeng. 95, 684–691.

Nishimura, Y., and Eguchi, T. (2006). Biosynthesis of archaeal membrane lipids: digeranylgeranylglycerophospholipid reductase of the thermoacidophilic archaeon *Thermoplasma acidophilum*. J. Biochem. *139*, 1073–1081.

Paddon, C.J., Westfall, P.J., Pitera, D.J., Benjamin, K., Fisher, K., McPhee, D., Leavell, M.D., Tai, A., Main, A., Eng, D., *et al.* (2013). High-level semi-synthetic production of the potent antimalarial artemisinin. Nature. *496*, 528–532.

Pare, P.W., and Tumlinson, J.H. (1999). Plant volatiles as a defense against insect herbivores. Plant Physiol. *121*, 325–332.

Peralta-Yahya, P.P., Ouellet, M., Chan, R., Mukhopadhyay, A., Keasling, J.D., and Lee, T.S. (2011). Identification and microbial production of a terpene-based advanced biofuel. Nat. Commun. 2, 483.

Peralta-Yahya, P.P., Zhang, F., Del Cardayre, S.B., and Keasling, J.D. (2012). Microbial engineering for the production of advanced biofuels. Nature *488*, 320–328.

Pfleger, B.F., Pitera, D.J., Smolke, C.D., and Keasling, J.D. (2006). Combinatorial engineering of intergenic regions in operons tunes expression of multiple genes. Nature Biotechnol. *24*, 1027–1032.

Phillips, M.A., Savage, T.J., and Croteau, R. (1999). Monoterpene synthases of loblolly pine (Pinus taeda) produce pinene isomers and enantiomers. Archiv. Biochem. Biophys. *372*, 197–204.

Phillips, M.A., Wildung, M.R., Williams, D.C., Hyatt, D.C., and Croteau, R. (2003). cDNA isolation, functional expression, and characterization of (+)-alpha-pinene synthase and (−)-alpha-pinene synthase from loblolly pine (Pinus taeda): stereocontrol in pinene biosynthesis. Archiv. Biochem. Biophys. *411*, 267–276.

Pitera, D.J., Paddon, C.J., Newman, J.D., and Keasling, J.D. (2007). Balancing a heterologous mevalonate pathway for improved isoprenoid production in *Escherichia coli*. Metab. Eng. *9*, 193–207.

Qi, L., Lucks, J.B., Liu, C.C., Mutalik, V.K., and Arkin, A.P. (2012). Engineering naturally occurring trans-acting non-coding RNAs to sense molecular signals. Nucleic Acids Res. *40*, 5775–5786.

Ramos, A.M., Crooijmans, R.P., Affara, N.A., Amaral, A.J., Archibald, A.L., Beever, J.E., Bendixen, C., Churcher, C., Clark, R., Dehais, P., *et al.* (2009). Design of a high density SNP genotyping assay in the pig using SNPs identified and characterized by next generation sequencing technology. PLoS ONE *4*, e6524.

Redding-Johanson, A.M., Batth, T.S., Chan, R., Krupa, R., Szmidt, H.L., Adams, P.D., Keasling, J.D., Lee, T.S., Mukhopadhyay, A., and Petzold, C.J. (2011). Targeted proteomics for metabolic pathway optimization: application to terpene production. Metab. Eng. *13*, 194–203.

Renninger, N.S., and Mcphee, D.J. (2010). Fuel compositions comprising farnesane and farnesane derivatives and method of making and using same, U.S. Patent, ed.

Ringer, K.L., Davis, E.M., and Croteau, R. (2005). Monoterpene metabolism. Cloning, expression, and characterization of (−)-isopiperitenol/(−)-carveol dehydrogenase of peppermint and spearmint. Plant Physiol, *137*, 863–872.

Ro, D.-K., Paradise, E.M., Ouellet, M., Fisher, K.J., Newman, K.L., Ndungu, J.M., Ho, K.A., Eachus, R.A., Ham, T.S., Kirby, J., *et al.* (2006). Production of the antimalarial drug precursor artemisinic acid in engineered yeast. Nature *440*, 940–943.

Rohmer, M., Seemann, M., Horbach, S., BringerMeyer, S., and Sahm, H. (1996). Glyceraldehyde 3-phosphate and pyruvate as precursors of isoprenic units in an alternative non-mevalonate pathway for terpenoid biosynthesis. J. Am. Chem. Soc. *118*, 2564–2566.

Rude, M.A., and Schirmer, A. (2009). New microbial fuels: a biotech perspective. Curr. Opin. Microbiol. *12*, 274–281.

Sasaki, D., Fujihashi, M., Iwata, Y., Murakami, M., Yoshimura, T., Hemmi, H., and Miki, K. (2011). Structure and mutation analysis of archaeal geranylgeranyl reductase. J. Mol. Biol. 409, 543–557.

Schmid, K., Ebner, R., Altenbuchner, J., Schmitt, R., and Lengeler, J.W. (1988). Plasmid-mediated sucrose metabolism in *Escherichia coli* K12: mapping of the scr genes of pUR400. Mol. Microbiol. *2*, 1–8.

Schmidt-Dannert, C., Umeno, D., and Arnold, F.H. (2000). Molecular breeding of carotenoid biosynthetic pathways. Nature Biotechnol. *18*, 750–753.

Temme, K., Hill, R., Segall-Shapiro, T.H., Moser, F., and Voigt, C.A. (2012a). Modular control of multiple pathways using engineered orthogonal T7 polymerases. Nucleic Acids Res. *40*, 8773–8781.

Temme, K., Zhao, D., and Voigt, C.A. (2012b). Refactoring the nitrogen fixation gene cluster from *Klebsiella oxytoca*. Proc. Natl. Acad. Sci. U.S.A. *109*, 7085–7090.

Tracy, N.I., Chen, D.C., Crunkleton, D.W., and Price, G.L. (2009). Hydrogenated monoterpenes as diesel fuel additives. Fuel *88*, 2238–2240.

Tracy, N.I., Crunkleton, D.W., and Price, G.L. (2010). Gasoline production from phytol. Fuel *89*, 3493–3497.

Tracy, N.I., Crunkleton, D.W., and Price, G.L. (2011). Catalytic cracking of squalene to gasoline-range molecules. Biomass Bioenerg. *35*, 1060–1065.

Voynova, N.E., Rios, S.E., and Miziorko, H.M. (2004). Staphylococcus aureus mevalonate kinase: isolation and characterization of an enzyme of the isoprenoid biosynthetic pathway. J. Bacteriol. *186*, 61–67.

Wagner, A.M., and Krab, K. (1995). The alternative respiration pathway in plants – role and regulation. Physiol. Plant. *95*, 318–325.

Wang, C.W., Oh, M.K., and Liao, J.C. (1999). Engineered isoprenoid pathway enhances astaxanthin production in *Escherichia coli*. Biotechnol. Bioeng. *62*, 235–241.

Wang, H.H., Isaacs, F.J., Carr, P.A., Sun, Z.Z., Xu, G., Forest, C.R., and Church, G.M. (2009). Programming cells by multiplex genome engineering and accelerated evolution. Nature *460*, 894–898.

Xu, Q., Eguchi, T., Mathews, I.I., Rife, C.L., Chiu, H.J., Farr, C.L., Feuerhelm, J., Jaroszewski, L., Klock, H.E., Knuth, M.W., *et al.* (2010). Insights into substrate specificity of geranylgeranyl reductases revealed by the structure of digeranylgeranylglycerophospholipid reductase, an essential enzyme in the biosynthesis of archaeal membrane lipids. J. Mol. Biol. *404*, 403–417.

Yamano, S., Ishii, T., Nakagawa, M., Ikenaga, H., and Misawa, N. (1994). Metabolic engineering for production of beta-carotene and lycopene in *Saccharomyces cerevisiae*. Biosci. Biotechnol. Biochemi. 58, 1112–1114.

Yang, J., Zhao, G., Sun, Y., Zheng, Y., Jiang, X., Liu, W., and Xian, M. (2012). Bio-isoprene production using exogenous MVA pathway and isoprene synthase in *Escherichia coli*. Bioresour. Technol. *104*, 642–647.

Yang, Y., Dec, J.E., Dronniou, N., and Simmons, B. (2010). Characteristics of Isopentanol as a Fuel for HCCI Engines. SAE Int. J. Fuels Lubricants 3, 725–741.

Yokota, A., and Sasajima, K. (1986). Formation of 1-deoxy-ketoses by pyruvate-dehydrogenase and acetoin dehydrogenase. Agric.]. Biol. Chem. *50,* 2517–2524.

Zhao, Y., Yang, J., Qin, B., Li, Y., Sun, Y., Su, S., and Xian, M. (2011). Biosynthesis of isoprene in *Escherichia coli* via methylerythritol phosphate (MEP) pathway. Appl. Microbiol. Biotechnol. l *90*, 1915–1922.

Zulak, K.G., and Bohlmann, J. (2010). Terpenoid biosynthesis and specialized vascular cells of conifer defense. J. Integr. Plant Biol. *52*, 86–97.

Engineering Microbial Fatty Acid Biosynthetic Pathways to Make Advanced Biofuels

7

Tingting Ning and Tiangang Liu

Abstract

Fatty acid-derived biofuels production received more and more concerns during the last decades of years, since they are closer to diesel fuels in characteristics of high combustion value, low water solubility and higher biodegradability than other existing biofuels. Thus the biosynthesis pathway of fatty acid was elucidated and the metabolic engineering works to produce free fatty acids, fatty alcohols, fatty acid methyl/ethyl esters, and fatty alkanes/alkenes were introduced. And finally future perspectives of this kind of work were provided.

7.1 Introduction

Fatty acids are composed of carboxylic acid with a long aliphatic tail (chain), which is either saturated or unsaturated. Most naturally occurring fatty acids have a chain containing an even number of carbon atoms, from 4 to 28. Fatty acids are integral components of living systems and play important roles in cell metabolism. Fatty acids have been studied from many perspectives. Fundamental research has focused on the function of fatty acids as structural components of cell membranes (Metz *et al.*, 2000; Silhavy *et al.*, 2006). In medicine, fatty acids are components of lipid signalling pathways that serves as drug targets (Kuhajda *et al.*, 1994; Vioque and Kolattukudy, 1997; Campbell and Cronan, 2001). From an economic perspective, fatty acids and their derivatives can be used as biofuels substitutes, and recently fatty acid metabolism has received significant attention as a route for producing high-value fuels (Miwa, 1971; Metz *et al.*, 2000; Peralta-Yahya and Keasling, 2010; Lennen and Pfleger, 2012).

7.2 Current status of biodiesel production

Due to depletion of the world's petroleum reserves and increasing environmental concerns, there is a great demand for alternative sources of cleaner renewable fuels. Biodiesel has recently been considered as the best candidate for a diesel fuel substitution because it can be used in any compression ignition engine without requiring mechanical modifications. Chemically, biodiesel is a mixture of methyl esters with long-chain fatty acids and is typically made from nontoxic, biological resources such as vegetable oils, animal fats, or even used cooking oils (Lardizabal *et al.*, 2000; Balat, 2011). Biodiesel can be used alone or blended with petrodiesel. Most current research on renewable biofuels is focused on producing biodiesel from vegetable oils. Natural oils and fats are the esters of glycerol and fatty acids. They are called glycerides or triglycerides. Several biodiesel production methods have been developed, among which transesterification using an alkali catalyst gives high level of triglyceride conversion to their corresponding methyl ester during short reaction times (Doan *et al.*, 2009; Leung *et al.*, 2010). This catalysed reaction is shown in Fig. 7.1.

The transesterification reaction is reversible and excess alcohol shifts the equilibrium to the product side. However, when the raw materials (oils or fats) contain a high percentage of free fatty acids (FFAs) or water, the alkali catalyst will react with the FFAs to form soaps and the water can hydrolyse the triglycerides into diglycerides and form more FFAs (Chouhan and Sarma, 2011; Hofvander *et al.*, 2011). These are undesirable reactions that reduce the yield of the biodiesel product. Although biodiesel is the main biofuel

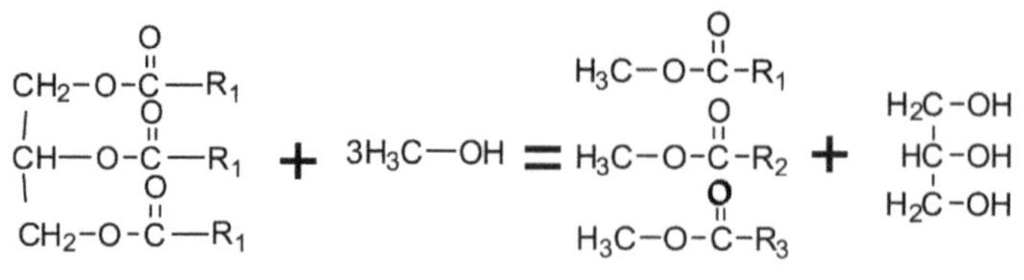

Figure 7.1 Transesterification reaction used for biodiesel production. R_1, R_2, and R_3 are long chains of fatty acid. Catalysed by alkali one mole of triglyceride and three moles of methanol will produce three moles methyl ester and one mole glycerol.

used in Europe, it has limitations in that it has only 91% of the energy content of D2 diesel and because crystals can form in the fuel at low temperatures, thus, it is difficult to transport with the current distribution infrastructure so there are geographical limits to its use (Leung *et al.*, 2010; Willis *et al.*, 2011). And the transesterification reaction is expensive and difficult to scale up. Therefore, the most desirable biodiesel production processes produce fuels from lignocellulose-derived carbohydrates rather than plant seed oils will be more helpful to achieve meeting sustainability goals.

7.3 Motivation for engineering fatty acid metabolism

Progress in metabolic engineering, and synthetic and systems biology, have used microbes to produce advanced biofuels from sustainable substrates that have similar properties to petroleum-based fuels. During these studies, the most extensive efforts have been focused on bioengineering of the fatty acid synthesis pathway.

In cells, fatty acids function mainly as components of the cell membrane; however, in the eyes of synthetic biologist fatty acids were considered to be attractive chemical feedstock for generating high-energy density liquid transportation fuels. Fatty acid methyl/ethyl esters, fatty alkenes/alkanes, and fatty alcohols are the most important biofuels, and are considered to be the most favourable substitutes for current petroleum-based fuels. The conversion of FFAs, fatty acyl-acyl carrier protein (ACP) and fatty acyl-CoA complexes to final commercially viable biofuel products is shown in Fig. 7.2. Because many high-value biofuel products can be generated by the fatty acid metabolism pathway, this approach has attracted significant attention during recent years. In the following discussion, we focus on state-of-the-art genetic engineering approaches to optimize conversion of fatty acids to biofuels.

7.4 Brief review of fatty acid metabolism

Fatty acid biosynthesis (FAB) is best understood in *E. coli*, in which almost every step has been intensively elucidated (Wettstein-Knowles *et al.*, 2006; Steen *et al.*, 2010). The pathway for FAB and β-oxidation are detailed shown in Fig. 7.3. De novo fatty acid synthesis is performed by fatty acid synthase (FAS), which catalyses cycles of multi-step chemical reactions that are essentially the same in all organisms. There are two principal classes of FAS. In type I system, the constituent catalytic components are covalently linked in multifunctional megasynthases, whereas in type II system, the catalytic components are freestanding monofunctional polypeptides (White *et al.*, 2005). In most bacterial, plant, and mitochondrial FASs belong to the type II FAS systems, while in animal, fungi and some bacterial, (e.g. corynebacteria) FASs belong to type I FAS system. FAS I and FAS II share similar elongation and reduction mechanisms, because the catalytic domains of the FAS II enzymes are largely homologous to their counterparts in FAS I multienzyme polypeptides and most engineering works were based on type II system. Here, we will focus on the mechanism of type II FAS.

Briefly, fatty acids are made via an iterative reduction cycle that operates on ACP (Tong, 2005). In each iteration, two carbons are added from malonyl-ACP to a growing acyl-chain and the resulting β-keto group is reduced to a saturated

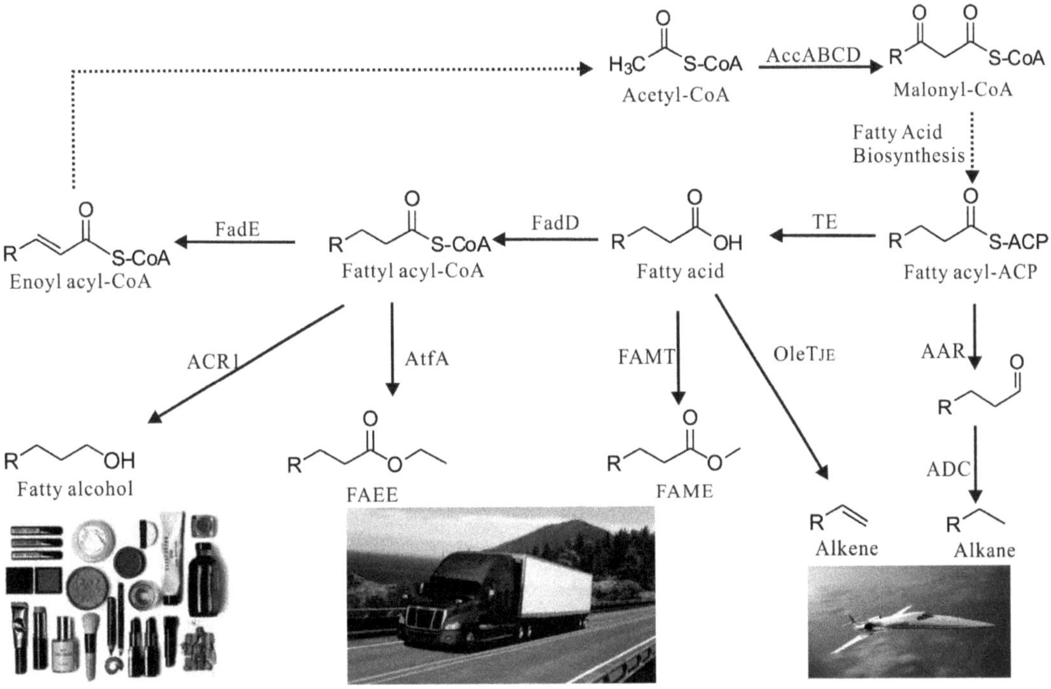

Figure 7.2 Biofuels derived from fatty acids. General synthesis pathways and applications of fatty alcohol, fatty acid methyl/ethyl ester, fatty alkene and fatty alkane are shown. AAR, acyl-ACP reductase; ACC, acetyl-CoA carboxylase; Acr1, acyl-CoA reductase; ADC, aldehyde decarbonylase; AtfA, wax-ester synthase; FadD, acyl-CoA synthetase; FadE, acyl CoA dehydrogenase; FAMT, fatty acid methyltransferase; OleTJE, *Jeotgalicoccus* sp. terminal olefin-forming fatty-acid decarboxylase; TE, acyl-ACP thioesterase. A colour version of this figure is available in the plate section at the back of the book.

methylene. The whole metabolic process can be divided into three parts; the first part is the synthesis of starter and extender units, the second step is the fatty acid elongation cycle, and finally the FFAs are activated for β-oxidation by acyl-CoA synthetase. The three parts are highlighted by different colours in Fig. 7.3.

The first step for fatty acid generation is the synthesis of starter and extender units. Acetyl-CoA is a central metabolite made from numerous carbon sources. As shown in Fig. 7.3a, glucose could be converted to pyruvate through glycolysis, and acetyl-CoA is formed from pyruvate through oxidative decarboxylation by the multienzyme complex pyruvate dehydrogenase. The acetyl group of acetyl-CoA is bonded as a thioester to the sulfhydryl portion of the β-mercaptoethylamine group. CoA thereby functions as a carrier of acetyl and other acyl groups.

Acetyl-CoA carboxylase (ACC) catalyses the first committed step of fatty acid biosynthesis, and is one of the reaction rate-limitation steps. ACC is a multi-subunit enzyme in most prokaryotes and in the chloroplasts of most plants and algae, whereas it is a large, multi-domain enzyme in the endoplasmic reticulum of most eukaryotes. ACC is a biotin-dependent enzyme that catalyses the irreversible carboxylation of acetyl-CoA to produce malonyl-CoA through its biotin carboxylase and carboxyl transferase catalytic activities. In *E. coli*, ACC consists of four subunits, biotin carboxyl carrier protein (AccB), biotin carboxylase (AccC), and acetyl CoA carboxytransferase (a heterodimer of AccA and AccD) (Tong, 2005). Malonyl-CoA is converted to malonyl-ACP by malonyl-CoA:ACP transacylase (FabD), and to the first β-ketoacyl-ACP in the fatty acid elongation cycle (acetoacetyl-ACP) catalysed by β-ketoacyl-ACP synthase III (FabH). Then acetoacetyl-ACP enters the elongation cycle (Magnuson *et al.*, 1993).

Then fatty acid chains are elongated on carrier proteins through repeated cycles of

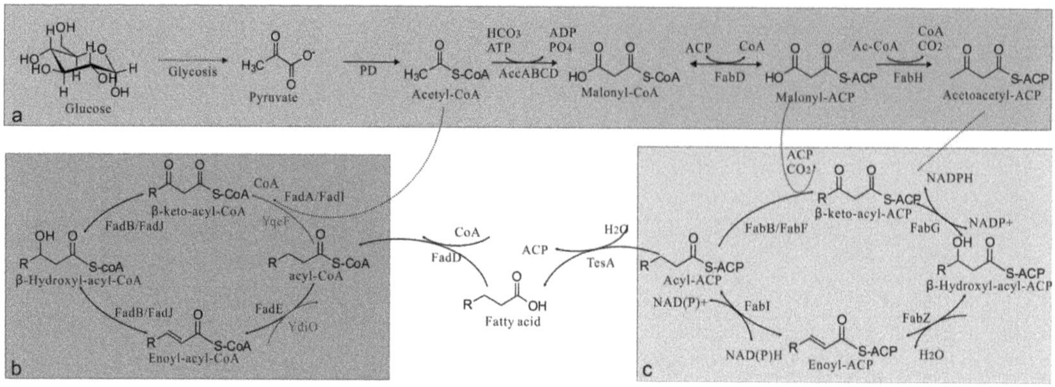

Figure 7.3 Metabolic pathways for fatty acid biosynthesis. The synthesis of starter and extender units is shown in the yellow box (a); the elongation cycle is shown in the green box (c), and the β-oxidation step is shown in the red box (b). The red in the β-oxidation cycle represent the critical enzymes involved in the reversal of β-oxidation. ACC, acetyl-CoA carboxylase; FabB, β-ketoacyl-ACP synthase I; FabD, malonyl-CoA:ACP transacylase; FabF, β-ketoacyl-ACP synthase II; FabG, β-ketoacyl-ACP reductase; FabH, β-ketoacyl-ACP synthase III; FabI, enoyl-acyl-ACP reductase; FabZ, β-hydroxyacyl-ACP dehydratase; FadB/FadJ, enoyl-CoA hydratase/3-hydroxyacyl-CoA dehydrogenase; FadD, acyl-CoA synthetase; FadE, acyl CoA dehydrogenase; FadA/FadI, acetyl-CoA-acyltransferase; PD, pyruvate dehydrogenase; TesA, *E. coli* acyl-ACP thioesterase; YdiO, enoyl-CoA reductase; YqeF, thiolase. A colour version of this figure is available in the plate section at the back of the book.

decarboxylative condensation, β-keto reduction, dehydration, and enol reduction. In detail, the first β-ketoacyl-ACP (acetoacetyl-ACP) is reduced by β-ketoacyl-ACP reductase (FabG) to yield 3-hydroxyacyl-ACP, which is in turn dehydrated by β-hydroxyacyl-ACP dehydratase (FabZ) to yield enoyl-ACPs that will be reduced by enoyl-ACP reductase (FabI). Then the resulting saturated acyl-ACP is condensed with malonyl-ACP by β-ketoacyl-ACP synthase I (FabB) or II (FabF) to elongate the fatty acid chain by two carbons. The elongation step is shown in Fig. 7.3c.

During chain elongation, the growing fatty acid is esterified to carrier protein ACP, and elongation is dependent on the structure and characteristics of the specific ACP. As does CoA, ACP contains a phosphopantetheine group that forms thioesters with acyl groups (Lambalot and Walsh, 1995). The phosphopantetheine phosphoryl group is esterified to a Ser OH group of ACP. In animals, ACP is part of a large multifunctional protein (Type I ACP), whereas in *E. coli* it is a 78-residue polypeptide (Type II ACP). The phosphopantetheine group is transferred from CoA to *apo*-ACP to form the active *holo*-ACP by ACP synthase. ('*apo*' means that it is in an inactive and unbound state, while '*holo*' means that it is bound by ligands). In their *holo*-form, ACPs play central roles in a broad range of biosynthetic pathways that depend on iterative acyl transfer steps, including polyketide, non-ribosomal peptide, and depsipeptide biosynthesis, as well as in the transacylation of oligosaccharides and proteins (Tong, 2005).

After elongation, long-chain fatty acids are released from the ACP by thioesterase (TE)-catalysed hydrolysis. Released FFAs are then dismembered through β-oxidation of fatty acyl-CoA in a cyclical reaction that involves four sequential steps. Before the β-oxidation cycle, the FFAs require activation by acyl-CoA synthetase (FadD). Then an acyl-CoA dehydrogenase (FadE) catalyses the activated acyl-CoA to form a trans-α, β double bond called enoyl-acyl-CoA, which is hydrated and dehydrogenated by an enoyl-CoA-hydratase (FadB or FadJ) to β-ketoacyl-CoA. Finally, β-ketoacyl-CoA thiolase (FadA or FadI) cleaves the C_α-C_β in a thiolysis reaction with CoA to form acetyl-CoA and a new acyl-CoA containing two fewer carbon atoms than the original molecule. The β-oxidation step is shown in Fig. 7.3b.

The pathway of fatty acid synthesis differs from that of fatty acid oxidation. Whereas fatty acid oxidation occurs in the mitochondria and utilizes fatty acyl-CoA esters, fatty acid biosynthesis occurs in the cytosol with the growing fatty acids esterified to ACP.

7.5 Regulation of fatty acid synthesis and degradation

The primary use for fatty acids in *E. coli* is in phospholipid biosynthesis, and to maintain bacterial growth. FAB is tightly regulated at multiple levels involving transcriptional, posttranslational, and enzymatic modulation. In *E. coli*, the key regulatory signal for controlling FAB is the long chain acyl-ACPs; *in vitro* studies confirmed that acyl-ACPs directly inhibit ACC and to a lesser degree FabH and FabI. Thus, overexpression of TE, which hydrolyses both acyl-ACPs and acyl-CoAs to generate FFAs and consequently depletes long chain acyl-ACPs, increases fatty acid production in many studies (Lu *et al.*, 2008; Liu *et al.*, 2010).

Transcriptional regulation of FAB in *E. coli* is mostly achieved by the action of two antagonistic repressors, FabR and FadR (DiRusso *et al.*, 1993). FabR regulates FAB by acting as a repressor of FabB and, to a lesser degree, FabA. FadR regulates fatty acid degradation in *E. coli* by binding to operators upstream of the fatty acid degradation operon, thereby repressing translation. When acyl-CoA is present in the cell it binds with FadR and de-represses transcription. This scheme is quite similar with that of the lactose repressor in *E. coli*. Research has shown that FadR also acts as a positive activator of FabA and FabB (Magnuson *et al.*, 1993; Handke *et al.*, 2011). The complex interactions between the two repressors indicate that *E. coli* has evolved to regulate fatty acid content necessary for proper membrane function under varying conditions. This reminds us that when attempting to increase the production of fatty acids or corresponding derived products, the experimental positive and negative controls must be considered in a deliberate manner.

7.6 Genetic engineering of bacteria to improve free fatty acid production

Because other products are derived from FFAs or their intermediates during the FAB pathway, the first step to meet commercialized biofuel demand is to increase the yield and productivity of FFAs. In the first report of a concerted metabolic engineering effort to maximize production of FFAs, all the engineering efforts were based on the knowledge of fatty acid regulation in *E. coli*. This study tested three modifications: elimination of β-oxidation (Δ*fadD*); overexpression of two TEs, *TesA'* and CcTE (FatB1) from *Cinnamomum camphorum*, and overexpression of *E. coli* ACC. Here, *TesA'* means a 'leaderless' version of TesA that is targeted to the cytosol (Lu *et al.*, 2008). An optimized combination of these modifications yielded a 19-fold increase in total fatty acids compared to the original strain. Subsequent modification increased FFA yield by altering expression levels of TesA' and CcTE in an ACC-overexpressing Δ*fadD* strain, elevating fatty acid production from 0.38 g/l to 0.94 g/l. And the difference between the strains used in these two studies is that in the first study, plant CcTE used the T7 promoter and *TesA'* used the pBAD promoter, while in the later study both CcTE and *TesA'* were controlled under the same pBAD promoter (Liu *et al.*, 2010). Subsequent studies have used similar sets of genetic modifications (Steen *et al.*, 2010; Peralta-Yahya *et al.*, 2012).

Overexpression of TEs is critical for producing FFAs in *E. coli*. This is because long chain acyl-ACPs inhibit FAB activity, and TE depletes long chain acyl-ACPs; therefore, overexpression of TE increases substrate flux through the FAB pathway. Different species' of TEs have varied substrate specificities. In plants, fatty acids are synthesized in the chloroplast, whereas lipid synthesis occurs in the cytoplasm. Because acyl-thioesters cannot cross the plastid membrane, TEs cleave acyl-ACPs to a desired length that enables transport. Plant TEs fall into two classes based on homology to enzymes from *Arabidopsis thaliana*. FatA isoforms prefer unsaturated acyl chains, while FatB isoforms target saturated acyl chains. The CcTE used in the above-detailed studies belongs to the FatB family. Fatty acyl-ACPs with chain lengths of C_{16}–C_{18} are the predominant acyl-ACPs; so most TEs exhibit the highest specificity towards C_{16}–C_{18} acyl-chains. To obtain medium-length acyl-chains, TEs from other oil seeds could be exploited (Doan *et al.*, 2009).

E. coli contains two native TEs, TE I and TE II. TE I, encoded by the *tesA* gene, was found to be specific for C12 to C18 acyl-CoA esters, but it was inactive for C6 to C10 acyl-CoA esters, while TE II, an enzyme encoded by the *tesB* gene that is composed of four identical subunits has broader substrate specificity and it can cleave C6 to C18 acyl-CoA esters (Zheng *et al.*, 2004). To date, a number of TEs have been expressed in *E. coli*, including TEs from *Cinnamomunm camphora* (14:0-ACP), *Umbellularis californica* (12:0-ACP), *Arabidopsis thaliana* (18:0-ACP), and *Cuphea hookeriana* (8:0 and 10:0-ACP). So, it is very convenient to control the length of the resulting biosynthetic fatty acid chain by choosing different plant TEs for transfection into *E. coli* (Doan *et al.*, 2009; Zhang *et al.*, 2011b).

By cytosolic expression of native *E. coli* TEs, combined with eliminating *fadD* and *fadE*, fatty acid production titre increased by roughly four-fold, from 0.32 g/l to 1.2 g/l (Steen *et al.*, 2010). Although concentrations of almost all key enzymes in the FAB pathway have been modified in efforts to maximize fatty acid production, the situation is much more complex. We appreciate that FAB is regulated at several levels, including transcriptional and enzymatic, and we know FadR is a key transcription factor that binds to specific DNA sequences and controls the expression of several FAB genes. One study upregulated FadR expression in an engineered *E. coli* host and increased fatty acid production to 5.27 g/l, or 73% of the theoretically possible yield (Zhang *et al.*, 2012b). To elucidate the mechanism by which FadR enhanced fatty acid yield so greatly, whole genome transcriptional analysis (microarray) and targeted proteomics was performed. The largest transcriptional upregulation in fatty acid pathway genes included *fabB*, *fabF*, and *accA*, but overexpression of any of these genes alone did not

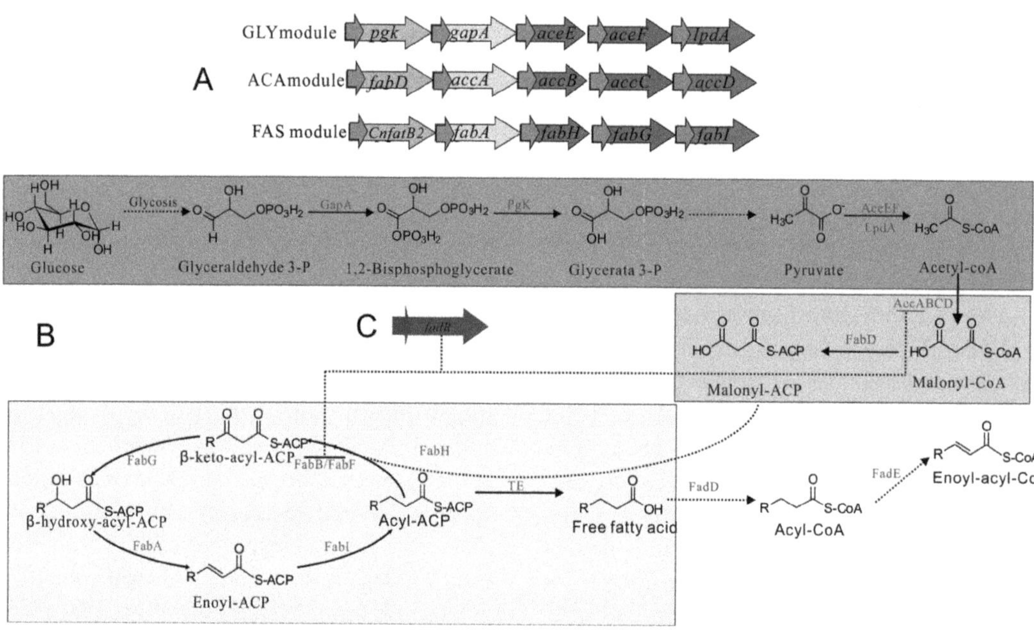

Figure 7.4 Summary of some important modifications on free fatty acid overproduction. The red, green and yellow boxes in panel (B) represent the three modules deconstructed respectively, and these three modules are shown in details at the upper panel (A). Genes overexpressed are depicted as green text. Genes knocked out are in red text. The *fadR* gene in the panel (C) shows overexpressing FadR upregulates the transcription of *fabF*, *fabB*, and accA. GapA, glyceraldehyde-3-phosphate dehydrogenase; PgK, phesosphoglycerate kinase; AceEF and LpdA, pyruvate dehydrogenase multi-enzyme; ACC, acetyl-CoA carboxylase; FabD, malonyl-CoA:ACPtransacylase; FabB, β-ketoacyl-ACP synthase I; FabD, malonyl-CoA:ACPtransacylase; FabF, β-ketoacyl-ACP synthase II; FabG, β-ketoacyl-ACP reductase; FabH, β-ketoacyl-ACP synthase III; FabI, enoyl-acyl-ACP reductase; FadD, acyl-CoA synthase; FadE, acyl CoA dehydrogenase. A colour version of this figure is available in the plate section at the back of the book.

result in a high yield comparable to *fadR* expression. This indicates that FadR enhances fatty acid production globally by regulating the expression levels of many genes to optimal levels. This is a good example for knowledge gained by combining systems biology and synthetic biology. The schematic of this regulation is shown in Fig. 7.4C.

This example also reminds us that the fatty acid synthesis pathway is an organized, unitary, and dynamic whole, so to obtain maximal production we must take a global view. Another study demonstrated such global control. These researchers used a modular engineering approach instead of increasing or knocking out single genes (Xu *et al.*, 2013). On the basis of central pathway architecture, *E. coli* fatty acid biosynthesis was deconstructed into three modules: (i) GLY module: the upstream glycolysis module encoded by *pgk, gapA, aceE, aceF,* and *lpdA*; (ii) ACA module: the intermediary acetyl-CoA activation module encoded by *fabD, accA, accB, accC,*and *accD*; and (iii) FAS module: the downstream fatty acid biosynthetic module encoded by *CnfatB2, fabA, fabH, fabG,* and *fabI.* These three modules were successfully expressed on the five compatible ePathBrick vectors, with varying promoter strengths, plasmid copy numbers, and antibiotic resistance markers (Xu *et al.*, 2012). An optimized combination of these three modules was determined as *m*GLY-*l*ACA-*h*FAS which means that the GLY module is expressed at a medium copy number, the ACA module is expressed at a low copy number, and the FAS module is expressed at a high copy number. The total fatty acid production reached to 1.42 g/l in this optimized strain. Fatty acid production was further improved by tuning translation rates, to 2.04 g/l. Different ribosome binding sites (RBSs) were chosen and used to replace the 5′-UTR region of the native T7 promoter, and the selected RBS 29 was combined with *m*GLY-*l*ACA-*h*FAS. This optimized strain (RBS29-*m*GLY-*l*ACA-*h*FAS) resulted in an increased final volumetric production (2.04 g/l) of total fatty acids, representing a 46% increase compared with the parental stain. After further optimizing cultivation conditions in a fed-batch 20-L fermenter, the engineered strain produced 8.6 g/l of FFAs, which is the highest production titre reported to date. The modular optimization scheme and other genetic modifications of FFA overexpression are summarized in panel A and B of Fig. 7.4.

An unconventional approach to synthesize biofuels involves reversing β-oxidation, and this approach is promising because it operates anabolically rather than in the usual catabolic direction (Dellomonaco *et al.*, 2011). Reversal of the β-oxidation cycle was engineered into *E. coli* and used in combination with endogenous dehydrogenases and TEs to synthesize *n*-alcohols, fatty acids, and other chemicals. This pathway is possible because of the reversibility of FadA activity and the two reactions catalysed by FadB. This pathway is advantageous because it eliminates the need for ATP in synthesizing malonyl-CoA; therefore, it could allow anaerobic production of higher-order alkyl-chain products. Briefly, thiolase (YqeF), FadA, FadB, and YdiO (enoyl-CoA reductase) were engineered into *E. coli,* and then FadM was used to release FFAs from the resulting fatty acyl-CoA. The results suggested that high levels of butanol and fatty acids could be produced (7 g/l). These studies highlight the complexity of modifying existing pathways and introducing new pathways into *E. coli,* while balancing the many other criteria desired for a suitable production host. The reversal of β-oxidation is shown in Fig. 7.3b, and the key steps were highlighted in red.

Based on clear understanding of its metabolism process, yeast is another host that has been extensively studied for biofuel synthesis. Two species, non-oleaginous *Saccharomyces cerevisiae* and oleaginous *Yarrowia lipolytica,* are the most popular candidates (Stoops *et al.*, 1978; Wakil, 1989; Lomakin *et al.*, 2007). In yeast, fatty acids biosynthesis is also initiated by the conversion of acetyl-CoA to malonyl-CoA via the enzyme acetyl-CoA carboxylase in the cytosol. This reaction requires acetyl-CoA, together with NADPH and ATP as key substrates and cofactors. Because yeasts are eukaryotic they contain many organelles that are not present in *E. coli,* so there are some differences between the FAB pathways between these two organisms. In yeast cells, there exist two different pathways for cytosol acetyl-CoA generation (Tehlivets *et al.*, 2007). In *S. cerevisiae,* the acetyl-CoA is produced from glycolysis of fermentable sugars through the pyruvate-acetaldehyde-acetate pathway via the activity of cytoplasmic acetyl-CoA synthases, while for oleaginous fungi, another source of cytosol acetyl-CoA production is from excess of citrate, which is exported to cytosol via the citrate transport protein and then cleaved by ATP-citrate lysase (ACL), an enzyme present in all oleaginous microorganisms studied so far. To accumulate citrate in the TCA cycle and then transport it out to generate cytosolic acetyl-CoA is a promising pathway to provide precursor for FAB. The mitochondrial enzyme-isocitrate dehydrogenase, which responsible for catalysing the conversion of isocitrate to α-ketoglutarate, is encoded by two distinct genes, *idh1* and *idh2*. Because ACL is absent in members of *S. cerevisiae*, by overexpressing a heterologous ACL and knocking down expression of *idh1* and *idh2*, 9-hexadecanoic acid production increased by 92% (Tang *et al.*, 2013). The oleaginous yeast *Yarrowia lipolytica* has excellent lipid accumulation capacity, commonly accumulating up to 36% of its dry cell weight in lipids, so whole-genome sequencing can give a lot of information on how to engineer fatty acid synthesis in these species (Nielsen, 2009; Santos and Riezman, 2012).

Both *E. coli* and yeast are heterotrophic, meaning that they rely on microorganisms that metabolize carbohydrate sources from land-based feedstocks. Increased food cost was resulted from the competition with land use for food crops. Therefore, the direct conversion of solar energy into liquid fuel using photosynthetic microorganisms is an attractive alternative to fossil fuels. There are several advantages to using organisms such as microalgae and cyanobacteria: their readily available genetic tools and sequenced genomes; their higher growth rate compared to plants; and their ability to thrive in areas that cannot support agriculture (Lu, 2010; Machado and Atsumi, 2012). Cyanobacteria are advantageous organisms for industrial applications because they grow and divide rapidly, have simple nutrient requirements (mainly water, sunlight, and CO_2), and are naturally transformable and thus have the potential to be genetically engineered. The production of fatty acids by genetically engineered cyanobacteria will be reviewed in Chapter 8.

7.7 Genetic engineering to improve fatty alcohol production

Because of the ionic nature of their carboxyl group, fatty acids cannot be used directly as biofuel, but they can be readily converted into non-ionic hydrophobic molecules, such as fatty alcohols, fatty acid alkyl esters, alkenes, and alkanes.

Fatty alcohols, usually with hydrocarbon chain length ranging from C_8 to C_{18}, are composed of a nonpolar, lipophilic carbon chain and a polar, hydrophilic hydroxyl group (Hofvander *et al.*, 2011). Fatty alcohols having higher selling prices than diesel, are widely used in industry, and are currently derived from plant oils or petrochemical sources (Kunst and Samuels, 2003). Their value was approximately $1500 per ton (2004 ICIS pricing), and they represent almost $3 billion annual market. These reasons motivate metabolic engineering efforts to construct fatty alcohol-producing biological strains. In plants, fatty alcohols are produced as components of the plant cuticle, a phenomenon that inspired scientists to develop new strains for biofuel production (de Azevedo Souza *et al.*, 2009; Doan *et al.*, 2009). It has been demonstrated that the alcohol is formed by a four-electron reduction of acyl-CoA, using NADH or NADPH as cofactors. Although the alcohol-generating fatty acid reduction reactions proceed through an aldehyde intermediate, free aldehyde is not released and these types of fatty acyl CoA/ACP reductases (FAR) employ integral membrane proteins (Vioque and Kolattukudy, 1997). Another kind of FARs performs two-electron reductions of fatty acyl-CoA, yielding free fatty aldehyde as a product. Researchers tried to find key enzymes for acyl-CoA reduction to fatty alcohol in fatty alcohol-rich plants such as Jojoba (*Simmondsia chinensis*) (Metz *et al.*, 2000) and pea leaves (*Pisum sativum* L.) (Vioque and Kolattukudy, 1997). In pea leaves, two distinct fatty acyl-CoA reductases have been identified, with one catalysing the four-electron reduction to the fatty alcohol and the other catalysing the two-electron reduction to the fatty aldehyde. Jojoba is native to the deserts of the American Southwest, and is unusual among higher plants in that its seed storage lipids are waxes rather than triglycerides (Miwa, 1971; Metz *et al.*, 2000). These waxes are esters of long-chain monounsaturated fatty acids and alcohols. In 2000, an alcohol-forming fatty acyl-CoA reductase from developing embryos was cloned, and when the cDNA of this jojoba FAR (mRNA accession number AF149917) was expressed in *E. coli*, accumulation of fatty alcohol was observed (Lardizabal *et al.*, 2000). Approximately 0.2 g/l of fatty alcohol could be produced when this FAR gene was coexpressed in *E. coli* with other genetic modifications. This result encouraged researchers to find more effective FAR genes in plants. Bioinformatics showed that the Arabidopsis (*Arabidopsis thaliana*) genome contains eight genes homologous to FAR genes from jojoba and other species. Expression of five *Arabidopsis* FAR homologues in *E. coli* resulted in production of alcohols from endogenous *E. coli* fatty acids (Doan *et al.*, 2009). However, the five FAR enzymes and the jojoba FAR produced distinctly different fatty alcohol profiles, indicating that they may be involved in the production of different types of alcohols in plant cells. Accumulation of fatty aldehyde is harmful to the cell, so identifying other reductases that can directly reduce fatty acyl-CoA to fatty alcohol would be beneficial. In 2010, a novel bacterial enzyme was identified from *Marinobacter aquaeolei* VT8 that catalyses the reduction of fatty acyl-CoA by four electrons to the corresponding fatty alcohol, eliminating the need for a separate fatty aldehyde reductase (Hofvander *et al.*, 2011). The enzyme, Maqu_2220, is shown to reduce fatty acyl-CoAs ranging from C8:0 to C20:4 to the corresponding fatty alcohols, with the highest rate found for palmitoyl-CoA (C16:0). Further studies reported that this enzyme, named with FALDR (Fatty Aldehyde Reductase), can also reduce acyl-CoAs and acyl-ACP to fatty alcohols, in addition to reducing fatty aldehydes, thereby performing both reduction steps (Willis *et al.*, 2011). Combining with other genetic modifications expression of this gene in *E. coli* lead to a remarkable level of 1.725 g/l of fatty alcohol under the fermentation condition (Liu *et al.*, 2013).

Thus, at least two pathways are available among both prokaryotes and eukaryotes for reducing activated acyl chains to fatty alcohols, and both of these reductase approaches have been used to produce fatty alcohols in different hosts.

To maximize the production of fatty alcohol, not only FARs should be overexpressed, but also combined with other modifications. Expressing of a FAR named *acr1* from *Acinetobacter calcoaceticus* BD413 in combination with a native aldehyde reductase (AHR) resulted in the production of medium-chain fatty alcohols up to 60 mg/l in a '*tesA*', *fadD*, Δ*fadE*' strain (Steen *et al.*, 2010). To engineer controlled fatty alcohol chain-length distributions, different plant TEs were used to replace *tesA*'. When *tesA*' was used, the majority of fatty alcohols is C_{14}-chain fatty alcohol, and when ucFatB is used, C_{12} is the predominant alcohol form, whereas chFatB3 increased production of C_{14}-chain fatty alcohol to about 80 mg/l (Silhavy *et al.*, 2006; Steen *et al.*, 2010). The genetic modification of fatty alcohols is summarized in Fig. 7.5.

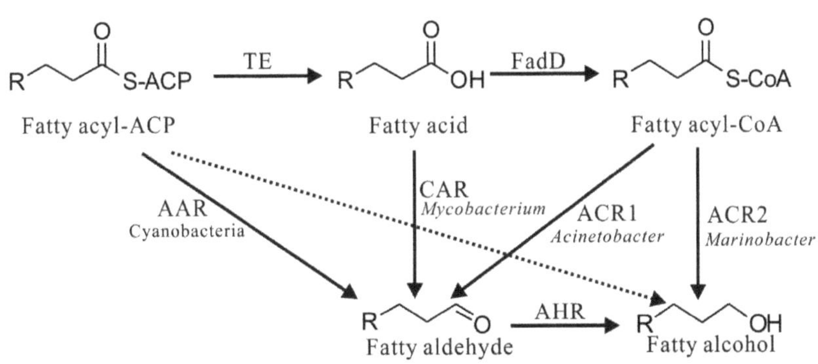

Figure 7.5 Metabolic pathway for fatty alcohol production. ACR1 and ACR2 are fatty acyl-CoA reductases from *Acinetobacter* and *Marinobacter*, respectively. AAR, acyl-ACP reductase; CAR, carboxylic acid reductase; AHR, aldehyde reductase.

Because $C_{16/18}$ alcohol can be employed as defoamers, solubility retarders, and consistency providing factors, another group tried to maximize the production of C_{16}/C_{18} fatty alcohol in *E. coli*. They first compared the substrate preferences of different FARs; the FAR gene from *S. chinensis*, the *acr1* gene from *Acinetobacter baylyi*, and the *cer4* gene from *A. thaliana*, respectively. They concluded that when FadD expression was combined with ACR1 or CER4 expression in *E. coli*, the engineered strains predominantly produced a $C_{12/14}$ alcohol, and combined expression of FadD with FAR produced a predominant $C_{16/18}$ alcohol. Secondly, they tried different TEs to provide the needed FFAs, and found that BTE from *Umbellularia californica* predominantly produces $C_{12/14}$ FFAs, while *TesA′* mainly produces C14-C18 FFAs. Finally, by coexpression of BTE, *fadD*, and *acr1*, 210.1 mg/l of C12/14 alcohol was obtained (Kuhajda *et al.*, 1994; Campbell and Cronan, 2001; Steen *et al.*, 2010). Further optimization of the expression levels of BTE, *fadD*, and *acr1* increased the $C_{12/14}$ alcohol production to 449.2 mg/l, accounting for 75.0% of the total fatty alcohol production (598.6 mg/l). In addition, coexpression of TE gene *tesA'*, *fadD*, and FAR produced 101.5 mg/l $C_{16/18}$ alcohol, with $C_{16/18}$ alcohol accounting for 89.2% of the total fatty alcohol production (Lennen and Pfleger, 2012; Peralta-Yahya *et al.*, 2012; Zheng *et al.*, 2012).

A unique pathway for fatty alcohol formation was reported that proceed via another fatty acid activation route that depends on a class of enzymes called carboxylic acid reductase (CAR) (Balat, 2011; Akhtar *et al.*, 2013). A central catalytic feature of CAR enzymes is the prosthetic group 4′-phosphopantetheine, which is covalently linked via a phosphodiester bond to a serine residue. The formation and insertion of this prosthetic group is mediated by a phosphopantetheinyl transferase. Three key steps are involved in its reaction mechanism: adenylylation of the bound fatty acid substrate to form an AMP-fatty acyl complex; formation of a thioester linkage between the fatty acyl moiety and the phosphopantetheine prosthetic group; and reduction of the thioester intermediate to the aldehyde. A putative CAR sequence from *Mycobacterium marinum* (UniProt accession number B2HN69, CAR) was characterized. Coexpression of a phosphopantetheinyl transferase, *Sfp*, from *Bacillus subtilis* alongside CAR and an aldehyde reductase (AHR) (slr1192) in *E. coli*, resulted a titre of 200 mg/l (C_{12}–C_{18}) (Leung *et al.*, 2010; Akhtar *et al.*, 2013). A three-fold increased fatty alcohol titre was obtained after further optimization by replacing slr1192 with *E. coli* YjgB (Ahr). Titres of 363 mg/l, was observed when engineered strains were cultivated in complex media or in minimal media with increased glucose and NH_4Cl content.

The same idea was transplanted into cyanobacteria. By overexpressing the FAR gene from jojoba, fatty alcohol production of 200 μg/l was obtained. Although the production efficiency described here is still low, the results show the potential of employing cyanobacteria-based platforms to produce fatty alcohols (Chouhan and Sarma, 2011; Tan *et al.*, 2011). It will be reviewed in Chapter 8.

A disadvantage of using *E. coli* as a host for fatty alcohol production is that the direct product of fatty acid synthase enzymes is in the form of fatty acyl-ACP. This enzyme-linked product needs to be hydrolysed by a TE to FFA and subsequently activated to fatty acyl-CoA by FadD before FAR can act on them. Because the product of *S. cerevisiae* fatty acid synthase is already in the form of fatty acyl-CoA, the correct form for the 'converting enzymes,' the yeast system is a more direct way to produce fatty alcohols (Leung *et al.*, 2010; Runguphan and Keasling, 2013). Moreover, *E. coli* is susceptible to phage infection, which could hamper production at industrial levels. *De novo* FAB in *S. cerevisiae* requires acetyl-CoA carboxylase (ACC; encoded by the ACC1 gene) and the fatty acid synthase complex (FAS; encoded by FAS1 and FAS2; the yeast FAS complex is a 2.6-MDa protein consisting of two non-identical, multi-functional subunits, α and β, organized as a hexamer (*a6β6*) (Wettstein-Knowles *et al.*, 2006; Tehlivets *et al.*, 2007). A primary distinction is that in *E. coli* approximately 10 genes encode for FAB while is *S. cerevisiae* C_{16}–C_{18} fatty acid biosynthesis enzymes are encoded by merely two genes (FAS1 and FAS2). Therefore, regulated engineering of the FAB pathway is more straightforward in *S. cerevisiae*.

In *S. cerevisiae*, by overexpression of *TesA'*, FFA production increased by eight-fold, to 5 mg/l.

In yeast, FFAs released from the FAS synthase complex are converted back to fatty acyl-CoAs by the fatty-acyl-CoA synthases FAA1 and FAA4. Deletion of FAA1 in a *TesA'* overexpressing yeast strain led to 164 mg/l of FFAs, and deletion of both FAA1 and FAA4 resulted in 207 mg/l FFAs. When all FAB genes were driven by a strong constitutive promoter P_{TEF1} (i.e. ACC, FAS1, and FAS2) coupled with *TesA'*, the production of FFAs increased to 400 mg/l, a 670-fold increase over the levels observed in the parent strain. Encouraged by this result, further engineering was done to transfer FFAs to fatty alcohol. By overexpressing an NADPH-dependent fatty acyl-CoA reductase from *Mus musculus* (mouse), 47.4 mg/l of fatty alcohol was obtained. Additional P_{Tef1}replacement of the native promoter improved production levels to 56.5 mg/l. At last, mFAR1 and P_{TEF1}-driven-ACC resulted in 93.4 mg/l production of fatty alcohol. Increasing the pool of cytosolic NADPH lead to a slight increase in fatty alcohol yield to 98 mg/l; this was achieved by overexpressing an NADP-dependent malic enzyme from the oleaginous fungus *M. alpina*. This is the highest production level of fatty alcohols reported in *S. cerevisiae* to date. The distribution of fatty alcohols was mainly C16:0 which accounted for 91.1% and $C_{18:0}$ accounted for about 8.9% (White *et al.*, 2005; Runguphan and Keasling, 2013).

7.8 Genetic engineering to improve fatty acid methyl/ethyl ester production

Biodiesel, a mixture of fatty acid methyl esters (FAMEs) or fatty acid ethyl esters (FAEEs), is closer in characteristics to diesel fuels than other fatty acid-derived biofuels. It is also non-toxic and biodegradable. Currently it is being produced from plant oils, animal fats, or FFA using traditional catalysis. The advantages of biodiesel as diesel fuel are liquid nature portability, ready availability, renewability, higher combustion efficiency, lower sulfur and aromatic content, higher octane number, and higher biodegradability. Pure biodiesel or biodiesel mixed in any ratio with petroleum-based diesel can be used in conventional diesel engines with no or only marginal modifications, and it can be distributed using the existing infrastructure. Biodiesel is already produced in a growing number of countries on large scale. How to improve the production of FAME/FAEE has become an increasingly important challenge during recent years.

Waxes are oxygen esters of primary fatty alcohols and fatty acids that are synthesized in a wide range of microorganisms and in specialized tissues of higher plants and animals, such as in the seed of jojoba. In all of these examples, the final step in wax biosynthesis is the transfer of an acyl chain from fatty acyl-CoA to a fatty alcohol. This reaction is catalysed by fatty acyl-CoA:fatty alcohol acyltransferase (wax synthase, WS). The jojoba WS gene was cloned in 2000, but this plant WS is not active in *E. coli* or yeast so it cannot be utilized to synthesize fatty acid methyl/ethyl esters. In 2003, a bifunctional WS/DGAT (WS/diacylglycerol acyltransferase) was found in a bacterium, *Acinetobacter calcoaceticus*. Different from WS, Acyl-CoA:DAG acyltransferase (DGAT) uses diacylglycerol (DAG) as a substrate and catalyses the acylation of DAG. This bifunctional acyltransferase is coded by the *atfA* gene. Heterologous expression of WS/DGAT in *S. cerevisiae* resulted also in the accumulation of fatty acid ethyl esters as well as fatty acid isoamyl esters. *In vitro* studies confirmed that WS/DGAT is capable of utilizing a broad range of alcohols as substrates including long-chain fatty alcohols such as hexadecanol as well as short-chain alcohols like ethanol or isoamyl alcohol (Kalscheuer, 2002; Tong, 2005). Although the production of FAEEs was very low in *S. cerevisiae*, these findings demonstrated the possibility of using this gene for biodiesel synthesis, even if further genetic modification is needed.

Because FAEEs are produced from acyl-CoA and ethanol, increasing the supply of ethanol and acyl-CoA are equivalently critical in maximizing FAEE production. Although *E. coli* is known to form ethanol during mixed-acid fermentation, ethanol synthesis was too inefficient to support detectable FAEE production under anaerobic condition. Heterologous expression of pyruvate decarboxylase (the *pdc* gene product) and alcohol dehydrogenase (the *adhB* gene product) from the strictly anaerobic ethanologenic Gram-negative bacterium *Zymomonas mobilis* into *E. coli* increased ethanol production. Thus FAEE biosynthesis was established by combining expression of *pdc, adhB* and *atfA*, and FAEE concentrations of 1.28 g/l and an FAEE content of 26% of the cellular dry mass were achieved by fed-batch fermentation using renewable carbon sources (Tong, 2005; Kalscheuer, 2006).

FAMEs and FAEEs have comparable chemical and physical fuel properties and engine performances. FAMEs are synthesized predominantly via the transesterification of triglycerides. Direct intracellular FAME synthesis in bacteria is an attractive alternative to current methods of biodiesel production since it can bypass the transesterification and subsequent purification steps. A novel methyltransferase that methylates fatty acids and 3-OH fatty acids using *S*-adenosylmethionine (SAM) as the methyl donor was identified and characterized (Magnuson *et al.*, 1993; Nawabi *et al.*, 2011). SAM is synthesized by the action of methionine adenosyltransferase (MAT, AdoMet), which catalyses the reaction between methionine and ATP. AdoMet in turn regulates methionine levels by interacting with the global methionine regulator MetJ. Expressing FAMT and a novel bacterial TE led to the production of FAME, and by deleting *metJ* SAM levels were increased, which in turn enhanced FAME production. Although the titres was quite low (16 mg/l), the author suggested that achieving higher yield of FAME will likely require improving TE and FAMT activities and optimizing SAM recycling.

To develop a cost-effective process that utilizes glycerol as the carbon source, a strategy that uses glycerol to produce ethanol was combined with acyltransferase expression in *S. cerevisiae*, and approximately 0.52 g/l FAEEs were produced, while nearly 17 g/l of glycerol was consumed. *FPS1*, a gene in the glycerol transport pathway, was first knocked out to prevent FPS1 from constitutively releasing glycerol, and deletion of GPD2 in the glycerol synthesis pathway resulted in a significant reduction of glycerol, which increased ethanol yield as a consequence (Lambalot and Walsh, 1995; Yu *et al.*, 2011).

Another group tried to modify all metabolic pathway involved in FAEE production and impressive results were obtained. Expression of *atfA* in an FFA-producing *E. coli* (*fadD, TesA'*) with exogenous feeding of ethanol produced a titre of 0.4 g/l of FAEE. Deletion of *fadE* contributed to the production of FAEE. As consistent with the FFA composition, different TEs resulted in different chain-length FAEEs, ranging from C12 to C18. To eliminate the need to feed exogenous ethanol to produce FAEEs, the FAEE-producing *E. coli* strain was further engineered to co-produce ethanol by expressing *pdc* and *adhB*. The addition of the minimal FAEE biosynthesis pathway (*TesA'* and *atfA*) into the ethanol-producing strain resulted in reduced FAEE production. But overexpressing of FAA2 (an acyl-CoA ligase from *S. cerevisiae*) resulted in an approximate 2.5-fold increase, and another 2.5-fold increase was achieved by overexpressing of *fadD*. Expression of an extra copy of *atfA* resulted in the production of 427 mg/l of FAEEs. Finally, by overlaying an organic phase of dodecane to prevent FAEE evaporation, 9.4% of the theoretical yield was obtained, or approximately about 674 mg/l. Higher titres were achieved by introducing *accBACD* genes into this strain in a scaled-up fed-batch fermentation, and nutritional and environmental conditions were optimized, yielding an FAEE titre of 922 mg/l (Tong, 2005; Steen *et al.*, 2010).

One approach to reduce the costs of advanced biofuel production is to engineer a single microorganism to both digest plant biomass and produce biofuels. However, large quantities of glycoside hydrolase enzymes are required to efficiently convert lignocellulose into fermentable sugars to feed the biofuel-producing microorganisms. Several microorganisms have been engineered to ferment model cellulosic and hemicellulosic substrates directly into ethanol or other fuels. Since *E. coli* has been engineered to biosynthesize perhaps the most chemically diverse range of chemicals of any organisms it will be very exciting if we can develop a cellulolytic strain that could also produce biofuels (Lu *et al.*, 2008; Liu *et al.*, 2010; Bokinsky and Peralta-Yahya, 2011). To overcome the obstacle that *E. coli* cannot secrete cellulases (Cel) in quantities required for industrial-scale lignocellulose hydrolysis, Cel enzyme from *Bacillus* and *Clostridium stercorarium* endoxylanase Xyn10B were fused with the protein OsmY, to enable protein export. To further hydrolyse cellodextrins into glucose, two β-glucosidases from *Cellvibrio japonicus*, *encel3A* or *cel3B*, were selected. To enable growth of *E. coli* on xylodextrins, the oligosaccharide products of xylan hydrolysis, *gly43F*, xylobiosidase genes from *C. japonicas* was chosen. These genes were put under the respective control of their two native promoters. This modification associated with previously described FAEE production strategies resulted in 100 mg/l FAEE from a minimal media formulated with ionic liquid pretreated switchgrass. This is the first step toward consolidated bioprocessing to produce advanced biofuels.

To close the gap between the theoretical yields and practical production, a promising way is to maximize the flux through FAB and balancing the rates of all modules involved in this pathway. A recent synthetic biology studied aimed at balancing the supply of ethanol and acyl-CoA. Expression of pathway genes at too low level creates bottlenecks, whereas expression at too high level diverts cellular resources to wasteful production of unnecessary RNAs, proteins, or intermediate metabolites. Furthermore, heterologous enzymes or pathway intermediates are sometimes toxic to the host. So, the ideal metabolic pathway would be dynamically regulated in response to the physiological state of the cell. Dynamic regulation would allow an organism to sense the concentration of critical pathway intermediates and dynamically regulate the production and consumption of the intermediates. A dynamic sensor-regulating system (DSRS) was designed to enhance the production of FAEE (DiRusso *et al.*, 1993; Zhang *et al.*, 2012a). Because we know that fatty acyl-CoA is a key intermediate during FAB, and FFAs also will inhibit the activity of enzymes involved, so a hybrid fatty acid/acyl-CoA biosensor was engineered. There is a naturally existed fatty acid-sensing protein FadR. So, the 17-bp FadR-binding DNA sequence from PfadBA was integrated into two locations in the phage promoters. In the absence of fatty acid, FadR is expected to bind to the 17-bp DNA sequences, which would interfere with RNA polymerase binding to the phage promoter and inhibit transcription. When fatty acid is present, FadD activates fatty acid to acyl-CoA and acyl-CoA in turn binds to FadR and releases FadR from the synthetic promoter, initiating downstream transcription. To eliminate leaky expression, the regulatory element of LacI was integrated to the acyl-CoA/fatty acid sensor. To realize the goal of dynamic regulation of the FAEE synthesis pathway, the biosensor-derived promoter was cloned upstream of the ethanol biosynthesis module (*adhB* and *pdc*) and wax-ester synthase module (*fadD* and *atfA*). At low intracellular fatty acid concentration, FadR appears as an *Apo*-form, which would bind DNA to repress production of ethanol and unnecessary acyl-CoA. When there is extra fatty acid present, the fatty acid would first be activated to acyl-CoA by the enzyme of the chromosomal *fadD*, acyl-CoA would then bind with FadR and turn FadR to its *holo*-form which would break away from the binding site and result in expression of genes that encode enzymes to produce ethanol, activate more FFA to fatty acyl-CoA, and convert ethanol and acyl-CoA to FAEE. After optimizing the combination of biosensor-derived promoters, 1.5 g/l FAEE was produced after 3 days incubation. The regulation mode of this DSRS system and all genetic modifications for FAEE and FAME synthesis are shown in Fig. 7.6.

7.9 Genetic engineering to improve fatty alkane/alkene production

Alkanes/alkenes with C_4–C_{23} carbon chain length possess higher energy density, hydrophobic properties, and are compatible with existing liquid fuel infrastructure including gasoline, petrodiesel, and jet fuels. Diesel and jet fuel currently account for more than half of the world refinery output destined for use in vehicles. Furthermore, diesel engines improve fuel efficiency in small passenger vehicles over gasoline engines, and have already been widely adopted in Europe. Medium- and long-chain alkanes/alkenes can potentially serve as replacements for diesel, rendering them an attractive target for microbial production from lignocellulosic feedstocks (Magnuson *et al.*, 1993; Knothe, 2010; Handke *et al.*, 2011).

Various organisms such as bacteria, yeasts, plants, and insects can produce alkanes/alkenes with different chain lengths. In higher plants, alkanes are mainly involved in synthesizing the epicuticula wax layer, whose primary function is to reduce water loss through the epidermis

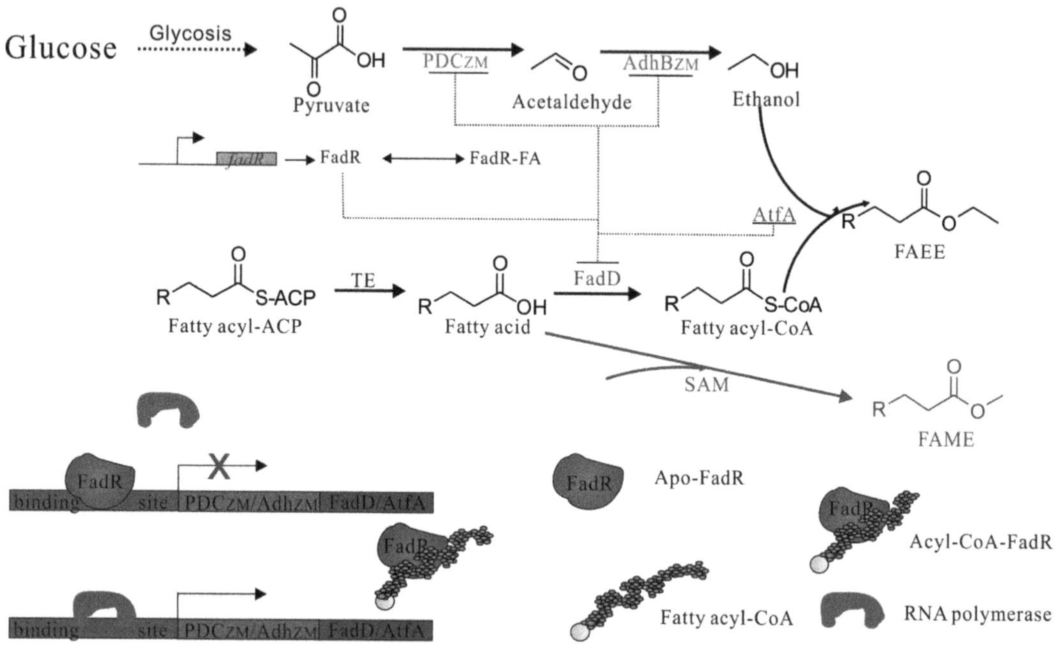

Figure 7.6 Genetic modifications for overproduction of fatty acid methyl/ethyl ester. PdcZM, pyruvate decarboxylase from *Zymomonas mobilis*. AdhB, alcohol dehydrogenase from *Zymomonas mobilis*. AtfA, wax-ester synthase from *Acinetobacter calcoaceticus.* FAMT, fatty acid methyltransferase. The dashed lines represent the DSRS regulation. FadR-FA represents FadR bound to acyl-CoA. When fatty acid is absent, FadR binds to the promoter and the transcription of genes in blue and magenta would be turned off. When there is abundant fatty acid, the fatty acid is catalysed to acyl-CoA by FadD, the acyl-CoA then binds with FadR, the Acyl-CoA-FadR breaks away from the DNA-binding site, and the transcription of the genes in blue and magenta are turned on. A colour version of this figure is available in the plate section at the back of the book.

(Post-Beittenmiller, 1996; Lu *et al.*, 2008). The role of hydrocarbons in microorganisms is still not completely understood. The first report on the production of C_{10}–C_{25} alkanes by sulfate-reducing bacteria grown in seawater-containing media supplemented with organic or fatty acids was made in 1944. But during the last decades years, most studies of alkane biosynthesis have focused on eukaryotic systems, such as plants including pea leaves, jojoba, Arabidopsis, etc. (Cheesbrough, 1984; Aarts *et al.*, 1995; Liu *et al.*, 2010). In 1995, a gene named CER (ECERIFERUM) was cloned in *Arabidopsis* and its role of conversing fatty aldehyde to fatty alkane was determined (Aarts *et al.*, 1995; Steen *et al.*, 2010; Peralta-Yahya *et al.*, 2012).

In 2010, by using a subtractive genome analysis method, two key enzymes responsible for fatty alkane/alkene formation were identified in cyanobacteria (Schirmer *et al.*, 2010; Steen *et al.*, 2010; Peralta-Yahya *et al.*, 2012). ACP can be reduced to aldehyde by an acyl-ACP reductase (AAR, PCC7942_orf1594 or Sll0209), and then aldehyde can be oxidized to alkane or alkene by an aldehyde deformylating oxygenase (ADO, PCC7942_orf1593 or Sll0208). Expression of PCC7942_orf1593 alone was indistinguishable from the *E. coli* negative controls, which was expected because *E. coli* does not naturally produce fatty aldehydes. Coexpression of PCC7942_orf 1593 and Orf 1594 in an *E. coli fadD*-deletion strain led to similar levels of alkane versus the wild-type strain, which demonstrated that this AAR preferred acyl-ACP but not acyl-CoA as substrate *in vivo*, which was confirmed by *in vitro* experiments. The acyl-ACP preference allows it to distinguish from fatty acyl-CoA reductases (FARs) that specifically reduce acyl-CoAs but not acyl-ACPs to fatty aldehydes (Doan *et al.*, 2009; Schirmer *et al.*, 2010).

Because acyl-ACP is the substrate for AAR, increasing the supply of ACP was hypothesized to increase alkane production. In 2010, the gene encoding acyl-ACP synthetase (AAS) in *Synechocystis sp.* PCC 6803 was cloned and designed as open reading frame *slr1609*. The slr1609-knockout mutant was incapable of importing exogenous fatty acids and secreted fatty acids released from membrane lipids into the medium. This suggests a remarkable role for this cyanobacterial AAS in recycling released fatty acids (Zheng *et al.*, 2004; Kaczmarzyk and Fulda, 2010). In another study, a series of *Synechocystis* sp. PCC6803 mutant strains have been constructed and confirmed. Alka (e)ne content in a *Synechocystis* mutant harbouring alkane biosynthetic genes overexpressed in both slr0168 and slr1556 gene loci (LX56) was 1.3% of cell dry weight, which was enhanced by 8.3 times compared with wild-type strain (0.14% of cell dry weight) cultivated in shake flasks (Doan *et al.*, 2009; Zhang *et al.*, 2011a; Wang *et al.*, 2013). it will be reviewed in Chapter 8.

For hydrocarbon biosynthesis, beside 'elongation-decarboxylation', another pathway mentioned as 'head-to-head condensation' also was described although not in depth. A metabolic pathway for alkene biosynthesis starting with acyl-CoA or acyl-ACP thioesters and involving decarboxylative Chainsen condensation was identified in *Micrococcus luteus* ATCC 4698 (also NCTC 2665), and this was catalysed by OleA (Beller *et al.*, 2010; Steen *et al.*, 2010). These researchers selected three candidate genes according to bioinformatics analysis, and purified protein from one of the three genes, Mlut_13230 (OleA), mixed with acyl-CoA, and cell lysates resulted in the formation of monoketone. The first step involves the formation of a β-ketoacyl-CoA, which is converted in the early step of a β-oxidation. Then OleA catalyses the decarboxylation of the β-ketoacyl-CoA and nucleophilic attack by the resulting carbanion on an acyl-CoA to form an aliphatic diketone. Following aliphatic diketone formation by OleA, alkene biosynthesis could follow a series of reductase and dehydratase reactions. In addition to carbon chain length, a key characteristic that distinguishes most intermediates in the proposed alkene biosynthesis pathway from those in the fatty acid biosynthesis pathway is the absence of an ACP thioester.

A third biosynthesis pathway to convert fatty acid intermediates into hydrocarbons involves a terminal olefin-forming P450 enzyme (Rude *et al.*, 2011; Zhang *et al.*, 2012b). This enzyme, *Jeotgalicoccus* sp. OleT ($OleT_{JE}$), was identified by purification from cell lysates, and its encoding gene was identified from a draft genome sequence of *Jeotgalicoccus* sp. ATCC 8456 using reverse genetics. Heterologous expression of the identified gene in *E. coli* demonstrated the new gene, *Jeotgalicoccus oleT* (*oleT*JE), was sufficient to confer olefin biosynthesis. Feeding studies suggested that the putative enzyme used an elongation-decarboxylation mechanism to convert fatty acyl-acyl carrier proteins (fatty acyl-ACPs) to α-olefins. A gene involved in the production of medium-chain α-olefins was identified in the cyanobacterium *Synechococcus sp.* strain PCC 7002. The gene encodes a large multidomain protein with homology to type I polyketide synthases, suggesting a route for hydrocarbon biosynthesis from fatty acids via a sulfonation-assisted decarboxylation mechanism (Mendez-Perez *et al.*, 2011; Xu *et al.*, 2013).

In 2013, a study suggested yet another pathway for alkane synthesis, which was distinguished from the pathway described in 2010. The synthesis of fatty aldehyde intermediates was not catalysed by AAR from acyl-ACP but by a carboxylic acid reductase (CAR) from fatty acid. By co-overexpressing all genes involved in the newly identified pathway [i.e. *TesA'*, Sfp (phosphopantetheinyl transferase from *B. subtilis*), CAR, and ADC (cyanobacterial aldehyde decarbonylase)], formation of undecane, tridecane, pentadecene and heptadecene with a titre of 2 mg/l was observed (Akhtar *et al.*, 2013). The pathway for fatty alkane/alkene synthesis is shown in Fig. 7.7.

The existing transport fuels are composed primarily of alkanes of various carbon chain lengths (C_n), branched hydrocarbons (iso-alkanes), and unsaturated hydrocarbons (n- alkenes). So the ideal biofuels are therefore n-alkanes, iso-alkanes, and n-alkenes that are chemically and structurally identical to the fossil fuels thus eliminating costly and time-consuming infrastructure and engine remodelling. In 2013, another group tried de

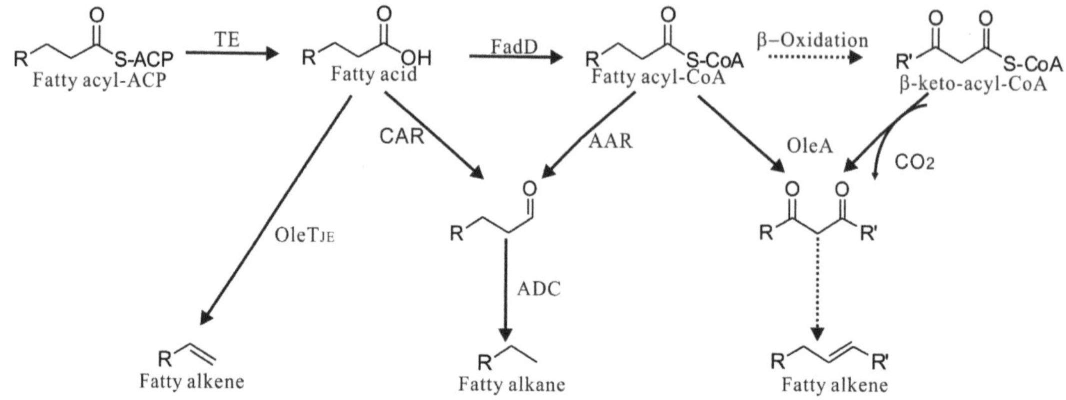

Figure 7.7 Alkane and alkene biosynthesis pathways. AAR, acyl-ACP reductase; CAR, carboxylic acid reductase. ADC, cyanobacterial aldehyde decarbonylase; $OleT_{JE}$, Olefin-forming fatty acid decarboxylase from *Jeotgalicoccus*; OleA, condensing-enzyme from *Micrococcus luteus*. The dashed line represents multi-enzyme involved alkene synthetic pathway.

novo design and assembly of synthetic metabolic pathways, to produce linear and branched-chain alkanes and alkenes of variable, but specified, Cn. At first, they confirmed the ability of a synthetic CEDDEC pathway that can converse free fatty acid to fatty alkane. The CEDDEC pathway involved luxC, luxE, and luxD from the *Photorhabdus luminescens*, these three genes composed the fatty acid reduction complex in this bioluminescent bacterium. Cn fatty aldehyde substrate then is catalysed to corresponding fatty alcohol by an aldehyde decarbonylase (NpAD) cloned from *Nostoc punctiforme*. As to control the chain length of the alkane FatB1 from *C. camphora* was used and as to produce branched-chain α-keto acid dehydrogenase complex (BCKD) and α-keto acyl-ACP synthase III (FabH2) from *B. subtilis* were implicated (Howard *et al.*, 2013). The characteristic of FabH2 that it has a broader specificity profile for fatty acid initiation than native FabH in *E. coli* was also utilized in another research that to expand the alkane product profile (Harger *et al.*, 2013).

7.10 Future perspectives

As detailed in this chapter, we can appreciate the significant sequential progress in bioengineered fuel production that has been achieved during recent decades. Nonetheless, the theoretical yield has never been attained, even in one case. But the great advances that have occurred encourage us to strive harder towards that target. The rapid progress in synthetic biology and molecular biology provide us with powerful tools to attempt this goal. We have also learned that there are several perspectives that we should take into account carefully.

As cells are an organic, dynamic, and unitary whole, every step in each metabolic pathway is not isolated but involves cross-talk with many other pathways. Therefore, engineering of each gene will often in turn modify other metabolic reactions or may cause the gene to be subject to feedback regulation by other genes. At the beginning of metabolic engineering studies, much research focused on modifying a single gene or several genes in one pathway; although some progress has been realized, production still lies well below the theoretical line. Some novel methods have appeared in the last 3 years that have significantly accelerated innovation in the microorganism-producing biofuel industry. These novel methods include targeted engineering, synthetic protein scaffolds and consolidated bioprocessing.

Targeted engineering refers to the method of optimizing the ratios of all enzymes, co-factors, metals, and substrates by reconstituting the whole metabolic pathway first and then using the obtained information to guide the engineering work *in vivo*. Almost all of the achievements to date are based on labour-intensive metabolic engineering work. Although we have made great strides in the field, there is much work to be done to identify potential new metabolic pathways, elucidate and modulate known pathways, and understand the interactions and cross-regulation that occurs between metabolic mechanisms. So to find the key enzymes or co-factors and to determine the optimized ratio of all genes involved as soon as possible will save us quantities of time and tell us the appropriate engineering directions to pursue. Targeted engineering affords us such convenience. Also for fatty acid overproduction, *in vitro* reconstitution and steady-state analysis of the fatty acid synthase was vital to understand all of the factors that control carbon flux from glucose to fatty acid (Dellomonaco *et al.*, 2011; Yu *et al.*, 2011). As elucidated in Fig. 7.8, the first step in targeted engineering is not to do genetic modification but to reconstitute the whole metabolic pathway *in vitro* using purified proteins and all co-factors, and then titrate all substrates, cofactors, and metals to determine the optimal molar ratio of these reagents. This is contrary to traditional metabolic engineering. This reconstituted system contains physiologically accurate ratios of all the macromolecular components responsible for transforming acetyl-CoA into fatty acids in *E. coli*. Then the roles of all enzyme and cofactors have been studied by *in vitro* steady-state analysis. Potential metabolic bottlenecks have been examined by titrating the amounts of substrates, intermediates, cofactors, and enzymes in this cell-free system (Stoops *et al.*, 1978; Wakil, 1989; Lomakin *et al.*, 2007; Yu *et al.*, 2011). This approach has provided useful direction for future genetic engineering. For example, this system demonstrated that fatty acid synthesis strongly depends on malonyl-CoA, so enhancing ACC activity is meaningful for our goals of increasing biofuel synthesis (Tehlivets *et al.*, 2007; Zha *et al.*, 2009). We have also learned that the ideal molar ratio of FabA:FabB:FabD:FabF:FabG:FabH:FabI:FabZ:holo-ACP:TesA for optimizing fatty acid synthesis is approximately 1:1:1:1:1:1:10:10:30:30. Then directed by this information further optimization for fatty acid production was made. For example the observation that FabI and especially FabZ increase FAS turnover *in vitro* guided us to express the full complement of reductive enzymes (FabG, FabZ,

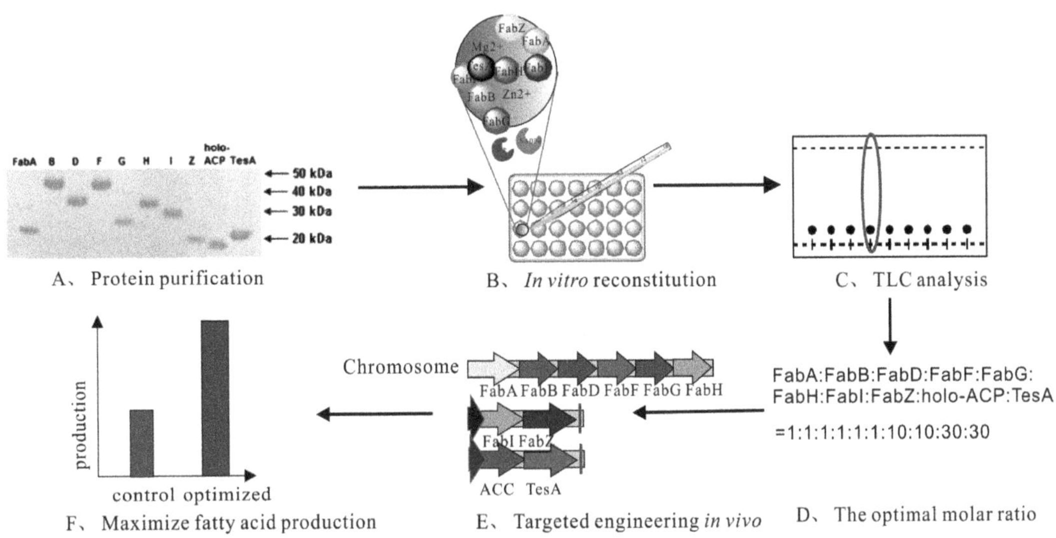

Figure 7.8 Schematic drawing of targeted engineering. (A) Protein purification; (B) *In vitro* reconstitution of the FAB pathway, which includes not only proteins, but also cofactors and metal ions; (C) Free fatty acid quantification using TLC to determine the optimal molar ratio; (D) The optimal molar ratio of different enzymes were determined; (E) Using the data from step D to guide *in vivo* genetic engineering; (F) Maximized free fatty acid production obtained. A colour version of this figure is available in the plate section at the back of the book.

and FabI) in the engineered strain, which was previously described. The resulting strain produces ca. 50% more fatty acids. And this method can be applied to other pathways engineering efforts beyond FAB. The schematic of targeted engineering is shown in Fig. 7.8.

Growing evidence indicates that many cellular reactions within metabolic pathways are catalysed not by free-floating 'soluble' enzymes, but via one or more membrane-associated multienzyme complexes, such as tryptophan biosynthesis (Conrado *et al.*, 2008; Tang *et al.*, 2013). The tryptophan biosynthesis pathway is mediated by 13 total enzyme reactions, seven of which constitute the polyaromatic branch. In fungi, this pathway includes a multifunctional enzyme known as the AROM complex that has evolved to link five distinct enzymatic activities into a single polypeptide. In this manner, complex small molecules can be synthesized with very low rates of loss of intermediates. This spatial organization reminds us that we can synthesize protein scaffolds to improve pathway efficiency, because the scaffold could prevent the loss of intermediates to diffusion or competing pathways. Some groups made valuable attempts to demonstrate this possibility. They used naturally existing domain/ligand to fuse with proteins in three pathways, and obtained improved titre production through this synthetic protein scaffold. A 77-fold improvement was achieved when this strategy was applied to the mevalonate biosynthetic pathway (Dueber *et al.*, 2009; Nielsen, 2009; Santos and Riezman, 2012). This result gave us useful inspiration that we employ similar approaches to further improve fatty acid derived biofuel production, in addition to complementary existing genetic engineering methods, such as manipulating promoter strengths, ribosome binding site strengths, plasmid copy number, or improving turnover activities of rate-limiting enzymes by directed evolution.

The progress obtained during these years, such as the production of FAEE and fatty alcohol, has a promising advantage over first generation products, such as corn ethanol and plant oil-derived biodiesel. However a primary goal in renewable energy is the production of these compounds directly from cellulosic plant biomass. As ethanol production from starch and simple sugar, the feedstock generally amounts to 60–80% of the fuel production cost. So, another term named 'consolidated bioprocessing', or CBP, is receiving more and more interest now. CBP was formerly referred as 'direct microbial conversion'. Two distinctly different approaches were developed to achieve the goal of increasing microbial biofuel production from cellusolic biomass. One is the use of highly cellulolytic organisms such as the thermophilic *Clostridium thermocellum*, either exclusively or in co-culture with other thermophilic, higher-producing sugar fermenters (Fischer *et al.*, 2008; Lu, 2010; Machado and Atsumi, 2012). These organisms permit cellulase production, cellulose hydrolysis, and fermentation to occur anaerobically in the same process vessel. Another approach tries to integrate the enzymes for sugar liberation into the biofuel-producing strain. Cellulosic plant biomass is the most abundant feedstock in the world, so some studies try to engineer the genes encoding endoxylanase catalytic domain (Xyn10B) from *Clostridium stercorarium* and a xylanase (Xsa) from *Bacteroides ovatus* into their previously engineered strains. To hydrolyse the hemicellulose in the medium into xylose, the hemicellulases were fused with OsmY (Qian *et al.*, 2008; Hofvander *et al.*, 2011). Although the production of biofuels was not as high as in glucose-feeding culture medium, the result demonstrated that it is possible that engineered *E. coli* have the ability to grow depending on cellulose and hemicellulose without addition of expensive enzymes (Kunst and Samuels, 2003; Steen *et al.*, 2010).

In summary, to date, the majority of improvements in biofuel production have been achieved by fine tuning discrete steps within the production pathway, or by improving the reactions that regulate carbon flux through the pathway. Now, the rapid development of system biology put more possibilities into the industrialization of microorganism-producing biofuels. Synthetic biology conceptualizes the cell as an assembly of parts that leads to a specific phenotype (de Azevedo Souza *et al.*, 2009; Doan *et al.*, 2009; Peralta-Yahya and Keasling, 2010), while system biology takes a system-level look of microbial metabolism and the effects of fuel toxicity. By integrating these two distinctly different but related and complementary research modes together, we may soon have engineered microorganisms that are capable of producing commercially feasible quantities of advanced biofuels.

References

Aarts, M.G., Keijzer, C.J., Stiekema, W.J., and Pereira, A. (1995). Molecular characterization of the CER1 gene of *Arabidopsis* involved in epicuticular wax biosynthesis and pollen fertility. Plant Cell 7, 2115–2127.

Akhtar, M.K., Turner, N.J., and Jones, P.R. (2013). Carboxylic acid reductase is a versatile enzyme for the conversion of fatty acids into fuels and chemical commodities. Proc. Natl. Acad. Sci. U.S.A. *110*, 87–92.

de Azevedo Souza, C., Kim, S.S., and Koch, S. (2009). A novel fatty Acyl-CoA Synthetase is required for pollen development and sporopollenin biosynthesis in Arabidopsis. Plant Cell *21*, 507–525.

Balat, M. (2011). Potential alternatives to edible oils for biodiesel production – a review of current work. Energy Conversion and Management 52, 1479–1492.

Beller, H.R., Goh, E.B., and Keasling, J.D. (2010). Genes involved in long-chain alkene biosynthesis in *Micrococcus luteus*. Appl. Environ. Microbiol. 76, 1212–1223.

Bokinsky, G., and Peralta-Yahya, P.P. (2011). Synthesis of three advanced biofuels from ionic liquid-pretreated switchgrass using engineered *Escherichia coli*. Proc. Natl. Acad. Sci. U.S.A. *108*, 19949–19954.

Campbell, J.W., and Cronan, J.E., Jr (2001). Bacterial fatty acid biosynthesis: targets for antibacterial drug discovery. Annu. Rev. Microbiol.55, 305–332.

Cheesbrough, T.M. (1984). Alkane biosynthesis by decarbonylation of aldehydes catalyzed by a particulate preparation from Pisum sativum. Proc. Natl. Acad. Sci. U.S.A.*81*, 6613–6617.

Chouhan, A.P.S., and Sarma, A.K. (2011). Modern heterogeneous catalysts for biodiesel production: A comprehensive review. Renewable and Sustainable Energy Reviews *15*, 4378–4399.

Conrado, R.J., Varner, J.D., and DeLisa, M.P. (2008). Engineering the spatial organization of metabolic enzymes: mimicking nature's synergy. Curr. Opin. Biotechnol. *19*, 492–499.

Dellomonaco, C., Clomburg, J.M., Miller, E.N., and Gonzalez, R. (2011). Engineered reversal of the [bgr]-oxidation cycle for the synthesis of fuels and chemicals. Nature *476*, 355–359.

DiRusso, C.O., Metzger, A.K., and Heimert, T.L. (1993). Regulation of transcription of genes required for fatty acid transport and unsaturated fatty acid biosynthesis in *Escherichia coli* by FadR. Mol. Microbiol. 7, 311–322.

Doan, T., Carlsson, A.S., Hamberg, M., and Bülow, L. (2009). Functional expression of five Arabidopsis fatty acyl-CoA reductase genes in *Escherichia coli*. Journal of Plant *166*, 787–796.

Dueber, J.E., Wu, G.C., Malmirchegini, G.R., Moon, T.S., Petzold, C.J., Ullal, A.V., Prather, K.L.J., and Keasling, J.D. (2009). Synthetic protein scaffolds provide modular control over metabolic flux. Nat. Biotechnol. *27*, 753–759.

Fischer, C.R., Klein-Marcuschamer, D., and Stephanopoulos, G. (2008). Selection and optimization of microbial hosts for biofuels production. Metab. Eng. *10*, 295–304.

Handke, P., Lynch, S.A., and Gill, R.T. (2011). Application and engineering of fatty acid biosynthesis in *Escherichia coli* for advanced fuels and chemicals. Metab. Eng. *13*, 28–37.

Harger, M., Zheng, L., Moon, A., Ager, C., An, J.H., Choe, C., Lai, Y.-L., Mo, B., Zong, D., Smith, M.D., *et al.* (2013). Expanding the Product Profile of a Microbial Alkane Biosynthetic Pathway. ACS Synth. Biol. *2*, 59–62.

Hofvander, P., Doan, T.T.P., and Hamberg, M. (2011). A prokaryotic acyl-CoA reductase performing reduction of fatty acyl-CoA to fatty alcohol. FEBS Lett. *585*, 3538–3543.

Howard, T.P., Middelhaufe, S., and Moore, K. (2013). Synthesis of customized petroleum-replica fuel molecules by targeted modification of free fatty acid pools in *Escherichia coli*. Proc. Natl. Acad. Sci. U.S.A. *110*, 7636–7641.

Kaczmarzyk, D., and Fulda, M. (2010). Fatty acid activation in cyanobacteria mediated by acyl-acyl carrier protein synthetase enables fatty acid recycling. Plant Physiol. *152*, 1598–1610.

Kalscheuer, R. (2002). A Novel Bifunctional Wax Ester Synthase/Acyl-CoA:Diacylglycerol Acyltransferase Mediates Wax Ester and Triacylglycerol Biosynthesis in Acinetobacter calcoaceticus ADP1. J. Biol. Chem.*278*, 8075–8082.

Kalscheuer, R. (2006). Microdiesel: *Escherichia coli* engineered for fuel production. Microbiology *152*, 2529–2536.

Knothe, G. (2010). Biodiesel and renewable diesel: a comparison. Progress in Energy and Combustion Science *36*, 364–373.

Kuhajda, F.P., Jenner, K., and Wood, F.D. (1994). Fatty acid synthesis: a potential selective target for antineoplastic therapy. Proc. Natl. Acad. Sci. U.S.A. *91*, 6379–6383.

Kunst, L., and Samuels, A.L. (2003). Biosynthesis and secretion of plant cuticular wax. Progr. Lipid Res. *42*, 51–80.

Lambalot, R.H., and Walsh, C.T. (1995). Cloning, overproduction, and characterization of the *Escherichia coli* holo-acyl carrier protein synthase. J. Biol.Chem. *270*, 24658–24661.

Lardizabal, K.D., Metz, J.G., and Sakamoto, T. (2000). Purification of a jojoba embryo wax synthase, cloning of its cDNA, and production of high levels of wax in seeds of transgenic Arabidopsis. Plant physiology *122*, 645–656.

Lennen, R.M., and Pfleger, B.F. (2012). Engineering *Escherichia coli* to synthesize free fatty acids. Trends in Biotechnology *30*, 659–667.

Leung, D.Y.C., Wu, X., and Leung, M.K.H. (2010). A review on biodiesel production using catalyzed transesterification. Applied Energy *87*, 1083–1095.

Liu, A., Tan, X., Yao, L., and Lu, X. (2013). Fatty alcohol production in engineered *E. coli* expressing *Marinobacter* fatty acyl-CoA reductases. Appl. Microbiol. Biotechnol. *97*, 7061–7071.

Liu, T., Vora, H., and Khosla, C. (2010). Quantitative analysis and engineering of fatty acid biosynthesis in *E. coli*. Metab. Eng. *12*, 378–386.

Lomakin, I.B., Xiong, Y., and Steitz, T.A. (2007). The crystal structure of yeast fatty acid synthase, a cellular machine with eight active sites working together. Cell *129*, 319–332.

Lu, X. (2010). A perspective: Photosynthetic production of fatty acid-based biofuels in genetically engineered cyanobacteria. Biotechnol. Advanc.*28*, 742–746.

Lu, X., Vora, H., and Khosla, C. (2008). Overproduction of free fatty acids in E. coli: Implications for biodiesel production. Metabolic Engineering *10*, 333–339.

Machado, I.M.P., and Atsumi, S. (2012). Cyanobacterial biofuel production. J. Biotechnol. *162*, 50–56.

Magnuson, K., Jackowski, S., Rock, C.O., and Cronan, J.E. (1993). Regulation of fatty acid biosynthesis in *Escherichia coli*. Microbiol. Rev. *57*, 522–542.

Mendez-Perez, D., Begemann, M.B., and Pfleger, B.F. (2011). Modular synthase-encoding gene involved in -olefin biosynthesis in *Synechococcus* sp. strain PCC 7002. Appl. Environ. Microbiol. *77*, 4264–4267.

Metz, J.G., Pollard, M.R., Anderson, L., Hayes, T.R., and Lassner, M.W. (2000). Purification of a jojoba embryo fatty acyl-coenzyme A reductase and expression of its cDNA in high erucic acid rapeseed. Plant Physiol. *122*, 635–644.

Miwa, T.K. (1971). Jojoba oil wax esters and derived fatty acids and alcohols: gas chromatographic analyses. J. AOC Soc. *48*, 259–264.

Nawabi, P., Bauer, S., Kyrpides, N., and Lykidis, A. (2011). Engineering *Escherichia coli* for biodiesel production utilizing a bacterial fatty acid methyltransferase. Appl. Environ. Microbiol. *77*, 8052–8061.

Nielsen, J. (2009). Systems biology of lipid metabolism: From yeast to human. FEBS Lett. *583*, 3905–3913.

Peralta-Yahya, P.P., and Keasling, J.D. (2010). Advanced biofuel production in microbes. Biotechnol. J. *5*, 147–162.

Peralta-Yahya, P.P., Zhang, F., and Del Cardayre, S.B. (2012). Microbial engineering for the production of advanced biofuels. Nature *488*, 320–328.

Post-Beittenmiller, D. (1996). Biochemistry and molecular biology of wax production in plants. Annual Review of Plant Biology *47*, 405–430.

Qian, Z.-G., Xia, X.-X., Choi, J.H., and Lee, S.Y. (2008). Proteome-based identification of fusion partner for high-level extracellular production of recombinant proteins in *Escherichia coli*. Biotechnol. Bioeng. *101*, 587–601.

Rude, M.A., Baron, T.S., Brubaker, S., Alibhai, M., Del Cardayre, S.B., and Schirmer, A. (2011). Terminal olefin (1-alkene) biosynthesis by a novel P450 fatty acid decarboxylase from *Jeotgalicoccus* species. Appl. Environ. Microbiol. *77*, 1718–1727.

Runguphan, W., and Keasling, J.D. (2013). Metabolic engineering of *Saccharomyces cerevisiae* for production of fatty acid-derived biofuels and chemicals. Metab. Eng. *27*, on line.

Santos, A., and Riezman, H. (2012). Yeast as a model system for studying lipid homeostasis and function. FEBS Lett. *586*, 2858–2867.

Schirmer, A., Rude, M.A., Li, X., and Popova, E. (2010). Microbial biosynthesis of alkanes. Science *329*, 559–562.

Silhavy, T.J., Ruiz, N., and Kahne, D. (2006). Advances in understanding bacterial outer-membrane biogenesis. Nat. Rev. Microbiol. *4*, 57–66.

Steen, E.J., Kang, Y., Bokinsky, G., Hu, Z., Schirmer, A., McClure, A., del Cardayre, S.B., and Keasling, J.D. (2010). Microbial production of fatty-acid-derived fuels and chemicals from plant biomass. Nature *463*, 559–562.

Stoops, J.K., Awad, E.S., Arslanian, M.J., and Gunsberg, S. (1978). Studies on the yeast fatty acid synthetase. J. Biol. Chem. *253*, 4464–4475.

Tan, X., Yao, L., Gao, Q., Wang, W., Qi, F., and Lu, X. (2011). Photosynthesis driven conversion of carbon dioxide to fatty alcohols and hydrocarbons in cyanobacteria. Metab. Eng. *13*, 169–176.

Tang, X., Feng, H., and Chen, W.N. (2013). Metabolic engineering for enhanced fatty acids synthesis in *Saccharomyces cerevisiae*. Metab. Eng. *16*, 95–102.

Tehlivets, O., Scheuringer, K., and Kohlwein, S.D. (2007). Fatty acid synthesis and elongation in yeast. Biochim. Biophys. Acta *1771*, 255–270.

Tong, L. (2005). Acetyl-coenzyme A carboxylase: crucial metabolic enzyme and attractive target for drug discovery. Cell. Mol. Life Sci. *62*, 1784–1803.

Vioque, J., and Kolattukudy, P.E. (1997). Resolution and Purification of an Aldehyde-Generating and an Alcohol-Generating Fatty Acyl-CoA Reductase from Pea Leaves (*Pisum sativum* L.). Archiv. Biochem. Biophys. *340*, 64–72.

Wakil, S.J. (1989). Fatty acid synthase, a proficient multifunctional enzyme. Biochemistry *28*, 4523–4530.

Wang, W., Liu, X., and Lu, X. (2013). Engineering cyanobacteria to improve photosynthetic production of alka (e) nes. Biotechnol. Biofuels 6:69.

Wettstein-Knowles, von, P., Olsen, J.G., McGuire, K.A., and Henriksen, A. (2006). Fatty acid synthesis. Role of active site histidines and lysine in Cys-His-His-type beta-ketoacyl-acyl carrier protein synthases. FEBS J. *273*, 695–710.

White, S.W., Zheng, J., Zhang, Y.-M., and Rock, C.O. (2005). The structural biology of type II fatty acid biosynthesis. Annu. Rev. Biochem. *74*, 791–831.

Willis, R.M., Wahlen, B.D., Seefeldt, L.C., and Barney, B.M. (2011). Characterization of a Fatty Acyl-CoA Reductase from Marinobacter aquaeolei VT8: A Bacterial Enzyme Catalyzing the Reduction of Fatty Acyl-CoA to Fatty Alcohol. Biochemistry *50*, 10550–10558.

Xu, P., Vansiri, A., Bhan, N., and Koffas, M. (2012). ePathBrick: a synthetic biology platform for engineering metabolic pathways in E. coli. ACS Synth. Biol. *1*, 256–266.

Xu, P., Gu, Q., Wang, W., Wong, L., Bower, A.G.W., Collins, C.H., and Koffas, M.A.G. (2013). Modular optimization of multi-gene pathways for fatty acids production in E. coli. Nat Comms *4*, 1409.

Yu, X., Liu, T., Zhu, F., and Khosla, C. (2011). In vitro reconstitution and steady-state analysis of the fatty acid synthase from *Escherichia coli*. Proc. Natl. Acad. Sci. U.S.A. *108*, 18643–18648.

Zha, W., Rubin-Pitel, S.B., Shao, Z., and Zhao, H. (2009). Improving cellular malonyl-CoA level in *Escherichia coli* via metabolic engineering. Metab. Eng. *11*, 192–198.

Zhang, F., Rodriguez, S., and Keasling, J.D. (2011a). Metabolic engineering of microbial pathways for advanced biofuels production. Curr. Opin. Biotechnol.22, 775–783.

Zhang, X., Li, M., Agrawal, A., and San, K.Y. (2011b). Efficient free fatty acid production in *Escherichia coli* using plant acyl-ACP thioesterases. Metab. Eng. *13*, 713–722.

Zhang, F., Carothers, J.M., and Keasling, J.D. (2012a). Design of a dynamic sensor-regulator system for production of chemicals and fuels derived from fatty acids. Nat Biotechnol.*30*, 354–359.

Zhang, F., Ouellet, M., Batth, T.S., Adams, P.D., Petzold, C.J., Mukhopadhyay, A., and Keasling, J.D. (2012b). Enhancing fatty acid production by the expression of the regulatory transcription factor FadR. Metab. Eng. *14*, 653–660.

Zheng, Y.N., Li, L.L., Liu, Q., Yang, J.M., and Wang, X.W. (2012). Optimization of fatty alcohol biosynthesis pathway for selectively enhanced production of C12/14 and C16/18 fatty alcohols in engineered *Escherichia coli*. Microb. Cell Factories,11, 65.

Zheng, Z., Gong, Q., Liu, T., Deng, Y., and Chen, J.C. (2004). Thioesterase II of *Escherichia coli* plays an important role in 3-hydroxydecanoic acid production. Appl. Environ. Microbiol. *70*, 3807–3813.

Biofuel Production by Genetically Engineered Cyanobacteria

8

Xuefeng Lu, Zhimin Li, Xiaoming Tan, Tao Zhu and Weihua Wang

Abstract

Cyanobacteria are a group of prokaryotes that perform plant-type oxygenic photosynthesis. In recent years, with the development of modern biotechnology, cyanobacteria have shown great potential for biofuel production directly from carbon dioxide and solar energy. Herein, the phylogenetic classification, cell structure and biotechnological and environmental applications of cyanobacteria will be introduced. After discussing the available molecular tools for the cyanobacterial genetic engineering and omics studies of cyanobacteria, biofuel production by engineered cyanobacteria is reviewed. At the end of this chapter, future trends for research on cyanobacterial biofuels and web resources are provided.

8.1 Overview of Cyanobacteria

8.1.1 Phylogenetic classification

Cyanobacteria are a versatile monophyletic group of prokaryotes that perform plant-type oxygenic photosynthesis. Traditionally, they are divided into five major subsections based on their morphological distinctions (Fig. 8.1). The strains of subsection I (e.g. Chroococcales, Prochlorophytes and Gloeobacterales) divide by binary fission, and those of subsection II reproduce by forming small, spherical baeocytes through internal multiple fission. These two taxonomic groups are unicellular. Subsections III, IV and V form filaments of varying morphological complexity. It was reported that the strains of subsection III (e.g. Oscillatoriales) had only vegetative cells, but in subsections IV (e.g. Nostocales) and V (e.g. Stigonematales), vegetative cells could differentiate into morphologically distinct heterocysts for nitrogen fixation and into akinetes (climate-resistant resting cells), depending on their growth conditions (Castenholz, 2001; Rippka *et al.*, 1979; Schirrmeister *et al.*, 2011; Shih *et al.*, 2013; Tomitani *et al.*, 2006).

Molecular phylogenetic analyses based on 16S rDNA indicate that cyanobacteria producing baeocytes (subsection II), heterocysts (subsections IV and V), and true-branching filaments (subsection V) are phylogenetically coherent (Giovannoni *et al.*, 1988; Turner *et al.*, 1999). To understand how and when cyanobacteria evolved, Tomitani *et al.* suggested that the clade of cyanobacteria marked by cell differentiation diverged once between 2450 and 2100 million

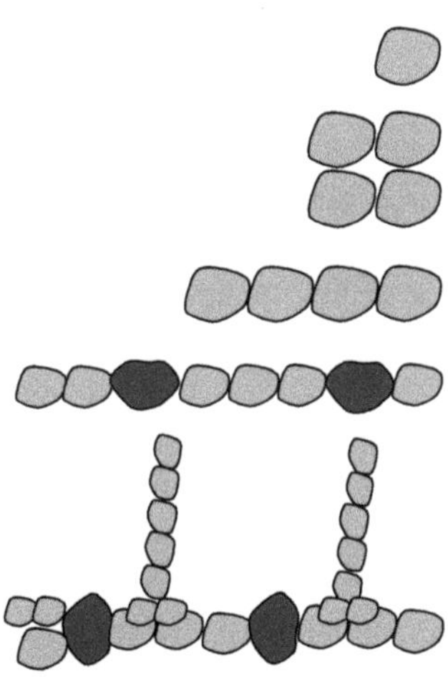

Figure 8.1 Morphological representation of the five cyanobacterial subsections (modified from Schirrmeister *et al.*).

Table 8.1 Representative genetic parts for the metabolic engineering of cyanobacteria

Genetic parts	Description	References
Promoters		
PrbcL	Different versions of the promoter for the large subunit of Rubisco in *Synechocystis* PCC6803	Huang *et al.* (2010), Qi *et al.* (2013)
PpetE	The promoter for the plastocyanin of *Synechocystis* PCC6803	Qi *et al.* (2013)
Ptrc	The *trc* promoter with lac operators	Huang *et al.* (2010)
PrnpB	The promoter for the RNaseP beta subunit of *Synechocystis* PCC6803	Huang *et al.* (2010)
PpsbD	The promoter for the D2 protein of the photosystem II reaction centre in *Synechocystis* PCC6803	Qi *et al.* (2013)
PatpB	The promoter for the ATP synthase beta subunit of *Synechocystis* PCC6803	Qi *et al.* (2013)
PpsbA1	The promoter for the D1 protein of photosystem II in *Synechocystis* PCC6803	Qi *et al.* (2013)
PpsbB	The promoter for the photosystem II core light harvesting protein of *Synechocystis* PCC6803	Qi *et al.* (2013)
PpsaD	The promoter for photosystem I subunit II of *Synechocystis* PCC6803	Qi *et al.* (2013)
Reporters		
Cerulean	The gene encoding cerulean fluorescent protein	Huang *et al.* (2010)
GFPmut3B	The gene encoding green fluorescent protein	Asayama (2012), Huang *et al.* (2010)
EYFP	The gene encoding yellow fluorescent protein	Huang *et al.* (2010), Landry *et al.* (2013)
LacZ	The gene encoding β-D-galactosidase	Qi *et al.* (2013)
Degradation tags		
SsrA tags	The native or modified protease degradation tags from the *ssrA* system	Huang *et al.* (2010), Landry *et al.* (2013)

years ago, basing this theory on the integration of phylogenetic analyses and geological evidence (Tomitani *et al.*, 2006). The evolutionary rates of cyanobacteria are very slow. Before eukaryotes evolved, cyanobacteria exhibited complex morphological diversity. Schirrmeister *et al.* conclude that a multicellular morphotype evolved early in the cyanobacterial lineage, and most of the morphological diversity exhibited in cyanobacteria today (including the majority of single-celled species, such as *Synechococcus* and *Prochlorococcus*) appears to arise from ancient multicellular lineages (Schirrmeister *et al.*, 2011). Cyanobacteria not only played a key role during the earth's evolution by converting the reducing atmosphere into an oxidizing one but are also of significant interest currently for biotechnological applications. As a promising platform for 'green' biotechnology, particularly for biofuel production, cyanobacteria have attracted considerable worldwide interest.

8.1.2 Cell structure

Cyanobacteria are primitive, single-celled, colonial or filamentous organisms characterized by a prokaryotic cell structure with highly differentiated membrane systems. They do not possess membrane-bound sub-cellular organelles or membrane-bound nuclei, and they are traditionally thought to be Gram-negative. However, the peptidoglycan layer found in cyanobacteria is considerably thicker than that of most Gram-negative bacteria (Hoiczyk and Hansel, 2000). The cyanobacterial cell envelope is composed of two membrane systems (plasma membrane and outer membrane) that are separated by a periplasmic space. In addition, cyanobacteria have an internal system of thylakoid membranes, in which the fully functional electron transfer chains of photosynthesis and respiration reside (Fig. 8.2). Each member of the membrane systems has specialized roles in the cyanobacterial cell. Understanding the organization, functionality, and dynamics of the membrane systems remains a great challenge in cyanobacterial cell biology. The interior of cyanobacterial cells usually contains several of the following components that are important to their lives, depending on the strain and growth conditions: (i) polygonal carboxysomes that contain the ribulose-1,5-bisphosphate carboxylase/oxygenase (RuBisCO) enzyme tightly packed in crystalline arrays; (ii) polyphosphate bodies; (iii) cyanophycin granules that consist of a non-ribosomally synthesized aspartic acid backbone and arginine side groups functioning as a dynamic nitrogen reservoir and possibly a carbon-storage compound; (iv) glycogenosomes; and (v) phycobilisomes (Allen, 1984) (Fig. 8.2).

In addition to the common structure and components, many species of cyanobacteria possess gas vesicles (e.g. *Microcystis, Anabaena, Calothrix* and *Planktothrix*) (Walsby, 1972, 1994) (Fig. 8.2). Gas vesicles are intracellular proteinaceous structures that function as floating devices to maintain cell buoyancy. They are the only subcellular structures that contain a gas-filled space in cyanobacteria (Pfeifer, 2012; Walsby, 1972, 1994). Gas vesicle proteins (Gvps) are encoded by a gene cluster of 8–14 genes, of which GvpA forms the main structure and GvpC stabilizes the structure. Gas vesicle protein gene expression can be regulated by ambient environmental factors (Mlouka *et al.*, 2004; Pfeifer, 2012).

During nitrogen starvation, the cyanobacterial strains of subsections IV (e.g. Nostocales) and V (e.g. Stigonematales) can form specialized nitrogen-fixing cells termed heterocysts. Compared to vegetative cells, heterocysts undergo many morphological and metabolic changes (Kumar *et al.*, 2010). To form heterocysts, the vegetative cells enlarge, granular inclusions decrease, cell envelopes become thicker, and polar plugs consisting of cyanophycin form at the poles adjacent to vegetative cells. Mature heterocysts contain photosystem I but no functional photosystem II, so they can generate ATP by cyclic photophosphorylation but cannot produce oxygen. These metabolic changes create a microaerobic environment and provide the appropriate conditions for the activity of the oxygen-sensitive nitrogenase. Heterocysts and vegetative cells are

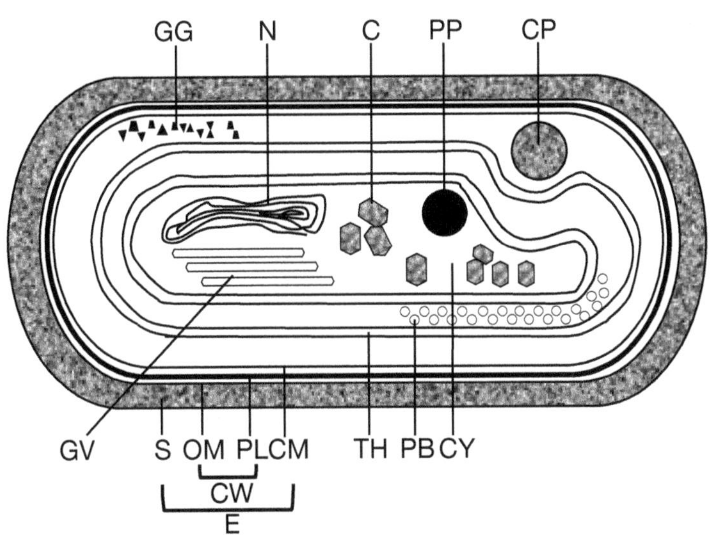

Figure 8.2 Schematic diagram of a cyanobacterial cell (S, external layer; OM, outer membrane; PL, peptidoglycan layer; CM, cytoplasmic membrane; CW, cell wall; CY, cytoplasm; E, cell envelope; GV, gas vesicle; N, nucleoplasmic region; C, carboxysome; PP, polyphosphate body; CP, cyanophycin granule; GG, glycogenosomes; TH, thylakoid; PB, phycobilisome) (modified from Castenholz, 2001).

mutually interdependent. Vegetative cells supply heterocysts with carbohydrates. In return, newly fixed nitrogen is distributed from heterocysts to nearby vegetative cells. Once a heterocyst has formed, it cannot revert to a vegetative cell. Certain heterocyst-forming bacteria can differentiate into resting cells called akinetes or motile cells called hormogonia, which serve as survival structures (Kumar *et al.*, 2010; Zhang *et al.*, 2006). It is important to thoroughly understand the cell structure to genetically engineer cyanobacterial cells that can synthesize biofuel molecules.

8.1.3 Biotechnological and environmental applications

Research on cyanobacteria in the last few decades has focused largely on their morphological, physiological and ecological traits, but relatively little research has been performed on their potential biotechnological and environmental applications. In recent years, these microorganisms have garnered extensive interest for their possible roles in helping mankind in various ways and due to the rapid progress in high throughput screening systems, genomics and bioinformatic tools. Cyanobacteria are under consideration as an alternative source of biofuels, human food, animal feed, bioactive chemicals, and biofertilizers and for the production of secondary metabolites, including exopolysaccharides, vitamins, toxins, enzymes and pharmaceuticals. They are also being developed to combat environmental pollution, such as their use in wastewater treatment and in the prevention of water bloom (Abed *et al.*, 2009).

Biofuels

Cyanobacteria are becoming more and more attractive as biofuel-producing microbial systems because they can utilize solar energy to directly convert carbon dioxide into biofuels in one single biological system. There are several advantages to the use of these microorganisms: (i) due to their natural diversity, cyanobacteria can thrive in a variety of environments, including areas that cannot support agriculture; (ii) cyanobacteria have a higher growth rate compared to microalgae and plants; and (iii) many cyanobacterial genomes have been sequenced (more than 126 strains), and several strains can be genetically engineered. Thus far, several biofuels, including ethanol, high alcohols (e.g. isobutanol, 1-butanol, 2-methyl-1-butanol, and diols), isoprene, and fatty acid-derived biofuels (fatty esters, fatty alcohols and fatty alkanes) (Angermayr *et al.*, 2009; Lu, 2010; Quintana *et al.*, 2011), have been successfully produced by model cyanobacterial strains. These bioenergy applications will be illustrated in detail in the appropriate section later in this chapter.

Food and feed

Several cyanobacterial strains are consumed as human food. Whole cells of *Arthrospira platensis* are widely accepted as health care products and have been successfully cultured in large scale and marketed in various forms (tablets, capsules, etc.). *Arthrospira platensis* has a protein content exceeding 60% of its whole cell weight; it is rich in beta-carotene, thiamine and riboflavin and is considered to be one of the richest sources of vitamin B12 (Spolaore *et al.*, 2006). *Nostoc commune* and *Aphanizomenon* sp. are available as food supplements to provide a protein-rich diet, because of their lack of toxicity and wealth of nutritional elements (Spolaore *et al.*, 2006; Yang *et al.*, 2011). Due to the increasing demand for protein sources in animal feed, the aquaculture industry is now actively investigating alternative nutrient sources. Biomasses of marine cyanobacteria *Phormidium valderianum* and *Arthrospira* sp. have been tested for their use as dietary supplements in aquacultures, but further studies are needed to determine the nutritional value of these products (Abed *et al.*, 2009; Sivakumar *et al.*, 2011; Thajuddin and Subramanian, 2005).

Bioactive compounds

Cyanobacteria are one of the richest sources of known and novel bioactive compounds, and people have started to pay attention to these chemicals in recent years. The modern high-throughout genome sequencing studies of cyanobacteria revealed that the genes responsible for these bioactive proteins, ribosomal and non-ribosomal peptides, and peptide-polyketide hybrid molecules were of significant diversity and novelty (Namikoshi and Rinehart, 1996).

These cyanobacterial secondary metabolites, especially the products of non-ribosomal polypeptide (NRP) or the mixed polyketide-NRP biosynthetic pathways (lipopeptides, macrolides, amides, etc.), show a wide range of biological activities, including anticancer, antibacterial, antiviral and protease inhibitory activities (Singh *et al.*, 2011). Cyanobacterial secondary metabolites may constitute a prolific source for potential lead compounds in further drug development. According to available data, over 4000 strains of marine and freshwater cyanobacteria have been screened, and a variety of bioactive compounds have been identified (Burja *et al.*, 2001; Uzair *et al.*, 2012). Prominent among these are toxins, such as hepatotoxins (e.g. microcystins and nodularins) and neurotoxins (e.g. anatoxins and saxitoxins). Microcystins are well-known hepatotoxic cyclic heptapeptides with more than 50 structural variations, which are synthesized not only in *Microcystis* but also in *Anabaena, Nostoc* and *Oscillatoria* and can form waterblooms. Cryptophycins, analogues of which have been either naturally isolated from *Nostoc* sp. or chemically synthesized, are potent anticancer agents. Curacin A, a bioactive compound isolated from *Lyngbya majuscula,* appeared to be a potent tubulin interactive compound, and its semi-synthetic derivatives with relatively increased solubility are undergoing preclinical evaluation. There are also some other anticancer compounds in cyanobacteria, such as dolastatin 10, synthadotin, apratoxin A, tolyporphin, and somocystinamide A. Antiviral compounds isolated from cyanobacteria include spirulan, nostoflan, cyanovirin-N, scytovirin N, and sulfoglycolipid. Noscomin, carbamidocyclophanes, ambiguine I isonitrile, and the phenolic compounds hapalindole T and pahayokolide A are some antibacterial compounds isolated from cyanobacteria. Antiprotozoal compounds isolated from cyanobacteria include viridamide A, symplocamide A, venturamides, dragomabin and ambigol C (Singh *et al.*, 2011). However, very few cyanobacterial compounds have entered clinical trials. These fields truly deserve more scientific attention and interdisciplinary research collaborations.

Biofertilizer

Cyanobacterial strains of subsections IV (e.g. Nostocales) and V (e.g. Stigonematales) are capable of fixing atmospheric nitrogen and converting it into an available form of ammonium. These species of cyanobacteria are major components of the nitrogen fixing biomass in paddy fields and play an important role in the maintenance of soil fertility, depending on the metabolic activity of the nitrogenase complex in the nitrogen-fixing heterocyst. Dried cyanobacteria have been used to inoculate soils to aid fertility; this was proven to increase grain and rice yield. Both free-living cyanobacteria and the *Azolla-Anabaena* symbiotic nitrogen-fixing complex have been studied as cyanobacterial biofertilizers (Vaishampayan *et al.*, 2001). In desert soils, nitrogen-fixing cyanobacteria were found to dominate the desert crusts on a global scale, and these microorganisms may eventually facilitate the vegetation of deserts (Garcia-Pichel and Pringault, 2001). In addition to providing access to nutrients, different biofertilizers also provide growth-promoting factors to crops. Compared to chemical fertilizers, cyanobacteria-based biofertilizers are environmentally friendly, cost effective and have great potential (Abed *et al.*, 2009).

Exopolysaccharides (EPSs)

The cyanobacterial EPSs, complex heteropolysaccharides consisting of six or more different kinds of monosaccharides, can be divided into two main groups: cell surface-associated EPSs and polysaccharides released into the surrounding environment (released polysaccharides, RPSs). The RPSs can be recovered easily from liquid cultures (De Philippis *et al.*, 2001). Their chemical properties show that such polysaccharides have abundant uronic acid subunits and non-saccharidic components, such as peptidic moieties, acetyl, pyruvyl and/or sulfate groups. Thus, these complex anionic heteropolymers show promise for their application as thickening or suspending agents and emulsifying or cation-chelating material (De Philippis *et al.*, 1998, 2001). Cyanobacterial EPSs are also suitable for a variety of biotechnological and environmental applications in wastewater treatment, brewing, cosmetology, pharmacology and as food additives.

Putative EPS biosynthesis genes in cyanobacteria are occasionally clustered, present in different regions of the genome, and often occur in multiple copies. To maximize the production of cyanobacterial EPSs, the gene regulatory networks involved in EPS biosynthesis need to be clarified, and optimal culture conditions influencing the synthesis and/or the characteristics of EPSs should be established.

Biodegradable plastic

Polyhydroxyalkanoates (PHAs) are crystalline thermoplastics with physical properties similar to those of polyethylene. They are linear head-to-tail polyesters and consist of 3-hydroxy fatty acid monomers. It is reported that there are at least 100 different 3-hydroxyalkanoic acids among the PHA constituents; the first PHA discovered was poly (3-hydroxy-butyrate) (PHB) (Steinbuchel and Valentin, 1995). In conditions of excess essential nutrients, several cyanobacterial strains can naturally accumulate PHAs under phototrophic and/or mixotrophic growth conditions, and the PHA production can amount to 7% of the cyanobacterial dry cell weight (DCW). The PHB biosynthetic pathway involves three distinct enzymes encoded by the genes of the *phbCAB* operon. The biosynthesis of PHA has been investigated in the recombinant cyanobacterial host strains of *Synechocystis* sp. PCC6803, *Synechococcus* sp. PCC7002 and *Synechococcus elongatus* PCC7942. The highest PHA yield was reported in genetically engineered *Synechococcus* sp. PCC7002, which yields 52% of the DCW (Akiyama *et al.*, 2011). PHA is an ideal biodegradable material that can replace chemical plastics in some applications, such as disposable bulk materials in packing films, containers, and paper coatings (Madison and Huisman, 1999). In addition, PHA is being developed for its application in the biomedical and biopharmaceutical fields (Chen, 2009).

Other applications

Phycocyanin (PC) is a blue, light-harvesting pigment that gives many cyanobacteria their bluish colour. Recent progress in PC biosynthesis and functionality has expanded its potential applications in biotechnology, diagnostics, food, and medicine (Eriksen, 2008). Cyanophycin is soluble in water under both acidic and alkali conditions and is of potential interest to biotechnology as a source of polyaspartic acid. The heterologous expression of cyanophycin synthetase allows the production of cyanophycin in a number of biotechnologically relevant microorganisms (Oppermann-Sanio and Steinbuchel, 2002). Cyanobacterial gas vesicles could be recombined and used as antigen presentation systems, and gas vesicles filled with oxygen and stabilized by glutaraldehyde treatment can be used to improve the gas supply in mammalian cell cultures (Pfeifer, 2012).

8.2 Molecular tools for the genetic engineering of cyanobacteria

8.2.1 Genetic transfer systems

The genetic modification of cyanobacteria relies on genetic transfer systems. Transformation, conjugation and electroporation were developed as genetic transfer systems for different cyanobacterial strains several decades ago.

Transformation

Most unicellular cyanobacterial species, such as *Synechocystis* sp., *Synechococcus* sp., etc., are naturally competent. Thus, transformation has been developed as the primary method of gene transfer in these species. Transformation was first applied to *Synechococcus* sp. PCC7943 in 1970 (Shestakov and Khyen, 1970). *Synechococcus* sp. PCC7942 (Golden and Sherman, 1984) and PCC7002 (Stevens and Porter, 1980), *Synechocystis* sp. PCC6308 (Devilly and Houghton, 1977) and PCC6803 (Grigorieva and Shestakov, 1982), *Thermosynechococcus elongatus* BP-1 (Onai *et al.*, 2004), etc. were subsequently shown to be transformable. Normally, the transformation of these unicellular cyanobacterial strains could be realized by simply mixing the cell suspensions with the plasmid DNA or PCR products for several hours without any other treatment. However, $CaCl_2$ treatment is needed to improve competency in the transformation of *Synechocystis* sp. PCC6308 (Devilly and Houghton, 1977).

The transformation was optimized in detail for some cyanobacteria, especially for *Synechocystis* sp. PCC6803 (Kufryk *et al.*, 2002; Williams, 1988; Zang *et al.*, 2007) and *Synechococcus* sp. PCC7942 (Golden and Sherman, 1984). The transformation efficiency depends on the concentration, form (linear or circular) and length of the incoming DNA, the incubation time of cells with DNA, and the recovery treatment after the incubation with DNA. Furthermore, the transformability of the host strain was increased by two orders of magnitude if the exonuclease RecJ was deleted in *Synechocystis* sp. PCC6803 (Kufryk *et al.*, 2002). In contrast, the deletion of ComA (Yoshihara *et al.*, 2001), PilT1 (Okamoto and Ohmori, 2002) or ComF (Nakasugi *et al.*, 2006), which are critical for phototaxis and the biogenesis of type IV piliin, led to the loss of natural competency in *Synechocystis* sp. PCC6803.

Although the protocol for transformation has been established for decades and some competence proteins have been identified, the mechanism of transformation in cyanobacteria is not yet fully understood. It was postulated that the mechanism might share some characteristics with other transformable bacteria (Porter, 1986).

Conjugation

Most filamentous cyanobacteria are not transformable. This may be due to the existence of extracellular nucleases in some heterocyst-forming filamentous strains of cyanobacteria (Wolk and Kraus, 1982). The introduction of foreign DNA into cyanobacteria through conjugation with *E. coli* was first described in 1984 (Wolk *et al.*, 1984) and has been extensively applied in many filamentous cyanobacteria (Elhai and Wolk, 1988; Taton *et al.*, 2012) and in some unicellular cyanobacteria (Asayama, 2012; Kreps *et al.*, 1990; Sode *et al.*, 1992). Typically, there are three plasmids involved in the conjugal transfer of DNA to cyanobacteria, including the conjugative, cargo and helper plasmids. Prior to conjugation, the cargo and helper plasmids were always introduced into the same *E. coli* host, and the cargo plasmid would be methylated by the methylases carried by the helper plasmid to protect the cargo plasmid from the restriction enzymes in the cyanobacterial host (Elhai *et al.*, 1997). Alternatively, these three plasmids could be maintained in the same donor *E. coli* host. During conjugation, the cargo plasmid is first nicked at the *bom* (*oriT*) site by the product of the *mob* gene, which is harboured by the helper or conjugative plasmid; the nicked single strand of the cargo plasmid is then transferred from the donor *E. coli* cell to the cyanobacterial cell by the gene products of *tra* harboured by the conjugative plasmid. The complementary strand is finally synthesized in the cyanobacterial cell to establish the double-stranded circular plasmid (Elhai and Wolk, 1988; Ippen-Ihler and Minkley, 1986; Koksharova and Wolk, 2002; Willetts and Wilkins, 1984).

Due to its specific gene-transfer mechanism, conjugation is more suitable than transformation for transferring large foreign DNA and obtaining merodiploids (Elhai and Wolk, 1988).

Electroporation

Electroporation is another method to introduce foreign DNA into cyanobacteria, during which the transient and reversible breakdown of the cell membrane can be induced by applying an external electric field (Gehl, 2003). *Anabaena* sp. strain M131 (Thiel and Poo, 1989), *Chlorogloeopsis fritschii* PCC6912, *Fischerella muscicola* PCC7414 (Toyomizu *et al.*, 2001), *Microcystis aeruginosa* PCC7806 (El Semary, 2010), *Spirulina platensis* (Toyomizu *et al.*, 2001) and *Synechocystis* sp. PCC6714 (Ludwig *et al.*, 2008) have been successfully transformed by electroporation. However, restriction enzymes contained by the host significantly reduce the efficiency of transformation (Thiel and Poo, 1989).

8.2.2 Construction of markerless mutants

Based on the established gene transfer systems, cyanobacteria can be easily modified by inserting antibiotic resistance markers into the targeted loci of the genome, generating a markered mutant (Golden *et al.*, 1987; Williams, 1988). However, this method depends on the selectable markers, and the multiple rounds of genetic engineering of cyanobacteria will be restricted by the limited number of available markers. Herein, the construction of markerless mutants by eliminating the selectable markers from the chromosomes will be necessary for the large-scale genetic modification and the future industrial application of cyanobacteria (Tan *et al.*, 2013). Negative selection systems and site-specific recombination systems were the two dominant methods for the construction of markerless mutant cyanobacteria.

Negative selection systems

By using negative selection systems, the targeted locus is first replaced by a drug marker coupled with a negative selection marker. By positively selecting cells with the corresponding antibiotic and then using negative selection to kill cells with the negative selection marker, markerless mutants can be obtained.

The *rps12* negative selection system, which is based upon the dominance of a streptomycin-sensitive *rps12* gene encoding a ribosomal S12 protein over a streptomycin-resistant *rps12-R43* allele with a single amino acid substitution, are feasible for constructing markerless mutant *Synechococcus* sp. PCC7942 (Matsuoka *et al.*, 2001).

Based on the sensitivity of *sacB*-expressing cells to sucrose, the *sacB* negative selection system was first described in 1996 (Vermaas, 1996) and has been successfully applied to the construction of marker-free mutants for the production of ethanol (Dexter and Fu, 2009) and fatty acids (Liu *et al.*, 2011b) in *Synechocystis* sp. PCC6803. However, this system cannot be applied to some sub-strains of *Synechocystis* sp. PCC6803 that are sucrose-sensitive and are unable to grow in the presence of sucrose, as indicated by Cheah *et al.* (2013).

The deletion of the *upp* gene encoding uracil phosphoribosyltransferase in most organisms leads to resistance to 5-fluorouracil, and markerless mutant cyanobacteria could be obtained using the native *upp* gene as the negative selection marker and the *upp* deletion mutant strain as the starting strain for genetic modifications (Aikens, 2012).

Very recently, a novel counter-selection system based on *mazF*, the gene product of which cleaves mRNA at the ACA triplet sequence (Zhang *et al.*, 2003), was developed for the construction of the gene mutants of *Synechocystis* sp. PCC6803 without drug markers (Cheah *et al.*, 2013).

Site-specific recombination systems

Some site-specific recombinases, such as FLP from *Saccharomyces cerevisiae* or Cre from *E. coli* bacteriophage P1, etc., which mediate precise DNA rearrangements between a pair of their target sites (Grindley *et al.*, 2006), have emerged as powerful molecular tools for genetic engineering in both prokaryotes and eukaryotes (Askora *et al.*, 2012; Lyznik *et al.*, 2003; Schweizer, 2003; Siegal and Hartl, 2000). In the most common scenario, foreign DNA fragments containing the drug marker flanked by the target sites of the site-specific recombinase are integrated into targeted chromosomal locations through homologous recombination, and the drug marker is excised by the subsequent site-specific recombination of the transgenic locus (Lyznik *et al.*, 2003). The Cre-loxP recombination system was first applied in the filamentous *Anabaena* sp. PCC7120 for generating the unmarked mutant (Zhang *et al.*, 2007). The FLP-FRT system was also shown to be feasible for constructing the marker-free mutants in both *Synechocystis* sp. PCC6803 and *Synechococcus* sp. PCC7942 (Tan *et al.*, 2013). In the future, these powerful tools will facilitate large-scale genetic modification and biofuel production in cyanobacteria.

8.2.3 Synthetic biology tools for the metabolic engineering of cyanobacteria

Synthetic biology can significantly advance metabolic engineering by contributing genetic engineering tools, and the availability of more tools could facilitate future applications (Keasling, 2012). Specifically, developing molecular tools compatible with multiple gene assembly and gene expression are especially crucial for the production of biofuels in cyanobacteria, which need system-level metabolic engineering (Tan *et al.*, 2012).

Expression vectors

Central to any genetic manipulation is the vector used to carry and/or harbour the transforming DNA in the host (Keasling, 2012). In the ideal expression vector, the core elements for gene expression, such as the promoter, the gene of interest, and the terminator, etc., should be easily assembled and replaced for the optimization of high efficiency gene expression. To this end, some RSF1010-based vectors, such as pPMQAK1 (Huang *et al.*, 2010), pAM500, and pAM461c (Asayama, 2012), and integrative vectors, such as pFQ9R, pFQ20, pXT37a and pXT37b (Qi *et al.*, 2013; Tan *et al.*, 2011), have been developed and tested for the heterologous expression of genes of interest in cyanobacteria. Among these platform vectors, three RSF1010-based vectors exist as self-replicative plasmids in the host after their conjugal transfer into cyanobacteria (Asayama, 2012; Huang *et al.*, 2010). They have a broad host range and could replicate in some different species of cyanobacteria, including *Synechocystis* sp. PCC6803, *Synechococcus* sp. PCC7942, *Pseudanabaena* sp. ABRG5-3, *Nostoc* sp. PCC7120 and *Nostoc punctiforme* ATCC 29133 (Asayama, 2012; Huang *et al.*, 2010). In contrast, the integrative vectors harbouring the homologous DNA fragments can integrate with the genomic DNA by homologous recombination rather than by replicating. Furthermore, to date pPMQAK1 is the only BioBrick-compatible vector for the metabolic engineering of cyanobacteria; its genetic parts can be easily assembled according to the BioBrick standard (Huang *et al.*, 2010).

Representative genetic parts

For the rational design of artificial genetic circuits, the behaviour of each individual part, including promoters, reporter genes, degradation tags, etc., has to be well characterized (Huang *et al.*, 2010). For this purpose, as shown in Table 8.1, a series of promoters, including constitutive and inducible ones, have been characterized and tested in cyanobacteria (Huang *et al.*, 2010; Qi *et al.*, 2013). The fluorescent proteins cerulean, GFPmut3B, EYFP (Huang *et al.*, 2010) and β-D-galactosidase (LacZ) (Qi *et al.*, 2013) have been demonstrated to work as reporter proteins in cyanobacteria. Moreover, luciferase has also been demonstrated previously as a good reporter in cyanobacteria (Aoki *et al.*, 2002). The effects of different protease degradation tags in manipulating the levels and turnover times of the EYFP protein have been evaluated in *Synechocystis* sp. PCC6803 (Huang *et al.*, 2010; Landry *et al.*, 2013).

8.3 Omics of Cyanobacteria

The technology-driven omics platforms have been widely used in cyanobacteria to understand the functional principles and dynamics of total cellular systems. The integration of multiple omics approaches has been applied to build a foundation for the global understanding of the biological processes of cyanobacteria (Schwarz *et*

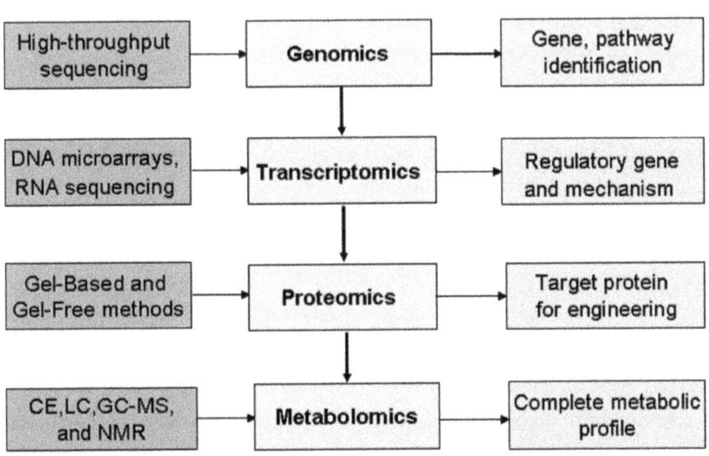

Figure 8.3 Overview of 'omics' technologies for systems biology research. CE, capillary electrophoresis; LC, liquid chromatography; GC, gas chromatography; MS, mass spectrometry; NMR, nuclear magnetic resonance. A colour version of this figure is available in the plate section at the back of the book.

al., 2011; Yoshikawa et al., 2013; Zhu et al., 2013). The multi-omics approach can be exploited to understand cellular physiology and metabolism and consequently to develop engineered cyanobacteria for biofuel production (Fig. 8.3).

8.3.1 Genomics

It is essential to acquire well-annotated genome information for further metabolic engineering and systems biology (transcriptomics, proteomics and metabolomics) research (Knoop et al., 2010). *Synechocystis* sp. PCC6803, the unicellular model species of cyanobacteria with a genome of 3.947 Mb and 3725 genes, was the first phototrophic organism for which the complete genome was sequenced in 1996 (Kaneko et al., 1996). Thus far, over 126 full genome sequences of cyanobacteria strains are available (Shih et al., 2013). The genome analysis of these cyanobacteria strains can provide detailed genetic information for biotechnological applications. The metabolic engineering of cyanobacteria strains can benefit greatly from a thorough analysis of the genomic sequences of previously sequenced cyanobacteria. Genome analysis will eventually contribute to the understanding of the metabolic capabilities of cyanobacteria species.

Most cyanobacterial genomes include a circular chromosome and several additional plasmids that can reach several hundred kbp in size. Large differences were also reported for chromosomal copy numbers. Real-time PCR for the measurement of genome copy numbers was optimized for its application in cyanobacteria. Experimentally determined ploidy levels suggest some marine cyanobacteria are mainly monoploid or diploid. Some cyanobacteria (e.g. *Synechococcus elongatus* PCC7942) were found to contain 3–4 genome copies per cell and are thus oligoploid (between 3 and 10 chromosome copies per cell). In contrast, some cyanobacteria are highly polyploid, such as the motile *Synechocystis* sp. PCC6803 strain with a maximum of 218 chromosome copies per cell in exponential phase (Griese et al., 2011).

A subtractive genome analysis was employed to identify cyanobacterial alkane biosynthetic pathway genes in 2010 (Schirmer et al., 2010). No alkane biosynthesis was observed in *Synechococcus* sp. PCC7002, so *Synechococcus* sp. PCC7002 and ten other alkane-producing cyanobacteria strains with sequenced genomes were analysed. The *Synechococcus* sp. PCC7002 genome was subtracted by the other 10 genomes to remove orthologues. Two candidate enzymes for alkane biosynthesis were identified. Acyl-acyl carrier protein (ACP) can be reduced to an aldehyde by an acyl-ACP reductase (AAR, EC 1.2.1.50), and then aldehyde can be converted to an alkane or alkene by an aldehyde-deformylating oxygenase (ADO) (Li et al., 2012). In 2011, another hydrocarbon-producing pathway was found in *Synechococcus* sp. PCC7002 by the basic local alignment search tool (BLAST) (Mendez-Perez et al., 2011). This route of hydrocarbon biosynthesis employed an elongation- decarboxylation mechanism converting fatty acyl-ACPs to medium-chain α-olefins by a type I polyketide synthase (PKS), Ols. These findings open the door to achieve the photosynthetic production of alkanes/alkenes with high efficiency by genetically engineering cyanobacteria.

8.3.2 Transcriptomics

DNA microarrays and the emerging RNA sequencing are widely employed in transcriptomics research (Sherstnev et al., 2012; Wang et al., 2009). The regulatory mechanisms can be elucidated by comparing transcriptome profiles between the wild-type and mutant strains or between the same strain cultured under different conditions. In addition, potential target genes for strain improvement can be provided by transcriptome analysis. Transcriptomics technologies have been applied in cyanobacteria to obtain global transcriptome profiles and to understand the mechanisms behind the observed expression and physiological changes.

RNA sequencing gives precise measurements of the dynamic genome-wide expression profile. A genome-wide map of 3,527 transcriptional start sites (TSSs) of *Synechocystis* sp. PCC6803 was established by differential RNA sequencing. One-third of all TSSs were located upstream of an annotated gene, and approximately 64% of all TSSs lead to antisense or noncoding RNAs in *Synechocystis* sp. PCC6803 genome. The global TSS map will facilitate the use of *Synechocystis* sp. PCC6803 as a model organism for further biofuel production by genetic engineering (Mitschke *et al.*, 2011). Combined with the chromatin immunoprecipitation (ChIP) procedure, transcriptomics technology can also be used to investigate genome-wide DNA-binding proteins and gene expression regulatory systems. A comprehensive analysis of the *Synechococcus elongatus* PCC7942 transcriptome was conducted by integrating data from RNA sequencing, tiling expression microarrays, and RNA polymerase (RNA pol) ChIP sequencing. The absolute transcript levels, operon identification, and high-resolution mapping of the 5′ and 3′ transcript ends of *Synechococcus elongatus* PCC7942 were reported for further studies of cyanobacterial transcription (Vijayan *et al.*, 2011).

Biohydrogen (H_2) is an appealing candidate for renewable energy. In *Synechocystis* sp. PCC6803, AbrB2 is a repressor of the hydrogenase-encoding operon. The transcriptome profiles of *abrB2* deletion mutants were compared with that of the wild-type strain by oligonucleotide DNA microarray analysis. The results showed that AbrB2 is a master negative regulator of H_2 production and also regulates metal transport, the oxidative stress response, and numerous plasmid genes of unknown function. A recent report promotes genetic engineering to develop powerful cell factories for the sustainable production of H_2 in cyanobacteria (Leplat *et al.*, 2013).

A quantitative transcriptomics approach with a next-generation RNA sequencing technology was applied to reveal the global metabolic responses to ethanol in *Synechocystis* sp. PCC6803. The results showed that ethanol stress induced genes involved in common stress responses, transportation and cell envelope modification. In addition, the cells can also utilize enhanced polyhydroxyalkanoate (PHA) accumulation and the glyoxalase detoxification pathway to defend against ethanol stress. These results showed that *Synechocystis* sp. PCC6803 most likely employed synergistic resistance mechanisms in response to ethanol stress (Wang *et al.*, 2012).

As energy-rich molecules, free fatty acids (FFAs) are ideal biofuel precursors. It is known that FFA production has negative physiological effects on cyanobacteria, ultimately limiting the high yields of FFAs (Ruffing and Jones, 2012). To identify genetic targets that may alleviate the toxic effects, RNA-seq analysis was used to investigate the response of *Synechococcus elongatus* PCC7942 to FFAs. This analysis indicated that a general stress response, nitrogen metabolism, PSII photosynthesis, and protein folding genes were activated during FFA production, while genes involved in PSI photosynthesis and carbon and hydrogen metabolism were repressed. The gene targets to reduce the toxic effects of FFA production were verified by mutagenesis experiments and comparative transcriptomics analysis (Ruffing, 2013).

8.3.3 Proteomics

Proteomics is important to understand biological processes because most cellular functions are accomplished by proteins. The coverage and sensitivity of gel-based proteomics analysis is not compatible with gel-free proteomics techniques that can detect several thousand proteins in a single run and that can also provide better detection for low-abundance proteins and membrane proteins (Abdallah *et al.*, 2012; Insenser *et al.*, 2010). Quantitative proteomic methods, such as stable isotope labelling-based isotope-coded affinity tags (ICAT) (Sethuraman *et al.*, 2004), isobaric tag for relative and absolute quantification (iTRAQ) (Mertins *et al.*, 2012), stable isotope labelling with amino acids in cell culture (SILAC) (Ong *et al.*, 2002) and label-free comparative quantitative proteomics, were developed recently for the precise quantification and identification of the proteome (Patel *et al.*, 2009). Proteomics methods have been extensively used to explore cyanobacterial metabolism and stress responses.

Nostoc punctiforme ATCC 29133 are capable of fixing atmospheric nitrogen and producing hydrogen through nitrogenase (EC 1.7.9.22). The disruption of the uptake hydrogenase gene in *N. punctiforme* gives rise to enhanced H_2 production but unchanged N_2 fixation rates. Quantitative proteomic approaches at both the filament and the heterocyst level were employed to investigate this metabolic redirection systematically. From the iTRAQ analysis, 1070 proteins were identified and quantified, and 239 proteins were differentially expressed in the uptake hydrogenase mutant compared to the wild type. The results indicate that the targets for biological engineering to improve H_2 production include components of the oxygen scavenging activity, the cell envelope of heterocysts, and proteins involved in reduced carbon transport from vegetative cells to heterocysts (Ekman *et al.*, 2011).

To understand the mechanisms of biofuel tolerance and engineer robust cyanobacterial hosts, the responses of *Synechocystis* sp. PCC6803 to ethanol (Qiao *et al.*, 2012), butanol (Tian *et al.*, 2013) and hexane (Liu *et al.*, 2012) were determined using a quantitative proteomics approach with iTRAQ LC–MS/MS technologies. Functional annotation and KEGG pathway enrichment analysis showed that the *Synechocystis* cells employed a synergistic resistance mechanism in dealing with biofuel stress, which includes the induction of heat-shock and transporter proteins, oxidative stress response proteins and proteins related to the sulfur relay system, photosynthesis and cell mobility, along with modifications of the cell membrane and envelope. The proteomic analysis provided the comprehensive molecular mechanisms against biofuel stress and also identified a list of potential target genes for further engineering stress-resistant hosts to produce biofuels efficiently.

8.3.4 Metabolomics

Metabolomics is the comprehensive measurement of the level of small-molecule metabolites and refers to the complete metabolic profile of the cell at specific times and/or under specific conditions. Three different sampling strategies (quenching, filtering, and centrifugation) to arrest the metabolic profile of two morphologically diverse cyanobacteria, the unicellular *Synechocystis* sp. PCC6803 and the filamentous *Nostoc* sp. PCC7120, were compared for gas chromatography-mass spectrometry (GC-MS) analysis. The results showed that fast filtering and centrifugation produced quite similar metabolite pool sizes, even for metabolites with predicted high turnover (Krall *et al.*, 2009).

An integrated, untargeted metabolic footprinting approach was employed to systematically evaluate the uptake and release of metabolites between *Synechococcus* sp. PCC 7002 and different growth media. *Synechococcus* was found to ingest a great diversity of metabolites that are dominant in and specific to its own metabolite extract, including histidine, betaine, γ-glutamyl phenylalanine and a hexosamine-based trisaccharide. Additionally, stable isotope probing was used to show that adenine, glutamate and glucosylglycerol are actively turned over as opposed to the hexosamine-based trisaccharide, which is taken up and maintained in its native state (Baran *et al.*, 2011).

Some cyanobacteria produce H_2 in low yields during the anaerobic metabolism of intracellular carbohydrates previously stored during aerobic photosynthesis. A mutant (the *ldhA* mutant) of the cyanobacterium *Synechococcus* sp. PCC7002 lacking the enzyme for the NADH-dependent reduction of pyruvate to D-lactate was constructed. Both nuclear magnetic resonance (NMR) spectroscopy and liquid chromatography-mass spectrometry (LC-MS) metabolomic methods have shown that auto-fermentation by the *ldhA* mutant resulted in higher concentrations of excreted acetate, alanine, succinate, and hydrogen (up to 5-fold) compared to those from the wild type. The increased NAD (P)H/NAD $(P)^+$ ratio during auto-fermentation in the *ldhA* mutant led to a considerable increase in H_2 production via an NADH-dependent, bidirectional [NiFe] hydrogenase (McNeely *et al.*, 2010).

Arthrospira platensis is a fast-growing halophilic cyanobacterium capable of accumulating glycogen, which has the potential to serve as a feedstock in biofuel production. The time-course profiling of widely targeted cyanobacterial metabolic intermediates demonstrated a global metabolic reprogramming that involves transient increases in the levels of some amino acids during the glycogen production phase induced by nitrate depletion. Additionally, *in vivo* labelling with $NaH^{13}CO_3$ enabled the direct measurement of metabolic intermediate turnover in *A. platensis*, revealing that under the conditions of nitrate depletion, glycogen was biosynthesized with carbon derived from amino acids released from proteins via gluconeogenesis. This dynamic metabolic profiling approach provided conclusive evidence of temporal alterations in the metabolic profile in cyanobacterial cells (Hasunuma *et al.*, 2013).

A metabolome study was performed with *Synechocystis* sp. PCC6803 and photorespiratory mutants shifted from high CO_2 (HC) into conditions of low CO_2 (LC). Shifting wild-type cells into LC conditions resulted in global metabolic reprogramming. A decrease in Calvin-Benson cycle activity and the increased usage of photorespiratory metabolism was indicated by the synergistic accumulation of the fumarate, malate, and 2-phosphoglycolate pools and a transient increase of 3-phosphoglycerate. The accumulation of 2-oxoglutarate with a concomitant decrease of glutamine demonstrated reduced nitrogen availability when cells were shifted to LC. Mutants defective in the photorespiratory metabolism accumulated glycolate and glycine. Metabolome profiling showed that carbon and nitrogen metabolism was coordinated to acclimate to LC conditions (Eisenhut *et al.*, 2008).

The *sigE* gene encodes a group 2 sigma factor for RNA polymerase and has been proposed to function in the transcriptional activation of sugar catabolic pathways in *Synechocystis sp.* PCC6803. Metabolome analysis with capillary electrophoresis-mass spectrometry (CE-MS) revealed that the metabolite levels of the tricarboxylic acid cycle and acetyl-CoA are significantly altered by SigE overexpression in *Synechocystis* sp. PCC6803 (Osanai *et al.*, 2005). CE-MS analysis also revealed that *sigE* overexpression significantly changes sugar catabolism and increases the production of polyhydroxybutyrate during nitrogen starvation in *Synechocystis sp.* PCC6803 (Osanai *et al.*, 2013).

The metabolomics analysis of the central metabolism of *Synechocystis* sp. PCC6803 cultured under autotrophic and mixotrophic conditions were performed by using capillary electrophoresis time-of-flight mass spectrometry (CE-TOFMS). The results showed that the oxidative pentose phosphate pathway and glycolysis are activated under mixotrophic conditions. The metabolism of *Synechocystis* sp. PCC6803 under photoheterotrophic conditions created by atrazine, a photosynthesis inhibitor, was investigated by metabolomics analysis. The results demonstrated that the glycolytic activity decreased due to the indirect effect of atrazine (Yoshikawa *et al.*, 2013). The above experimental examples and successful research experiences with cyanobacterial metabolomics analysis indicate that it will provide further information for understanding the metabolism of cyanobacteria and the sustainable production of biofuels.

8.4 Biofuel production by genetically engineered cyanobacteria

Concerns about the depletion of fossil fuels and global climate change have made the research and development of renewable biofuels increasingly attractive. Cyanobacteria, which use sunlight as their energy source and directly convert carbon dioxide into biofuels, have been proposed as microbial systems to produce a series of biofuel and biochemical products. Furthermore, as discussed in the previous sections, the development of the molecular tools of genetic engineering and the extensive omics studies of cyanobacteria make the construction and modification of biosynthetic pathways to produce biofuels in cyanobacteria relatively easier and more efficient. In this section, the proof of concept for the production of biofuels, including ethanol, higher alcohols, isoprene-derived biofuels and fatty acid-derived biofuels, by genetically engineering cyanobacteria will be discussed (Fig. 8.4).

8.4.1 Ethanol

As described in the previous chapters, most ethanol is currently produced from agricultural feedstocks such as sugar cane, corn and potato, and this practice is believed to cause the food price soar due to the large amount of arable land required for planting crops. In contrast, cyanobacterial ethanol production utilizes a consolidated bioprocessing (CBP) method to directly produce ethanol from carbon dioxide and solar energy. In this process, the central metabolite pyruvate is first converted to acetaldehyde by pyruvate decarboxylase (Pdc), and then acetaldehyde is converted to ethanol by alcohol dehydrogenase (Adh).

In 1999, Deng *et al.* reported the first case of ethanol production by genetic engineering in *Synechococcus* sp. PCC7942, in which the pyruvate decarboxylase and alcohol dehydrogenase II genes from *Zymomonas mobilis* were expressed at high

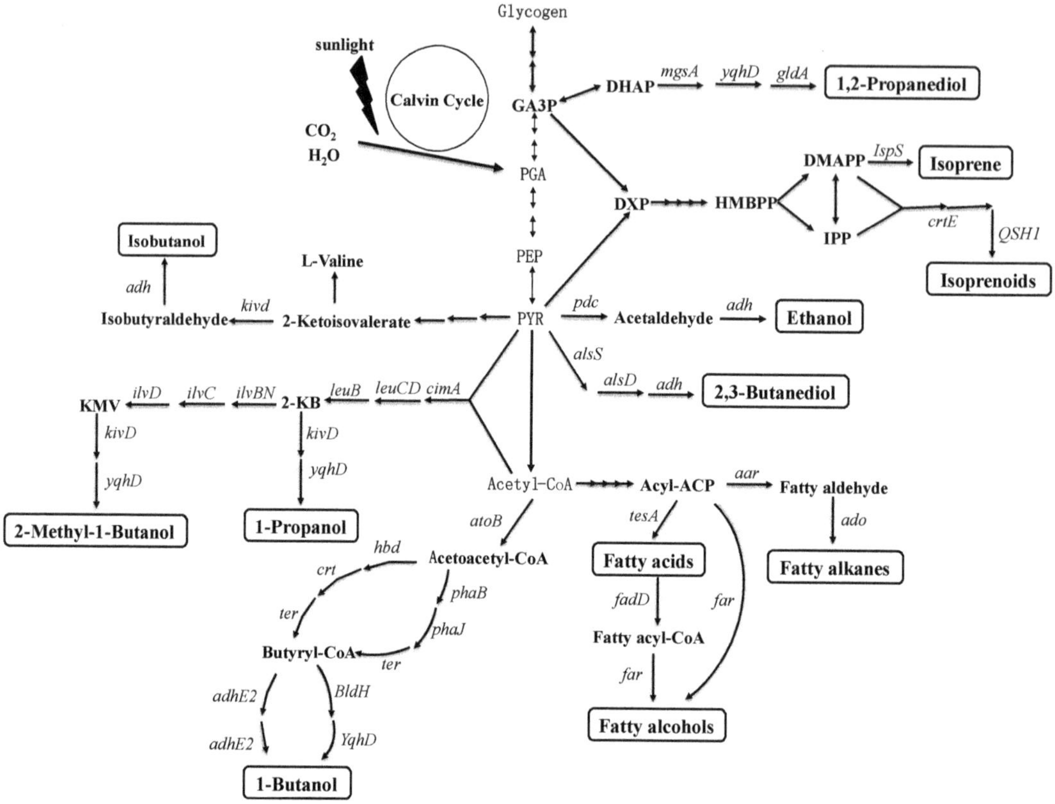

Figure 8.4 Schematic illustration of biofuel production by genetically engineered cyanobacteria. The molecules discussed in this chapter are in black boxes. 3PGA, 3-phosphoglycerate; GA3P, glyceraldehyde 3-phosphate; DHAP, dihydroxyacetone phosphate; DXP, deoxyxylulose 5-phosphate; HMBPP, hydroxymethylbutenyl diphosphate; DMAPP, dimethylallyl pyrophosphate; IPP, isopentenyl diphosphate; PEP, phosphoenolpyruvate; PYR, pyruvate; 2-KB, 2-ketobutyrate; KMV, 2-keto-3-methylvalerate. *mgsA*, methylglyoxal synthase; *yqhD*, aldehyde reductase; *gldA*, glycerol dehydrogenase; *IspS*, isoprene synthase; *crtE*, geranylgeranyl pyrophosphate synthase; *QSH1*, β-caryophyllene synthase; *pdc*, pyruvate decarboxylase; *adh*, alcohol dehydrogenase; *alsS*, acetolactate synthase; *alsD*, 2-acetolactate decarboxylase; *kivd*, ketoacid decarboxylase; *cimA*, citramalate synthase; *leuCD*, isopropyl malate isomerase; *leuB*, β-isopropylmalate dehydrogenase; *ilvBN*, acetohydroxyacid synthase; *ilvC*, acetohydroxy acid isomeroreductase; *ilvD*, dihydroxy-acid dehydratase; *aar*, acyl-ACP reductase; *ado*, aldehyde decarbonylase/oxygenase; *tesA*, thioesterase; *fadD*, fatty acyl-CoA synthetase; *far*, fatty acyl-CoA reductase; *atoB*, acetyl-CoA acetyltransferase; *hbd*, 3-hydroxybutyryl-CoA dehydrogenase; *crt*, crotonase; *ter*, trans-2-enoyl-CoA reductase; *adhE2*, bifunctional aldehyde/alcohol dehydrogenase; *phaB*, acetoacetyl-CoA reductase; *phaJ*, enoyl-CoA hydratase; *BldH*, butyraldehyde dehydrogenase.

levels under the control of the *rbcLS* promoter. As a result, the amount of ethanol accumulated to concentrations up to 5 mM (0.23 g/l) in the culture media after 4 weeks of growth (Deng and Coleman, 1999). In 2009, Dexter *et al.* integrated the same *pdc-adh II* cassette into the genome of *Synechocystis* sp. PCC6803 at the *psbA2* locus. An average ethanol yield of 5.2 mM/OD_{730} unit/l was obtained in the computerized photobioreactor system where the mutant strain was cultivated (Dexter and Fu, 2009). In the above-mentioned cyanobacterial ethanol production, the *Adh II* gene from obligately ethanol-producing *Zymomonas mobilis* was cloned into the chromosome of cyanobacteria and overexpressed to convert acetaldehyde to ethanol. Vidal *et al.* reported that the *slr1192* gene from *Synechocystis* sp. PCC6803 encodes a NADP-dependent alcohol dehydrogenase that exhibits superior catalytic efficiency for aldehyde reduction (Vidal *et al.*, 2009). Based on this finding, Algenol Biofuels constructed the mutant strains of *Synechocystis* sp. PCC6803 with overexpressed endogenous *slr1192* rather than *Adh II* from *Z. mobilis* and integrated pyruvate decarboxylase from *Z. mobilis* under the control of different promoters, resulting in ethanol accumulation of 3.6 g/l in the culture media after 38 days of cultivation (Dehring, 2012). In 2012, Gao *et al.* adopted a similar strategy to construct a *Synechocystis* sp. PCC6803 mutant strain in which *pdc* from *Z. mobilis* and *slr1192* were incorporated at two different sites of the chromosome through homologous recombination, and the biosynthetic pathway of poly-β-hydroxybutyrate was disrupted. As a result, the amount of ethanol reached 5.5 g/l in the culture media after 26 days of growth, which is the highest yield ever reported (Gao *et al.*, 2012).

8.4.2 Higher alcohols

Currently, up to 10% ethanol (in gasoline) is being used in most cars in the USA; the compatibility of higher ethanol blends with existing infrastructure still needs to be validated (Dugar and Stephanopoulos, 2011). Therefore, the development of higher alcohols (C_3–C_6) has attracted considerable interest recently because of their higher energy content, lower solubility in aqueous culture media and compatibility with existing infrastructure. Atsumi *et al.* implemented a synthetic approach in *E. coli* to convert 2-keto acids, intermediates in the amino acid biosynthesis pathway, to aldehydes by 2-keto acid decarboxylases (KDC) and then to higher alcohols by alcohol dehydrogenases (ADH) (Atsumi *et al.*, 2008). Decarboxylation was considered as a driving force for the product formation in this process (Lan and Liao, 2012). Later, the same group utilized the same strategy to produce higher alcohols in cyanobacteria (Atsumi *et al.*, 2009). Isobutanol (450 mg/l) was produced directly from CO_2 by genetically engineering *Synechococcus elongatus* PCC7942, and the productivity was increased by overexpressing ribulose 1,5-bisphosphate carboxylase/oxygenase (Rubisco), a key enzyme in the Calvin cycle. This pioneering work evidenced the biochemical feasibility of converting CO_2 to higher alcohols by using a synthetic approach. Isobutanol production from *Synechocystis* sp. PCC6803 has also been recently reported (Varman *et al.*, 2013). Varman *et al.* integrated two heterologous genes (*kivd* and *adhA*) from the Ehrlich pathway into the genome of *Synechocystis* sp. PCC6803, the engineered strain can accumulate 90 mg/l of isobutanol from 50 mM bicarbonate in a gas-tight shaking flask (Varman *et al.*, 2013). In 2011, Lan *et al.* reported the production of 1-butanol from CO_2 by integrating a CoA-dependent pathway into *Synechococcus elongatus* PCC7942 (Lan and Liao, 2011). The *hbd, crt,* and *adhE2* genes from *Clostridium acetobutylicum,* the *ter* gene from *Treponema denticola,* and the *AtoB* gene from *E. coli* were introduced into wild-type *Synechococcus elongatus* PCC7942, and 14.5 mg/l 1-butanol was ultimately produced under dark anoxic conditions rather than oxic conditions, because oxygen was shown to inhibit 1-butanol synthesis. Furthermore, the 1-butanol production was enhanced 5-fold when a polyhistidine tag was attached to *T. denticola ter.* This modification was believed to increase the protein expression and then increase enzyme activity (Doray *et al.*, 2001). Because the photosynthetic production of 1-butanol is desired for oxygenic cyanobacteria, the same group coupled the ATP-driven acetyl-CoA condensation to malonyl-CoA and decarboxylation to obtain the acetoacetyl-CoA precursor. Moreover, NADPH-utilizing enzymes *phaB, phaJ, bldH* and *yqhD* were used to replace the aforementioned NADH-utilizing enzymes *hbd, crt* and *adhE2.* With these modifications, 1-butanol production increased to 30 mg/l under photosynthetic conditions (Lan and Liao, 2012). Recently, Shen *et al.* integrated the heterologous citramalate pathway into *Synechococcus elongatus* PCC7942 to produce 2-ketobutyrate, which is the precursor of the native isoleucine biosynthesis pathway (Shen and Liao, 2012). Similar to the production of isobutanol and 1-butanol, 2-methyl-1-butanol (200 mg/l) was produced from 2-keto-3-methylvalerate with the aid of a ketoacid decarboxylase (KDC) and an alcohol dehydrogenase (ADH) (Shen and Liao, 2012) after ten days cultivation. At the same time, small amounts of 1-propanol, 1-butanol, isobutanol and 3-methyl-1-butanol were also produced as byproducts due to the versatile substrate selectivity of KDC and ADH.

Diols are important chemical intermediates because they can be used to make polymers. Diols are usually produced by the chemical processing of petroleum. However, glucose/glycerol-based diol production by microbial fermentation has achieved significant progress in recent years (Altaras and Cameron, 1999; Li *et al.*, 2010; Nakamura and Whited, 2003; Yim *et al.*, 2011). More recently, the photosynthetic production of diols by cyanobacteria has been reported. Oliver *et al.* integrated the *als*, *aldc* and *adh* genes into the chromosome of *Synechococcus elongatus* PCC7942 to produce 2,3-butanediol via acetoin (Oliver *et al.*, 2013). By screening *als*, *aldc* and *adh* from different sources, the authors could achieve 2.38 g/l 2,3-butanediol from CO_2 after 20 days. Another example is the photosynthetic production of 1,2-propanediol (Li and Liao, 2013). DHAP, an intermediate of the Calvin cycle, was used as the precursor. By introducing three genes coding for methylglyoxal synthase (*mgsA*), glycerol dehydrogenase (*gldA*) and aldehyde reductase (*yqhD*) into *Synechococcus elongatus* PCC7942, 22 mg/l 1,2-propanediol was produced from CO_2. The optimization of the secondary alcohol dehydrogenases to be NADPH-dependent increased the yield by approximately 7-fold to 150 mg/l (Li and Liao, 2013).

8.4.3 Isoprene-derived biofuels

Isoprene is a volatile hydrocarbon that can be used to make rubber. Isoprenoid-based compounds are suitable for the direct replacement of diesel fuels. The precursors for isoprenoid synthesis are isopentenyl pyrophosphate (IPP) and dimethylallyl diphosphate (DMAPP), which can be derived from two different pathways: the mevalonate (MVA) pathway and the methylerythritol phosphate (MEP) pathway. In nature, many plants produce isoprene from IPP by isoprene synthase. However, cyanobacteria are not endowed with the ability to produce isoprene, due to the lack of an isoprene synthase gene (*IspS*). Lindberg *et al.* (2010) reported isoprene production in genetically engineered *Synechocystis* sp. PCC6803 by introducing an *IspS* gene from *Pueraria montana* (kudzu vine). More recently, it was reported that the production of the sesquiterpene β-caryophyllene was achieved by expressing the *Artemisia annua* β-caryophyllene synthase gene in *Synechocystis* sp. PCC6803 through the native MEP pathway in *Synechocystis* sp. PCC6803 (Reinsvold *et al.*, 2011).

8.4.4 Fatty acid-derived biofuels

Fatty acid-derived biofuels, including fatty acid alcohol esters, fatty alcohols and fatty alkanes, are ideal alternatives for petroleum-based fuels. Fatty acid metabolism in bacteria has been investigated extensively (Cronan and Thomas, 2009). The biosynthesis of fatty acid-based biofuels in cyanobacteria includes two steps: the overproduction of fatty acids and either the trans-esterification of fatty acids to form fatty acid ethyl esters or the reduction of fatty acids to form fatty alcohols or fatty alkanes (Lu, 2010). Recently, the production of free fatty acids (FFAs) by genetically engineered *Synechocystis* sp. PCC6803 was reported by Liu *et al.* (Liu *et al.*, 2011b). The authors codon-optimized the acyl–acyl carrier protein (acyl-ACP) thioesterase gene and made six successive genetic modifications in wild type *Synechocystis* sp. PCC6803 (SD100). Lastly, the free fatty acid secretion yield was increased to 197 mg/l of culture media in the engineered strain SD277 (Liu *et al.*, 2011b). Later, the same group developed a CO_2-limitation-inducible green recovery strategy to generate free fatty acids from membrane lipids (Liu *et al.*, 2011a). In this strategy, lipase genes controlled by CO_2-limitation-inducible promoters were integrated in an FFA-secretion strain, and the FFA yield increased by two-fold in strain SD239 after CO_2 limitation (Liu *et al.*, 2011a). Ruffing and colleagues reported the production of FFA in genetically engineered *Synechococcus elongatus* PCC7942 using a similar strategy, in which they deleted the acyl-ACP synthetase gene (Synpcc7942_0918) and simultaneously overexpressed a truncated thioesterase (*'tesA*) from *E. coli* (Ruffing and Jones, 2012). Unlike the previously described higher chain alcohols, fatty alcohols are alcohols containing long alkyl chains. The production of fatty alcohols by cyanobacteria was first reported by Tan *et al.* (Tan *et al.*, 2011). Seven different *far* genes were separately integrated into the genome of *Synechocystis* sp. PCC6803. The highest yield of fatty alcohols (200 μg/L) was obtained when the *far* gene from jojoba was expressed under the control of the P_{rbc} promoter (Tan *et al.*, 2011). Alkanes are the major components of gasoline, diesel and jet fuels. As early as the 1960s, the production of fatty alkanes in 11 cyanobacterial species was first reported (Winters *et al.*, 1969). Some initial investigations were performed to elucidate the biosynthesis of alkanes in two filamentous cyanobacteria (Fehler and Light, 1972; Han *et al.*, 1969). However, there have been no other reports regarding the biosynthesis of alkanes in cyanobacteria in the following 40 years. Tan *et al.* (2011) demonstrated that the overexpression of acetyl-CoA carboxylase would increase the alkane production by 56% in *Synechocystis* sp. PCC6803. Recently, the biosynthesis pathway of alkanes in cyanobacteria was identified by Schirmer *et al.* (2010). In this biosynthesis pathway, fatty acyl-ACP was first converted to fatty aldehyde by acyl-ACP reductase (AAR), and then the fatty aldehyde was deformylated to an alkane by aldehyde-deformylating oxygenase (ADO) (Li *et al.*, 2012; Li *et al.*, 2011; Warui *et al.*, 2011; Zhang *et al.*, 2013). The overexpression of AAR and ADO in both *slr0168* and *slr1556* gene loci led to an 8-fold increase of alkane production in *Synechocystis* sp. PCC6803 compared to the wild type (Wang *et al.*, 2013). Liu *et al.* analysed the hydrocarbon profiles and phylogenetic clusters of 19 freshwater cyanobacterial species and found that the classification of species based on hydrocarbon profiles is similar to that based on phylogenetic clusters (Liu *et al.*, 2013). Furthermore, the topology of 16S rDNA showed a general congruence with that of AAR but not with that of ADO (Liu *et al.*, 2013).

8.5 Future trends

Compared to other biofuel-producing microbial systems, cyanobacteria possess some promising advantages due to their intrinsic properties described above. However, as yet, there is no commercial application of biofuel production by cyanobacteria, while corn ethanol is in commercial production and cellulosic ethanol production is in a demonstration stage. The low biosynthetic efficiency, high processing cost and difficult scalability of biofuel production by cyanobacteria are still hurdles to overcome. As discussed in this chapter, advances in biotechnology may help to circumvent these hurdles. First, with the discovery of new pathways based on genome sequencing, new biofuel molecules can be synthesized by cyanobacteria, or the yield of existing biofuel molecules may be increased. At the same time, the comprehensive pathway information will help to systematically modify the biofuel-producing pathways and enhance the specific biofuel productivity. Second, the development of efficient genetic engineering tools with synthetic biology will facilitate multiple gene manipulation in cyanobacteria. Third, more efforts in the optimization of large-scale photobioreactor systems will scale up biofuel production by cyanobacteria and lower the production cost. Briefly, research progress made at the molecular, cellular, process and equipment levels will eventually make commercial cyanobacterial biofuel production feasible.

Acknowledgements

Research work in Xuefeng Lu's group was supported by grants from the National Basic Research Program of China (973: 2011CBA00907), Knowledge Innovation Program of the Chinese Academy of Sciences (KSCX2-EW-G-1–4), the Shandong Province Natural Science Fund for Distinguished Young Scholars, and the Shandong Province 'Taishan Scholarship'.

Web resources

Culture collection

ANACC (Australian National Algae Culture Collection) http://www.csiro.au/Organisation-Structure/National-Facilities/Australian-National-Algae-Culture-Collection.aspx

ATCC (American Type Culture Collection) http://www.atcc.org/

CCAP (Culture Centre of Algae and Protozoa) http://www.ccap.ac.uk/

CPCC (Canadian Phycological Culture Centre) http://www.phycol.ca/

FACHB (Freshwater Algae Culture Collection of the Institute of Hydro biology) http://algae.ihb.ac.cn/

NCMA (National Center for Marine Algae and Microbiota) https://ncma.bigelow.org/

NIES (National Institute for Environmental Studies) http://mcc.nies.go.jp/

PCC (The Pasteur Culture Collection of Cyanobacteria) http://www.pasteur.fr/ip/easysite/pasteur/en/research/collections/crbip/general-informations-concerning-the-collections/iv-the-open-collections/iv-iii-pasteur-culture-collection-of-cyanobacteria

SAG (The Culture Collection of Algae at Goettingen University) http://www.uni-goettingen.de/de/184982.html

UTEX (The Culture Collection of Algae) http://web.biosci.utexas.edu/utex/

Database

Biobike, powerful tools to manipulate and analyse biological data of cyanobacteria: http://biobike.csbc.vcu.edu:8003/biologin

Cyanobase, the Genome Database for Cyanobacteria: http://genome.microbedb.jp/cyanobase

Cyano2Dbase, a protein-gene linkage map of the unicellular cyanobacterium *Synechosystis* sp. strain PCC6803: http://wiki.annotation.jp/Cyano2Dbase

CyanoMutants, an open annotation system for the genes and mutants information: http://genome.microbedb.jp/cyanobase/mutants/

CYORF (Cyanobacteria Gene Annotation Database): http://cyano.genome.jp/

GEBA-Cyano, a collaboration project between the JGI (DOE Joint Genome Institute) and PCC (the Pasteur Culture Collection) for sequencing the genomes of 50 phylogenetically diverse cyanobacterial strains: http://genome.jgi.doe.gov/programs/bacteria-archaea/GEBA-Cyano.jsf

References

Abdallah, C., Dumas-Gaudot, E., Renaut, J., and Sergeant, K. (2012). Gel-based and gel-free quantitative proteomics approaches at a glance. Int. J. Plant Genom. *2012*, 494572.

Abed, R.M.M., Dobretsov, S., and Sudesh, K. (2009). Applications of cyanobacteria in biotechnology. J. Appl. Microbiol. *106*, 1–12.

Aikens, J., Turner, R.J. (2012). Transgenic photosynthetic microorganisms and photobioreactor. US Patent, Application Number: US20120301563.

Akiyama, H., Okuhata, H., Onizuka, T., Kanai, S., Hirano, M., Tanaka, S., Sasaki, K., and Miyasaka, H. (2011). Antibiotics-free stable polyhydroxyalkanoate (PHA) production from carbon dioxide by recombinant cyanobacteria. Bioresource Technol. *102*, 11039–11042.

Allen, M.M. (1984). Cyanobacterial cell inclusions. Annu. Rev. Microbiol. *38*, 1–25.

Altaras, N.E., and Cameron, D.C. (1999). Metabolic engineering of a 1,2-propanediol pathway in *Escherichia coli*. Appl. Environ. Microbiol. *65*, 1180–1185.

Angermayr, S.A., Hellingwerf, K.J., Lindblad, P., and de Mattos, M.J. (2009). Energy biotechnology with cyanobacteria. Curr. Opin. Biotechnol. *20*, 257–263.

Aoki, S., Kondo, T., and Ishiura, M. (2002). A promoter-trap vector for clock-controlled genes in the cyanobacterium *Synechocystis* sp. PCC 6803. J. Microbiol. Methods *49*, 265–274.

Asayama, M. (2012). Overproduction and easy recovery of target gene products from cyanobacteria, photosynthesizing microorganisms. Appl. Microbiol. Biotechnol. *95*, 683–695.

Askora, A., Abdel-Haliem, M.E., and Yamada, T. (2012). Site-specific recombination systems in filamentous phages. Mol. Genet. Genomics *287*, 525–530.

Atsumi, S., Hanai, T., and Liao, J.C. (2008). Non-fermentative pathways for synthesis of branched-chain higher alcohols as biofuels. Nature *451*, 86–89.

Atsumi, S., Higashide, W., and Liao, J.C. (2009). Direct photosynthetic recycling of carbon dioxide to isobutyraldehyde. Nat. Biotechnol. *27*, 1177–1180.

Baran, R., Bowen, B.P., and Northen, T.R. (2011). Untargeted metabolic footprinting reveals a surprising breadth of metabolite uptake and release by Synechococcus sp. PCC 7002. Mol. Biosyst. *7*, 3200–3206.

Burja, A.M., Banaigs, B., Abou-Mansour, E., Burgess, J.G., and Wright, P.C. (2001). Marine cyanobacteria-a prolific source of natural products. Tetrahedron *57*, 9347–9377.

Castenholz, R.W. (2001). Phylum BX. Cyanobacteria. In: D.R. Boone, G. Garrity and r.W. Castenholz eds. Bergey´s Manual of Systematic Bacteriology. New York: Springer, pp. 473–599.

Cheah, Y.E., Albers, S.C., and Peebles, C.A. (2013). A novel counter-selection method for markerless genetic modification in *Synechocystis* sp. PCC 6803. Biotechnol. Prog. *29*, 23–30.

Chen, G.Q. (2009). A microbial polyhydroxyalkanoates (PHA) based bio- and materials industry. Chem. Soc. Rev. *38*, 2434–2446.

Cronan, J.E., and Thomas, J. (2009). Bacterial Fatty Acid Synthesis and Its Relationships with Polyketide Synthetic Pathways. Complex Enzymes in Microbial Natural Product Biosynthesis, Part B: Polyketides, Aminocoumarins and Carbohydrates *459*, 395–433.

Dehring, U., Kramer, D., and Ziegler, K. (2012). *US Pat.*, 8 163 516 B2.

Deng, M.D., and Coleman, J.R. (1999). Ethanol synthesis by genetic engineering in cyanobacteria. Appl. Environ. Microbiol. *65*, 523–528.

De Philippis, R., Margheri, M.C., Materassi, R., and Vincenzini, M. (1998). Potential of unicellular cyanobacteria from saline environments as exopolysaccharide producers. Appl. Environ. Microbiol. *64*, 1130–1132.

De Philippis, R., Sili, C., Paperi, R., and Vincenzini, M. (2001). Exopolysaccharide-producing cyanobacteria and their possible exploitation: A review. J. Appl. Phycol. *13*, 293–299.

Devilly, C.I., and Houghton, J.A. (1977). A Study of Genetic Transformation in Gloeocapsa alpicola. J. Gen. Microbiol. *98*, 277–280.

Dexter, J., and Fu, P.C. (2009). Metabolic engineering of cyanobacteria for ethanol production. Energy Environ. Sci. *2*, 857–864.

Doray, B., Chen, C.D., and Kemper, B. (2001). N-terminal deletions and His-tag fusions dramatically affect expression of cytochrome P4502C2 in bacteria. Arch. Biochem. Biophys. *393*, 143–153.

Dugar, D., and Stephanopoulos, G. (2011). Relative potential of biosynthetic pathways for biofuels and bio-based products. Nat. Biotechnol. *29*, 1074–1078.

Eisenhut, M., Huege, J., Schwarz, D., Bauwe, H., Kopka, J., and Hagemann, M. (2008). Metabolome Phenotyping of Inorganic Carbon Limitation in Cells of the Wild Type and Photorespiratory Mutants of the Cyanobacterium Synechocystis sp Strain PCC 6803. Plant Physiol. *148*, 2109–2120.

Ekman, M., Ow, S.Y., Holmqvist, M., Zhang, X., van Wagenen, J., Wright, P.C., and Stensjo, K. (2011). Metabolic adaptations in a H2 producing heterocyst-forming cyanobacterium: potentials and implications for biological engineering. J. Proteome Res. *10*, 1772–1784.

Elhai, J., and Wolk, C.P. (1988). Conjugal transfer of DNA to cyanobacteria. Methods Enzymol. *167*, 747–754.

Elhai, J., Vepritskiy, A., Muro-Pastor, A.M., Flores, E., and Wolk, C.P. (1997). Reduction of conjugal transfer efficiency by three restriction activities of *Anabaena* sp. strain PCC 7120. J. Bacteriol. *179*, 1998–2005.

El Semary, A. (2010). Optimized electroporation-induced transformation in Microcystis aeruginosa PCC7806. Biotechnologie, Agronomie, Société et Environnement *14*, 149–152.

Eriksen, N.T. (2008). Production of phycocyanin-a pigment with applications in biology, biotechnology, foods and medicine. Appl. Microbiol. Biotechnol. *80*, 1–14.

Fehler, S.W.G., and Light, R.J. (1972). Biosynthesis of Methylheptadecanes in Anabaena-Variabilis – in-Vitro Incorporation of S-[Methyl-C-14] Adenosylmethionine. Biochemistry *11*, 2411–2416.

Gao, Z.X., Zhao, H., Li, Z.M., Tan, X.M., and Lu, X.F. (2012). Photosynthetic production of ethanol from carbon dioxide in genetically engineered cyanobacteria. Energy Environ. Sci. *5*, 9857–9865.

Garcia-Pichel, F., and Pringault, O. (2001). Microbiology: Cyanobacteria track water in desert soils. Nature *413*, 380–381.

Gehl, J. (2003). Electroporation: theory and methods, perspectives for drug delivery, gene therapy and research. Acta Physiologica Scandinavica *177*, 437–447.

Giovannoni, S.J., Turner, S., Olsen, G.J., Barns, S., Lane, D.J., and Pace, N.R. (1988). Evolutionary relationships among cyanobacteria and green chloroplasts. J. Bacteriol. *170*, 3584–3592.

Golden, S.S., and Sherman, L.A. (1984). Optimal conditions for genetic transformation of the cyanobacterium Anacystis nidulans R2. J. Bacteriol. *158*, 36–42.

Golden, S.S., Brusslan, J., and Haselkorn, R. (1987). Genetic engineering of the cyanobacterial chromosome. Methods Enzymol. *153*, 215–231.

Griese, M., Lange, C., and Soppa, J. (2011). Ploidy in cyanobacteria. FEMS Microbiol. Lett. 323, 124–131.

Grigorieva, G., and Shestakov, S. (1982). Transformation in the cyanobacterium Synechocystis sp. 6803. FEMS Microbiol. Lett. *13*, 367–370.

Grindley, N.D., Whiteson, K.L., and Rice, P.A. (2006). Mechanisms of site-specific recombination. Annu. Rev. Biochem. *75*, 567–605.

Han, J., Chan, H.W.S., and Calvin, M. (1969). Biosynthesis of Alkanes in Nostoc Muscorum. J. Am. Chem. Soc. *91*, 5156–5159.

Hasunuma, T., Kikuyama, F., Matsuda, M., Aikawa, S., Izumi, Y., and Kondo, A. (2013). Dynamic metabolic profiling of cyanobacterial glycogen biosynthesis under conditions of nitrate depletion. J. Exp. Bot. *64*, 2943–2954.

Hoiczyk, E., and Hansel, A. (2000). Cyanobacterial cell walls: News from an unusual prokaryotic envelope. J. Bacteriol. *182*, 1191–1199.

Huang, H.H., Camsund, D., Lindblad, P., and Heidorn, T. (2010). Design and characterization of molecular tools for a Synthetic Biology approach towards developing cyanobacterial biotechnology. Nucleic Acids Res. *38*, 2577–2593.

Insenser, M.R., Hernaez, M.L., Nombela, C., Molina, M., Molero, G., and Gil, C. (2010). Gel and gel-free proteomics to identify Saccharomyces cerevisiae cell surface proteins. J. Proteomics 73, 1183–1195.

Ippen-Ihler, K.A., and Minkley, E.G. (1986). The Conjugation System of F, The Fertility Factor of Escherichia Coli. Annu. Rev. Genet. *20*, 593–624.

Kaneko, T., Sato, S., Kotani, H., Tanaka, A., Asamizu, E., Nakamura, Y., Miyajima, N., Hirosawa, M., Sugiura, M., Sasamoto, S., *et al.* (1996). Sequence analysis of the genome of the unicellular cyanobacterium *Synechocystis* sp. strain PCC6803. II. Sequence determination of the entire genome and assignment of potential protein-coding regions. DNA Res. *3*, 109–136.

Keasling, J.D. (2012). Synthetic biology and the development of tools for metabolic engineering. Metab. Eng. *14*, 189–195.

Knoop, H., Zilliges, Y., Lockau, W., and Steuer, R. (2010). The Metabolic Network of Synechocystis sp. PCC 6803: Systemic Properties of Autotrophic Growth. Plant Physiol. *154*, 410–422.

Koksharova, O., and Wolk, C. (2002). Genetic tools for cyanobacteria. Appl. Microbiol. Biotechnol. *58*, 123–137.

Krall, L., Huege, J., Catchpole, G., Steinhauser, D., and Willmitzer, L. (2009). Assessment of sampling strategies for gas chromatography-mass spectrometry (GC-MS) based metabolomics of cyanobacteria. Journal of chromatography B, *877*, 2952–2960.

Kreps, S., Ferino, F., Mosrin, C., Gerits, J., Mergeay, M., and Thuriaux, P. (1990). Conjugative Transfer and Autonomous Replication of a Promiscuous Incq Plasmid in the Cyanobacterium Synechocystis Pcc-6803. Mol. Gen. Genet. *221*, 129–133.

Kufryk, G.I., Sachet, M., Schmetterer, G., and Vermaas, W.F. (2002). Transformation of the cyanobacterium *Synechocystis* sp. PCC 6803 as a tool for genetic mapping: optimization of efficiency. FEMS Microbiol. Lett. *206*, 215–219.

Kumar, K., Mella-Herrera, R.A., and Golden, J.W. (2010). Cyanobacterial Heterocysts. Csh. Perspect. Biol. 2: a000315.

Lan, E.I., and Liao, J.C. (2011). Metabolic engineering of cyanobacteria for 1-butanol production from carbon dioxide. Metab. Eng. *13*, 353–363.

Lan, E.I., and Liao, J.C. (2012). ATP drives direct photosynthetic production of 1-butanol in cyanobacteria. Proc. Natl. Acad. Sci. USA *109*, 6018–6023.

Landry, B.P., Stockel, J., and Pakrasi, H.B. (2013). Use of Degradation Tags To Control Protein Levels in the Cyanobacterium Synechocystis sp Strain PCC 6803. Appl. Environ. Microbiol. *79*, 2833–2835.

Leplat, C., Champeimont, R., Saenkham, P., Cassier-Chauvat, C., Jean-Christophe, A., and Chauvat, F. (2013). Genome-wide transcriptome analysis of hydrogen production in the cyanobacterium *Synechocystis*: Towards the identification of new players. Int. J. Hydrogen Energ. *38*, 1866–1872.

Li, H., and Liao, J.C. (2013). Engineering a cyanobacterium as the catalyst for the photosynthetic conversion of CO2 to 1,2-propanediol. Microbial cell factories *12:4*.

Li, N., Norgaard, H., Warui, D.M., Booker, S.J., Krebs, C., and Bollinger, J.M., Jr. (2011). Conversion of fatty aldehydes to alka (e)nes and formate by a cyanobacterial aldehyde decarbonylase: cryptic redox by an unusual dimetal oxygenase. J. Am. Chem. Soc. *133*, 6158–6161.

Li, N., Chang, W.C., Warui, D.M., Booker, S.J., Krebs, C., and Bollinger, J.M. (2012). Evidence for Only Oxygenative Cleavage of Aldehydes to Alk (a/e) nes and Formate by Cyanobacterial Aldehyde Decarbonylases. Biochemistry *51*, 7908–7916.

Li, Z.-J., Jian, J., Wei, X.-X., Shen, X.-W., and Chen, G.-Q. (2010). Microbial production of meso-2,3-butanediol by metabolically engineered *Escherichia coli* under low oxygen condition. Appl. Microbiol. Biotechnol. *87*, 2001–2009.

Lindberg, P., Park, S., and Melis, A. (2010). Engineering a platform for photosynthetic isoprene production in cyanobacteria, using Synechocystis as the model organism. Metab. Eng. *12*, 70–79.

Liu, A., Zhu, T., Lu, X., and Song, L. (2013). Hydrocarbon profiles and phylogenetic analysis of diversified cyanobacterial species. Appl. Energ. *111*, 383–393.

Liu, J., Chen, L., Wang, J., Qiao, J., and Zhang, W. (2012). Proteomic analysis reveals resistance mechanism against biofuel hexane in Synechocystis sp. PCC 6803. Biotechnol. Biofuels 5, 68.

Liu, X., Fallon, S., Sheng, J., and Curtiss, R., 3rd (2011a). CO2-limitation-inducible Green Recovery of fatty acids from cyanobacterial biomass. Proc. Natl. Acad. Sci. USA *108*, 6905–6908.

Liu, X., Sheng, J., and Curtiss, R., 3rd (2011b). Fatty acid production in genetically modified cyanobacteria. Proc. Natl. Acad. Sci. USA *108*, 6899–6904.

Lu, X. (2010). A perspective: photosynthetic production of fatty acid-based biofuels in genetically engineered cyanobacteria. Biotechnol. Adv. *28*, 742–746.

Ludwig, A., Heimbucher, T., Gregor, W., Czerny, T., and Schmetterer, G. (2008). Transformation and gene replacement in the facultatively chemoheterotrophic, unicellular cyanobacterium Synechocystis sp. PCC6714 by electroporation. Appl. Microbiol. Biotechnol. *78*, 729–735.

Lyznik, L.A., Gordon-Kamm, W.J., and Tao, Y. (2003). Site-specific recombination for genetic engineering in plants. Plant cell reports *21*, 925–932.

McNeely, K., Xu, Y., Bennette, N., Bryant, D.A., and Dismukes, G.C. (2010). Redirecting Reductant Flux into Hydrogen Production via Metabolic Engineering of Fermentative Carbon Metabolism in a Cyanobacterium. Appl. Environ. Microbiol. *76*, 5032–5038.

Madison, L.L., and Huisman, G.W. (1999). Metabolic engineering of poly (3-hydroxyalkanoates): From DNA to plastic. Microbiol. Mol. Biol. Rev. *63*, 21–53.

Matsuoka, M., Takahama, K., and Ogawa, T. (2001). Gene replacement in cyanobacteria mediated by a dominant streptomycin-sensitive *rps12* gene that allows selection of mutants free from drug resistance markers. Microbiology *147*, 2077–2087.

Mendez-Perez, D., Begemann, M.B., and Pfleger, B.F. (2011). Modular Synthase-Encoding Gene Involved in alpha-Olefin Biosynthesis in Synechococcus sp Strain PCC 7002. Appl. Environ. Microbiol. 77, 4264–4267.

Mertins, P., Udeshi, N.D., Clauser, K.R., Mani, D.R., Patel, J., Ong, S.E., Jaffe, J.D., and Carr, S.A. (2012). iTRAQ Labeling is Superior to mTRAQ for Quantitative Global Proteomics and Phosphoproteomics. Mol. Cell. Proteomics *11*, 6.

Mitschke, J., Georg, J., Scholz, I., Sharma, C.M., Dienst, D., Bantscheff, J., Voss, B., Steglich, C., Wilde, A., Vogel, J., *et al.* (2011). An experimentally anchored map of transcriptional start sites in the model cyanobacterium Synechocystis sp PCC6803. Proc. Natl. Acad. Sci. USA *108*, 2124–2129.

Mlouka, A., Comte, K., Castets, A.M., Bouchier, C., and de Marsac, N.T. (2004). The gas vesicle gene cluster from *Microcystis aeruginosa* and DNA rearrangements that lead to loss of cell buoyancy. J. Bacteriol. *186*, 2355–2365.

Nakamura, C.E., and Whited, G.M. (2003). Metabolic engineering for the microbial production of 1,3-propanediol. Curr. Opin. Biotechnol. *14*, 454–459.

Nakasugi, K., Svenson, C.J., and Neilan, B.A. (2006). The competence gene, comF, from Synechocystis sp. strain PCC 6803 is involved in natural transformation, phototactic motility and piliation. Microbiology *152*, 3623–3631.

Namikoshi, M., and Rinehart, K.L. (1996). Bioactive compounds produced by cyanobacteria. J. Ind. Microbiol. Biot. *17*, 373–384.

Okamoto, S., and Ohmori, M. (2002). The Cyanobacterial PilT Protein Responsible for Cell Motility and Transformation Hydrolyzes ATP. Plant Cell Physiol. *43*, 1127–1136.

Oliver, J.W.K., Machado, I.M.P., Yoneda, H., and Atsumi, S. (2013). Cyanobacterial conversion of carbon dioxide to 2,3-butanediol. Proc. Natl. Acad. Sci. USA *110*, 1249–1254.

Onai, K., Morishita, M., Kaneko, T., Tabata, S., and Ishiura, M. (2004). Natural transformation of the thermophilic cyanobacterium Thermosynechococcus elongatus BP-1: a simple and efficient method for gene transfer. Mol. Genet. Genomics *271*, 50–59.

Ong, S.E., Blagoev, B., Kratchmarova, I., Kristensen, D.B., Steen, H., Pandey, A., and Mann, M. (2002). Stable isotope labeling by amino acids in cell culture, SILAC, as a simple and accurate approach to expression proteomics. Mol. Cell. Proteomics *1*, 376–386.

Oppermann-Sanio, F.B., and Steinbuchel, A. (2002). Occurrence, functions and biosynthesis of polyamides in microorganisms and biotechnological production. Naturwissenschaften *89*, 11–22.

Osanai, T., Kanesaki, Y., Nakano, T., Takahashi, H., Asayama, M., Shirai, M., Kanehisa, M., Suzuki, I., Murata, N., and Tanaka, K. (2005). Positive regulation of sugar catabolic pathways in the cyanobacterium *Synechocystis* sp PCC 6803 by the group 2 sigma factor sigE. J. Biol. Chem. *280*, 30653–30659.

Osanai, T., Numata, K., Oikawa, A., Kuwahara, A., Iijima, H., Doi, Y., Tanaka, K., Saito, K., and Hirai, M.Y. (2013). Increased Bioplastic Production with an RNA Polymerase Sigma Factor SigE during Nitrogen Starvation in Synechocystis sp. PCC 6803. DNA Res. 20, 525–535.

Patel, V.J., Thalassinos, K., Slade, S.E., Connolly, J.B., Crombie, A., Murrell, J.C., and Scrivens, J.H. (2009). A Comparison of Labeling and Label-Free Mass Spectrometry-Based Proteomics Approaches. J. Proteome Res. *8*, 3752–3759.

Pfeifer, F. (2012). Distribution, formation and regulation of gas vesicles. Nat. Rev. Microbiol. *10*, 705–715.

Porter, R.D. (1986). Transformation in cyanobacteria. Crit. Rev. Microbiol. *13*, 111–132.

Qi, F., Yao, L., Tan, X., and Lu, X. (2013). Construction, characterization and application of molecular tools for metabolic engineering of Synechocystis sp. Biotechnol. Lett. 35, 1655–1661.

Qiao, J.J., Wang, J.X., Chen, L., Tian, X.X., Huang, S.Q., Ren, X.Y., and Zhang, W.W. (2012). Quantitative iTRAQ LC-MS/MS Proteomics Reveals Metabolic Responses to Biofuel Ethanol in Cyanobacterial Synechocystis sp PCC 6803. J. Proteome Res. *11*, 5286–5300.

Quintana, N., Van der Kooy, F., Van de Rhee, M.D., Voshol, G.P., and Verpoorte, R. (2011). Renewable energy from cyanobacteria: Energy production optimization by metabolic pathway engineering. Appl. Microbiol. Biotechnol. *91*, 471–490.

Reinsvold, R.E., Jinkerson, R.E., Radakovits, R., Posewitz, M.C., and Basu, C. (2011). The production of the sesquiterpene beta-caryophyllene in a transgenic strain of the cyanobacterium Synechocystis. J. Plant Physiol. *168*, 848–852.

Rippka, R., Deruelles, J., Waterbury, J.B., Herdman, M., and Stanier, R.Y. (1979). Generic assignments, strain histories and properties of pure cultures of cyanobacteria. J. Gen. Microbiol. *111*, 1–61.

Ruffing, A.M. (2013). RNA-Seq analysis and targeted mutagenesis for improved free fatty acid production in an engineered cyanobacterium. Biotechnol. Biofuels *6*, 113.

Ruffing, A.M., and Jones, H.D.T. (2012). Physiological effects of free fatty acid production in genetically engineered Synechococcus elongatus PCC 7942. Biotechnol. Bioeng. *109*, 2190–2199.

Schirmer, A., Rude, M.A., Li, X., Popova, E., and del Cardayre, S.B. (2010). Microbial biosynthesis of alkanes. Science *329*, 559–562.

Schirrmeister, B.E., Antonelli, A., and Bagheri, H.C. (2011). The origin of multicellularity in cyanobacteria. BMC evolutionary biology *11*, 45.

Schwarz, D., Nodop, A., Huge, J., Purfurst, S., Forchhammer, K., Michel, K.P., Bauwe, H., Kopka, J., and Hagemann, M. (2011). Metabolic and Transcriptomic Phenotyping of Inorganic Carbon Acclimation in the Cyanobacterium Synechococcus elongatus PCC 7942. Plant Physiol. *155*, 1640–1655.

Schweizer, H.P. (2003). Applications of the *Saccharomyces cerevisiae* Flp-FRT system in bacterial genetics. J. Mol. Microbiol. Biotechnol. *5*, 67–77.

Sethuraman, M., McComb, M.E., Huang, H., Huang, S.Q., Heibeck, T., Costello, C.E., and Cohen, R.A. (2004). Isotope-coded affinity tag (ICAT) approach to redox proteomics: Identification and quantitation of oxidant-sensitive cysteine thiols in complex protein mixtures. J. Proteome Res. *3*, 1228–1233.

Shen, C.R., and Liao, J.C. (2012). Photosynthetic production of 2-methyl-1-butanol from CO2 in cyanobacterium Synechococcus elongatus PCC7942 and characterization of the native acetohydroxyacid synthase. Energy Environ. Sci. *5*, 9574–9583.

Sherstnev, A., Duc, C., Cole, C., Zacharaki, V., Hornyik, C., Ozsolak, F., Milos, P.M., Barton, G.J., and Simpson, G.G. (2012). Direct sequencing of Arabidopsis thaliana RNA reveals patterns of cleavage and polyadenylation. Nat. Struct. Mol. Biol. *19*, 845–852.

Shestakov, S.V., and Khyen, N. (1970). Evidence for genetic transformation in blue-green alga Anacystis nidulans. Mol. Gen. Genet. *107*, 372–375.

Shih, P.M., Wu, D.Y., Latifi, A., Axen, S.D., Fewer, D.P., Talla, E., Calteau, A., Cai, F., de Marsac, N.T., Rippka, R., *et al.* (2013). Improving the coverage of the cyanobacterial phylum using diversity-driven genome sequencing. Proc. Natl. Acad. Sci. USA *110*, 1053–1058.

Siegal, M.L., and Hartl, D.L. (2000). Application of Cre/loxP in Drosophila. Site-specific recombination and transgene coplacement. Methods Mol. Biol. *136*, 487–495.

Singh, R.K., Tiwari, S.P., Rai, A.K., and Mohapatra, T.M. (2011). Cyanobacteria: An emerging source for drug discovery. J. Antibiot. *64*, 401–412.

Sivakumar, N., Sundararaman, M., and Selvakumar, G. (2011). Efficacy of microalgae and cyanobacteria as a live feed for juveniles of shrimp *Penaeus monodon*. Afr. J. Biotechnol. *10*, 11594–11599.

Sode, K., Tatara, M., Takeyama, H., Burgess, J.G., and Matsunaga, T. (1992). Conjugative Gene-Transfer in Marine Cyanobacteria – Synechococcus Sp, Synechocystis Sp and Pseudanabaena Sp. Appl. Microbiol. Biotechnol. 37, 369–373.

Spolaore, P., Joannis-Cassan, C., Duran, E., and Isambert, A. (2006). Commercial applications of microalgae. J. Biosci. Bioeng. *101*, 87–96.

Steinbuchel, A., and Valentin, H.E. (1995). Diversity of bacterial polyhydroxyalkanoic acids. FEMS Microbiol. Lett. *128*, 219–228.

Stevens, S.E., and Porter, R.D. (1980). Transformation in Agmenellum quadruplicatum. Proc. Natl. Acad. Sci. USA 77, 6052–6056.

Tan, X., Yao, L., Gao, Q., Wang, W., Qi, F., and Lu, X. (2011). Photosynthesis driven conversion of carbon dioxide to fatty alcohols and hydrocarbons in cyanobacteria. Metab. Eng. 13, 169–176.

Tan, X., Wang, W., and Lu, X. (2012). Production of Photosynthetic Biofuels by Genetically Engineering Cyanobacteria. Curr. Chem. Biol. 6, 26–31.

Tan, X., Liang, F., Cai, K., and Lu, X. (2013). Application of the FLP/FRT recombination system in cyanobacteria for construction of markerless mutants. Appl. Microbiol. Biotechnol. 14, 6373–6382.

Taton, A., Lis, E., Adin, D.M., Dong, G., Cookson, S., Kay, S.A., Golden, S.S., and Golden, J.W. (2012). Gene transfer in Leptolyngbya sp. strain BL0902, a cyanobacterium suitable for production of biomass and bioproducts. PLoS One 7, e30901.

Thajuddin, N., and Subramanian, G. (2005). Cyanobacterial biodiversity and potential applications in biotechnology. Curr. Sci. India 89, 47–57.

Thiel, T., and Poo, H. (1989). Transformation of a filamentous cyanobacterium by electroporation. J. Bacteriol. 171, 5743–5746.

Tian, X.X., Chen, L., Wang, J.X., Qiao, J.J., and Zhang, W.W. (2013). Quantitative proteomics reveals dynamic responses of Synechocystis sp PCC 6803 to next-generation biofuel butanol. J. Proteomics 78, 326–345.

Tomitani, A., Knoll, A.H., Cavanaugh, C.M., and Ohno, T. (2006). The evolutionary diversification of cyanobacteria: Molecular-phylogenetic and paleontological perspectives. Proc. Natl. Acad. Sci. USA 103, 5442–5447.

Toyomizu, M., Suzuki, K., Kawata, Y., Kojima, H., and Akiba, Y. (2001). Effective transformation of the cyanobacterium Spirulina platensis using electroporation. J. Appl. Phycol. 13, 209–214.

Turner, S., Pryer, K.M., Miao, V.P.W., and Palmer, J.D. (1999). Investigating deep phylogenetic relationships among cyanobacteria and plastids by small submit rRNA sequence analysis. J. Eukaryot. Microbiol. 46, 327–338.

Uzair, B., Tabassum, S., Rasheed, M., and Rehman, S.F. (2012). Exploring marine cyanobacteria for lead compounds of pharmaceutical importance. The Scientific World Journal 2012, 179782.

Vaishampayan, A., Sinha, R.P., Hader, D.P., Dey, T., Gupta, A.K., Bhan, U., and Rao, A.L. (2001). Cyanobacterial biofertilizers in rice agriculture. Bot. Rev. 67, 453–516.

Varman, A.M., Xiao, Y., Pakrasi, H.B., and Tang, Y.J.J. (2013). Metabolic Engineering of Synechocystis sp Strain PCC 6803 for Isobutanol Production. Appl. Environ. Microbiol. 79, 908–914.

Vermaas, W. (1996). Molecular genetics of the cyanobacterium *Synechocystis* sp. PCC 6803: Principles and possible biotechnology applications. J. Appl. Phycol. 8, 263–273.

Vidal, R., Lopez-Maury, L., Guerrero, M.G., and Florencio, F.J. (2009). Characterization of an alcohol dehydrogenase from the Cyanobacterium Synechocystis sp. strain PCC 6803 that responds to environmental stress conditions via the Hik34-Rre1 two-component system. J. Bacteriol. 191, 4383–4391.

Vijayan, V., Jain, I.H., and O'Shea, E.K. (2011). A high resolution map of a cyanobacterial transcriptome. Genome Biol. 12: R47.

Walsby, A.E. (1972). Structure and function of gas vacuoles. Bacteriol. Rev. 36, 1–32.

Walsby, A.E. (1994). Gas vesicles. Microbiol. Rev. 58, 94–144.

Wang, J.X., Chen, L., Huang, S.Q., Liu, J., Ren, X.Y., Tian, X.X., Qiao, J.J., and Zhang, W.W. (2012). RNA-seq based identification and mutant validation of gene targets related to ethanol resistance in cyanobacterial Synechocystis sp PCC 6803. Biotechnol. Biofuels 5, 89.

Wang, W., Liu, X., and Lu, X. (2013). Engineering cyanobacteria to improve photosynthetic production of alka (e)nes. Biotechnol. Biofuels 6, 69.

Wang, Z., Gerstein, M., and Snyder, M. (2009). RNA-Seq: a revolutionary tool for transcriptomics. Nat. Rev. Genet. 10, 57–63.

Warui, D.M., Li, N., Norgaard, H., Krebs, C., Bollinger, J.M., Jr., and Booker, S.J. (2011). Detection of formate, rather than carbon monoxide, as the stoichiometric coproduct in conversion of fatty aldehydes to alkanes by a cyanobacterial aldehyde decarbonylase. J. Am. Chem. Soc. 133, 3316–3319.

Willetts, N., and Wilkins, B. (1984). Processing of plasmid DNA during bacterial conjugation. Microbiol. Rev. 48, 24–41.

Williams, J.G.K. (1988). Construction of specific mutations in photosystem II photosynthetic reaction center by genetic engineering methods in *Synechocystis* 6803. Methods Enzymol. 167, 766–778.

Winters, K., Parker, P.L., and Vanbaale, C. (1969). Hydrocarbons of Blue-Green Algae – Geochemical Significance. Science 163, 467–468.

Wolk, C.P., and Kraus, J. (1982). Two approaches to obtaining low, extracellular deoxyribonuclease activity in cultures of heterocyst-forming cyanobacteria. Arch. Microbiol. 131, 302–307.

Wolk, C.P., Vonshak, A., Kehoe, P., and Elhai, J. (1984). Construction of Shuttle Vectors Capable of Conjugative Transfer from Escherichia-Coli to Nitrogen-Fixing Filamentous Cyanobacteria. Proc. Natl. Acad. Sci.-Biol. 81, 1561–1565.

Yang, Y., Park, Y., Cassada, D.A., Snow, D.D., Rogers, D.G., and Lee, J. (2011). In vitro and in vivo safety assessment of edible blue-green algae, *Nostoc commune* var. *sphaeroides* Kutzing and *Spirulina plantensis*. Food Chem. Toxicol. 49, 1560–1564.

Yim, H., Haselbeck, R., Niu, W., Pujol-Baxley, C., Burgard, A., Boldt, J., Khandurina, J., Trawick, J.D., Osterhout, R.E., Stephen, R., *et al.* (2011). Metabolic engineering of *Escherichia coli* for direct production of 1,4-butanediol. Nat. Chem. Biol. 7, 445–452.

Yoshihara, S., Geng, X., Okamoto, S., Yura, K., Murata, T., Go, M., Ohmori, M., and Ikeuchi, M. (2001). Mutational Analysis of Genes Involved in Pilus Structure, Motilityand Transformation Competency in the Unicellular Motile CyanobacteriumSynechocystis sp. PCC6803. Plant Cell Physiol. *42*, 63–73.

Yoshikawa, K., Hirasawa, T., Ogawa, K., Hidaka, Y., Nakajima, T., Furusawa, C., and Shimizu, H. (2013). Integrated transcriptomic and metabolomic analysis of the central metabolism of Synechocystis sp PCC 6803 under different trophic conditions. Biotechnol. J. *8*, 571–580.

Zang, X., Liu, B., Liu, S., Arunakumara, K.K., and Zhang, X. (2007). Optimum conditions for transformation of *Synechocystis* sp. PCC 6803. J. Microbiol. (Seoul, Korea) *45*, 241–245.

Zhang, C.C., Laurent, S., Sakr, S., Peng, L., and Bedu, S. (2006). Heterocyst differentiation and pattern formation in cyanobacteria: a chorus of signals. Mol. Microbiol. *59*, 367–375.

Zhang, J., Lu, X., and Li, J.J. (2013). Conversion of fatty aldehydes into alk (a/e)nes by in vitro reconstituted cyanobacterial aldehyde-deformylating oxygenase with the cognate electron transfer system. Biotechnol. Biofuels *6*, 86.

Zhang, Y.L., Zhang, J.J., Hoeflich, K.P., Ikura, M., Qing, G.L., and Inouye, M. (2003). MazF cleaves cellular mRNAs specifically at ACA to block protein synthesis in *Escherichia coli*. Mol. Cell *12*, 913–923.

Zhang, Y., Pu, H., Wang, Q., Cheng, S., Zhao, W., Zhang, Y., and Zhao, J. (2007). PII is important in regulation of nitrogen metabolism but not required for heterocyst formation in the cyanobacterium *Anabaena* sp. PCC 7120. J. Biol. Chem. *282*, 33641–33648.

Zhu, H., Ren, X., Wang, J., Song, Z., Shi, M., Qiao, J., Tian, X., Liu, J., Chen, L., and Zhang, W. (2013). Integrated OMICS guided engineering of biofuel butanol-tolerance in photosynthetic Synechocystis sp. PCC 6803. Biotechnol. Biofuels 6, 106.

Index

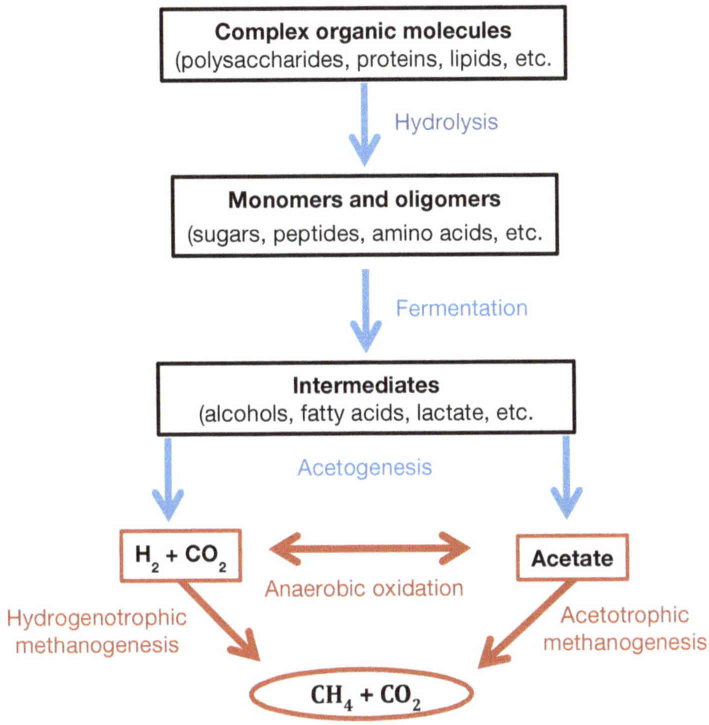

Figure 2.1 Stepwise decomposition of organic matter into biogas.

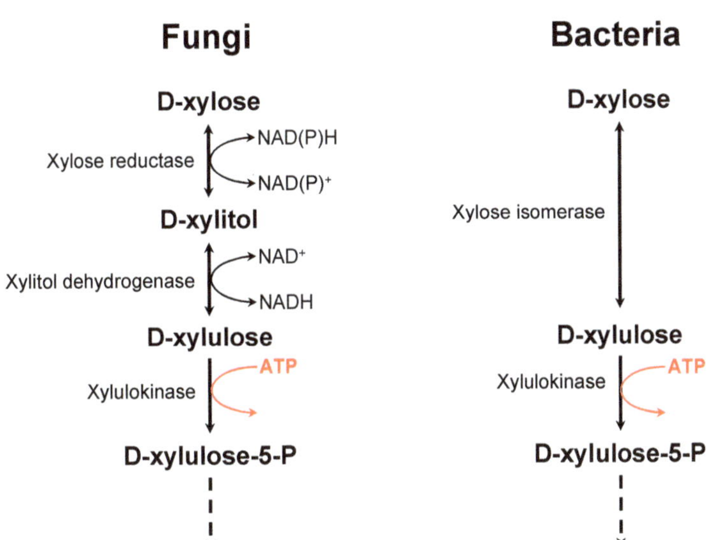

Figure 3.1 Metabolic pathways for the utilization of xylose by fungi and bacteria.

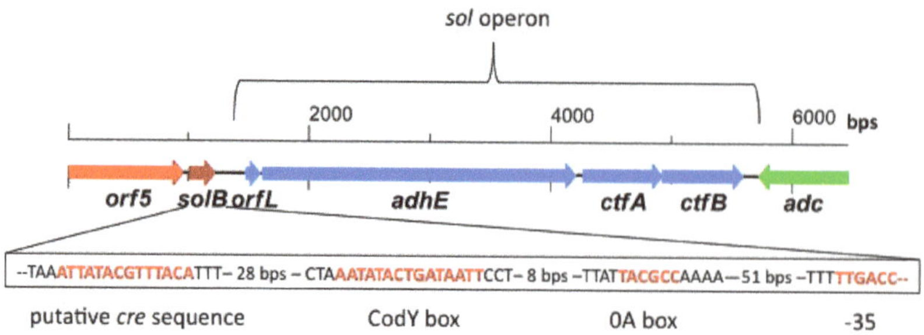

Figure 4.2 Binding sites for regulators upstream of the *sol* operon of *C. acetobutylicum.* Consensus motifs for binding of Spo0A~P (0A), CodY, and CcpA (*cre* sequence) upstream of the *sol* operon promoter are shown (–35 region of this promoter is also indicated). *orf5* encodes a glycosylase/deglycosylase, *solB* a small regulatory RNA. The function of *orfL* is still unknown.

Figure 4.4 LanzaTech's gas fermentation demonstration plant with BaoSteel in Shanghai.

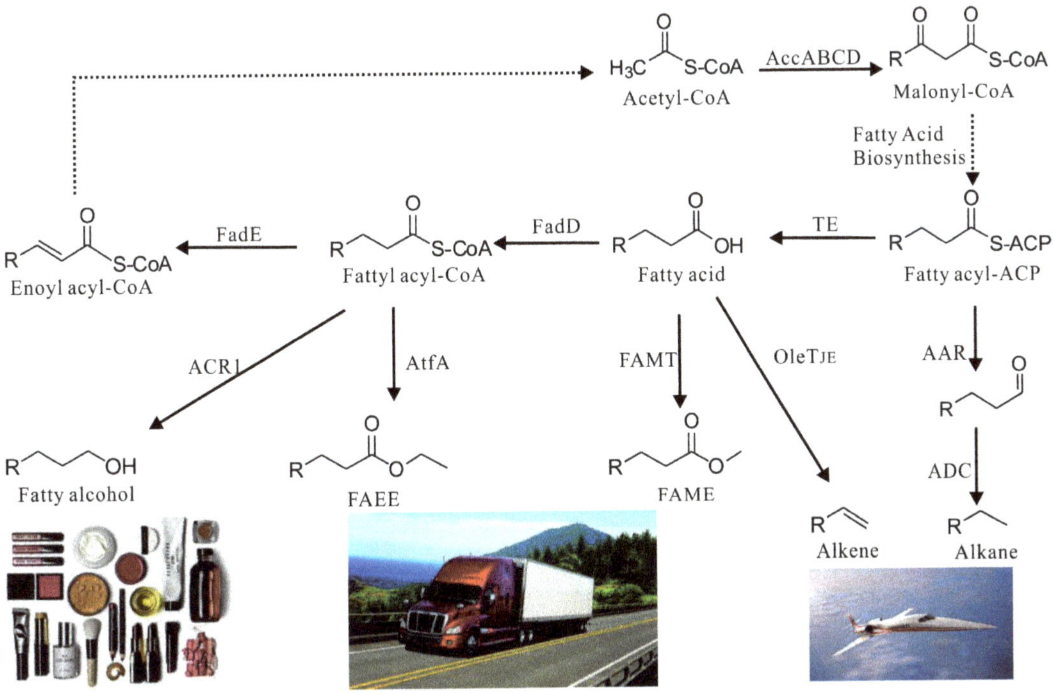

Figure 7.2 Biofuels derived from fatty acids. General synthesis pathways and applications of fatty alcohol, fatty acid methyl/ethyl ester, fatty alkene and fatty alkane are shown. AAR, acyl-ACP reductase; ACC, acetyl-CoA carboxylase; Acr1, acyl-CoA reductase; ADC, aldehyde decarbonylase; AtfA, wax-ester synthase; FadD, acyl-CoA synthetase; FadE, acyl CoA dehydrogenase; FAMT, fatty acid methyltransferase; OleTJE, *Jeotgalicoccus* sp. terminal olefin-forming fatty-acid decarboxylase; TE, acyl-ACP thioesterase.

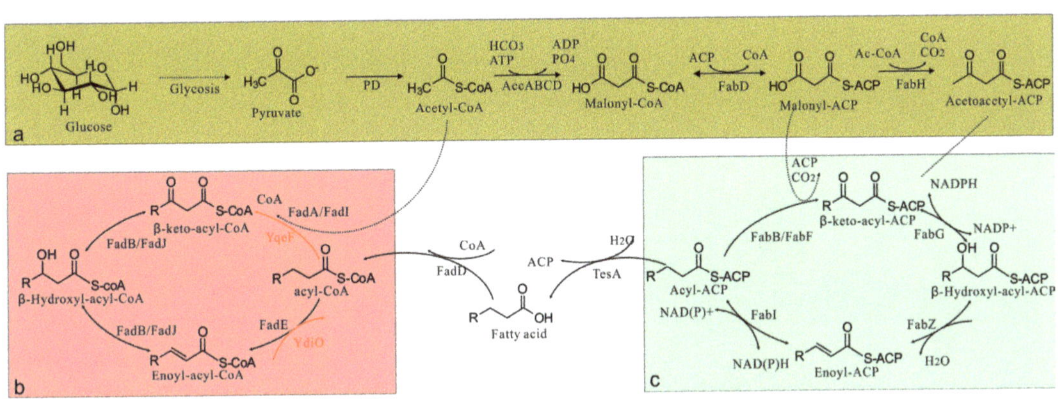

Figure 7.3 Metabolic pathways for fatty acid biosynthesis. The synthesis of starter and extender units is shown in the yellow box (a); the elongation cycle is shown in the green box (c), and the β-oxidation step is shown in the red box (b). The red in the β-oxidation cycle represent the critical enzymes involved in the reversal of β-oxidation. ACC, acetyl-CoA carboxylase; FabB, β-ketoacyl-ACP synthase I; FabD, malonyl-CoA:ACP transacylase; FabF, β-ketoacyl-ACP synthase II; FabG, β-ketoacyl-ACP reductase; FabH, β-ketoacyl-ACP synthase III; FabI, enoyl-acyl-ACP reductase; FabZ, β-hydroxyacyl-ACP dehydratase; FadB/FadJ, enoyl-CoA hydratase/3-hydroxyacyl-CoA dehydrogenase; FadD, acyl-CoA synthetase; FadE, acyl CoA dehydrogenase; FadA/FadI, acetyl-CoA-acyltransferase; PD, pyruvate dehydrogenase; TesA, *E. coli* acyl-ACP thioesterase; YdiO, enoyl-CoA reductase; YqeF, thiolase.

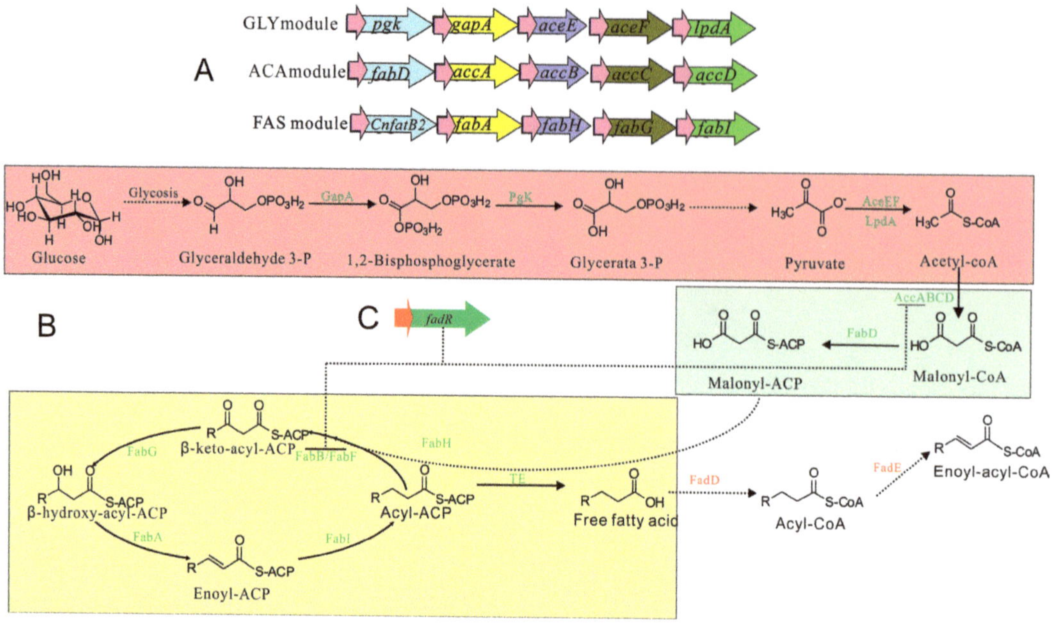

Figure 7.4 Summary of some important modifications on free fatty acid overproduction. The red, green and yellow boxes in panel (B) represent the three modules deconstructed respectively, and these three modules are shown in details at the upper panel (A). Genes overexpressed are depicted as green text. Genes knocked out are in red text. The *fadR* gene in the panel (C) shows overexpressing FadR upregulates the transcription of *fabF*, *fabB*, and accA. GapA, glyceraldehyde-3-phosphate dehydrogenase; PgK, phesosphoglycerate kinase; AceEF and LpdA, pyruvate dehydrogenase multi-enzyme; ACC, acetyl-CoA carboxylase; FabD, malonyl-CoA:ACPtransacylase; FabB, β-ketoacyl-ACP synthase I; FabD, malonyl-CoA:ACPtransacylase; FabF, β-ketoacyl-ACP synthase II; FabG, β-ketoacyl-ACP reductase; FabH, β-ketoacyl-ACP synthase III; FabI, enoyl-acyl-ACP reductase; FadD, acyl-CoA synthase; FadE, acyl CoA dehydrogenase.

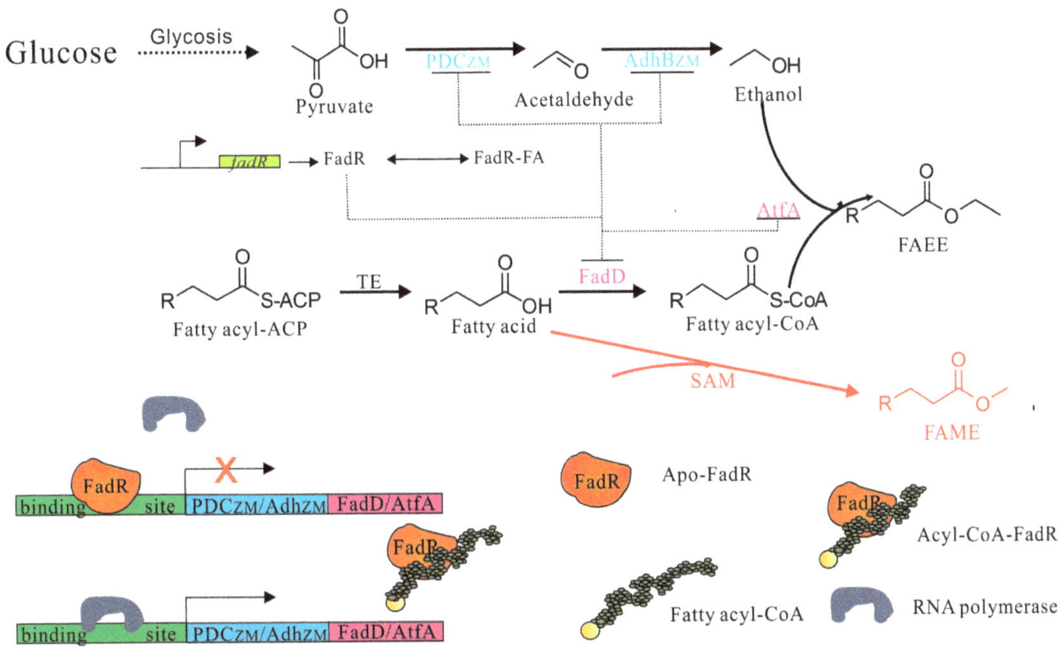

Figure 7.6 Genetic modifications for overproduction of fatty acid methyl/ethyl ester. PdcZM, pyruvate decarboxylase from *Zymomonas mobilis*. AdhB, alcohol dehydrogenase from *Zymomonas mobilis*. AtfA, wax-ester synthase from *Acinetobacter calcoaceticus*. FAMT, fatty acid methyltransferase. The dashed lines represent the DSRS regulation. FadR-FA represents FadR bound to acyl-CoA. When fatty acid is absent, FadR binds to the promoter and the transcription of genes in blue and magenta would be turned off. When there is abundant fatty acid, the fatty acid is catalysed to acyl-CoA by FadD, the acyl-CoA then binds with FadR, the Acyl-CoA-FadR breaks away from the DNA-binding site, and the transcription of the genes in blue and magenta are turned on.

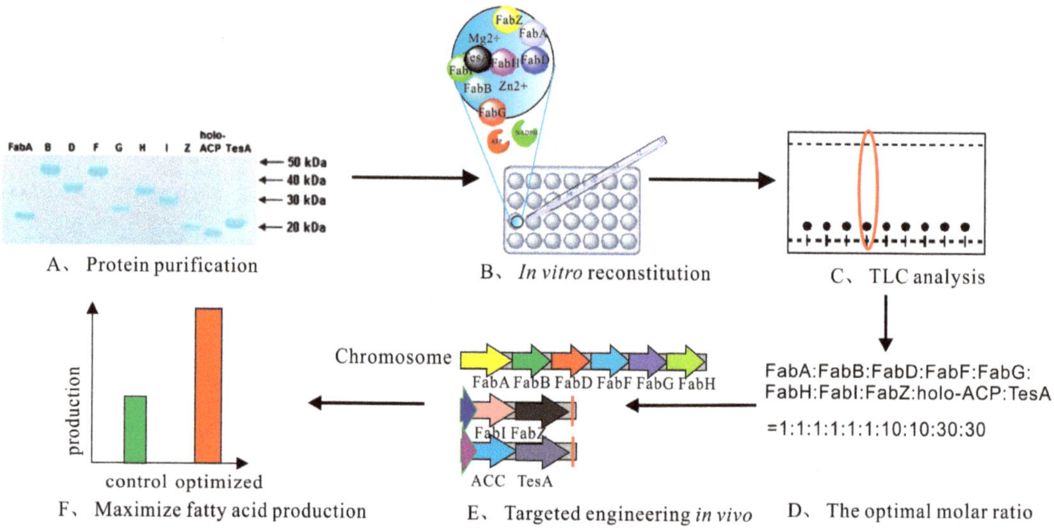

Figure 7.8 Schematic drawing of targeted engineering. (A) Protein purification; (B) *In vitro* reconstitution of the FAB pathway, which includes not only proteins, but also cofactors and metal ions; (C) Free fatty acid quantification using TLC to determine the optimal molar ratio; (D) The optimal molar ratio of different enzymes were determined; (E) Using the data from step D to guide *in vivo* genetic engineering; (F) Maximized free fatty acid production obtained.

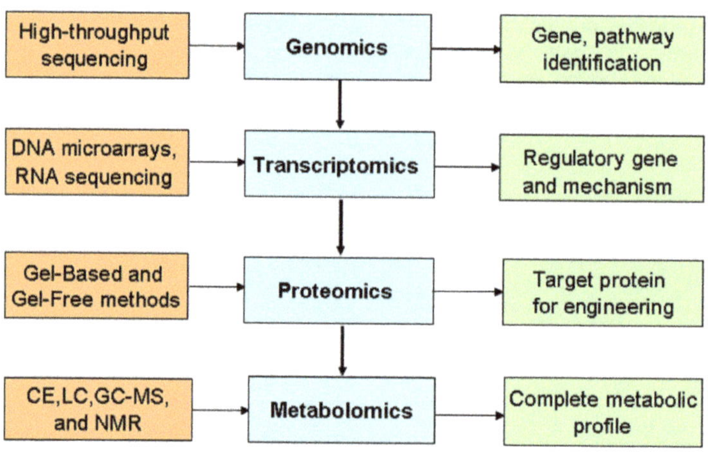

Figure 8.3 Overview of 'omics' technologies for systems biology research. CE, capillary electrophoresis; LC, liquid chromatography; GC, gas chromatography; MS, mass spectrometry; NMR, nuclear magnetic resonance.